The Foundations o
A Fundamental Course with 347 Exercises and Detailed Solutions

Richard Mikula

BrownWalker Press
Boca Raton

The Foundations of Real Analysis:
A Fundamental Course with 347 Exercises and Detailed Solutions

Copyright © 2015 Richard Mikula
All rights reserved. No part of this book may be reproduced or transmitted in any form or by any means, electronic or mechanical, including photocopying, recording, or by any information storage and retrieval system, without written permission from the publisher.

BrownWalker Press
Boca Raton, Florida
USA • 2015

ISBN-10: 1-62734-565-5
ISBN-13: 978-1-62734-565-1

www.brownwalker.com

Cover design by Marija Milanovic

Macrovector used under license from Shutterstock.com

Publisher's Cataloging-in-Publication Data

Mikula, Richard, 1975-
The foundations of real analysis : a fundamental course with 347 exercises and detailed solutions / Richard Mikula.
pages cm
Includes bibliographical references and index.
ISBN: 978-1-62734-565-1 (pbk.)
1. Functions of real variables. 2. Mathematical analysis. 3. Transcendental functions—Textbooks. 4. Trigonometry—Textbooks. 5. Calculus—Textbooks. I. Title.
QA300 .M55 2015
515`.8—dc23

2015951048

Contents

1 **The Set of Real Numbers**   11
   1.1 Introduction 11
   1.2 Fields and the Set of Rational Numbers 11
     1.2.1 Fields in the Algebraic Sense 11
     1.2.2 The Rational Numbers and the Integers, a Rigorous Discussion 19
     1.2.3 Homework Exercises 27
   1.3 Further Properties of the Real Numbers 29
     1.3.1 Positivity and Ordering on Fields 29
     1.3.2 The Completeness Axiom of the Set of Real Numbers . 31
     1.3.3 Some Further Properties of the Set of Real Numbers . 34
     1.3.4 Homework Exercises 43
   1.4 The Construction of the Real Numbers 46
     1.4.1 Dedekind Cuts and the Construction of the Real Numbers 46
     1.4.2 The Uniqueness of a Complete Ordered Field up to Isomorphism 56
   1.5 The Complex Numbers 60
     1.5.1 Homework Exercises 62

2 **Elementary Point-Set Topology**   65
   2.1 Euclidean Spaces 65
     2.1.1 Intervals in Euclidean Space 67
     2.1.2 Balls or Disks in Euclidean Space 68
     2.1.3 Convexity in Euclidean Space 69
   2.2 Metric Spaces 70
   2.3 Open Sets and Closed Sets 70
     2.3.1 Open Sets 70

		2.3.2	Closed Sets	71

        2.3.2 Closed Sets 71
        2.3.3 Unions and Intersections of Open and Closed Sets . . . 73
        2.3.4 Homework Exercises 74
  2.4 Compactness 76
        2.4.1 Homework Exercises 84
  2.5 Connectedness 84
  2.6 The Cantor Set 85
        2.6.1 Perfect Subsets of Euclidean Space 85
        2.6.2 The Construction of the Cantor Set 86
        2.6.3 Homework Exercises 87

3 Sequences and Series of Real Numbers   89
  3.1 Limits of Sequences 89
        3.1.1 Homework Exercises 92
        3.1.2 Properties of Limits 92
        3.1.3 Bounded Monotone Sequences 95
        3.1.4 Homework Exercises 97
  3.2 Subsequences 98
        3.2.1 Homework Exercises 99
  3.3 Limit Superior and Limit Inferior 100
        3.3.1 Homework Exercises 101
  3.4 Cauchy Sequences 103
        3.4.1 Homework Exercises 105
  3.5 Series 106
        3.5.1 Geometric and Telescoping Series 109
        3.5.2 Series of Non-negative Terms 111
        3.5.3 The Root Test and the Ratio Test 114
        3.5.4 Homework Exercises 118
        3.5.5 The Number e 120
        3.5.6 Alternating Series 122
        3.5.7 Absolute and Conditional Convergence 125
        3.5.8 The Algebra of Series 127
        3.5.9 Rearrangements of Series 128
        3.5.10 Power Series 133
        3.5.11 Homework Exercises 136

4 Limits and Continuity — 139
- 4.1 Limits of Functions 139
  - 4.1.1 Properties of Limits 140
  - 4.1.2 Homework Exercises 143
- 4.2 Continuous Functions 143
  - 4.2.1 The Exponential and Logarithmic Functions 146
  - 4.2.2 A Topological Characterization of Continuity 152
  - 4.2.3 Continuity and Compactness 153
  - 4.2.4 Continuity and Connectedness 154
  - 4.2.5 Uniform Continuity 154
  - 4.2.6 Discontinuities 156
  - 4.2.7 Semicontinuity, lim sup and lim inf 157
- 4.3 Monotone Functions 158
- 4.4 Infinite Limits and Limits at Infinity 160
- 4.5 Homework Exercises 162

5 Differentiation — 167
- 5.1 Definitions and Basic Differentiability Results 167
  - 5.1.1 Derivatives of Some Functions 169
  - 5.1.2 Approximation by Linear Functions 173
  - 5.1.3 The Product and Quotient Rule 173
  - 5.1.4 The Chain Rule 175
  - 5.1.5 The Trigonometric Functions 177
  - 5.1.6 Homework Exercises 190
- 5.2 Extrema for Functions 191
- 5.3 The Mean Value Theorem 192
  - 5.3.1 The Inverse Function Theorem 194
  - 5.3.2 The Continuity of Derivatives and the Intermediate Value Property 197
  - 5.3.3 L'Hôpital's Rule 198
  - 5.3.4 Homework Exercises 200
- 5.4 Higher Order Derivatives 201
  - 5.4.1 Homework Exercises 201
- 5.5 Monotonicity, Concavity and Convexity 202
  - 5.5.1 Some Monotonicity Results 202
  - 5.5.2 The First and Second Derivative Tests 202
  - 5.5.3 Concavity and Convexity 204
  - 5.5.4 Homework Exercises 208

	5.6	Taylor's Theorem 208
		5.6.1 Some Examples 210
		5.6.2 Homework Exercises 211
	5.7	Newton's Method 211
		5.7.1 Homework Exercises 214

6 The Riemann Integral 217
  6.1 Riemann Sums and the Riemann Integral 217
      6.1.1 The Existence of Riemann Integrable Functions . . . 219
      6.1.2 Linearity and Comparative Properties of the Integral . 220
      6.1.3 A Cauchy Convergence Criterion for Integrability . . . 222
      6.1.4 Some Examples 224
      6.1.5 Further Properties of Riemann Integrable Functions . . 226
      6.1.6 Upper and Lower Sums and the Darboux Integral . . . 232
      6.1.7 A Sufficient Condition for Integrability 238
      6.1.8 Additive Properties of the Riemann Integral 239
      6.1.9 Homework Exercises 243
  6.2 The Fundamental Theorem of Calculus 244
      6.2.1 Integral Representation of Some Transcendental Functions 247
      6.2.2 Homework Exercises 248
  6.3 Functions of Bounded Variation 249
      6.3.1 Homework Exercises 260
  6.4 Rectifiable Curves and Arclength 260
      6.4.1 Homework Exercises 263
  6.5 Lebesgue's Integrability Criterion 264
      6.5.1 Sets of Measure Zero 264
      6.5.2 The Oscillation of a Function 265
      6.5.3 Lebesgue's Criterion for Riemann Integrability 266
      6.5.4 Homework Exercises 269

7 Sequences and Series of Functions 271
  7.1 Point-wise Convergence 271
      7.1.1 Homework Exercises 273
  7.2 Uniform Convergence 274
      7.2.1 Uniform Convergence and Continuity 276
      7.2.2 Uniform Convergence and Integration 278
      7.2.3 Uniform Convergence and Differentiation 279

	7.2.4	Homework Exercises 281
7.3	Series of Functions 283	
	7.3.1	Uniform Convergence of Power Series 284
	7.3.2	Homework Exercises 287
7.4	Equicontinuity 287	
	7.4.1	Properties of Convergent Sequences of Functions 287
	7.4.2	The Ascoli-Arzelà Theorem and Compactness 290
	7.4.3	Homework Exercises 292
7.5	Uniform Approximation by Polynomials 293	
	7.5.1	The Weierstrass Approximation Theorem 294

8 Functions of Several Real Variables   297

8.1	The Algebra and Topology of $\mathbb{R}^n$ 297	
	8.1.1	$\mathbb{R}^n$ as a Normed Linear Space 298
	8.1.2	Homework Exercises 305
8.2	Linear Transformations and Affine Functions 306	
	8.2.1	Linear Transformations on Vector Spaces 306
	8.2.2	Homework Exercises 311
	8.2.3	Affine Functions 311
8.3	Differentiation 312	
	8.3.1	The Chain Rule 315
	8.3.2	The Gradient and Directional Derivatives 316
	8.3.3	Taylor's Theorem, the Second Derivative and Extreme Values 320
	8.3.4	Homework Exercises 325
8.4	The Inverse Function and Implicit Function Theorems 327	
	8.4.1	The Contraction Mapping Principle 327
	8.4.2	The Inverse Function Theorem 328
	8.4.3	The Implicit Function Theorem 333
	8.4.4	Homework Exercises 338

9 The Lebesgue Integral   339

9.1	Lebesgue Outer Measure and Measurable Sets 339	
	9.1.1	The Lebesgue Outer Measure 339
	9.1.2	The Lebesgue Measure 342
	9.1.3	Homework Exercises 352
9.2	Lebesgue Measurable Functions 353	
	9.2.1	Sequences of Measurable Functions 355

- 9.2.2 Homework Exercises 360
- 9.3 The Riemann and Lebesgue Integrals 362
  - 9.3.1 The Riemann Integral in $\mathbb{R}^n$ 362
  - 9.3.2 The Lebesgue Integral of Non-Negative Functions in $\mathbb{R}^n$ 373
  - 9.3.3 Lebesgue Integral for a Measurable Function of Any Sign 382
  - 9.3.4 Homework Exercises 389
- 9.4 Iterated Integration and Fubini's Theorem 392
  - 9.4.1 Convolutions 397
  - 9.4.2 Homework Exercises 398

A Preliminary Materials 401
- A.1 Sets and Related Notation 401
- A.2 Notation in Symbolic Logic 402
- A.3 The Basics in Symbolic Logic 402
- A.4 Quantified Sentences in Symbolic Logic 405
  - A.4.1 Negations of Quantified Sentences 407
- A.5 Equivalence Relations 408
- A.6 The Natural Numbers and Induction 409
  - A.6.1 The Well-Ordering Principle 411
  - A.6.2 Inductive Sets 412
  - A.6.3 How the Principle of Mathematical Induction is Used in Proofs 413
  - A.6.4 The Principle of Complete Induction 413

B Homework Solutions 415
- B.1 Homework Solutions 1.2.3 415
- B.2 Homework Solutions 1.3.4 418
- B.3 Homework Solutions 1.5.1 421
- B.4 Homework Solutions 2.3.4 422
- B.5 Homework Solutions 2.4.1 425
- B.6 Homework Solutions 2.6.3 426
- B.7 Homework Solutions 3.1.1 426
- B.8 Homework Solutions 3.1.4 427
- B.9 Homework Solutions 3.2.1 429
- B.10 Homework Solutions 3.3.1 430
- B.11 Homework Solutions 3.4.1 434
- B.12 Homework Solutions 3.5.4 435
- B.13 Homework Solutions 3.5.11 437

CONTENTS

B.14 Homework Solutions 4.1.2 439
B.15 Homework Solutions 4.5 440
B.16 Homework Solutions 5.1.6 447
B.17 Homework Solutions 5.3.4 448
B.18 Homework Solutions 5.4.1 449
B.19 Homework Solutions 5.5.4 450
B.20 Homework Solutions 5.6.2 451
B.21 Homework Solutions 5.7.1 453
B.22 Homework Solutions 6.1.9 453
B.23 Homework Solutions 6.2.2 457
B.24 Homework Solutions 6.3.1 459
B.25 Homework Solutions 6.4.1 461
B.26 Homework Solutions 6.5.4 462
B.27 Homework Solutions 7.1.1 463
B.28 Homework Solutions 7.2.4 464
B.29 Homework Solutions 7.3.2 468
B.30 Homework Solutions 7.4.3 469
B.31 Homework Solutions 8.1.2 471
B.32 Homework Solutions 8.2.2 474
B.33 Homework Solutions 8.3.4 476
B.34 Homework Solutions 8.4.4 480
B.35 Homework Solutions 9.1.3 482
B.36 Homework Solutions 9.2.2 485
B.37 Homework Solutions 9.3.4 489
B.38 Homework Solutions 9.4.2 494

Chapter 1

The Set of Real Numbers

1.1 Introduction

Real Analysis must begin with the set of real numbers $\mathbb{R}$, along the the two binary operations addition $+$ and multiplication $\cdot$, which makes $(\mathbb{R},+,\cdot)$ a field in the algebraic sense. In fact $(\mathbb{R},+,\cdot)$ can be called the only *complete, ordered field*, with "only" being up to algebraic isomorphism of rings.

It is the completeness property of $\mathbb{R}$ that is essential to calculus, and many of its results. Note that $(\mathbb{Q},+,\cdot)$ is an ordered field, where $\mathbb{Q}$ being the set of rational numbers $\{\frac{m}{n} : m, n \text{ integers, and } n \neq 0\}$. However $\mathbb{Q}$ is not complete. To see this, take for instance terms in the sequence

$$1, 1.4, 1.41, 1.414, 1.4142, 1.41421, 1.414213, 1.4142135, \cdots$$

that is the sequence of terms $\{a_n\}_{n=0}^{\infty}$ which have the decimal expansion for $\sqrt{2}$ out to the $(10)^{-n}th$ place. Clearly all the terms in this sequence are rational, however, if the sequence converges, it must converge to $\sqrt{2} \notin \mathbb{Q}$. Thus $\mathbb{Q}$ is not complete.

1.2 Fields and the Set of Rational Numbers

1.2.1 Fields in the Algebraic Sense

Let $\mathbb{F}$ be a set that has the following binary relations $+ : \mathbb{F} \times \mathbb{F} \to \mathbb{F}$ and $\cdot : \mathbb{F} \times \mathbb{F} \to \mathbb{F}$, called addition and multiplication respectively. We will assume that the triple $(\mathbb{F},+,\cdot)$ satisfies the following list of properties:

For any $x, y, z \in \mathbb{F}$

1. **Commutativity of Addition:**
$$x + y = y + x$$

2. **Associativity of Addition:**
$$x + (y + z) = (x + y) + z$$

3. **Existence of Zero:** There is an element $0 \in \mathbb{F}$, called **zero**, which satisfies
$$0 + x = x$$
for all $x \in \mathbb{F}$.

4. **Additive Inverse:** For each $x \in \mathbb{F}$ there is a $-x \in \mathbb{F}$ so that
$$x + (-x) = 0.$$

5. **Commutativity of Multiplication:**
$$x \cdot y = y \cdot x$$

6. **Associativity of Multiplication:**
$$x \cdot (y \cdot z) = (x \cdot y) \cdot z$$

7. **Existence of Unity**[1]**:** There is an element $1 \in \mathbb{F}$, called **one**, so that
$$1 \cdot x = x$$
for all $x \in \mathbb{F}$.

8. **Multiplicative Inverse:** For each non-zero $x \in \mathbb{F}$ there is a $x^{-1} \in \mathbb{F}$ so that
$$x \cdot x^{-1} = 1$$

---
[1] Unity in a ring refers to 1.

1.2. FIELDS AND THE SET OF RATIONAL NUMBERS

9. **Distributive Property:** $+$ and $\cdot$ are related by

$$x \cdot (y + z) = (x \cdot y) + (x \cdot z).$$

Needless to say $1, 0 \in \mathbb{F}$ may not be the $1, 0$ in the set of real numbers.

Such an ordered triple $(\mathbb{F}, +, \cdot)$ is called a **field**.

The Set of Natural Numbers

There is an important set of numbers called the **natural numbers**[2], denoted by $\mathbb{N}$, and which satisfies the property that $1 \in \mathbb{N}$, and if $n \in \mathbb{N}$, then $n + 1 \in \mathbb{N}$. Moreover, there does not exists an $m \in \mathbb{N}$ so that $m + 1 = 1$. Here of course $\mathbb{N}$ has two binary operations $+, \cdot$.[3] Moreover, the set of natural numbers is closed under $+$ and $\cdot$ and satisfies the following list of properties for any $k, m, n \in \mathbb{N}$:

1. **Associativity:**

$$k + (m + n) = (k + m) + n, \quad k \cdot (m \cdot n) = (k \cdot m) \cdot n$$

2. **Commutativity:**

$$k + m = m + k, \quad k \cdot m = m \cdot k$$

3. **Existence of a Unity:** There is an element $1 \in \mathbb{N}$, called **one**, so that

$$1 \cdot k = k$$

for any $k \in \mathbb{N}$.

---

[2] The natural numbers $\mathbb{N}$ can actually be characterized by the following five axioms: A1: $1 \in \mathbb{N}$. A2: $\forall x \in \mathbb{N}$ there exists a unique $(x + 1) \in \mathbb{N}$, called the successor of x. A3: $\forall x \in \mathbb{N}$, $x + 1 \neq 1$. That is 1 is not the successor of any natural number. A4: If $x + 1 = y + 1$, then $x = y$. A5 (the axiom of induction): Suppose $A \subseteq \mathbb{N}$ satisfies: $1 \in A$; $x \in A$ implies $(x + 1) \in A$. Then $A = \mathbb{N}$. For a complete construction of the properties of $\mathbb{N}$ see for instance [7].

[3] Although we are using the same notation for $+, \cdot$ for both our field $\mathbb{F}$ and $\mathbb{N}$, as well as for the element $1 \in \mathbb{F}$ and $1 \in \mathbb{N}$, they unrelated in general.

4. **Cancellation Property:**

$$k + m = k + n \Rightarrow m = n$$

$$k \cdot m = k \cdot n \Rightarrow m = n$$

5. **The Distributive Property:**

$$k \cdot (m + n) = (k \cdot m) + (k \cdot n)$$

6. For a given pair of natural numbers m, n one and only one of the following holds:

   - $m = n$,
   - $m = n + k$ for some $k \in \mathbb{N}$,
   - $n = m + k$ for some $k \in \mathbb{N}$.

This in turn allows us to define an ordering on $\mathbb{N}$. The following are **order properties:**

(a) $m < n$ if and only if there exists $j \in \mathbb{N}$ so that $n = m + j$.

(b) $m \leq n$ if and only if $m < n$ or $m = n$.

(c) $k < m$ and $m < n$ implies $k < n$.

(d) $m \leq n$ and $n \leq m$ implies $m = n$

(e) $m < n$ implies $m + k < n + k$ and $k \cdot m < k \cdot n$.

Moreover, we also have the following axiom for the set $\mathbb{N}$ of natural numbers:

The Axiom of Induction: Let $S \subseteq \mathbb{N}$. Suppose S satisfies

1. $1 \in S$.

2. $n \in S$ implies $n + 1 \in S$.

Then $S = \mathbb{N}$.

From this characterization of $\mathbb{N}$ in terms of the above properties, we see that $\mathbb{N} = \{1, 2, 3, 4, \cdots\}$.

1.2. FIELDS AND THE SET OF RATIONAL NUMBERS

The Set of Integers

From the set of natural numbers we may define the set of **integers**, denoted by[4] $\mathbb{Z}$, that has the same binary operations $+, \cdot$ and is closed under these operations.[5] $\mathbb{Z}$ contains $\mathbb{N}$ as a subset, and also contains an element 0 so that $m + 0 = m$ for all $m \in \mathbb{Z}$. Moreover, for any $m \in \mathbb{Z}$ we have $m = 0$, $m \in \mathbb{N}$ or $-m \in \mathbb{N}$, where $m + (-m) = 0$. Thus we see that $\mathbb{Z} = \{\cdots, -3, -2, -2, 0, 1, 2, 3, 4, \cdots\}$. $\mathbb{Z}$ satisfies all the field axioms except the axiom which says that every non-zero element x has a multiplicative inverse x^{-1}. Moreover, the set of integers also satisfies the **cancellation property**[6], and so $\mathbb{Z}$ cannot have any zero divisors, that is

$$x \cdot y = 0 \quad \Rightarrow \quad x = 0, \quad \text{or} \quad y = 0.$$

Moreover, on $\mathbb{Z}$ we may define an ordering via the relation $<$, where $m < n$ if and only if $n + (-m) \in \mathbb{N}$.

The Set of Rational Numbers

From the set of integers, we may construct set set of **rational numbers**, denoted by $\mathbb{Q}$, which are defined by

$$\mathbb{Q} := \{a \cdot b^{-1} : a, b \in \mathbb{Z}, b \neq 0\}. \tag{1.1}$$

We may think of this as the algebraic extension of the integral domain $\mathbb{Z}$ to a field of fractions. In a later section we shall give a rigorous construction of $\mathbb{Q}$ from $\mathbb{Z}$, as well as a rigorous construction of $\mathbb{Z}$ from $\mathbb{N}$.

Notational Conventions for Fields

The following is a list of standard notational conventions:

1. We often write xy instead of $x \cdot y$.

2. We define **subtraction** by

$$x - y := x + (-y).$$

---

[4] $\mathbb{Z}$ for the German *Zahl*, which means number.
[5] In a later section we shall give a rigorous argument for the construction of $\mathbb{Z}$ from $\mathbb{N}$.
[6] $a \cdot b = a \cdot c$ and $a \neq 0$ implies $b = c$

3. We often write $\frac{1}{x}$ instead of x^{-1}.

4. We define **division** by
$$x \div y \text{ or } \frac{x}{y} := x \cdot y^{-1}$$
for $x \in \mathbb{F}$, and $y \neq 0$ in $\mathbb{F}$.

5. We define **powers** of a element x by $x^1 := x$, $x^2 := x \cdot x$, $x^3 := x \cdot x^2 = x \cdot x \cdot x$ and for n a positive integer
$$x^{n+1} := x \cdot x^n.$$
For n a positive integer, we define x^{-n} as $(x^{-1})^n$ or equivalently $(x^n)^{-1}$. Since both ways to define this will agree with one another.

6. We define **multiples** of an element x by $1x := x$, $2x := x + x$, $3x := x + 2x = x + x + x$, and for a positive integer n by
$$(n+1)x = x + nx.$$
Moreover, because of the distributive property, we have
$$n \cdot x = nx.$$
That is,
$$\underbrace{(1 + 1 + \cdots + 1)}_{n \text{ times}} \cdot x = \underbrace{x + x + \cdots + x}_{n \text{ times}}.$$
Note as well
$$\underbrace{(1 + 1 + \cdots + 1)}_{n \text{ times}} \cdot x$$
where here $1 \in \mathbb{F}$.

7. **Order of Operations:**

    (a) Operations in parentheses are performed first, working with the inner most first. In the innermost parentheses, or in the absence of parentheses, follow the remaining steps.

    (b) Powers are computed first, going from left to right.

1.2. FIELDS AND THE SET OF RATIONAL NUMBERS

(c) Multiplication and division are performed next (with equal weight) going from left to right.

(d) Addition and subtraction are performed after all multiplications and divisions have been performed. Addition and subtraction (which hold equal weight) are performed from left to right.

Thus, we see that we may write the distributive property as

$$x(y+z) = xy + xz.$$

Some Consequences of the Field Axioms

The following result shows that additive and multiplicative inverses for a fixed element x are unique, and the elements 1 and 0 are unique in a field $\mathbb{F}$.

Proposition 1.2.1 *The elements 0 and 1 of a field are unique. Moreover, for a given element $x \in \mathbb{F}$, the additive inverse $-x$ is unique, and for $x \neq 0$ the multiplicative inverse x^{-1} is unique.*

Proof: Suppose that there are two elements $0_1, 0_2 \in \mathbb{F}$ such that $0_1 + x = x$ and $0_2 + x = x$ for any $x \in \mathbb{F}$. Then $0_1 + 0_2 = 0_2$ by using $x = 0_2$, and $0_1 + 0_2 = 0_1$ by using $x = 0_1$. Thus, $0_1 = 0_2$. Next, we suppose there are two elements $1_1, 1_2 \in \mathbb{F}$ such that $1_1 \cdot x = x$ and $1_2 \cdot x = x$. Then $1_1 \cdot 1_2 = 1_2$ by using $x = 1_2$ and on the other hand $1_1 \cdot 1_2 = 1_1$ by using $x = 1_1$. Thus, $1_1 = 1_2$.

Now, given $x \in \mathbb{F}$, suppose there are two elements $y, z \in \mathbb{F}$ so that $x + y = 0$ and $x + z = 0$. Then, by associativity we have

$$y = y + 0 = y + (x + z) = (y + x) + z = 0 + z = z.$$

Given $x \neq 0$, suppose there are two elements $y, z \in \mathbb{F}$ so that $xy = 1$ and $xz = 1$. Then, by associativity we have

$$y = y \cdot 1 = y(xz) = (yx)z = 1 \cdot z = z.$$

□

Since additive and multiplicative inverses are unique,

$$-(-x) = x, \quad \frac{1}{\frac{1}{x}} = (x^{-1})^{-1} = x \qquad (1.2)$$

Another important result to note is the **cancellation property:**

Proposition 1.2.2 *If $x + y = x + z$, then $y = z$. Moreover, if $x \neq 0$ and $xy = xz$, then $y = z$.*

Proof: If $x + y = x + z$, by adding $-x$ to both sides we get

$$y = 0 + y = (-x + x) + y = -x + (x + y)$$
$$= -x + (x + z) = (-x + x) + z = 0 + z = z.$$

Moreover, if $x \neq 0$ and $xy = xz$, by multiplying both sides by x^{-1} we get

$$y = 1 \cdot y = (x^{-1}x)y = x^{-1}(xy) = x^{-1}(xz) = (x^{-1}x)z = 1 \cdot z = z.$$

□

The next result relates $(-1) \cdot x$ to $-x$ and also gives us the property $0 \cdot x = 0$, neither of which are axioms of our field $\mathbb{F}$.

Proposition 1.2.3 *For any $x \in \mathbb{F}$ we have $(-1) \cdot x = -x$ and $0 \cdot x = 0$. Moreover, $(-1)^2 = 1$.*

Proof: $0 \cdot x = (0 + 0) \cdot x = 0 \cdot x + 0 \cdot x$ and thus by the cancellation property we have

$$0 \cdot x = 0.$$

To show $(-1) \cdot x = -x$ we will show $(-1) \cdot x$ is the additive inverse $-x$, and thus by uniqueness of inverses, we get the result. To see this, note

$$(-1) \cdot x + x = (-1) \cdot x + 1 \cdot x = (-1 + 1) \cdot x = 0 \cdot x = 0.$$

For the last result, we will show $(-1)^2$ is the additive inverse of -1.

$$-1 + (-1)^2 = (-1) \cdot 1 + (-1)^2 = (-1) \cdot (1 + (-1)) = (-1) \cdot 0 = 0.$$

□

1.2. FIELDS AND THE SET OF RATIONAL NUMBERS

A consequence of the above proposition is that for any $x, y \in \mathbb{F}$ we have

$$(-x) \cdot y = -(x \cdot y) = x \cdot (-y) \tag{1.3}$$

To see this, we use the property $-x = (-1) \cdot x$, and thus

$$(-x) \cdot y = ((-1) \cdot x) \cdot y = (-1) \cdot (x \cdot y) = -(x \cdot y).$$

To see the other equality, we use commutativity, and reverse the roles of x and y.

Moreover, we also have

$$(-x) \cdot (-y) = x \cdot y,$$

since

$$(-x) \cdot (-y) = (-1)^2 \cdot x \cdot y.$$

The next result tells us that there are no zero divisors in a field.

Proposition 1.2.4 *Suppose that $x \cdot y = 0$, then either $x = 0$ or $y = 0$.*

Proof: Suppose $x \cdot y = 0$ and $x \neq 0$. Then multiplying $x \cdot y = 0$ by $\frac{1}{x}$ we get

$$0 = \frac{1}{x} \cdot 0 = \frac{1}{x} \cdot (x \cdot y) = (\frac{1}{x} \cdot x) \cdot y = 1 \cdot y = y.$$

□

1.2.2 The Rational Numbers and the Integers, a Rigorous Discussion

In this section we will discuss the construction of the set of integers from the set of natural numbers, and then a construction of the set of rational numbers from the set of integers. We will think of the set of rational numbers as an algebraic extension of the set of integers (as an integral domain) to a field.

The Construction of the Integers from the Natural Numbers

Consider the set $\mathbb{N} \times \mathbb{N}$. On this set we define a relation[7] R by

$$(a,b) \sim_R (c,d) \quad \Longleftrightarrow \quad a+d = b+c. \tag{1.4}$$

We claim that R is an equivalence relation on $\mathbb{N} \times \mathbb{N}$. Clearly $(a,b) \sim_R (a,b)$ since $a+b = b+a$, and thus R is reflexive. Likewise, if $(a,b) \sim_R (c,d)$, then $a+d = b+c$. By commutativity of addition on $\mathbb{N}$, we know that $c+b = d+a$, and so $(c,d) \sim_R (a,b)$. Hence R is a symmetric relation on $\mathbb{N} \times \mathbb{N}$. Next, we suppose that $(a,b) \sim_R (c,d)$ and $(c,d) \sim_R (e,f)$. Then $a+d = b+c$ and $c+f = d+e$. Adding f to the first equation $a+d = b+c$ we get $a+d+f = b+c+f$, and then using the second equation $c+f = d+e$ we get $b+c+f = b+d+e$, and so $a+d+f = b+d+e$. By the cancellation property of $\mathbb{N}$, we may conclude that $a+f = b+e$, and so $(a,b) \sim_R (e,f)$. Thus we see that R is a transitive relation on $\mathbb{N} \times \mathbb{N}$. This allows us to define $\mathcal{Z}$ to be the set of equivalence classes defined by this relation on $\mathbb{N} \times \mathbb{N}$. Let $[a,b]$ denote the equivalence class of $(a,b) \in \mathbb{N} \times \mathbb{N}$ with respect to this equivalence relation R.

On $\mathcal{Z}$ we define an operation $+$ by $[a,b] + [c,d] := [a+c, b+d]$. Likewise, we define an operation $\cdot$ by $[a,b] \cdot [c,d] := [ac+bd, ad+bc]$. Now it is our job to show that on $\mathcal{Z}$ the operations $+$ and $\cdot$ as defined are well-defined. Moreover, $\mathcal{Z}$ is closed with respect to $+$ and $\cdot$. On $\mathcal{Z}$, $+$ is commutative and associative, $\cdot$ is commutative and associative, and $+$ and $\cdot$ satisfy a distributive property.

First we address the well-defined nature of our definitions of $+$ and $\cdot$. Suppose $[a,b] = [a',b']$, and $[c,d] = [c',d']$. In other words, $(a,b) \sim_R (a',b')$ and $(c,d) \sim_R (c',d')$. We claim that $[a,b] + [c,d] = [a',b'] + [c',d']$ and $[a,b] \cdot [c,d] = [a',b'] \cdot [c',d']$. In order to see this, we first observe that for any $n \in \mathbb{N}$ we have that $[a,b] = [a+n, b+n]$, since $a + (b+n) = b + (a+n)$. Now,

$$[a,b] + [c,d] = [a+c, b+d]$$
$$= [a+c+a'+c', b+d+a'+c']$$
$$= [a+c+a'+c', a'+b+c'+d]$$

---
[7] Our motivation is (a,b) should be thought of as $a-b$, since we may obtain $\mathbb{Z}$ from $\mathbb{N}$ by taking all possible differences of two natural numbers.

1.2. FIELDS AND THE SET OF RATIONAL NUMBERS   21

$$= [a + c + a' + c', a + c + b' + d']$$
$$= [a' + c', b' + d']$$
$$= [a', b'] + [c', d'].$$

Likewise,
$$[a, b] \cdot [c, d] = [ac + bd, ad + bc]$$
$$= [ac + bd + ac' + bd', ac' + ad + bd' + bc]$$
$$= [ac + bd + ac' + bd', a(c' + d) + b(d' + c)]$$
$$= [ac + bd + ac' + bd', a(c + d') + b(d + c')]$$
$$= [ac + bd + ac' + bd', ac + bd + c'b + d'a]$$
$$= [ac' + bd', c'b + d'a]$$
$$= [ac' + bd' + a'c' + b'd', c'a + c'b + d'b' + d'a]$$
$$= [ac' + bd' + a'c' + b'd', c'(a' + b) + d'(b' + a)]$$
$$= [ac' + bd' + a'c' + b'd', c'(a + b') + d'(b + a')]$$
$$= [ac' + bd' + a'c' + b'd', ac' + bd' + a'd' + b'c']$$
$$= [a'c' + b'd', a'd' + b'c']$$
$$= [a', b'] \cdot [c', d'].$$

We leave it as an exercise to the reader to show the associativity and commutativity of both $+$ and $\cdot$ as defined on $\mathcal{Z}$.

Next, we address the distributive property, which relates the operations $+$ and $\cdot$.
$$[a, b] \cdot ([c, d] + [e, f]) = [a, b] \cdot [c + e, d + f]$$
$$[a(c + e) + b(d + f), a(d + f) + b(c + e)]$$
$$= [(ac + bd) + (ae + bf), (ad + bc) + (af + be)]$$
$$= [ac + bd, ad + bc] + [ae + bf, af + be]$$
$$= ([a, b] \cdot [c, d]) + ([a, b] \cdot [e, f]).$$

Next, we will show how to embed $\mathbb{N}$ into $\mathcal{Z}$, and we will do this via the identification of $\mathbb{N}$ with the elements of $\mathcal{Z}$ in the subset $\mathcal{P} := \{[m, n] : m > n\}$. First, we claim that $[m, n] \in \mathcal{P}$ has a representation of the form $[n + k, n]$.

This is clear from the fact that $m > n$ in $\mathbb{N}$ means that $m = n + k$ for some $k \in \mathbb{N}$. We claim that $\mathcal{P}$ is isomorphic to $\mathbb{N}$ relative to the operations $+, \cdot$ as defined on each set. Since we can think of elements of $\mathcal{P}$ as elements of the form $[n+k, n]$, we define the mapping $k \mapsto [n+k, n]$, where $n \in \mathbb{N}$ is any fixed natural number. Since $[k+m, m] = [k+n, n]$ for all $m, n \in \mathbb{N}$, what we are defining is well-defined and the mapping $k \mapsto [n+k, n]$ is thus independent of the natural number n. This is an isomorphism of rings since $[k+m, m] = [j+n, n]$ if and only if $k = j$. Likewise, $[k+n, n] + [j+m, m] = [k+j+m+n, m+n]$ and $[k+n, n] \cdot [j+m, m] = [(k+n)(j+m) + nm, (k+n)m + n(j+m)] = [jk+l, l]$ where $l = nj + km + 2nm$.

We define the zero of $\mathcal{Z}$ to be the element $[m, m]$, where $m \in \mathbb{N}$ is any fixed integer. Clearly $[j, k] + [m, m] = [j+k, k+m] = [j, k]$, $[j, k] \cdot [m, m] = [jm+km, jm+km] = [m, m]$. We define the negative of an element $[j, k]$ to be $[k, j]$ since $[j, k] + [k, j] = [j+k, k+j] = [m, m]$. From this we see that we can identify the collection of negatives of elements of $\mathcal{P}$ with the collection of elements of the form $[j, j+k]$, and such an element can be thought of as the negative of the natural number k. Through this, we identify the collection of integers, commonly denoted by $\mathbb{Z}$ with the integral domain $\mathcal{Z}$.

The Construction of the Rational Numbers from the Integers

In the future we will write a rational number in the form $\frac{m}{n}$ where $m, n \in \mathbb{Z}$ and $n \neq 0$. We will define this collection from $\mathbb{Z} \times \mathbb{Z}$ by defining binary operations $+, \cdot$ from those on $\mathbb{Z}$.

Let
$$Q = \{(a, b) : a, b \in \mathbb{Z}, b \neq 0\}.$$

We need to identify ordered pairs that would give the same value of $\frac{a}{b}$, where we think of (a, b) as $\frac{a}{b}$ or ab^{-1}. To do this, we define an equivalence relation $\sim$ on Q, which is given by

$$(a, b) \sim (c, d) \iff ad = bc.$$

Clearly $(a, b) \sim (a, b)$ since $ab = ba$. Thus $\sim$ is reflexive.

$$(a, b) \sim (c, d) \iff ad = bc \iff da = cb \iff (c, d) \sim (a, b).$$

1.2. FIELDS AND THE SET OF RATIONAL NUMBERS

Thus we see $\sim$ is symmetric. Moreover, if $(a,b) \sim (c,d)$ and $(c,d) \sim (e,f)$, then we know $ad = cb$ and $cf = de$. Now,

$$ad = cb \Rightarrow adf = cbf$$

and

$$cf = de \Rightarrow cfb = deb.$$

Since $\mathbb{Z}$ is commutative we know

$$adf = deb,$$

or simply

$$(af)d = (be)d.$$

Since $d \neq 0$, by the cancellation property we may conclude that

$$af = be,$$

and thus

$$(a,b) \sim (e,f).$$

Hence we have shown $\sim$ is transitive.

We have established $\sim$ is an equivalence relation on Q, and hence we may view Q as a disjoint union of equivalence classes. Let $[a,b]$ denote the equivalence class of (a,b). Let $\mathcal{Q}$ be the set of equivalence classes. On this set we define $+, \cdot$ by:

$$[a,b] + [c,d] = [ad+bc, bd]$$

and

$$[a,b] \cdot [c,d] = [ac, bd].$$

To understand these definitions, simply remember the formulae:

$$\frac{a}{b} + \frac{c}{d} = \frac{ad+bc}{bd}, \quad \frac{a}{b} \cdot \frac{c}{d} = \frac{ac}{bd}.$$

However, before we proceed any further, we must make certain these definitions of $+, \cdot$ on $\mathcal{Q}$ are well-defined. That is if $[a,b] = [r,s]$ and $[c,d] = [t,u]$, then

$$[ad+bc, bd] = [a,b] + [c,d]$$

$$= [r,s] + [t,u] = [ru + st, su]$$

and
$$[ac, bd] = [a,b] \cdot [c,d] = [r,s] \cdot [t,u] = [rt, su].$$

That is we actually need to show
$$[ad + bc, bd] = [ru + st, su]$$

and
$$[ac, bd] = [rt, su]$$

whenever $(a,b) \sim (r,s)$ and $(c,d) \sim (t,u)$.

Taking the condition $as = br$ and multiplying both sides by cu gives
$$ascu = brcu.$$

However, $cu = dt$ and thus
$$ascu = bdrt,$$

which implies
$$[ac, bd] = [rt, su].$$

Taking the condition $as = br$ and multiplying both sides by du gives
$$adsu = brdu.$$

Taking the condition $cu = dt$, and multiplying both sides by bs gives us
$$bcsu = dtbs.$$

Adding these together gives
$$(ad + bc)su = (ru + st)bd,$$

and thus
$$[ad + bc, bd] = [ru + st, su].$$

Thus we see that the binary operations $+, \cdot$ on $\mathcal{Q}$ are well-defined. It is left as an exercise to check that addition and multiplication are associative and commutative, and the distributive property holds.

1.2. FIELDS AND THE SET OF RATIONAL NUMBERS

In $\mathcal{Q}$ the element $[0,1]$ is the zero element, and the element $[1,1]$ is the unity. The negative of $[a,b]$ is $[-a,b]$ and the multiplicative inverse of a non-zero element $[a,b]$ is $[b,a]$. Thus we see that $\mathcal{Q}$ is a field. Moreover, we can identify $\mathbb{Z}$ with the subset $\{[a,1] : a \in \mathbb{Z}\}$ and it is clear that we may view $\mathcal{Q}$ as an extension of $\mathbb{Z}$, since

$$[a,1] + [b,1] = [a+b,1]$$

and

$$[a,1] \cdot [b,1] = [ab,1].$$

Moreover, any non-zero $a \in \mathbb{Z}$ has $[a,1]$ as an invertible element in $\mathcal{Q}$.

This field $(\mathcal{Q}, +, \cdot)$ is what we usually called the field of rational numbers $(\mathbb{Q}, +, \cdot)$. We instead use familiar the notation $\frac{a}{b}$ for $[a,b]$.

Order on the Set of Rational Numbers

We may define an ordering on the set of integers, which extends to the set of rational numbers as follows: The **positive** integers are the natural numbers, and the **negative** integers are the integers whose additive inverses are positive. The intersection between these two collections is empty. We say that an integer is **non-negative** if it is positive or zero, and **non-positive** if it is negative or zero.

Note that from its definition, and the property that $-n = (-1) \cdot n$ holds for any $n \in \mathbb{Z}$, as well as $(-1)^2 = 1$, we get that the set of positive integers are closed under $\cdot$, a positive times a negative is negative, and a negative times a negative is a positive.

We extend the notions of positivity and negativity to set of the rational numbers. We say that a rational number $\frac{m}{n}$ is positive if m, n are either both positive or both negative. A rational number $\frac{m}{n}$ is negative if one of $\{m, n\}$ is positive, and the other is negative. From these definitions, the properties of the set of positive rational numbers and negative rational numbers satisfy: the positive rational numbers are closed under $\cdot$, a positive times a negative is negative, and a negative times a negative is a positive.

We then define < on $\mathbb{Q}$ by $x < y$ if and only if $y - x$ is positive. $x \leq y$ if $x < y$ or $x = y$. We often write $y > x$ in place of $x < y$, and $y \geq x$ instead of $x \leq y$. Also, we often write $x < y < z$ to mean $x < y$ and $y < z$. Using this notion of order, we get that $(\mathbb{Q}, +, \cdot)$ along with < is an ordered field.

Distance and the Absolute Value

We then the so-called **absolute value** of x, denoted by $|x|$, by

$$|x| = \begin{cases} x & \text{if } x \text{ is positive} \\ 0 & \text{if } x \text{ is zero} \\ -x & \text{if } x \text{ is negative} \end{cases} \tag{1.5}$$

The following proposition gives the **triangle inequality**:

Proposition 1.2.1 *For $x, y \in \mathbb{Q}$ we have*

$$|x + y| \leq |x| + |y| \tag{1.6}$$

Proof: Let $x, y \in \mathbb{Q}$. From the definition of $|x|$, we see

$$-|x| \leq x \leq |x|, \quad -|y| \leq y \leq |y|,$$

and thus

$$-|x| - |y| \leq x + y \leq |x| + |y|.$$

Since $|x + y|$ either equals $x + y$ or $-(x + y)$, it follows that

$$|x + y| \leq |x| + |y|.$$

□

The absolute value allows us to define a notion of distance on $\mathbb{Q}$ by: for x, y in $\mathbb{Q}$, the **distance** between x and y is defined to be $|x - y|$.

Note: *The properties of the absolute value and < for $\mathbb{Q}$ that were discussed in this section and ones assigned in homework exercises below can be shown to apply to the set of real numbers.*

1.2. FIELDS AND THE SET OF RATIONAL NUMBERS

1.2.3 Homework Exercises

1. Suppose $n \in \mathbb{N}$, use the properties of the natural numbers to show that there does not exist a natural number m so that $n < m < n+1$.

2. If $r \neq 0$ is rational and i is irrational, show that $r \cdot i$ and $r + i$ are irrational.

3. Prove that there cannot exist a rational number whose square is 3.

4. Show that for $a \neq 0$ in any field $\mathbb{F}$ that
$$\frac{1}{-a} = -\frac{1}{a}.$$

5. Show that for any $a, b \in \mathbb{F}$ a field, with $b \neq 0$, we have
$$\frac{-a}{b} = -\frac{a}{b} = \frac{a}{-b}.$$

6. Show that for any $a, b \in \mathbb{F}$ a field, with $a, b \neq 0$, we have
$$\frac{1}{a} \cdot \frac{1}{b} = \frac{1}{a \cdot b}.$$

7. Let $a \in \mathbb{F}$ a field, show that $a^2 = a$ implies $a = 0$ or 1.

8. For $x \in \mathbb{F}$ and $m, n \in \mathbb{N}$ use the definition of x^n to prove the following **properties of exponents**
$$x^m \cdot x^n = x^{m+n}$$
and
$$(x^n)^m = x^{nm}.$$
Hint: Use mathematical induction on one of the natural numbers.

9. For any two rational numbers x, y, use the triangle inequality to prove
$$||x| - |y|| \leq |x - y|.$$

10. Use the definition of the absolute value to show that for any rational number x
$$|x| = |-x|.$$

11. Let $d(x,y) := |x-y|$ for $x, y \in \mathbb{Q}$, show that $d(\cdot, \cdot)$ satisfies the following properties:

    (a) $d(x,y) \geq 0$ and equals 0 if and only if $x = y$.
    (b) $d(x,y) = d(y,x)$
    (c) $d(x,z) \leq d(x,y) + d(y,z)$.

12. Prove the following density property of the rational numbers: Given any two rational numbers x, y, with $x < y$, there is a rational number z satisfying $x < z$ and $z < y$.

13. Prove that for any two rational numbers x, y,
$$|x \cdot y| = |x| \cdot |y|.$$

14. Let $y > 0$ be a rational number, and x and rational number. Prove the following:

    (a) $|x| < y$ if and only if $-y < x < y$.
    (b) $|x| \leq y$ if and only if $-y \leq x \leq y$.

15. For any non-zero rational number x, show that
$$|x^{-1}| = |x|^{-1}.$$

16. For any two rational numbers a, b with $b \neq 0$ show that
$$\left|\frac{a}{b}\right| = \frac{|a|}{|b|}.$$

17. Let $\{a_1, a_2, \cdots, a_n\} \subseteq \mathbb{Q}$. Use an induction argument to show
$$\left|\sum_{k=1}^{n} a_k\right| \leq \sum_{k=1}^{n} |a_k|.$$

18. Suppose $x < y$, where $x, y \in \mathbb{Q}$, show that the following holds
$$x < \frac{x+y}{2} < y.$$

19. Let x, y, z be rational numbers with $x \leq y$, prove the following:
$$x + z \leq y + z.$$

20. Let x, y, z be rational numbers with $x \leq y$ and $z > 0$, prove the following:
$$x \cdot z \leq y \cdot z.$$

21. Let x, y, z be rational numbers with $x \leq y$ and $z < 0$, prove the following:
$$x \cdot z \geq y \cdot z.$$

22. Suppose $a > 1$ is a rational number. Show that $0 < \frac{1}{a} < 1$.

23. Let $a, b \in \mathbb{Q}$, show that
$$a \cdot b \leq \frac{a^2 + b^2}{2}.$$

24. Show that equality holds in the triangle inequality $|x + y| \leq |x| + |y|$ if and only if $x \cdot y \geq 0$.

1.3 Further Properties of the Real Numbers

In this section we will discus further properties of the set of real numbers as a complete, ordered field.

1.3.1 Positivity and Ordering on Fields

An **ordered field** is a field $\mathbb{F}$ which is an ordered set under an ordering defined by $<$. Here $<$ satisfies the following properties for any $x, y, z \in \mathbb{F}$:

1. **Trichotomy Property:** For each x, exactly one of the following is true
$$x = 0, \quad 0 < x, \quad \text{or} \quad x < 0.$$

2. $x < y$ implies $x + z < y + z$.

3. $0 < x$, $0 < y$ implies $0 < x \cdot y$.

4. $x < y$ and $y < z$ implies $x < z$.[8]

We define the subset $\mathcal{P}$ of $\mathbb{F}$ to be the collection $\{x : 0 < x\}$. Note that $\mathcal{P}$ is closed under $+$ and $\cdot$. We call $\mathcal{P}$ the collection of **positive** elements of $\mathbb{F}$.

Note the following proposition, which lists some general properties of an ordered field. Here of course $x < y$ may be written as $y > x$, and we may define $\leq$ as $<$ or $=$ and $\geq$ as $>$ or $=$.

Proposition 1.3.1 *For any ordered field $\mathbb{F}$ we have the following:*

1. $x > 0$ implies $-x < 0$.

2. $x > 0, y > 0$ implies $x + y > 0$.

3. $x > 0$ and $y < z$ implies $x \cdot y < x \cdot z$.

4. $x < 0$ and $y < z$ implies $x \cdot y > x \cdot z$.

5. For $x \neq 0$ we have $x^2 > 0$. In particular, since $1^2 = 1$ we have $1 > 0$.

6. For $x > 1$ we have $0 < \frac{1}{x} < 1$.

Proof:

1. Adding $-x$ to both sides of $x > 0$ gives us
$$0 = -x + x > -x + 0 = -x.$$

2. $x > 0$ implies $x + y > 0 + y = y$. Using $x + y > y$ and $y > 0$, together we get $x + y > 0$.

3. By adding $-y$ to both sides of $y < z$, we get $z - y > 0$. Then multiplying by $x > 0$, and using that $\mathcal{P}$ is closed under $\cdot$, we get $x(z-y) > 0$. Then, the distributive property tells us that $xz - xy > 0$. Adding xy to both sides gives the result.

4. Using $y < z$ and $-x > 0$, we get
$$(-x)y < (-x)z.$$
However $(-x)y = -(xy)$ and $(-x)z = -(xz)$. Then adding $xy + xz$ to both sides we get the result.

---
[8] We often write $x < y$ and $y < z$ as $x < y < z$.

1.3. FURTHER PROPERTIES OF THE REAL NUMBERS

5. If $x > 0$, then since $\mathcal{P}$ is closed under $\cdot$, we get $x^2 > 0$. However $-x = (-1) \cdot x$ and thus $(-x)^2 = (-1)^2 \cdot x^2 = 1 \cdot x^2 = x^2$ Thus if $x < 0$, then $-x > 0$ and $(-x)^2 > 0$. This gives the desired result.

6. If $x > 1$, then $x > 0$. Suppose $\frac{1}{x} < 0$, then

$$1 = \frac{1}{x} \cdot x < \frac{1}{x} \cdot 1 = \frac{1}{x} < 0.$$

This contradicts the fact that $1 > 0$. Hence $\frac{1}{x} > 0$. By taking $x > 1$, and multiplying both sides by $\frac{1}{x}$ we get

$$1 = \frac{1}{x} \cdot x > \frac{1}{x} \cdot 1 = \frac{1}{x}.$$

$\square$

This gives us an alternate **trichotomy property**, which says that for any $x \in \mathbb{F}$ one and only one of the following is true

$$x \in \mathcal{P}, \quad -x \in \mathcal{P}, \quad \text{or} \quad x = 0.$$

The elements x which satisfy $-x \in \mathcal{P}$ satisfy $x < 0$, and are referred to as the **negative** elements of $\mathbb{F}$ and are denoted by $\mathcal{N}$. Thus we see that $\mathbb{F}$ is the disjoint union of the sets $\mathcal{P}, \mathcal{N}$ and $\{0\}$.

Moreover, if $x > 0$ and $y < 0$, we have $x(-y) > 0$. This implies $xy < 0$. Thus a negative times a positive is negative. If $x, y < 0$, then $(-x) \cdot (-y) > 0$. However $(-x) \cdot (-y) = (-1)^2 xy = xy$. Thus a negative times a negative is positive.

1.3.2 The Completeness Axiom of the Set of Real Numbers

Boundedness and the Least Upper Bound Property

To define what it means to be complete, we must first introduce some terminology. We say that a subset S of an ordered field $\mathbb{F}$ is **bounded from above** if there exists $B \in \mathbb{F}$ so that

$$x \leq B, \quad \forall x \in S.$$

We say that a S is **bounded from below** if there exists $b \in \mathbb{F}$ so that

$$b \leq x, \qquad \forall x \in S.$$

A set S is **bounded** if it is bounded from above and bounded from below. Note that the **bounds** B or b don't have to be elements of the set S. The element B is called an **upper bound** for S, and the element b is called a **lower bound** for S. If S is not bounded, it is said to be **unbounded.**

We define the **supremum** or **least upper bound** of a set S, denoted by $\sup S$, to be an element B^* so that

$$x \leq B^*, \qquad \forall x \in S$$

and for any other upper bound B of S,

$$B^* \leq B.$$

Similarly, we define the **infimum** or **greatest lower bound** of a set S, denoted by $\inf S$, to be an element b_* so that

$$b_* \leq x, \qquad \forall x \in S$$

and for any other lower bound b of S,

$$b \leq b_*.$$

Clearly if either $\sup S$ or $\inf S$ exists, then it is unique.

The so-called **completeness axiom** of $\mathbb{R}$ says that every nonempty subset S of $\mathbb{R}$ that is bounded from above has a supremum $\sup S \in \mathbb{R}$.

If S is bounded from below, then $-S = \{-s : s \in S\}$ is bounded from above, and one can use this to show $\inf S \in \mathbb{R}$ exists. To do this show

$$\inf(-S) = -\sup S, \quad \sup(-S) = -\inf S.$$

In the case where S is bounded and non-empty, clearly we have

$$\inf S \leq \sup S.$$

1.3. FURTHER PROPERTIES OF THE REAL NUMBERS 33

In the context of $\mathbb{R}$, if S is not bounded from below, we say that $\inf S = -\infty$, and if S is not bounded from above we say that $\sup S = \infty$. Thus, if S is unbounded, either $\inf S = -\infty$ or $\sup S = \infty$.

For any non-empty finite set S, there exists $\inf S, \sup S \in S$. Moreover, $\inf S = \min S$ and $\sup S = \max S$.

Also we note that for any $x \in \mathbb{R}$, x satisfies

$$-\infty < x < \infty.$$

Some Observations about the Supremum and Infimum

Another observation to make is that for any $S \subset \mathbb{R}$ which is bounded from above, we have that the set $S + a := \{s + a : s \in S\}$ satisfies

$$\sup(S + a) = \sup S + a.$$

To see this, we note that for $s \in S$ we have $s \leq \sup S$ and thus

$$s + a \leq \sup S + a \quad \forall s \in S.$$

Hence we see that
$$\sup(S + a) \leq \sup S + a.$$

Suppose B is an upper bound for the set $S + a$. Then any $s \in S$ satisfies $s + a \leq B$. Consequently, $s \leq B - a$, for all $z \in S$. Using $B = \sup(S + a)$ we get

$$s \leq \sup(S + a) - a, \quad \forall s \in S$$

and thus
$$\sup S \leq \sup(S + a) - a.$$

Together, we have
$$\sup S + a = \sup(S + a).$$

Next, suppose that we have two sets $A, B \subset \mathbb{R}$ which satisfy $a \leq b$ for any $a \in A$ and any $b \in B$. Then from this, fixing $b \in B$ we get $\sup A \leq b$. Thus $\sup A$ is a lower bound for B and thus

$$\sup A \leq \inf B.$$

1.3.3 Some Further Properties of the Set of Real Numbers

The Archimedean Property

The completeness of $\mathbb{R}$ and the lack of an upper bound on $\mathbb{N}$ allows us to deduce the so-called **Archimedean property**:

Proposition 1.3.2 *For $x \in \mathbb{R}$, there is an $n \in \mathbb{N}$ so that $x < n$.*

Proof: If this proposition we not true, then there is an $x \in \mathbb{R}$ so that for all $n \in \mathbb{N}$ we have $n \leq x$. Thus the set of natural numbers has a supremum $\sup \mathbb{N} \in \mathbb{R}$. Now, $\sup \mathbb{N} - 1$ is not an upper bound on $\mathbb{N}$, and thus there is an $m \in \mathbb{N}$ so that $m > \sup \mathbb{N} - 1$. This implies $m + 1 > \sup \mathbb{N}$. However, the inductive property of $\mathbb{N}$ implies that $m + 1 \in \mathbb{N}$, and this is a contradiction since $m + 1 > \sup \mathbb{N}$.
□

A consequence of this result is

$$\inf\{\frac{1}{n} : n \in \mathbb{N}\} = 0. \tag{1.7}$$

Since $0 < \frac{1}{n} \leq 1$ we know $x := \inf\{\frac{1}{n} : n \in \mathbb{N}\} \geq 0$. If $x > 0$, then $\frac{1}{2} \cdot x < \frac{1}{n}$ for all $n \in \mathbb{N}$. However, consider the real number $(\frac{1}{2} \cdot x)^{-1}$. By the Archimedean property, there is an $N \in \mathbb{N}$ so that $N > (\frac{1}{2} \cdot x)^{-1}$. Multiplying both sides of this inequality by the positive number $\frac{1}{N} \cdot \frac{1}{2} \cdot x$, we get $\frac{1}{2}x > \frac{1}{N}$, and this is a contradiction. Thus $x = 0$.

This guarantees that any positive real number y is not a lower bound for $\{\frac{1}{n} : n \in \mathbb{N}\}$. Thus, there is an $n \in \mathbb{N}$ so that

$$0 < \frac{1}{n} < y.$$

Another useful consequence of the Archimedean property is the following:

Corollary 1.3.3 *Given any positive real number x, there is a natural number n so that*

$$n - 1 \leq x < n.$$

1.3. FURTHER PROPERTIES OF THE REAL NUMBERS

Proof: Consider the set $\{m \in \mathbb{N} : x < m\}$. By the Archimedean property, this set is non-empty. By the Well-Ordering property of $\mathbb{N}$, we may conclude that this set has a least element, say n. Then $n - 1$ is not in this set, and we get the desired result.
□

The Supremum of a Function

Suppose $f : D \to \mathbb{R}$. We say that f is **bounded** if the set $\{f(x) : x \in D\}$ is a bounded set. We will often use the notation

$$\sup_{x \in D} f(x) = \sup\{f(x) : x \in D\}, \tag{1.8}$$

and

$$\inf_{x \in D} f(x) = \inf\{f(x) : x \in D\}. \tag{1.9}$$

The Existence of the nth Roots of Positive Real Numbers

At this point we shall show the existence of nth **roots** of positive real numbers x, which we shall denote as either $x^{\frac{1}{n}}$ or $\sqrt[n]{x}$, and if $n = 2$ as $x^{\frac{1}{2}}$ or $\sqrt{x}$.

Theorem 1.3.4 *For any $x > 0$ there exists a unique positive real number, denoted as either $\sqrt[n]{x}$ or $x^{\frac{1}{n}}$, which satisfies*

$$(\sqrt[n]{x})^n = x.$$

Proof: First we shall address uniqueness of such n^{th} roots. To see this, suppose that
$$0 < y < z.$$
Then for any $n \in \mathbb{N}$ we will show by induction on n that
$$0 < y^n < z^n.$$

The $n = 1$ case being our initial assumption. Suppose that for some $k \in \mathbb{N}$ that
$$0 < y^k < z^k,$$

upon multiplying this inequality by y we get

$$0 < y^{k+1} < z^k \cdot y.$$

However, $y < z$, and thus multiplying this by z^k we get

$$yz^k < z^{k+1}.$$

This gives us the desired inequality.

Suppose that we have two numbers y, z with $y < z$ so that $y^n = z^n$. Then using this inequality, we would get a contradiction. Hence, n^{th} roots of positive numbers must be unique, provided they exist.

Fix $x > 0$. Then we define S to be the set

$$\{t \in \mathbb{R} : t^n < x\}.$$

Since $t := \frac{x}{x+1} > 0$ is less than x, and is less than 1, we have $t^n \leq t < x$, and thus S is non-empty. If $t > 1 + x$, since $t > 1$ we have $t^n \geq t > x$, and hence S is bounded from above. Thus $\sup S$ exists and is a positive real number. Let y denote this quantity. To show that $y^n = x$, we will show that $y^n < x$ and $y^n > x$ can not hold.

The identity

$$\frac{b^n - a^n}{b - a} = b^{n-1} + a \cdot b^{n-2} + a^2 \cdot b^{n-3} + \cdots + a^{n-2} \cdot b + a^{n-1}$$

gives us

$$\frac{b^n - a^n}{b - a} < nb^{n-1}, \tag{1.10}$$

whenever $0 < a < b$.

If $y^n < x$, then we may choose an h satisfying $0 < h < 1$ and $h < \frac{x - y^n}{n \cdot (y+1)^{n-1}}$. Then using the above inequality (1.10) with $a = y$ and $b = y + h$ gives us

$$\frac{(y+h)^n - y^n}{h} < n(y+h)^{n-1} < n(y+1)^{n-1}.$$

1.3. FURTHER PROPERTIES OF THE REAL NUMBERS

This implies that
$$(y+h)^n - y^n < hn(y+1)^{n-1} < x - y^n.$$

Therefore we get
$$(y+h)^n < x,$$
and thus $y + h \in S$. This is a contradiction to the definition of $y = \sup S$. Hence $y^n < x$ is impossible.

Next we suppose $y^n > x$. By letting
$$k := \frac{y^n - x}{ny^{n-1}},$$
we have $k > 0$. Since $n \geq 2$ we have $y^n < ny^n$, and thus $0 < y^n - x < y^n < ny^n$. This implies
$$0 < k = \frac{y^n - x}{ny^{n-1}} < y.$$

If $z \geq y - k > 0$, then by using $a = y - k$ and $b = y$ in the above inequality (1.10), we see
$$y^n - z^n \leq y^n - (y-k)^n < kny^{n-1} = y^n - x.$$

Hence
$$z^n > x,$$
and in particular $z \notin S$. Thus $y - k$ is an upper bound for S, and this contradicts the definition of $y = \sup S$.

Thus, we must have
$$(\sup S)^n = y^n = x.$$

☐

Hence, we see that for each $x > 0$
$$y = \sqrt[n]{x}, \quad \text{or} \quad x^{\frac{1}{n}}$$
is a well-defined quantity.

Suppose $x, z > 0$, then we have the following **property of exponents for n^{th} roots**:
$$\sqrt[n]{x} \cdot \sqrt[n]{z} = \sqrt[n]{x \cdot z}$$
or
$$x^{\frac{1}{n}} \cdot z^{\frac{1}{n}} = (x \cdot z)^{\frac{1}{n}}.$$
To see this, we note that $(x \cdot z)^{\frac{1}{n}}$ exists and is unique. Moreover, since $(ab)^k = a^k b^k$ for $k \in \mathbb{N}$, we have
$$(x^{\frac{1}{n}} \cdot z^{\frac{1}{n}})^n = (x^{\frac{1}{n}})^n \cdot (z^{\frac{1}{n}})^n = x \cdot z.$$
Thus we get the desired identity.

The Existence of an Irrational Number

Here we will show that the number $\sqrt{2} \in \mathbb{R}$ is not an element of $\mathbb{Q}$. Thus the set $\mathbb{R} \setminus \mathbb{Q}$ is non-empty.

On the set of integers $\mathbb{Z}$, we say that an integer n is **even** if there is an integer k so that $n = 2k$. We say that an integer n is **odd** if there is an integer j so that $n = 2j + 1$. Since $\frac{1}{2}$ is not an integer, we get that no integer can be both even and odd. Moreover, every integer is either even or odd. One can show that the square of an even is even, and the square of an odd is odd.

We suppose that the number $\sqrt{2} \in \mathbb{Q}$. Then there are positive integers m, n so that
$$\sqrt{2} = \frac{m}{n}.$$
Moreover, we can assume that m, n have no common factors. In particular, both are not even. Then $2 = \frac{m^2}{n^2}$ and thus
$$2n^2 = m^2.$$
From this we may conclude that m is even, and thus $m = 2k$, for some positive $k \in \mathbb{Z}$. Hence $2n^2 = (2k)^2$ or
$$n^2 = 2k^2$$

1.3. FURTHER PROPERTIES OF THE REAL NUMBERS

and thus n is even as well. This is a contradiction. Hence

$$\sqrt{2} \notin \mathbb{Q}.$$

Thus we have established that the set of irrational numbers $\mathbb{R} \setminus \mathbb{Q}$ is non-empty.

The Density of the Rational Numbers and the Irrational Numbers in the Set of Real Numbers

We will now show that between any two real numbers, one can always find a rational number, and one can always find an irrational number. To see this we will use that a non-zero rational times an irrational is irrational. Thus, any rational number times $\sqrt{2}$ is irrational.

Theorem 1.3.5 *Let $x, y \in \mathbb{R}$ with $x < y$. Then there exists $r \in \mathbb{Q}$ so that*

$$x < r < y.$$

Proof: If $x < 0 < y$ we are done, since $0 \in \mathbb{Q}$. Thus, we may assume $0 \leq x < y$ or $x < y \leq 0$. Moreover, if $x = 0$ in the first case, we can look at the two numbers $\frac{y}{2}, y$ instead, noting that they satisfy $0 < \frac{y}{2} < y$. If $y = 0$ in the second case, we can look at the two numbers $\frac{x}{2}, x$ instead, noting that they satisfy $x < \frac{x}{2} < 0$. Thus it is only necessary to consider the cases

$$0 < x < y \quad \text{or} \quad x < y < 0.$$

If $0 < x < y$, then $y - x > 0$. Since $\inf\{\frac{1}{n} : n \in \mathbb{N}\} = 0$, we may find $n \in \mathbb{N}$ so that $\frac{1}{n} < y - x$. Thus

$$nx + 1 < ny.$$

Since $nx > 0$, by the Archimedean property (1.3.2) we know that we can find a natural number m so that

$$m - 1 \leq nx < m.$$

Therefore,

$$m \leq nx + 1 < ny.$$

Together we get
$$nx < m < ny,$$
and thus
$$x < \frac{m}{n} < y.$$

In the case $x < y < 0$. We apply the result to $-x, -y$ which satisfy $0 < -y < -x$ to find $r \in \mathbb{Q}$ with $0 < -y < r < -x$. Then
$$x < -r < y.$$

□

Corollary 1.3.6 *Let $x, y \in \mathbb{R}$ with $x < y$. Then there exists $i \in \mathbb{R} \setminus \mathbb{Q}$ so that*
$$x < i < y.$$

Proof: Applying the density theorem above to a pair of real numbers $\frac{x}{\sqrt{2}}, \frac{y}{\sqrt{2}}$, we may find a rational number[9] r satisfying
$$\frac{x}{\sqrt{2}} < r < \frac{y}{\sqrt{2}},$$
giving us
$$x < r \cdot \sqrt{2} < y.$$

□

Decimal Expansions and Real Numbers

Here we will discuss the decimal expansion of positive real numbers. Since any positive real number x has a natural number n satisfying $n - 1 \leq x < n$, we have that $x = (n - 1) + r$, where $r \in [0, 1)$. We write $n - 1$ as $D_0 + D_1 \cdot (10) + D_2 \cdot (10)^2 + \cdots D_k \cdot (10^k)$, and this can be done in a unique way via repeated use of the Euclidean Algorithm: First write $n - 1 = q_0 \cdot 10 + D_0$, then write $q_0 = q_1 \cdot 10 + D_1$, and then write $q_1 = q_2 \cdot 10 + D_2$, etc. This

---

[9] Note that if $x < 0$ and $y > 0$ we may instead use the pair $0, y$ or the pair $x, 0$ to insure that $r \neq 0$.

1.3. FURTHER PROPERTIES OF THE REAL NUMBERS

process will end when $q_{k-1} = 0 \cdot 10 + D_k$, and D_k is the final digit.[10]

Then by corollary 1.3.3 there is a whole number d_1 so that $d_1 \leq 10r < d_1 + 1$, with $d_1 \in \{0, 1, 2, 3, 4, 5, 6, 7, 8, 9\}$ because $r < 1$. Then

$$\frac{d_1}{10} \leq r < \frac{d_1 + 1}{10}.$$

Since $0 \leq 10r - d_1 < 1$, we can use the corollary 1.3.3 to find a $d_2 \in \{0, 1, 2, 3, 4, 5, 6, 7, 8, 9\}$ so that $d_2 \leq 10^2 r - 10 d_1 < d_2 + 1$. Hence,

$$\frac{d_1}{10} + \frac{d_2}{10^2} \leq r < \frac{d_1}{10} + \frac{d_2 + 1}{10^2}.$$

Continuing in this matter, we may find $d_3, \cdots, d_m \in \{0, 1, 2, 3, 4, 5, 6, 7, 8, 9\}$ so that

$$\frac{d_1}{10} + \frac{d_2}{10^2} + \cdots + \frac{d_m}{10^m} \leq r < \frac{d_1}{10} + \frac{d_2}{10^2} + \cdots + \frac{d_m + 1}{10^m}.$$

Let S be the set of real numbers

$$\{\frac{d_1}{10} + \frac{d_2}{10^2} + \cdots + \frac{d_j}{10^j} : j \in \mathbb{N}\},$$

then

$$\sup S = r.$$

In such a case, we say that r has a **decimal expansion** given by

$$0.d_1 d_2 d_3 \cdots d_m \cdots.$$

The decimal expansion of x is given by

$$D_k \cdots D_2 D_1 D_0 . d_1 d_2 d_3 \cdots d_m \cdots.$$

For negative values of x, we find the decimal expansion for $|x|$, and place a minus sign in front of it.

It can be shown that up to examples like $0.19999 \cdots = 0.20000 \cdots$, the decimal expansion is unique.

---

[10]The Euclidean algorithm: For any two natural numbers $a < b$, we may write

$$b = q \cdot a + r,$$

where $r \in \{0, 1, \cdots a - 1\}$ is unique, and $q \in \mathbb{N}$ is unique.

Cardinality of the Real Numbers, the Rational Numbers and the Irrational Numbers

We say that a non-empty set S is **countable** if it is either **finite** – meaning there is an $n \in \mathbb{N}$ so that there is a bijection between S and $\{1, 2, 3, \cdots, n\}$ – or there is a bijection between S and $\mathbb{N}$. In the case where there is a bijection between S and $\mathbb{N}$, we say that S is **countably infinite**. If a set is not countable, we say that it is **uncountable**.

It can be show that if S, T are countable, then $S \cup T$ and $S \times T$ are countable sets. Moreover, if $S \subseteq T$, and T is countable, then S is countable as well.

We will now show that the set of rational numbers is countable. Since the set of rational numbers is the union of the sets of the positive rational numbers, the negative rational numbers and $\{0\}$, it suffices to show that the set of positive rational numbers are countable.[11] Moreover, by our construction of $\mathbb{Q}$, the countability of $\mathbb{N} \times \mathbb{N}$ implies that $\mathbb{Q}$ is countable. To show that $\mathbb{N} \times \mathbb{N}$ is countable[12], we simply use the typical "snaking argument" of listing the ordered pairs

$$
\begin{array}{cccccc}
(1,1) & (1,2) & (1,3) & (1,4) & (1,5) & \cdots \\
(2,1) & (2,2) & (2,3) & (2,4) & (2,5) & \cdots \\
(3,1) & (3,2) & (3,3) & (3,4) & (3,5) & \cdots \\
(4,1) & (4,2) & (4,3) & (4,4) & (4,5) & \cdots \\
(5,1) & (5,2) & (5,3) & (5,4) & (5,5) & \cdots \\
\vdots & \vdots & \vdots & \vdots & \vdots & \ddots
\end{array}
$$

as

$$(1,1), (2,1), (1,2), (1,3), (2,2), (3,1), (4,1),$$
$$(3,2), (2,3), (1,4), (1,5), (2,4), \cdots$$

We see that $\mathbb{Q}$ is a countable set.

We will now show that $\mathbb{R}$ is uncountable, and we will do that by showing the interval $[0, 1]$ is uncountable, because if a set has an uncountable subset,

---

[11] Clearly there is a bijection $f(x) = -x$ between the positive rational numbers and the negative rational numbers.

[12] Or more generally, $S \times T$, where T, S are countable.

1.3. FURTHER PROPERTIES OF THE REAL NUMBERS

it too must be uncountable. Suppose that the interval $[0,1]$ is countable. Let the following be a list of the decimal expansions of all the numbers in this interval

$$0.a_{11}a_{12}a_{13}a_{14}\cdots a_{1n}\cdots$$
$$0.a_{21}a_{22}a_{23}a_{24}\cdots a_{2n}\cdots$$
$$0.a_{31}a_{32}a_{33}a_{34}\cdots a_{3n}\cdots$$
$$0.a_{41}a_{42}a_{43}a_{44}\cdots a_{4n}\cdots$$
$$0.a_{51}a_{52}a_{53}a_{54}\cdots a_{5n}\cdots$$
$$\vdots$$
$$0.a_{k1}a_{k2}a_{k3}a_{k4}\cdots a_{kn}\cdots$$
$$\vdots$$

Where here $a_{ij} \in \{0,1,2,3,4,5,6,7,8,9\}$ for any $i,j \in \mathbb{N}$. Consider the real number in $[0,1]$ whose decimal expansion is given by

$$0.a_{11}a_{22}a_{33}a_{44}\cdots.$$

We define another real number in $[0,1]$ by

$$0.d_1d_2d_3d_4d_5\cdots$$

where $d_i \in \{0,1,2,3,4,5,6,7,8,9\}$ satisfies

$$d_i = a_{ii} + 2 \quad \text{mod } 10.$$

If the number $0.d_1d_2d_3d_4\cdots$ were in the list, it is in the nth place for some $n \in \mathbb{N}$. Then d_n must be a_{nn}, and by arrangement this is impossible. [13]

Thus, we see that $\mathbb{R}$ is uncountable. The the set of irrational numbers is uncountable as well, because if $\mathbb{R} \setminus \mathbb{Q}$ were countable, then the union $\mathbb{Q} \cup (\mathbb{R} \setminus \mathbb{Q}) = \mathbb{R}$ would also be countable.

1.3.4 Homework Exercises

1. Define the set
$$-S := \{-s : s \in S\},$$
where S is some given non-empty set. Show the following:
$$\inf(-S) = -\sup S, \quad \sup(-S) = -\inf S.$$

---

[13] Note that this construction rules out the problem caused by lack of uniqueness in decimal representation, since $0.45999999\cdots$ is the same as $0.460000\cdots$.

2. Show that if a set S contains one of its upper bounds, then this upper bound is the supremum for the set.

3. Suppose $A \subseteq B \subset \mathbb{R}$ are non-empty sets, show the following:
$$\inf B \leq \inf A \leq \sup A \leq \sup B.$$

4. Show that every non-empty finite subset of the real numbers contains its supremum and its infimum.

5. Show that
$$\sup\{1 - \frac{1}{n} : n \in \mathbb{N}\} = 1.$$

6. Show that there exists a real number x such that $x^3 = 3$.

7. Show that any real number x satisfying $x^3 = 3$ is not rational.

8. For $a > 0$ and S bounded, let $aS = \{as : s \in S\}$. Show that
$$a \sup S = \sup(aS), \quad a \inf S = \inf(aS).$$

9. For $a < 0$ and S bounded, let $aS = \{as : s \in S\}$. Show that
$$a \sup S = \inf(aS), \quad a \inf S = \sup(aS).$$

10. Prove that $\sqrt{3}$ is irrational.

11. Prove that $\sqrt[3]{5}$ is irrational.

12. For $b > 0$ and $m, n \in \mathbb{N}$ show that
$$(b^{\frac{1}{n}})^m = (b^m)^{\frac{1}{n}},$$
and this allows us to define $b^{\frac{m}{n}}$ to be this quantity. Show that if $i, j, n, m \in \mathbb{N}$ satisfy $\frac{i}{j} = \frac{m}{n}$, then
$$b^{\frac{i}{j}} = b^{\frac{m}{n}}.$$

13. The above exercise allows us to define b^r, for $b > 0$ and r a positive rational number. For $b > 0$ and for positive rational numbers r, s prove the following properties of exponents
$$b^r \cdot b^s = b^{r+s}, \quad (b^r)^s = b^{r \cdot s} = (b^s)^r.$$

1.3. FURTHER PROPERTIES OF THE REAL NUMBERS

14. A real number x is said to be **algebraic** if it a root of a polynomial equation

$$a_n x^n + a_{n-1} x^{n-1} + \cdots + a_2 x^2 + a_1 x + a_0 = 0 \qquad (1.11)$$

where $a_0, a_1, \cdots, a_{n-1}, a_n \in \mathbb{Z}$ and $a_n \neq 0$. For instance $\sqrt{2}$ is algebraic since it satisfies $x^2 - 2 = 0$. Numbers which are not algebraic are called **transcendental numbers**.[14] Moreover, it is well-known that the irrational numbers π, e are transcendental numbers.[15]

(a) Show that $\sqrt[7]{\frac{2}{5}}$ is algebraic.

(b) Show that every element of $\mathbb{Q}$ is algebraic.

(c) Let $\mathbb{A}$ be the collection of algebraic real numbers. Show that $\mathbb{A}$ is countable. *Hint: For each polynomial equation with integer coefficients (1.11) assign the quantity $n + |a_n| + |a_{n-1}| + \cdots + |a_2| + |a_1| + |a_0|$. For each positive integer m, there are only finitely many equations which are of the form*

$$m = n + |a_n| + |a_{n-1}| + \cdots + |a_2| + |a_1| + |a_0|.$$

---

[14] The mathematician Euler was probably the first person to define what a transcendental number is in the modern sense of the definition. The first number proven to be transcendental was the number $\sum_{n=1}^{\infty} \frac{1}{10^{n!}} = 0.11000100000000000000000010\cdots$, and it was proven to be transcendental in 1844 by the mathematician Joseph Liouville.

[15] The symbol e was first used by Leonard Euler in 1727. In 1737 Euler showed e is irrational, and in 1873 the mathematician Charles Hermite proved that e was transcendental.

π is the symbol used today to represent the ratio of a circumference of a circle to its diameter, namely

$$3.141592653589793238462\cdots.$$

The symbol was first used in 1707 by the mathematician William Jones, and was popularized by Euler who started using this notation in 1736.

Johann Heinrich Lambert conjectured that e and π were both transcendental and proved that π is irrational in 1761. It was proven that π was transcendental in 1882 by the mathematician Ferdinand von Lindemann. For a complete proof, see for instance the **Lindemann-Weierstrass Theorem** in [3] section 4.12.

1.4 The Construction of the Real Numbers

We now will show how to construct the set of real numbers $\mathbb{R}$ from the set of rational numbers $\mathbb{Q}$ so that $\mathbb{Q} \subseteq \mathbb{R}$, with $\mathbb{R}$ inheriting the same operations $+, \cdot$ and the ordering $<$, but so that $\mathbb{R}$ is a complete, ordered field.

There are several approaches that are commonly used to construct the real numbers out of the set of rational numbers. One approach uses the collection of all Cauchy sequences of rational numbers, with Cauchy sequences being identified that would have a common limit, if one did exist.[16] This approach was first taken by the mathematician Georg Cantor, and published in 1872.

Another common approach to construct $\mathbb{R}$ from $\mathbb{Q}$ involves what are commonly referred to as **Dedekind cuts**. This approach was first taken by the mathematician Richard Dedekind, and was also published in 1872. Below we will discuss this method. A third approach that was employed to construct the real numbers was by using decimal expansions,[17] and this was done by the mathematician Karl Weierstrass, also in the mid to late nineteenth century.

1.4.1 Dedekind Cuts and the Construction of the Real Numbers

In this section we will suppose that the set of rational numbers $\mathbb{Q}$ are at our disposal, and from the field $(\mathbb{Q}, +, \cdot)$ we will produce the field $(\mathbb{R}, +, \cdot)$.

The idea involved is very simple. We will think of a real number α in terms of a set of rational numbers, namely the set A of rational numbers which are less than α. However, without the prior existence of the set of real numbers, we cannot describe this set A in this matter. So we think about the ways we may describe such a set A, and the following works: If $x \in A$, and $y \in \mathbb{Q}$ satisfies $y < x$, then $y \in A$. A should be non-empty. A should not be all of $\mathbb{Q}$, and so there is a $z \in \mathbb{Q}$ so that $x < z$ for all $x \in A$. Also, if $x \in A$, there is a $y \in A$ so that $x < y$, and so A should not contain a greatest member.

---

[16] For a construction of $\mathbb{R}$ using Cauchy Sequences see for instance [13] sections 2.1 - 2.3.
[17] For a brief discussion on this approach see for instance [13] section 2.4.1.

1.4. THE CONSTRUCTION OF THE REAL NUMBERS

This motivates us to define a **Dedekind cut** A, or simply a **cut** A by

1. $A \subseteq \mathbb{Q}$.
2. $x \in A$, and $y \in \mathbb{Q}$ with $y < x$ implies $y \in A$.
3. A is not empty.
4. $A \neq \mathbb{Q}$.
5. $x \in A$ implies there is a $y \in A$ satisfying $x < y$.

Note that we actually have the following as well, which follow from the definition of a cut:

1. $x \in A$ and $y \notin A$ implies $x < y$.
2. $y \notin A$ and $y < z$ implies $z \notin A$.

We will actually so that the collection of Dedekind Cuts $\mathcal{R}$ with the appropriate operations $+, \cdot$, defined on this collection, can be identified with the real numbers $\mathbb{R}$. A rational number x will have its natural counter part $x_\mathcal{R} := \{y \in \mathbb{Q} : y < x\}$, which is a cut.

We note that throughout our discussion a cut will be denoted by a capital Roman letter, a rational number by a lower case Roman letter, and if necessary, a real number by a script Greek letter.

Ordering on the Collection of Dedekind Cuts

On the collection of cuts we define our ordering by A is a proper subset of B, denoted by
$$A \subset B.$$
Given two cuts A, B, we will now show that one and only one of the following holds
$$A \subset B, \quad A = B, \quad \text{or} \quad B \subset A.$$
To see this, suppose the first two do not hold. That is, A is not a subset of B. Then there is an $y \in A$ with $y \notin B$. Then for any $x \in B$, from the first property that follows from the definition of a cut, we have
$$x \in B, \quad \text{and} \quad y \notin B \quad \Rightarrow \quad x < y.$$
Then by the second property in the definition of a cut, we get $x \in A$, and thus $B \subset A$. Hence $\subset$ defines an ordering on $\mathcal{R}$.

The Ordered Set $\mathcal{R}$ Satisfies the Completeness Property

Let $\mathcal{A}$ be a non-empty subset of $\mathcal{R}$, with $B \in \mathcal{R}$ being an upper bound for $\mathcal{A}$. Define
$$C := \cup_{D \in \mathcal{A}} D.$$
We will now show that C is a cut and $C = \sup \mathcal{A}$.

Since $\mathcal{A}$ is non-empty, there exists a non-empty $E \in \mathcal{A}$. Since $E \subseteq C$ we have C is non-empty. Since B is an upper bound for $\mathcal{A}$, we have $D \subseteq B$ for all D in $\mathcal{A}$, and hence
$$C \subseteq B.$$
Thus $C \neq \mathbb{Q}$. We see that properties 1, 3 and 4 of the definition of a Dedekind Cut hold for C.

Now, let $y \in C$, then $y \in F$ for some $F \in \mathcal{A}$. For $x \in \mathbb{Q}$ with $x < y$ we have $x \in F$, and hence $x \in C$. Thus property 2 holds. Likewise, for this arbitrary $y \in C$, we see that since $y \in F$, for some $F \in \mathcal{A}$, we have that there is a $z \in F$ with $y < z$, and thus $z \in C$ as well. Hence property 5 holds for C, and thus $C \in \mathcal{R}$.

It is also clear that C is an upper bound for $\mathcal{A}$, since $G \in \mathcal{A}$ has $G \subseteq C$, by the definition of C.

Suppose $H \in \mathcal{R}$ satisfies $H \subset C$. Then there is an $x \in C$ with $x \notin H$. $x \in C$ implies $x \in J$ for some J in $\mathcal{A}$. Thus we see that neither $J \subset H$ nor $H = J$ is true. Since H, J are cuts, we must have $H \subset J$ holds. Thus H is not an upper bound for $\mathcal{A}$, and therefore, $C = \sup \mathcal{A}$.

Addition on $\mathcal{R}$

For $A, B \in \mathcal{R}$ we define
$$A + B = \{a + b : a \in A, b \in B\}.$$
We need to show $A + B \in \mathcal{R}$.

Since A, B are non-empty, we have $A + B$ is non-empty. Take $x \notin A$ and $y \notin B$, then $a < x$ and $b < y$ for any $a \in A, b \in B$. Thus $a + b < x + y$

1.4. THE CONSTRUCTION OF THE REAL NUMBERS 49

for any $a + b \in A + B$. Thus $A + B \neq \mathbb{Q}$, and so wee see $A + B$ satisfies properties 1, 3 and 4 for Dedekind cuts

Let $z \in A + B$. Then $z = a + b$ for some $a \in A$, $b \in B$. If $y < z$ is a rational number, then $y < a + b$ and so $y - b < a$. This implies $y - b \in A$. Hence $y = (y - b) + b \in A + B$. Thus property 2 in the definition of a cut holds for $A + B$.

Now, choose $x > a$ in A. Then $z < x + b$ and $x + b \in A + B$. Thus property 5 holds for a cut. Thus $A + B$ is a Dedekind Cut.

By the definition of $A + B$, we have

$$A + B = \{a + b : a \in A, b \in B\} = \{b + a : a \in A, b \in B\} = B + A.$$

Thus + is a commutative operation. Likewise, it is associative, since

$$(A + B) + C = \{(a + b) + c : a + b \in A + B, c \in C\}$$
$$= \{(a + b) + c : a \in A, b \in B, c \in C\}$$
$$= \{a + (b + c) : a \in A, b \in B, c \in C\}$$
$$= \{a + (b + c) : a \in A, b + c \in B + C\}.$$

The set of negative rational numbers

$$0_\mathcal{R} := \{x \in \mathbb{Q} : x < 0\}$$

will serve as our zero for this form of addition. We will now show $0_\mathcal{R}$ is a cut and it satisfies the property $0_\mathcal{R} + A = A$, for any $A \in \mathcal{R}$. Clearly $0_\mathcal{R} \subset \mathbb{Q}$, is not empty, and is not all of $\mathbb{Q}$. Now if $x \in 0_\mathcal{R}$, then $x < 0$. Given any $y \in \mathbb{Q}$ with $y < x$, we get $y < 0$, and thus $y \in 0_\mathcal{R}$. Also, given any $x \in 0_\mathcal{R}$, $x < \frac{x}{2} < 0$, and $y = \frac{x}{2} \in 0_\mathcal{R}$. Thus $0_\mathcal{R} \in \mathcal{R}$. Since any $x \in 0_\mathcal{R}$ is negative, we have for any $a \in A$, $a + x < a$, and thus $a + x \in A$. Thus $A + 0_\mathcal{R} \subseteq A$. We will now show $A \subseteq A + 0_\mathcal{R}$. Let $a \in A$, pick $x \in A$ so that $a < x$. Then $a - x < 0$ so $a - x \in 0_\mathcal{R}$. Then $a = x + (a - x)$, and clearly $x + (a - x) \in A + 0_\mathcal{R}$. Thus we get $A \subseteq A + 0_\mathcal{R}$.

Next we need to show that there is a cut $-A$ so that $A + (-A) = 0_\mathcal{R}$. We define $-A$ by

$$-A := \{x \in \mathbb{Q} : -x \notin A, \text{ but } -x \text{ is not the least element of } \mathbb{Q} \setminus A\},$$

or simply

$$-A = \{x \in \mathbb{Q} : \exists r > 0, r \in \mathbb{Q}, \text{ such that } -r - x \notin A\}.$$

We will now show $-A \in \mathcal{R}$ and $A + (-A) = 0_\mathcal{R}$.

If $x \notin A$, then $y = -x - 1$ has $x = -y - 1 \notin A$. Thus $y \in (-A)$, and we see that $-A$ is not empty. If $a \in A$, then $-a \notin (-A)$, So $-A \neq \mathbb{Q}$. Thus properties 1,3 and 4 hold for $-A$ in the definition of a Dedekind Cut. Pick $c \in (-A)$. Then we may find a positive rational r so that $-r - c \notin A$. Pick any rational $b < c$. Then $-b - r > -c - r$ and so $-b - r \notin A$. So $b \in (-A)$, and thus property 2 for being a cut holds for $-A$. Let $d = c + \frac{r}{2}$. Then $d > c$ and $-d - \frac{r}{2} = -c - r \notin A$, so $d \in (-A)$. Therefore, property 5 for being a cut holds for $-A$. Hence $-A \in \mathcal{R}$.

Next we will show $A + (-A) = 0_\mathcal{R}$. Let $a \in A$ and $b \in (-A)$. Then $-b \notin A$, and $a < -b$. Thus $a + b < 0$ which implies $a + b \in 0_\mathcal{R}$, and so $A + (-A) \subseteq 0_\mathcal{R}$. Now suppose $y \in 0_\mathcal{R}$. Let $z = -\frac{y}{2}$, and so $z > 0$. We claim that there is a positive integer n so that $nz \in A$ but $(n+1)z \notin A$. Let $N = \{k \in \mathbb{N} : kz \notin A\}$. This set is a non-empty subset of the naturals since there is a $q \in \mathbb{Q}$ so that $q \notin A$, and by the Archimedean Property of $\mathbb{Q}$, there is a $j \in \mathbb{N}$ so that $j - 1 \leq \frac{q}{z} < j$. Thus $(j-1)z \leq q < jz$, and $jz > q$; so $jz \notin A$ as well. Hence N is non-empty as well. So by the well ordering property of $\mathbb{N}$, N has a smallest element m. Thus $mz \notin A$, but $(m-1)z \in A$. Let $n = m - 1$ to see the claim, and we define $x = -(n+2)z$. Then $x \in (-A)$ since $-x - z = (n+1)z \notin A$, and $y = nz + x \in A + (-A)$. Hence $0_\mathcal{R} \subseteq A + (-A)$, and so we see $A + (-A) = 0_\mathcal{R}$.

Thus $+$ as defined on $\mathcal{R}$ satisfies the first four field axioms.

Further Properties of the Ordering on $\mathcal{R}$

Before we go on to define multiplication on $\mathcal{R}$, we will show that property two of an ordered field holds. That is for any $A, B, C \in \mathcal{R}$, we have $A \subset B$ implies $A + C \subset B + C$. If $A \subset B$, then $a \in A$ has $a \in B$, and $\exists b \in B$ so that $b \notin A$. Now, $A + C = \{a + c : a \in A, c \in C\} \subseteq \{b + c : b \in B, c \in C\} = B + C$. If $A + C = B + C$, then by adding $-C$ to both sides we get $A = B$, and so $A + C \subset B + C$ must hold.

1.4. THE CONSTRUCTION OF THE REAL NUMBERS

Moreover, we also know that for any $A \in \mathcal{R}$ one and only one of the following is true:

$$A \subset 0_\mathcal{R}, \quad A = 0_\mathcal{R}, \quad \text{or} \quad 0_\mathcal{R} \subset A.$$

Multiplication on $\mathcal{R}$

We will first restrict ourselves to positive cuts when we define multiplication. Let $\mathcal{R}^+$ denote the set $\{A \in \mathcal{R} : 0_\mathcal{R} \subset A\}$.

Suppose $A, B \in \mathcal{R}^+$. We define AB by

$$AB := \{p \in \mathbb{Q} : \exists a \in A, b \in B; a, b > 0; p \leq ab\}.$$

We define $1_\mathcal{R}$ by

$$1_\mathcal{R} := \{q \in \mathbb{Q} : q < 1\}.$$

We now show that $AB \in \mathcal{R}^+$ whenever $A, B \in \mathcal{R}^+$. For $A, B \in \mathcal{R}^+$ we have that there exists $a, b > 0$ so that $a \in A$ and $b \in B$. Moreover $ab \in AB$ so AB is non-empty, and is a subset of $\mathbb{Q}$. Also, there exist rational numbers x, y so that $x \notin A$, $y \notin B$ and $a < x$, $b < y$ for any $a \in A$, $b \in B$. Then any $p \in AB$ satisfies $p < xy$, and thus $AB \neq \mathbb{Q}$. So properties 1, 3 and 4 hold in the definition of a cut for AB. Suppose $p \in AB$, and let $q < p$ be a rational number. Then there is an $a \in A$ and $b \in B$ so that $p \leq ab$, and thus $q \leq ab$ as well, so $q \in AB$. Next, suppose $r \in AB$. Then there is an $a \in A$, $b \in B$ so that $r \leq ab$. Now there is an $a' \in A$ and $b' \in B$ so that $a < a'$ and $b < b'$. Then $r < a'b'$. Moreover, $a'b' \in AB$. Thus we see that properties 2 and 5 also hold. Thus $AB \in \mathcal{R}$. Moreover, $AB \in \mathcal{R}^+$ since it will contain all negative rational numbers, and at least one positive rational.

For $A, B \in \mathcal{R}^+$ we have

$$AB = \{p \in \mathbb{Q} : \exists a \in A, \exists b \in B, a > 0, b > 0 \text{ such that } p \leq ab\}$$

$$= \{p \in \mathbb{Q} : \exists b \in B, \exists a \in A, b > 0, a > 0 \text{ such that } p \leq ba\} = BA.$$

Therefore $\cdot$ is commutative. Next we shall show that $\cdot$ is associative. That is, for any $A, B, C \in \mathcal{R}^+$, we have $(AB)C = A(BC)$.

Let $p \in (AB)C$. Then there exists positive rational numbers $d \in AB$ and $c \in C$ so that $p \leq dc$. Since $d \in AB$, there exists positive rational numbers

$a \in A$ and $b \in B$ so that $d \leq ab$ and thus $p \leq abc = a(bc)$. Now $bc \in BC$ and thus we see that $p \in A(BC)$. Hence $(AB)C \subseteq A(BC)$. Now let $q \in A(BC)$. Then there are positive rational numbers $a \in A$ and $e \in BC$ so that $q \leq ae$. Since $e \in BC$ there are positive rational numbers $b \in B$ and $c \in C$ so that $e \leq bc$ and thus $q \leq abc = (ab)c$. Since $ab \in AB$, we see that $q \in (AB)C$. Thus $A(BC) \subseteq (AB)C$. So together we have

$$A(BC) = (AB)C.$$

Now we will show $1_\mathcal{R} \cdot A = A$ for all $A \in \mathcal{R}^+$. Let $x \in 1_\mathcal{R} \cdot A$. There there exists a positive rational number $r \in 1_\mathcal{R}$ and a positive rational $a \in A$ so that $x \leq r \cdot a$. However $r < 1$ and so $x < a$, and thus $x \in A$. Hence $1_\mathcal{R} \cdot A \subseteq A$. Now suppose $x \in A$. Since the negative rational numbers are a subset of every element of $\mathcal{R}^+$, we may assume $x \geq 0$. Then there exists a $z \in A$ with $z > x$. Thus $x = \frac{x}{z} z$. Moreover $\frac{x}{z}$ is a rational number satisfying $0 \leq \frac{x}{z} < 1$ and so it is in $1_\mathcal{R}$, and $z \in A$. Thus $z \in 1_\mathcal{R} \cdot A$. Hence $A \subseteq 1_\mathcal{R} \cdot A$, and thus we have established

$$1_\mathcal{R} \cdot A = A.$$

For $A \in \mathcal{R}^+$, let A^{-1} be the subset of $\mathbb{Q}$ defined by

$$A^{-1} := \{x \in \mathbb{Q} : x \leq 0, \text{ or } x > 0 \text{ and } \frac{1}{x} \in \mathbb{Q} \setminus A, \frac{1}{x} \neq \min(\mathbb{Q} \setminus A)\}.$$

We will now show $A^{-1} \in \mathcal{R}^+$ and $A^{-1} \cdot A = 1_\mathcal{R}$. Clearly by definition A^{-1} is a non-empty subset of $\mathbb{Q}$. Thus properties 1 and 3 of the definition of a cut hold for A^{-1}. Now we will establish property 4, namely $A^{-1} \neq \mathbb{Q}$. For if $z \in A$, $z > 0$, then $\frac{1}{z} > 0$ is not an element of A^{-1}. Suppose $x \in A^{-1}$. If $x \leq 0$, then any $y < x$ will be an element of A^{-1} as well, so we assume $x > 0$. Let $y < x$ be a rational number. If $y \leq 0$ we clearly have $y \in A^{-1}$. Thus we shall assume $y > 0$ as well. $y < x$ if and only if $\frac{1}{x} < \frac{1}{y}$. Since $\frac{1}{x} \notin A$, we also have $\frac{1}{y} \notin A$ and $\frac{1}{y}$ is not the smallest element of $\mathbb{Q} \setminus A$. Thus $y \in A^{-1}$. Hence property 2 of a cut holds for A^{-1}. Finally, we establish property 5. Suppose $x \in A^{-1}$. We need to show that there is a $y > x$ in A^{-1}. If $x < 0$, then clearly there is such a y since all non-positive rational numbers are in A^{-1}. Thus we assume $x \geq 0$. We will show that A^{-1} must contain a positive element $y > x$. First suppose $x = 0$. Since A is bounded from above, there is an rational number $r > 0$ so that $r \notin A$. Moreover, any rational $s > r$ also

1.4. THE CONSTRUCTION OF THE REAL NUMBERS 53

is not in A. Thus $\frac{1}{s}$ is a positive element of A^{-1}. If $x > 0$, then we know $\frac{1}{x} \notin A$. Let $0 < z < \frac{1}{x}$ satisfy $z \notin A$, and z is rational – since $\frac{1}{x}$ is not the smallest element of $\mathbb{Q} \setminus A$, we know such an element exists. Let $y \in \mathbb{Q}$ satisfy $z < y < \frac{1}{x}$. Since $z \notin A$ we know $y \notin A$. Moreover $\frac{1}{y} > 0$ and in fact $\frac{1}{y} > x$. Clearly $\frac{1}{y} \in A^{-1}$. Hence we have established $A^{-1} \in \mathcal{R}$. Moreover, since A^{-1} contains at least one positive element, we know $A^{-1} \in \mathcal{R}^+$.

We will now show that $A^{-1} \cdot A = 1_\mathcal{R}$ for any $A \in \mathcal{R}^+$. By definition $A^{-1} \cdot A = \{p \in \mathbb{Q} : \exists a, b > 0, a \in A, b \in A^{-1}, p \leq ab\}$. We will show that $A^{-1} \cdot A \subseteq 1_\mathcal{R}$ by showing that for any positive $a \in A$ and any positive $b \in A^{-1}$, we have $ab < 1$. Recall that for positive $b \in A^{-1}$, $\frac{1}{b} \notin A$ and is not the smallest element of $\mathbb{Q} \setminus A$, and thus all $a \in A$ satisfy $a < \frac{1}{b}$ giving $ab < 1$. Next suppose $r \in 1_\mathcal{R}$. If $r \leq 0$, then clearly $r \in A \cdot A^{-1}$, since every element of $\mathcal{R}^+$ contains all the non-positive rational numbers. Thus we may assume $0 < r < 1$. We must establish that $r \leq ab$ for some positive $a \in A$, $b \in A^{-1}$. Let $z = \frac{1}{r}$. Since $z - 1 > 0$, and

$$z^n = (1 + (z-1))^n \geq 1 + n(z-1),$$

it follows that the numbers

$$z, z^2, z^3, z^4, z^5, \cdots$$

cannot all be in A, since A is bounded from above. Assume first that $z \in A$. By the well-ordering principle, there exists a natural number m so that $z^m \in A$, but $z^{m+1} \notin A$. Let $a = z^m$ and $b = \frac{1}{z^{m+1}}$. Clearly $\frac{1}{b} \notin A$. If $\frac{1}{b}$ is not the least element of $\mathbb{Q} \setminus A$, we are done since $r = ab \in A \cdot A^{-1}$. If $\frac{1}{b}$ is the least element of $\mathbb{Q} \setminus A$, then instead we choose $a^* > a$ in A and define $b^* := \frac{a}{a^*}b$. Then $\frac{1}{b^*} > \frac{1}{b}$ and so $\frac{1}{b^*} \in \mathbb{Q} \setminus A$ is not the least element. Hence $b^* \in A^{-1}$. Moreover, $r = a^*b^* \in A \cdot A^{-1}$. In the case where $z \notin A$, we use that

$$1 > \frac{1}{z} > \frac{1}{z^2} > \frac{1}{z^3} > \frac{1}{z^4} > \cdots > 0$$

cannot all fail to be in A, since eventually the terms in this list will become smaller than some positive element of A. Thus $\frac{1}{z^k} \in A$, for some $k \in \mathbb{N}$. Once again, by the well ordering principle, we let k be the smallest natural number so that $\frac{1}{z^k} \in A$, but $\frac{1}{z^{k-1}} \notin A$. In the case where $k > 1$, We define $a = \frac{1}{z^k}$ and $b = z^{k-1}$. If $\frac{1}{z^{k-1}}$ is not the smallest element of $\mathbb{Q} \setminus A$, then we get $r = ab \in A \cdot A^{-1}$. Otherwise we take $a^* > a$ in A and define $b^* = \frac{a}{a^*}b$. Then $\frac{1}{b^*} > \frac{1}{b}$ and thus $b^* \in A^{-1}$, and $r = a^*b^* = ab \in A \cdot A^{-1}$. If $k = 1$, then we have two cases to consider. If $1 \in A$ or $1 \notin A$. If $1 \notin A$, take $a > \frac{1}{z}$ in A and define $b = \frac{1}{az}$. Then $\frac{1}{b} = az > 1$, and thus is not the smallest element of $\mathbb{Q} \setminus A$. Thus $b \in A^{-1}$ and $r = ab \in A \cdot A^{-1}$. Finally, we assume $1 \in A$. Here, we pick $a > 1$ in A and define $b := \frac{1}{az}$. Clearly $\frac{1}{b} = az > z \notin A$ and thus $b \in A^{-1}$. Moreover, $r = ab \in A \cdot A^{-1}$. Thus in all cases, $r \in A \cdot A^{-1}$ and so we get $1_\mathcal{R} = A \cdot A^{-1}$.

We now define multiplication for general elements of $\mathcal{R}$. Let $A, B \in \mathcal{R}$, and define
$$|A| = \begin{cases} A & \text{if } 0_\mathcal{R} \subseteq A, \\ -A & \text{if } A \subset 0_\mathcal{R}. \end{cases}$$

Let
$$A \cdot B = \begin{cases} 0 & \text{if } A = 0 \text{ or } B = 0, \\ |A| \cdot |B| & \text{if } A, B \subset 0_\mathcal{R} \text{ or } 0_\mathcal{R} \subset A, B, \\ -(|A| \cdot |B|) & \text{if } A \subset 0_\mathcal{R}, 0_\mathcal{R} \subset B \text{ or } 0_\mathcal{R} \subset A, B \subset 0_\mathcal{R}. \end{cases}$$

Clearly, we see that for $A \subset 0_\mathcal{R}$, we have $A^{-1} = -(|A|)^{-1}$. Thus $A \cdot A^{-1} = 1_\mathcal{R}$.

The Distributive Property for $\mathcal{R}$

Finally, we will establish the distributive property
$$A \cdot (B + C) = (A \cdot B) + (A \cdot C)$$
for any $A, B, C \in \mathcal{R}$.

First, we suppose $A, B, C \in \mathcal{R}^+$. Let $x \in A(B + C)$. Then if $x \leq 0$ we are done, since $\{x \in \mathbb{Q} : x \leq 0\}$ is a subset of every element of $\mathcal{R}^+$. Now there exists $a > 0$ in A and $b + c > 0$ in $B + C$ so that $x \leq a(b + c)$. However, $a(b + c) = (ab) + (ac)$ and thus $x \leq (ab) + (ac)$. Since $ab \in AB$ and $ac \in AC$, and $(ab) + (ac) \in (AB) + (AC)$ we get $x \in (AB) + (AC)$. Now suppose $x \in (AB) + (AC)$, and as before we may assume $x > 0$. There are $y \in AB$ and $z \in AC$ so that $x = y + z$. There exists positive $a \in A$ and $b \in B$ so that $y \leq ab$ and there are positive $a' \in A$, $c \in C$ so that $z \leq a'c$. Thus, $x \leq ab + a'c$. Either $a \leq a'$ or $a' \leq a$. Without loss of generality, let's suppose $a \leq a'$. Then $b' = \frac{a}{a'}b \leq b$ and thus $b' \in B$.

1.4. THE CONSTRUCTION OF THE REAL NUMBERS

Therefore $ab + a'c = a'b' + a'c = a'(b' + c) \in A \cdot (B + C)$ and so $x \in A(B+C)$. Thus we have established the distributive property for elements of $\mathcal{R}^+$.

We now consider the case where $A, B, C \in \mathcal{R}$ are not necessarily in $\mathcal{R}^+$. If $A = 0_\mathcal{R}$, then $A(B+C) = 0_\mathcal{R} = 0_\mathcal{R} + 0_\mathcal{R} = (0_\mathcal{R} B) + (0_\mathcal{R} C) = (AB) + (AC)$. If $B = 0_\mathcal{R}$, then $A(B+C) = (AC) = 0_\mathcal{R} + (AC) = (A \cdot 0_\mathcal{R}) + (AC) = (AB) + (AC)$. The case where $C = 0_\mathcal{R}$ is similar. Next suppose $0_\mathcal{R} \subset A, C$; $B \subset 0_\mathcal{R}$ and $0_\mathcal{R} \subseteq B + C$. Then $AC = A[(B+C) + (-B)] = [A(B+C)] + [A(-B)] = [A(B+C)] - (AB)$, and thus $(AB) + (AC) = A(B+C)$. If $0_\mathcal{R} \subset A, C$; $B \subset 0_\mathcal{R}$ and $B + C \subset 0_\mathcal{R}$. Then $-(AB) = A(-B) = A[-(B+C) + C] = (A[-(B+C)]) + (AC) = [-A(B+C)] + (AC)$. Thus $A(B+C) = (AB) + (AC)$. If $0_\mathcal{R} \subset A, B$ and $C \subset 0_\mathcal{R}$ the proof is similar, we simply interchange B and C in the above case. If $0_\mathcal{R} \subset A$ and $B, C \subset 0_\mathcal{R}$, then $A[-(B+C)] = A[(-B) + (-C)] = (A(-B)) + (A(-C)) = -((AB) + (AC))$, and the result follows. If $A \subset 0_\mathcal{R}$ simply apply the result above to $-A$.

The Embedding of $\mathbb{Q}$ in $\mathcal{R}$

Recall that for $x \in \mathbb{Q}$ we may associate with x the cut $x_\mathcal{R} = \{p \in \mathbb{Q} : p < x\}$. We define $\mathcal{Q} = \{x_\mathcal{R} \in \mathcal{R} : x \in \mathbb{Q}\}$. Then the mapping $x \mapsto x_\mathcal{R}$ satisfies the following properties:

1. $x_\mathcal{R} + y_\mathcal{R} = (x + y)_\mathcal{R}$

2. $x_\mathcal{R} \cdot y_\mathcal{R} = (x \cdot y)_\mathcal{R}$

3. $x_\mathcal{R} \subset y_\mathcal{R}$ if and only if $x < y$.

Thus the mapping $x \mapsto x_\mathcal{R}$ is an isomorphism that preserves order.

To prove the first property, suppose $p \in x_\mathcal{R} + y_\mathcal{R}$. Then $p = a + b$ where $a < x$ and $b < y$. Thus $p < x + y$ and so $p \in (x+y)_\mathcal{R}$. Conversely, suppose $q \in (x+y)_\mathcal{R}$. Then $q < x+y$. Let $t = \frac{x+y-q}{2}$, and let $x' = x - t$ and $y' = y - t$. Then $x' \in x_\mathcal{R}$ and $y' \in y_\mathcal{R}$ and $q = x' + y'$, and thus $q \in x_\mathcal{R} + y_\mathcal{R}$.

To prove the third property, we first suppose $x < y$. Then $x \notin x_\mathcal{R}$, but $x \in y_\mathcal{R}$, and thus $x_\mathcal{R} \subset y_\mathcal{R}$. If $x_\mathcal{R} \subset y_\mathcal{R}$, then there is a $p \in y_\mathcal{R}$ such that $p \notin x_\mathcal{R}$. Hence $x \leq p < y$ and thus $x < y$.

For $x \in \mathbb{Q}$, $(-x)_\mathcal{R} = -x_\mathcal{R}$. To see this note that if $a \in x_\mathcal{R}$ and $b \in (-x)_\mathcal{R}$, then $a < x$, $b < -x$ and thus $a + b < x + (-x) = 0$. Hence $x_\mathcal{R} + (-x)_\mathcal{R} \subseteq 0_\mathcal{R}$. If $q \in 0_\mathcal{R}$, then $q < 0$. Writing $q = (x + \frac{q}{2}) + (-x + \frac{q}{2})$ we see $x + \frac{q}{2} \in x_\mathcal{R}$ and $-x + \frac{q}{2} \in (-x)_\mathcal{R}$, and thus we see $0_\mathcal{R} \subseteq x_\mathcal{R} + (-x)_\mathcal{R}$. Together we get $x_\mathcal{R} + (-x)_\mathcal{R} = 0_\mathcal{R}$, and so $(-x)_\mathcal{R} = -x_\mathcal{R}$.

To prove the second property we will consider cases. Observe that from property 3 x, y are positive if and only if $0_\mathcal{R} \subset x_\mathcal{R}, y_\mathcal{R}$, and x, y are negative if and only if $x_\mathcal{R}, y_\mathcal{R} \subset 0_\mathcal{R}$. If $x, y < 0$ then $xy = |x||y| = (-x)(-y)$. If $x_\mathcal{R}, y_\mathcal{R} \subset 0_\mathcal{R}$, then $x_\mathcal{R} \cdot y_\mathcal{R} = |x_\mathcal{R}| \cdot |y_\mathcal{R}| = (-x_\mathcal{R}) \cdot (-y_\mathcal{R})$. Thus if $x, y < 0$ we may instead use $-x, -y > 0$. So we will only consider $x, y > 0$, since $x, y < 0$ will be handled under this case.

Suppose that $p \in x_\mathcal{R} \cdot y_\mathcal{R}$. Then $p \leq ab$ for some positive $a \in x_\mathcal{R}$ and some positive $b \in y_\mathcal{R}$. Moreover, $a < x$ and $b < y$, and thus $p \leq ab < xy$. Hence $p \in (xy)_\mathcal{R}$. If $q \in (xy)_\mathcal{R}$ then $q < xy$. Since $\mathbb{N}$ has no upper bound, we can find an $n \in \mathbb{N}$ so that $n > \max\{\frac{x+y}{xy-q}, \frac{1}{x}, \frac{1}{y}\}$, and for such an n we have $q \leq (x - \frac{1}{n})(y - \frac{1}{n}) < xy$. Moreover $x - \frac{1}{n} \in x_\mathcal{R}$ and $y - \frac{1}{n} \in y_\mathcal{R}$, and so we see that $q \in x_\mathcal{R} \cdot y_\mathcal{R}$.

Finally we consider the case where x or y is negative and the other is positive. Without loss of generality we assume $x > 0$ and $y < 0$. Here we use $-(x_\mathcal{R} y_\mathcal{R}) = x_\mathcal{R}(-y_\mathcal{R}) = (x(-y))_\mathcal{R} = (-(xy))_\mathcal{R} = -(xy)_\mathcal{R}$.

1.4.2 The Uniqueness of a Complete Ordered Field up to Isomorphism

Let $\mathbb{F}_1, \mathbb{F}_2$ be two fields and let $f : \mathbb{F}_1 \to \mathbb{F}_2$ be a bijection that satisfies

$$f(x + y) = f(x) + f(y), \quad f(x \cdot y) = f(x) \cdot f(y) \quad \forall x, y \in \mathbb{F}_1.$$

We say that f is an **isomorphism** between two $\mathbb{F}_1$ and $\mathbb{F}_2$. In addition, if $\mathbb{F}_1$ and $\mathbb{F}_2$ are ordered fields, we shall also require

$$x < y \Rightarrow f(x) < f(y) \quad \forall x, y \in \mathbb{F}_1.$$

In such a case $\mathbb{F}_1$ is said to be **isomorphic** to $\mathbb{F}_2$ and this is denoted by $\mathbb{F}_1 \cong \mathbb{F}_2$. It is an exercise in algebra to show that $\cong$ is an equivalence relation

1.4. THE CONSTRUCTION OF THE REAL NUMBERS

on rings.

We will show that if $\mathbb{F}$ is a complete ordered field, then $\mathbb{F} \cong \mathbb{R}$. To avoid confusion, elements of $\mathbb{F}$ will have a $'$ after them.

We will now describe an isomorphism $f : \mathbb{R} \to \mathbb{F}$ which preserves order. For $n \in \mathbb{N}$ f is defined by

$$f(0) = 0', \quad f(1) = 1'$$

$$f(n) = \underbrace{1' + 1' + \cdots 1'}_{n \text{ times}}$$

which we denote as n'. That is, $f(n) = n'$. We also define

$$f(-n) = -n'.$$

Then we have

$$f(m+n) = m' + n', \quad f(m \cdot n) = m' \cdot n', \quad \forall m, n \in \mathbb{Z}.$$

$f : \mathbb{Z} \to \mathbb{F}$ as defined is injective. For rational numbers $\frac{m}{n}$ so that $m, n \in \mathbb{Z}$, $n \neq 0$ we define

$$f(\frac{m}{n}) = \frac{m'}{n'} = m' \cdot (n')^{-1}.$$

We claim that this definition of f on $\mathbb{Q}$ is well-defined. Note that $\frac{m}{n} = \frac{k}{j}$ if and only if $m \cdot j = n \cdot k$, and so $m' \cdot j' = k' \cdot n'$ and thus $m' \cdot (n')^{-1} = k' \cdot (j')^{-1}$. For integers m, n, q with $n, q \neq 0$ we have $\frac{m}{n} + \frac{p}{q} = \frac{mq+np}{nq}$, and thus

$$f(\frac{m}{n} + \frac{p}{q}) = f(\frac{mq + np}{nq})$$

$$= (mq + np)' \cdot ((nq)')^{-1}$$

$$= (m'q' + n'p')(n'q')^{-1}$$

$$= (m'q' + n'p')((n')^{-1}(q')^{-1})$$

$$m'(n')^{-1} + p'(q')^{-1}$$

$$= f(\frac{m}{n}) + f(\frac{p}{q}).$$

Also,
$$f(\frac{m}{n} \cdot \frac{p}{q}) = f(\frac{mp}{nq})$$
$$= (mp)' \cdot ((nq)')^{-1}$$
$$= m'p'(n'q')^{-1}$$
$$= m'p'(n')^{-1}(q')^{-1}$$
$$= m'(n')^{-1} \cdot p'(q')^{-1}$$
$$= f(\frac{m}{n}) \cdot f(\frac{p}{q}).$$

Thus restricted to $\mathbb{Q}$, f is an isomorphism onto its image.

Next, we will show f preserves order. We will do this by showing that for x a positive rational $x' > 0'$. By definition $f(n) > 0'$ for $n \in \mathbb{N}$, and thus for naturals m, n, $f(\frac{m}{n}) = m' \cdot (n')^{-1}$ is a product of elements of elements of $\mathbb{F}$ which are $> 0'$, and thus $f(\frac{m}{n}) > 0'$ as well.

Now we define $f(x)$ for any $x \in \mathbb{R}$. For $x \in \mathbb{R}$, let $S_x = \{f(r) : r \in \mathbb{Q}, \ r < x\}$. Clearly $S_x \neq \emptyset$ since the rational numbers are dense in $\mathbb{R}$. Also, S_x is bounded above by $f(y)$ where $y > x$ is any rational number. Since $\mathbb{F}$ is a complete ordered field, $\sup S_x$ exists, and let us define $f(x) = \sup S_x$ for $x \notin \mathbb{Q}$.

We will now show $f(r) = \sup S_r$ for $r \in \mathbb{Q}$. Let $r = \frac{m}{n}$, $m, n \in \mathbb{Z}$ and $n \neq 0$. Thus $f(r) = m' \cdot (n')^{-1}$. One can easily show that theorems (1.3.2) and (1.3.5) hold for $\mathbb{F}$, where the rational numbers in $\mathbb{F}$ is the image of $\mathbb{Q}$ under f. Thus for any $a', b' \in \mathbb{F}$ with $a' < b'$, we may find an $p \in \mathbb{Q}$ so that
$$a' < f(p) < b'.$$
Thus we see that every element $q' \in S_r$ satisfies $q' < f(r)$ and thus $\sup S_r \leq f(r)$. If $\sup S_r < f(r)$, then we may find $s \in \mathbb{Q}$ so that $\sup S_r < f(s) < f(r)$, and thus $s < r$. However, since $s < r$ we must have $f(s) \leq \sup S_r$, and this is a contradiction. Thus we see that $\sup S_r = f(r)$ for any rational r.

We will now show $f : \mathbb{R} \to \mathbb{F}$ as defined is an isomorphism. Let $x, y \in \mathbb{R}$ satisfy $x < y$. Then $S_x \subseteq S_y$ and thus $f(x) = \sup S_x \leq \sup S_y = f(y)$.

1.4. THE CONSTRUCTION OF THE REAL NUMBERS

By the density of the rational numbers in $\mathbb{R}$, there exists rational numbers r, s which satisfy $x < r < s < y$, and thus $f(x) \le f(r) < f(s) \le f(y)$. Thus we see f preserves order on $\mathbb{R}$ and thus f is injective. We will now show f is surjective. Let $a' \in \mathbb{F}$ and Let $A_{a'} = \{r \in \mathbb{Q} : f(r) < a'\}$. Since $f : \mathbb{Q} \to \mathbb{F}$ is an isomorphism onto its image which preserves order we may find rational elements $p', q' \in f(\mathbb{Q})$ so that $p' < a' < r'$. Letting $p, q \in \mathbb{Q}$ be so that $f(p) = p'$ and $f(q) = q'$ we see $p \in A_{a'}$ and q is an upper bound for $A_{a'}$. Let $x \in \mathbb{R}$ be defined by $x = \sup A_{a'}$. We claim that $f(x) = a'$. Suppose $f(x) < a'$, then there is an $r \in \mathbb{Q}$ so that $f(x) < f(r) < a'$ and thus $x < r$. Moreover, $r > x$ and $r \in A_{a'}$, and this is a contradiction. If $a' < f(x)$, then there is a rational s so that $a' < f(s) < f(x)$, which implies that $s < x$. Hence s is an upper bound for $A_{a'}$ which is smaller than x, which is a contradiction. Thus $f(x) = a'$, and so f is a surjection from $\mathbb{R}$ to $\mathbb{F}$. Hence we see that $f : \mathbb{R} \to \mathbb{F}$ is a bijection.

Suppose there exists $x, y \in \mathbb{R}$ so that $f(x+y) \ne f(x) + f(y)$. If $f(x+y) < f(x) + f(y)$ then we may find $r \in \mathbb{Q}$ so that $f(x+y) < f(r) < f(x) + f(y)$. This means that $x + y < r$. Thus r may be written as $r = r_x + r_y$ where $r_x > x$ and $r_y > y$, $r_x, r_y \in \mathbb{Q}$. Thus $f(r) = f(r_x + r_y) = f(r_x) + f(r_y) > f(x) + f(y)$ which is a contradiction. If $f(x+y) > f(x) + f(y)$ then we may find $r \in \mathbb{Q}$ so that $f(x) + f(y) < f(r) < f(x+y)$. This means that $r < x+y$. Thus r may be written as $r = r_x + r_y$, where $r_x < x$ and $r_y < y$. Thus $f(r) = f(r_x + r_y) = f(r_x) + f(r_y) < f(x) + f(y)$ which is a contradiction. Thus $f(x+y) = f(x) + f(y)$ for all $x, y \in \mathbb{R}$.

If there exists $x, y \in \mathbb{R}$ so that $f(xy) \ne f(x)f(y)$. By linearity of f, we see that $0' = f(0) = f(z - z) = f(z) + f(-z)$ and thus $f(z) = -f(z)$. Moreover, since $-(xy) = (-x)y = x(-y)$ we see that we may assume $x, y > 0$. If $f(xy) < f(x)f(y)$ we may find $r \in \mathbb{Q}$ so that $f(xy) < f(r) < f(x)f(y)$. Hence $xy < r$. We may find positive rational numbers r_x, r_y so that $x < r_x$ and $y < r_y$, with $r = r_x r_y$. Hence $f(r) = f(r_x r_y) = f(r_x)f(r_y) < f(x)f(y)$. However, $f(x) < f(r_x)$ and $f(y) < f(r_y)$ and so $f(x)f(y) < f(r_x)f(r_y)$, and this is a contradiction. If $f(xy) > f(x)f(y)$ we may find $r \in \mathbb{Q}$ so that $f(x)f(y) < f(r) < f(xy)$. Hence $r < xy$. We may find positive rational numbers r_x, r_y so that $x > r_x$ and $y > r_y$, and $r = r_x r_y$. Hence $f(r) = f(r_x r_y) = f(r_x)f(r_y) > f(x)f(y)$. However, $f(x) > f(r_x)$ and $f(y) > f(r_y)$, and so $f(x)f(y) > f(r_x)f(r_y)$ and this is a contradiction. Hence $f(xy) = f(x)f(y)$ for all $x, y \in \mathbb{R}$.

Hence we see that $\mathbb{R} \cong \mathbb{F}$.

1.5 The Complex Numbers

In this section we shall define the field of **complex numbers**, denoted by $\mathbb{C}$, as an extension of the field of real numbers $\mathbb{R}$. This shall be done in a way that initially avoids discussion of $\sqrt{-1}$.

We may think of the field $\mathbb{C}$ as the collection of all ordered pairs of real numbers with binary operations $+$ and $\cdot$ defined by:

$$(a,b) + (c,d) = (a+c, b+d), \quad (a,b) \cdot (c,d) = (ac - bd, ad + bc),$$

$$\forall a,b,c,d \in \mathbb{R}.$$

Moreover, we can identify $\mathbb{R}$ with the collection $\{(x,0) : x \in \mathbb{R}\}$ since $(x,0) + (y,0) = (x+y, 0)$ and $(x,0) \cdot (y,0) = (xy, 0)$. We claim that $\mathbb{C}$ as so defined is indeed a field. In fact, $\mathbb{C}$ will be a complete field, but we cannot define an ordering on $\mathbb{C}$. Moreover, we will be able to define a notion of distance or absolute value, which will allow us to discuss issues of density and other important notions for convergence that will be discussed later in this text.

Clearly $\mathbb{C}$ is closed under $+$ and $\cdot$, and addition is commutative and associative since

$$(a,b) + (c,d) = (a+c, b+d) = (c+a, d+b) = (c,d) + (a,b)$$

and

$$[(a,b) + (c,d)] + (e,f) = (a+c, b+d) + (e,f)$$
$$= ((a+c) + e, (b+d) + f)$$
$$= (a + (c+e), b + (d+f))$$
$$= (a,b) + [(c,d) + (e,f)].$$

We also have $(0,0)$ is our zero element, and $(1,0)$ is our unity, since

$$(a,b) + (0,0) = (a,b), \quad (a,b) \cdot (1,0) = (a,b) = (1,0) \cdot (a,b).$$

1.5. THE COMPLEX NUMBERS

Given $(a, b) \in \mathbb{C}$ we see that $-(a, b) = (-a, -b)$ since $(a, b) + (-a, -b) = (0, 0)$.

Multiplication is commutative since
$$(a, b) \cdot (c, d) = (ac - bd, ad + bc) = (ca - db, da + cb) = (c, d) \cdot (a, b).$$

Multiplication is associative since
$$[(a, b) \cdot (c, d)] \cdot (e, f) = (ac - bd, ad + bc) \cdot (e, f)$$
$$= ((ac - bd)e - (ad + bc)f, (ac - bd)f + (ad + bc)e)$$
$$= (ace - bde - adf - bcf, acf - bdf + ade + bce)$$
$$= (a(ce - df) - b(cf + de), a(cf + de) + b(ce - df))$$
$$= (a, b) \cdot (ce - df, cf + de)$$
$$= (a, b) \cdot [(c, d) \cdot (e, f)].$$

For $(a, b) \neq (0, 0)$ we have $a^2 + b^2 \neq 0$ and thus the complex number $(\frac{a}{a^2+b^2}, -\frac{b}{b^2+c^2})$ exists. Moreover,

$$(a, b) \cdot (\frac{a}{a^2 + b^2}, -\frac{b}{a^2 + b^2})$$
$$= (a \cdot \frac{a}{a^2 + b^2} - b \cdot \frac{-b}{a^2 + b^2}, a \cdot \frac{-b}{a^2 + b^2} + b \cdot \frac{a}{a^2 + b^2})$$
$$= (1, 0).$$

Thus for $(a, b) \neq (0, 0)$ we have
$$(a, b)^{-1} = (\frac{a}{a^2 + b^2}, -\frac{b}{a^2 + b^2}).$$

Finally, we shall address the distributive property.
$$(a, b) \cdot [(c, d) + (e, f)] = (a, b) \cdot (c + e, d + f)$$
$$= (a(c + e) - b(d + f), a(d + f) + b(c + e))$$
$$= (ac + ae - bd - bf, ad + af + bc + be)$$
$$= (ac - bd, ad + bc) + (ae - bf, af + be)$$

$$= [(a,b) \cdot (c,d)] + [(a,b) \cdot (e,f)].$$

Hence we have established that $(\mathbb{C}, +, \cdot)$ is a field.

For a given complex number (a,b), we define the **complex conjugate** of (a,b) as
$$\overline{(a,b)} = (a, -b),$$
for which we have
$$\overline{(a,b)} \cdot (a,b) = (a^2 + b^2, 0),$$
which we may think of as the real number $a^2 + b^2$. Thus, we define the an absolute value or norm on the set $\mathbb{C}$ by
$$|(a,b)| = \sqrt{a^2 + b^2}$$
which we may think of as the real number corresponding to the complex number
$$\sqrt{\overline{(a,b)} \cdot (a,b)}.$$

1.5.1 Homework Exercises

1. For $(a,0) \in \mathbb{C}$ show that $(a,0)^{-1}$ exists if and only if $a \neq 0$. Moreover, show $(a,0)^{-1} = (\frac{1}{a}, 0)$.

2. For $i = (0,1) \in \mathbb{C}$, show that $i^2 = (0,1) \cdot (0,1) = (-1, 0)$.

3. Show that for $a, b \in \mathbb{R}$ that $(a,0) + (b,0) \cdot (0,1) = (a,b)$ and thus we may use the familiar notation
$$a + bi = (a,b)$$
for any complex number $(a,b) \in \mathbb{C}$.

4. Use the definition of the complex conjugate to prove the following:

    (a) $\overline{(a,b) + (c,d)} = \overline{(a,b)} + \overline{(c,d)}$
    (b) $\overline{(a,b) \cdot (c,d)} = \overline{(a,b)} \cdot \overline{(c,d)}$
    (c) $(a,b) + \overline{(a,b)} = (2a, 0)$ and $(a,b) - \overline{(a,b)} = (0, 2b)$.

5. Use the definition of the complex absolute value or norm to prove the following:

(a) $|\overline{(a,b)}| = |(a,b)|$

(b) $|(a,b) \cdot (c,d)| = |(a,b)| \cdot |(c,d)|$.

(c) The Triangle Inequality
$$|(a,b) + (c,d)| \leq |(a,b)| + |(c,d)|.$$

6. Suppose $(a,b) \in \mathbb{C}$, show that the following complex number
$$(c,d) = (\sqrt{\frac{\sqrt{a^2+b^2}+a}{2}}, \sqrt{\frac{\sqrt{a^2+b^2}-a}{2}})$$
has the property that $(c,d) \cdot (c,d) = (a,b)$ if $b > 0$ and $\overline{(c,d)} \cdot \overline{(c,d)} = (a,b)$ if $b \leq 0$. Conclude that every non-zero complex number has a square root.

Chapter 2

Elementary Point-Set Topology

2.1 Euclidean Spaces

Let $n \in \mathbb{N}$, $n > 1$. We let $\mathbb{R}^n$ denote the collections of ordered n-tuples

$$\{(x_1, x_2, \cdots, x_n) : x_1, x_2, \cdots, x_n \in \mathbb{R}\},$$

and a general element of this set $(x_1, x_2, \cdots, x_n)$ is often denoted by $\vec{x}$.

On this space $\mathbb{R}^n$ we may define two operations, vector addition $+$ and scalar multiplication, and they are defined by

$$\vec{x} + \vec{y} = (x_1, x_2, \cdots, x_n) + (y_1, y_2, \cdots, y_n) = (x_1 + y_1, x_2 + y_2, \cdots, x_n + y_n).$$

and

$$c\vec{x} = c(x_1, x_2, \cdots, x_n) = (cx_1, cx_2, \cdots, cx_n)$$

for any $\vec{x}, \vec{y} \in \mathbb{R}^n$ and $c \in \mathbb{R}$.

We may also define the so-called **dot product**, or **inner product** on $\mathbb{R}^n$ by

$$\vec{x} \cdot \vec{y} = (x_1, x_2, \cdots, x_n) \cdot (y_1, y_2, \cdots, y_n) := \sum_{k=1}^{n} x_k y_k.$$

Some properties of the dot product are: for any $c \in \mathbb{R}$ and any $\vec{x}, \vec{y}, \vec{z} \in \mathbb{R}^n$

1. $\vec{x} \cdot \vec{y} = \vec{y} \cdot \vec{x}$

2. $(c\vec{x}) \cdot \vec{y} = c(\vec{x} \cdot \vec{y})$.

3. $(\vec{x} + \vec{y}) \cdot \vec{z} = (\vec{x} \cdot \vec{z}) + (\vec{y} \cdot \vec{z})$.

Proof: To see the first, notice that

$$\vec{x} \cdot \vec{y} = \sum_{k=1}^{n} x_k y_k = \sum_{k=1}^{n} y_k x_k = \vec{y} \cdot \vec{x}.$$

For the second, notice that

$$(c\vec{x}) \cdot \vec{y} = \sum_{k=1}^{n} (cx_k) y_k = c \sum_{k=1}^{n} x_k y_k = c(\vec{x} \cdot \vec{y}).$$

For the third property, notice that

$$(\vec{x} + \vec{y}) \cdot \vec{z} = \sum_{k=1}^{n} (x_k + y_k) z_k = \sum_{k=1}^{n} x_k z_k + \sum_{k=1}^{n} y_k z_k = (\vec{x} \cdot \vec{z}) + (\vec{y} \cdot \vec{z}).$$

□

The dot product also allows us to define a **norm** or **magnitude** on $\mathbb{R}^n$ by

$$\|\vec{x}\| = \sqrt{\vec{x} \cdot \vec{x}}.$$

Some properties of $\|\cdot\|$ are the following: Let $\vec{x}, \vec{y} \in \mathbb{R}^n$ and $c \in \mathbb{R}$, then

1. $\|\vec{x}\| \geq 0$ and $\|\vec{x}\| = 0$ if and only if $\vec{x} = \vec{0} = (0, 0, \cdots, 0)$.

2. $\|c\vec{x}\| = |c| \cdot \|\vec{x}\|$.

3. **The Cauchy-Schwarz inequality**: $|\vec{x} \cdot \vec{y}| \leq \|\vec{x}\| \cdot \|\vec{y}\|$.

4. **The triangle inequality**: $\|\vec{x} + \vec{y}\| \leq \|\vec{x}\| + \|\vec{y}\|$.

We may also define the **distance** between $\vec{x}$ and $\vec{y}$ to be $\|\vec{x} - \vec{y}\|$ and the **angle** θ between the $\vec{x}$ and $\vec{y}$ to be the smallest positive θ so that $\cos \theta = (\frac{1}{\|\vec{x}\|}\vec{x}) \cdot (\frac{1}{\|\vec{y}\|}\vec{y})$. $\mathbb{R}^n$ along with addition, scalar multiplication and the dot product is called n dimensional **Euclidean space**.

We will now prove the above properties of $\|\cdot\|$: The first is obvious from the definition of $\|\vec{x}\|$. The second is clear since $\sqrt{\sum_{k=1}^{n}(cx_k)^2} = \sqrt{c^2 \sum_{k=1}^{n} x_k^2} =$

2.1. EUCLIDEAN SPACES

$|c|\sqrt{\sum_{k=1}^{n} x_k^2}$.

To see the Cauchy-Schwarz inequality, we consider the quantity $\|\vec{x}-t\vec{y}\|^2$, which satisfies

$$0 \leq \|\vec{x} - t\vec{y}\|^2 = (\vec{x} - t\vec{y}) \cdot (\vec{x} - t\vec{y}) = (\vec{x} \cdot \vec{x}) - 2t(\vec{x} \cdot \vec{y}) + t^2(\vec{y} \cdot \vec{y})$$
$$= \|\vec{x}\|^2 - 2t(\vec{x} \cdot \vec{y}) + t^2\|\vec{y}\|^2.$$

We then use $t = \frac{1}{\|\vec{y}\|^2}(\vec{x} \cdot \vec{y})$ to get

$$0 \leq \|\vec{x}\|^2 - \frac{1}{\|\vec{y}\|^2}(\vec{x} \cdot \vec{y})^2,$$

and thus

$$(\vec{x} \cdot \vec{y})^2 \leq \|\vec{x}\|^2 \|\vec{y}\|^2.$$

To prove the triangle inequality, we notice that

$$\|\vec{x} + \vec{y}\|^2 = (\vec{x} + \vec{y}) \cdot (\vec{x} + \vec{y}) = \|\vec{x}\|^2 + 2(\vec{x} \cdot \vec{y}) + \|\vec{y}\|^2$$
$$\leq \|\vec{x}\|^2 + 2\|\vec{x}\| \cdot \|\vec{y}\| + \|\vec{y}\|^2 = (\|\vec{x}\| + \|\vec{y}\|)^2,$$

and thus we get the desired result.
□

The name for a set S with a norm $\|\cdot\|$ which has the same properties as discussed above for $\mathbb{R}^n$ is usually called a **normed linear space**. A set S with an inner product or dot product that has the properties as described above for $\mathbb{R}^n$ is usually referred to as an **inner product space**. Note that inner product spaces are always normed linear spaces by defining the norm $\|x\|$ to be $\sqrt{x \cdot x}$. Moreover, you can recapture the inner product from its norm by noticing $x \cdot y = \frac{1}{4}(\|x + y\|^2 - \|x - y\|^2)$. However, it should also be noted that not all normed linear spaces are inner product spaces.

2.1.1 Intervals in Euclidean Space

Given any $a < b$ in $\mathbb{R}$ we define a bounded **interval** to be set of the form

$$(a, b) := \{x \in \mathbb{R} : a < x < b\},$$

$$[a, b) := \{x \in \mathbb{R} : a \leq x < b\},$$
$$(a, b] := \{x \in \mathbb{R} : a < x \leq b\},$$
$$[a, b] := \{x \in \mathbb{R} : a \leq x \leq b\}. \tag{2.1}$$

An non-bounded interval in $\mathbb{R}$ is a set of the form

$$(a, \infty) := \{x \in \mathbb{R} : x > a\},$$
$$[a, \infty) := \{x \in \mathbb{R} : x \geq a\},$$
$$(-\infty, a) := \{x \in \mathbb{R} : x < a\},$$
$$(-\infty, a] := \{x \in \mathbb{R} : x \leq a\}, \tag{2.2}$$

and of course
$$\mathbb{R} = (-\infty, \infty).$$

Later we shall refer to (a, b) as an **open interval**, $[a, b]$ as a **closed interval**, and $(a, b]$ or $[a, b)$ as a half-open and half-closed interval.

In $\mathbb{R}^n$ we have the counterpart, which will be referred to an an interval as well
$$\{\vec{x} \in \mathbb{R}^n : a_i < x_i < b_i; i = 1, 2, \cdots, n\},$$
where $a_1 < b_1, \cdots, a_n < b_n$ are any fixed real numbers and $\vec{x} = (x_1, x_2, \cdots, x_n)$. Of course we may replace any of the strict inequalities in this definition by a non-strict inequality and the resulting set is still called an interval. If the inequalities are all strict, it will be referred to as an open interval, and if they are all not strict, it will be referred to as a closed interval.

2.1.2 Balls or Disks in Euclidean Space

Let $\vec{x} \in \mathbb{R}^n$ be any element, and let $r > 0$. An **open ball** in $\mathbb{R}^n$ is a set of the form
$$B_r(\vec{a}) := \{\vec{x} \in \mathbb{R}^n : \|\vec{x} - \vec{a}\| < r\}. \tag{2.3}$$

A **closed ball** in $\mathbb{R}^n$ is a set of the form
$$\overline{B_r(\vec{a})} := \{\vec{x} \in \mathbb{R}^n : \|\vec{x} - \vec{a}\| \leq r\}. \tag{2.4}$$

The **boundary** of either $B_r(\vec{a})$ or $\overline{B_r(\vec{a})}$ is the set
$$\partial B_r(\vec{a}) := \{\vec{x} \in \mathbb{R}^n : \|\vec{x} - \vec{a}\| = r\}. \tag{2.5}$$

2.1. EUCLIDEAN SPACES

In the case of $\mathbb{R}^2$, $\partial B_r(\vec{a})$ is a circle. Also, in the case $n = 2$, the set $B_r(\vec{a})$ may be referred to as an **open disk** and the set $\overline{B_r(\vec{a})}$ may be referred to as a **closed disk**.

2.1.3 Convexity in Euclidean Space

A set E is said to be **convex** if for any $\vec{x}, \vec{y} \in E$ and any $0 \leq \lambda \leq 1$ we have

$$\lambda \vec{x} + (1 - \lambda)\vec{y} \in E. \tag{2.6}$$

We note that this definition carries over to any setting where the background set has addition and scalar multiplication by elements of $\mathbb{R}$ defined; in particular, with vector spaces over $\mathbb{R}$.

Theorem 2.1.1 *Balls and Intervals in $\mathbb{R}^n$ are convex.*

Proof: We will examine the case where our ball or interval is closed, and this can be modified easily in the case where either is not closed, since the inequalities need not be strict in the argument below. Consider first a closed ball $B := \overline{B_r(\vec{a})}$. Then for $\vec{x}, \vec{y} \in B$ and any $0 \leq \lambda \leq 1$ we have

$$\|\lambda \vec{x} + (1-\lambda)\vec{y} - \vec{a}\| = \|\lambda \vec{x} - \lambda \vec{a} + (1-\lambda)\vec{y} - (1-\lambda)\vec{a}\|$$

$$\leq \lambda \|\vec{x} - \vec{a}\| + (1-\lambda)\|\vec{y} - \vec{a}\| \leq r.$$

Also, let $I = \{\vec{x} : a_i \leq x_i \leq b_i, \quad i = 1, 2, \cdots, n\}$. Then for $\vec{x}, \vec{y} \in I$ and $0 \leq \lambda \leq 1$ we have

$$a_i \leq x_i \leq b_i, \quad a_i \leq y_i \leq b_i, \quad i = 1, 2, \cdots, n,$$

and thus

$$\lambda a_i \leq \lambda x_i \leq \lambda b_i, \quad (1-\lambda)a_i \leq (1-\lambda)y_i \leq (1-\lambda)b_i, \quad i = 1, 2, \cdots, n.$$

Hence we get

$$a_i \leq \lambda x_i + (1-\lambda)y_i \leq b_i, \quad i = 1, 2, \cdots, n.$$

□

2.2 Metric Spaces

Let M be a set, and let $d : M \times M \to [0, \infty)$ be a function which satisfies the following properties for each $x, y, z \in M$:

1. $d(x, y) \geq 0$
2. $d(x, y) = 0 \iff x = y$
3. $d(x, y) = d(y, x)$
4. $d(x, z) \leq d(x, y) + d(y, z)$.

d is called a **distance** function on M, and (M, d) is called a **metric space**. The inequality $d(x, z) \leq d(x, y) + d(y, z)$ is often called the **triangle inequality**.

In a metric space (M, d) we often call $x \in M$ a **point** in M.

A set S is said to be **bounded** in a metric space (M, d) if there is a $B > 0$ so that
$$d(x, y) \leq B \quad \forall x, y \in S.$$
If S is not bounded, it is said to be **unbounded**. The **diameter** of a set S, denoted $\text{diam}(S)$, is defined to be the quantity
$$\text{diam}(S) := \sup\{d(x, y) : x, y \in S\}. \tag{2.7}$$

Clearly a set is bounded if and only if the diameter of the set is finite.

Note that an inner product space or any normed linear space is a metric space by defining $d(x, y) = \|x - y\|$.

2.3 Open Sets and Closed Sets

2.3.1 Open Sets

In a metric space (M, d), for $x \in M$ and $r > 0$ we define
$$B_r(x) := \{y \in M : d(x, y) < r\}. \tag{2.8}$$

2.3. OPEN SETS AND CLOSED SETS

This shall be called the **open ball** in M with center x and radius r.

A set S is said to be **open** in M if for each $x \in S$ there is an $r > 0$ so that $B_r(x) \subseteq S$. For any set $S \subseteq M$, $x \in S$ is called an **interior point** of S if there is an $r > 0$ so that $B_r(x) \subseteq S$. Thus open sets are sets for which all points as interior points. The **interior** of a set S, denoted by S°, is the set of all interior points of S. Thus S is open if and only if $S = S^\circ$.

Theorem 2.3.1 *Every open ball is an open set.*

Proof: Let $B_R(x)$ be an open ball in M, and let $y \in B_R(x)$. Let $d(x,y) = r < R$ and let $h = \min\{r, R-r\}$. Consider the ball $B_h(y)$, and let $z \in B_h(y)$. Then
$$d(x,z) \leq d(x,y) + d(y,z) < r + h \leq r + R - r = R.$$
Therefore $z \in B_R(x)$, and thus $B_h(y) \subseteq B_R(x)$.
□

Theorem 2.3.2 *Let $I = \{\vec{x} = (x_1, x_2, \cdots, x_n) \in \mathbb{R}^n : a_i < x_i < b_i, \ i = 1, 2, \cdots, n\}$. Then I is open in $\mathbb{R}^n$.*

Proof: Let $\vec{x} \in I$. We define
$$r := \min\{x_1 - a_1, b_1 - x_1, x_2 - a_2, b_2 - x_2, \cdots, x_n - a_n, b_n - x_n\}.$$
Claim that $B_r(\vec{x}) \subseteq I$. Let $\vec{y} \in B_r(\vec{x})$. Then for any $i = 1, 2, \cdots, n$ we have $|x_i - y_i| < r$. Hence $-r + x_i < y_i < x_i + r$ and thus
$$a_i = -(x_i - a_i) + x_i \leq -r + x_i < y_i < x_i + r \leq x_i + b_i - x_i = b_i.$$
Hence $y \in I$, and this implies that $B_r(\vec{x}) \subseteq I$. Thus we conclude that I is open in $\mathbb{R}^n$.
□

2.3.2 Closed Sets

We say that $S \subset M$ is **closed** if and only if its **complement** $S^c := M \setminus S$ is open. Thus a set is open if and only if its complement is closed, since $(S^c)^c = S$. Clearly M is both open and closed, since $M^c = \emptyset$ trivially satisfies every point is an interior point.

A point $x \in M$ is called a **limit point** or an **accumulation point** of a set S if for every open ball $B_r(x)$ we have $[B_r(x) \setminus \{x\}] \cap S \neq \emptyset$. If a point $x \in S$ is not a limit point of S, it is said to be an **isolated point** of S. A set S is said to be **perfect** if it is both closed and every point of S is a limit point of S. A set S is said to be **dense** in M if every point of M is a limit point of S, or a point in S. Given a point $x \in M$ a set N is called a **neighbourhood** of x if N is a connected[1] open set containing x. For instance, for $r > 0$ $B_r(x)$ is a neighbourhood of x.

Theorem 2.3.3 *A set S is closed if and only if S contains all its limit points.*

Proof: Suppose S is closed. Then S^c is open. Let $x \in S^c$, then there is an $r > 0$ so that $B_r(x) \subseteq S^c$, and thus x is not a limit point of S. Hence S contains all its limit points.

Conversely, suppose S contains all its limit points. Then $x \in S^c$ is not a limit point, and thus there is an $r > 0$ so that $B_r(x) \cap S = \emptyset$. Thus $B_r(x) \subseteq S^c$, and so we see that S^c is open.
□

Theorem 2.3.4 *If x is a limit point of a set S then any open ball $B_r(x)$ contains infinitely many points of S.*

Proof: Suppose that there is an open ball $B_R(x)$ so that $[B_R(x) \setminus \{x\}] \cap S = \{s_1, s_2, \cdots s_N\}$. We define

$$r = \min\{d(x, s_1), d(x, s_2), \cdots, d(x, s_N)\}.$$

Then for any $y \in B_{\frac{r}{2}}(x)$ we have $d(x, y) < \frac{r}{2}$. This implies $s_i \notin B_{\frac{r}{2}}(x)$ for any $i = 1, 2, \cdots, N$, and thus $[B_{\frac{r}{2}}(x) \setminus \{x\}] \cap S = \emptyset$. This is a contradiction, since x is a limit point of S.
□

Theorem 2.3.5 *If $S \subseteq M$ is a finite set, then S has no limit points.*

We define the **closure** of a set S to be $S \cup \{\text{limit points of } S\}$, and let $\overline{S}$ denote the closure of S. Clearly a closed set S satisfies $\overline{S} = S$. We define the **boundary** of S, denoted by ∂S, to be the set $\overline{S} \setminus S^\circ$.

---

[1] The definition of connected will be given later in this chapter.

2.3. OPEN SETS AND CLOSED SETS

2.3.3 Unions and Intersections of Open and Closed Sets

Let $\{S_\alpha : \alpha \in \Delta\}$ be an indexed family of sets. The following result is often called **DeMorgan's laws** for sets.

Theorem 2.3.6
$$(\bigcap_{\alpha \in \Delta} S_\alpha)^c = \bigcup_{\alpha \in \Delta} S_\alpha^c,$$

$$(\bigcup_{\alpha \in \Delta} S_\alpha)^c = \bigcap_{\alpha \in \Delta} S_\alpha^c.$$

Proof: Suppose $x \in (\bigcap_{\alpha \in \Delta} S_\alpha)^c$, then $x \notin \bigcap_{\alpha \in \Delta} S_\alpha$. Thus there is some $\beta \in \Delta$ so that $x \notin S_\beta$ and thus $x \in S_\beta^c$. Hence $x \in \bigcup_{\alpha \in \Delta} S_\alpha^c$. Conversely, suppose $x \in \bigcup_{\alpha \in \Delta} S_\alpha^c$. Then for some $\gamma \in \Delta$, $x \in S_\gamma^c$ and so $x \notin S_\gamma$. Hence $x \notin \bigcap_{\alpha \in \Delta} S_\alpha$, and so $x \in (\bigcap_{\alpha \in \Delta} S_\alpha)^c$.

Suppose $x \in (\bigcup_{\alpha \in \Delta} S_\alpha)^c$, then $x \notin \bigcup_{\alpha \in \Delta} S_\alpha$. Thus for all $\beta \in \Delta$, $x \notin S_\beta$ and so for all $\beta \in \Delta$, $x \in S_\beta^c$. Hence $x \in \bigcap_{\alpha \in \Delta} S_\alpha^c$. Conversely, suppose $x \in \bigcap_{\alpha \in \Delta} S_\alpha^c$. Then for all $\gamma \in \Delta$, $x \in S_\gamma^c$, and so $x \notin S_\gamma$ for all $\gamma \in \Delta$. Hence $x \notin \bigcup_{\alpha \in \Delta} S_\alpha$, and so $x \in (\bigcup_{\alpha \in \Delta} S_\alpha)^c$.
□

If $\{S_\alpha : \alpha \in \Delta\}$ is a family of open sets, then $\bigcup_{\alpha \in \Delta} S_\alpha$ is open since $x \in \bigcup_{\alpha \in \Delta} S_\alpha$ implies $x \in S_\beta$ for some $\beta \in \Delta$. Thus there exists an $r > 0$ so that $B_r(x) \subseteq S_\beta \subseteq \bigcup_{\alpha \in \Delta} S_\alpha$. Thus $\bigcap_{\alpha \in \Delta} S_\alpha^c$ is closed. Moreover $\{S_\alpha^c : \alpha \in \Delta\}$ is a family of closed sets, and hence the intersection of any family of closed sets is closed and the union of any family of open sets is open.

For any finite collection of open sets $\{S_1, S_2, \cdots, S_n\}$ we have $\bigcap_{i=1}^n S_i$ is open. To see this let x be in this set, then for each i there is a positive r_i so that $B_{r_i}(x) \subseteq S_i$. Take $r := \min\{r_1, r_2, \cdots, r_N\}$, and then we see that $B_r(x) \subseteq \bigcap_{i=1}^n S_i$. Likewise, by taking complements, we see that the union of finitely many closed sets is closed. It is left as an exercise to show that the intersection of a general family of open sets may not be open, and the union of a general family of closed sets may not be closed. You simply need to look at $\mathbb{R}$ and intervals to see this.

2.3.4 Homework Exercises

1. Construct a bounded subset of $\mathbb{R}$ with exactly 1 limit point, and this point is not in the set.

2. Construct a bounded subset of $\mathbb{R}$ with exactly 1 limit point, and this point is in the set.

3. Construct a bounded subset of $\mathbb{R}$ with exactly 2 limit points, and these points are not in the set.

4. Construct a bounded subset of $\mathbb{R}$ with exactly 2 limit points, and these points are in the set.

5. Construct a bounded subset of $\mathbb{R}$ with exactly 3 limit points, and these points are not in the set.

6. Construct a bounded subset of $\mathbb{R}$ with exactly 3 limit points, and these points are in the set.

7. In the exercises above. Are the sets open, closed or neither? Also, find the complements of these sets. Are they open, closed or neither?

8. Consider the indexed family of open sets $\{(-\frac{1}{n}, \frac{1}{n}), n \in \mathbb{N}\}$. Find the following sets. Are they open, closed or neither?

    (a) $\bigcup_{i=1}^{N} (-\frac{1}{i}, \frac{1}{i})$
    (b) $\bigcup_{i=1}^{\infty} (-\frac{1}{i}, \frac{1}{i})$
    (c) $\bigcap_{i=1}^{N} (-\frac{1}{i}, \frac{1}{i})$
    (d) $\bigcap_{i=1}^{\infty} (-\frac{1}{i}, \frac{1}{i})$

9. Consider the indexed family of closed sets $\{[-\frac{1}{n}, \frac{1}{n}], n \in \mathbb{N}\}$. Find the following sets. Are they open, closed or neither?

    (a) $\bigcup_{i=1}^{N} [-\frac{1}{i}, \frac{1}{i}]$
    (b) $\bigcup_{i \in \mathbb{N}} [-\frac{1}{i}, \frac{1}{i}]$
    (c) $\bigcap_{i=1}^{N} [-\frac{1}{i}, \frac{1}{i}]$
    (d) $\bigcap_{i \in \mathbb{N}} [-\frac{1}{i}, \frac{1}{i}]$

2.3. OPEN SETS AND CLOSED SETS

10. Consider the indexed family of closed sets $\{[n, n+1], n \in \mathbb{Z}\}$. Find the following sets. Are they open, closed or neither?

    (a) $\bigcup_{i=-N}^{N}[i, i+1]$
    (b) $\bigcup_{i \in \mathbb{Z}}[i, i+1]$
    (c) $\bigcap_{i=-N}^{N}[i, i+1]$
    (d) $\bigcap_{i \in \mathbb{Z}}[i, i+1]$

11. Let S be an infinite set and define $d: S \times S \to [0, \infty)$ by

    $$d(x, y) := \begin{cases} 1 & \text{if } x \neq y, \\ 0 & \text{if } x = y. \end{cases}$$

    (a) Show that (S, d) is a metric space.
    (b) What are the bounded sets?
    (c) What are the unbounded sets?
    (d) What are the open sets?
    (e) What are the closed sets?

12. Let $S \subset \mathbb{R}$ be a non-empty bounded set. Show $\sup S, \inf S \in \overline{S}$. Concluded that if S is closed, then S contains $\inf S$ and $\sup S$.

13. On $\mathbb{R}^n$ we define $d(\vec{x}, \vec{y}) = \max\{|x_1 - y_1|, |x_2 - y_2|, \cdots, |x_n - y_n|\}$. Is $(\mathbb{R}^n, d)$ a metric space?

14. On $\mathbb{R}^n$ we define $d_1(\vec{x}, \vec{y}) = |x_1 - y_1| + |x_2 - y_2| + \cdots + |x_n - y_n|$. Show that $(\mathbb{R}^n, d_1)$ is a metric space. Show that the usual metric $d(\vec{x}, \vec{y}) = \|\vec{x} - \vec{y}\|$ and d_1 are equivalent. That is, show that there are real numbers $C, K \geq 1$ so that for any $\vec{x}$ and $\vec{y}$:

    $$\frac{1}{C}d_1(\vec{x}, \vec{y}) \leq d(\vec{x}, \vec{y}) \leq Cd_1(\vec{x}, \vec{y}),$$

    $$\frac{1}{K}d(\vec{x}, \vec{y}) \leq d_1(\vec{x}, \vec{y}) \leq Kd(\vec{x}, \vec{y}).$$

    In such a case we say that the two metrics define the same topology.

15. Let (M, d) be a metric space. Define for $x, y \in M$ define $\delta(x, y) = \frac{d(x,y)}{1+d(x,y)}$. Show that (M, δ) is also a metric space. Conclude that if M is a set for which (M, d) is a metric space, then there are infinitely many different metrics one can define on M.

2.4 Compactness

For a metric space (M, d) we say that a collection of open sets $\{\mathcal{O}_\alpha : \alpha \in \Delta\}$ is an **open cover** for $S \subseteq M$ if

$$S \subseteq \bigcup_{\alpha \in \Delta} \mathcal{O}_\alpha. \tag{2.9}$$

A set $C \subseteq M$ is said to be **compact** if any open cover has a finite subcover. Here a **subcover** of an open cover is a subfamily of the indexed family of open sets that is itself an open cover. A **finite subcover** is a subcover that consists of a finite number of sets in the family of sets. If M itself is compact, we say that (M, d) is a **compact metric space**.

We see that C is compact if and only if for any open cover $\{\mathcal{O}_\alpha : \alpha \in \Delta\}$ there are $\{\alpha_1, \alpha_2 \cdots, \alpha_N\} \subseteq \Delta$ so that

$$C \subseteq \mathcal{O}_{\alpha_1} \cup \mathcal{O}_{\alpha_2} \cup \cdots \cup \mathcal{O}_{\alpha_N}.$$

The next result shows that compact sets are closed and bounded sets.

Theorem 2.4.1 *Compact sets are closed and bounded.*

Proof: Let C be a compact subset of a metric space. We will show C^c is open. Fix $x \in C^c$, and consider any $y \in C$. Let U_y be a neighbourhood of x and V_y a neighbourhood of y whose diameters are less than $\frac{d(x,y)}{2}$. Note that the collection $\{V_y, y \in C\}$ is an open cover of C and since C is compact there are finitely many $y_i \in C$ so that

$$C \subseteq V_{y_1} \cup V_{y_2} \cup V_{y_2} \cup \cdots \cup V_{y_N}.$$

Then the set $U = U_{y_1} \cap U_{y_2} \cap \cdots \cap U_{y_N}$ is an open neighbourhood of x and the set U satisfies $U \cap V_{y_i} = \emptyset$, for $i = 1, 2, \cdots, N$. Thus $U \cap C = \emptyset$. So U is a neighbourhood of x that is entirely contained in C^c, and so C^c is open, which implies C is closed.

We will now show that C is bounded. Let $y, z \in C$. Consider the open cover of C given by

$$\{B_1(p) : p \in C\}.$$

2.4. COMPACTNESS

By compactness, there is a finite subcover of C given by

$$\{B_1(p_i) : p_1, p_2, \cdots, p_m \in C\}$$

for some $p_1, p_2, \cdots, p_m \in C$. Clearly y and z are each in one of the open balls $B_1(p_1), B_1(p_2), \cdots, B_1(p_m)$. If they are in the same one, then $d(y, z) < 2$. Otherwise, without loss of generality suppose $y \in B_1(p_1)$, and let $z \in B_1(p_j)$ for some $1 < j \leq m$. Then

$$d(y, z) \leq d(y, p_1) + d(p_1, p_2) + d(p_2, p_3) + \cdots + d(p_{j-1}, p_j) + d(p_j, z)$$
$$\leq 2 + d(p_1, p_2) + d(p_2, p_3) + \cdots + d(p_{m-1}, p_m),$$

which is independent of y, z, and clearly finite.
□

So we see that compact subsets of metric spaces must be closed and bounded. The converse in general is not true,[2] but the converse is true in Euclidean space. The next result tells us that closed subsets of compact sets are compact.

Theorem 2.4.2 *Closed subsets of compact sets are compact.*

Proof: Let $K \subseteq C$ be a closed subset of a compact set C. Let $\{U_\alpha : \alpha \in \Delta\}$ be any open cover of K. Consider the open cover $\{U_\alpha : \alpha \in \Delta\} \cup \{K^c\}$ of C. Then there is a finite subcover of this cover that covers C, since C is compact. Moreover, this subcover take away $\{K^c\}$ is a finite subcover of K. Hence K is compact.
□

Now we observe what is usually referred to as the **finite intersection property of compact sets**.

Theorem 2.4.3 *Let $\{K_\alpha : \alpha \in \Delta\}$ be a collection of compact sets, such that for every finite collection $\{\alpha_1, \alpha_2, \cdots, \alpha_N\} \subseteq \Delta$ we have*

$$K_{\alpha_1} \cap K_{\alpha_2} \cap \cdots \cap K_{\alpha_N} \neq \emptyset.$$

Then

$$\bigcap_{\alpha \in \Delta} K_\alpha \neq \emptyset.$$

---

[2] The closed unit ball in l_2 is closed and bounded, but not compact. Here l_2 is the collection of all sequences $\{a_n : n \in \mathbb{N}\}$ so that $\sum_{n=1}^\infty a_n^2$ converges. See [8] section IV.3 for more details.

Proof: Suppose that for some $K_\beta \in \{K_\alpha : \alpha \in \Delta\}$ no point of K_β is in every element of $\{K_\alpha : \alpha \in \Delta\}$. Then for each $x \in K_\beta$ there is a $K_{\alpha_x} \in \{K_\alpha : \alpha \in \Delta\}$ so that $x \in K_{\alpha_x}^c$. Thus the collection $\{K_\alpha^c : \alpha \in \Delta\}$ is an open cover of K_β. By compactness, there is a finite subcover $\{K_{\alpha_1}^c, K_{\alpha_2}^c, \cdots, K_{\alpha_m}^c\}$ that covers K_β. Hence

$$K_\beta \subseteq K_{\alpha_1}^c \cup K_{\alpha_2}^c \cup \cdots \cup K_{\alpha_m}^c = (K_{\alpha_1} \cap K_{\alpha_2} \cap \cdots \cap K_{\alpha_m})^c.$$

This means that
$$K_\beta \cap K_{\alpha_1} \cap K_{\alpha_2} \cap \cdots \cap K_{\alpha_m} = \emptyset.$$

However, this is a contradiction to our assumption.
□

This also gives us a useful corollary:

Corollary 2.4.4 *Let $\{K_n : n \in \mathbb{N}\}$ be a collection of compact sets satisfying*

$$\cdots \subseteq K_{n+1} \subseteq K_n \subseteq \cdots \subseteq K_2 \subseteq K_1.$$

Then
$$\bigcap_{i=1}^{\infty} K_i \neq \emptyset.$$

Proof: Clearly for any $m \in \mathbb{N}$ we have

$$\bigcap_{i=1}^{m} K_i = K_m,$$

and in general for any $i_1 < i_2 < \cdots < i_k$ we have

$$\bigcap_{j=1}^{k} K_{i_j} = K_{i_k}.$$

Thus we may apply the above theorem to get the result.
□

The next result tells us that compact sets with infinitely many points in them must contain at least one limit point.

2.4. COMPACTNESS

Theorem 2.4.5 *If S is an infinite subset of a compact set K, then S has a limit point in K.*

Proof: Suppose the contrary. Then no point of K is a limit point of S. Then for each $x \in K$, there is an open neighbourhood N_x of x that satisfies

$$N_x \cap S = \begin{cases} \emptyset & \text{if } x \notin S, \\ \{x\} & \text{if } x \in S. \end{cases}$$

Then the collection $\{N_x : x \in K\}$ is an open cover of K, but this cannot have a finite subcover, since no finite subcover will cover S. This is a contradiction to the compactness of K.
□

Another important observation to make is the following result about nested intervals of $\mathbb{R}$ having a nonempty intersection.

Theorem 2.4.6 *Let $\{I_n : n \in \mathbb{N}\}$ be a collection of closed bounded intervals in $\mathbb{R}$ that satisfy the nesting property:*

$$\cdots \subseteq I_{n+1} \subseteq I_n \subseteq \cdots \subseteq I_2 \subseteq I_1.$$

Then

$$\bigcap_{j=1}^{\infty} I_j \neq \emptyset.$$

Proof: Let $I_n := [a_n, b_n]$ and let $S = \{a_1, a_2, \cdots, a_n, \cdots\}$. Then S is non-empty and is bounded from above by b_1. Thus $\sup S$ exists. Notice for any $m, n \in \mathbb{N}$ we have

$$a_n \leq a_{m+n} \leq b_{m+n} \leq b_m,$$

and so $\sup S \leq b_m$ for all $m \in \mathbb{N}$. Hence $\sup S \in I_m$ for all $m \in \mathbb{N}$.
□

This result extends easily to $\mathbb{R}^n$.

Theorem 2.4.7 *Let $\{I_j : j \in \mathbb{N}\}$ be a collection of closed bounded intervals in $\mathbb{R}^n$ that satisfy the nesting property:*

$$\cdots \subseteq I_{m+1} \subseteq I_m \subseteq \cdots \subseteq I_2 \subseteq I_1.$$

Then
$$\bigcap_{j=1}^{\infty} I_j \neq \emptyset.$$

Proof: Let
$$I_j = \{\vec{x} = (x_1, x_2, \cdots, x_n) : a_{ij} \leq x_i \leq b_{ij}, \quad i = 1, 2, \cdots, n\}$$
be a collection of nested intervals as in the hypothesis of the theorem. Let $I_{m,j} = [a_{mj}, b_{mj}]$. For each $m = 1, 2, \cdots n$ we have $\{I_{m,j}\}_{j=1}^{\infty}$ is a nested sequence of intervals in $\mathbb{R}$, which satisfy the hypothesis of theorem (2.4.6). Thus for each $m = 1, 2, \cdots n$ we have that there are real numbers z_m so that $z_m \in \bigcap_{j \in \mathbb{N}} I_{m,j}$, and thus $\vec{z} := (z_1, z_2, \cdots, z_n) \in \bigcap_{j \in \mathbb{N}} I_j$.
□

As a result of theorem (2.4.1) we see that the interval $(0, 1]$ is not compact. However, the set $[0, 1]$ is compact. This will follow once we discuss a theorem often called the Heine-Borel theorem, which is given below.

The Heine-Borel Theorem 2.4.8 *A subset of $\mathbb{R}^n$ is compact if and only if it is closed and bounded.*

Proof: Clearly theorem (2.4.1) shows one part of this equivalence, and so it is only necessary to show that closed and bounded implies compact. We will show that any interval in $\mathbb{R}^n$ of the form
$$\{\vec{x} = (x_1, x_2, \cdots, x_n) : -R \leq x_1 \leq R, i = 1, 2, \cdots, n\}$$
is compact. That is any closed and bounded interval is compact. Since any bounded set is a proper subset of such an interval, this and theorem (2.4.2) will give us the result.

Suppose E is a bounded subset of $\mathbb{R}^n$. Then there exists a $B > 0$ so that
$$\sup\{\|\vec{x} - \vec{y}\| : \vec{x}, \vec{y} \in E\} < B.$$
Fix any $\vec{z} \in E$, then for any $\vec{y} \in E$ we have
$$\|\vec{y}\| \leq \|\vec{y} - \vec{z}\| + \|\vec{z}\| < B + \|\vec{z}\| =: D.$$

2.4. COMPACTNESS

Consider the interval J given by $J = \{\vec{x} = (x_1, x_2, \cdots, x_n) : -D \leq x_i \leq D, \ i = 1, 2, \cdots, n\}$. Then clearly $E \subseteq J$ since for $\vec{y} = (y_1, y_2, \cdots, y_n) \in E$ we have

$$|y_j| \leq \sqrt{y_1^2 + y_2^2 + \cdots + y_j^2 + \cdots + y_n^2} \leq D$$

for any $j = 1, 2, 3, \cdots, n$. Thus it remains to show that such closed and bounded intervals are compact.

Suppose that J is not compact. Then there is an open cover $\{G_\alpha : \alpha \in \Delta\}$ of J so that no finite collection $\{G_{\alpha_1}, G_{\alpha_2}, \cdots, G_{\alpha_N}\}$, $\{\alpha_1, \alpha_2, \cdots, \alpha_N\} \subseteq \Delta$, completely covers J.

Note that any $\vec{x}, \vec{y} \in J$ satisfy $\|\vec{x}-\vec{y}\| = \sqrt{\sum_{i=1}^n (y_i - x_i)^2} \leq \sqrt{\sum_{j=1}^n (2D)^2} = 2\sqrt{n}D$. Let $I_0 := J$. We partition I_0 into 2^n non-overlapping cubes $\{J_k : k = 1, 2, 3, \cdots, 2^n\}$ each with edge length equal to D. Note as well that any $\vec{x}, \vec{y} \in J_k$ satisfy $\|\vec{x} - \vec{y}\| \leq 2\sqrt{n}\frac{D}{2}$. Observe that at least one of the J_k's cannot be covered with finitely many elements of our open cover. Let I_1 denote such a J_k.

Subdivide I_1 into 2^n closed subintervals whose interiors are pairwise disjoint, each of which is an n-dimensional cube of edge length $\frac{D}{2}$. Then $\vec{x}, \vec{y}$ in any one of these subintervals satisfies $\|\vec{x} - \vec{y}\| \leq 2\sqrt{n}\frac{D}{4}$. Moreover, at least one of these subintervals cannot be covered by finitely many elements of our open cover. Let I_2 denote this subinterval.

Subdivide I_2 into 2^n subintervals which are also n-dimensional cubes each with an edge length of $\frac{D}{4}$, and continue this process. This gives rise to a nested collection of intervals $\{I_0, I_1, I_2, \cdots\}$ which satisfy

1. $\cdots \subseteq I_m \subseteq \cdots \subseteq I_2 \subseteq I_1 \subseteq I_0$

2. For any non-negative integer k, I_k cannot be covered by finitely many elements of our open cover $\{G_\alpha : \alpha \in \Delta\}$.

3. For any non-negative integer k, for any $\vec{x}, \vec{y} \in I_k$ we have $\|\vec{x} - \vec{y}\| \leq 2\sqrt{n}\frac{D}{2^k}$.

Then by theorem (2.4.7),

$$\bigcap_{k=0}^{\infty} I_k \neq \emptyset.$$

Let $\vec{z} \in \bigcap_{k=0}^{\infty} I_k$. Then for some $\beta \in \Delta$, $\vec{z} \in G_\beta$. Since G_β is open, there is an $r > 0$ so that $B_r(\vec{z}) \subseteq G_\beta$. Let $m \in \mathbb{N}$ be such that $2\sqrt{n}\frac{D}{2^m} < \frac{r}{2}$. Then $I_m \subseteq B_r(\vec{z}) \subseteq G_\beta$. This is a contradiction, since we have that there is no finite subcover for any I_k obtained from the covering $\{G_\alpha : \alpha \in \Delta\}$. Hence J is compact.

□

The next result is the converse of theorem (2.4.5) in the setting of $\mathbb{R}^n$.

Theorem 2.4.9 *Suppose K has the property that every infinite subset S of K has a limit point in K, then K is compact.*

Proof: Without loss of generality K has infinitely many elements in it, since otherwise there is nothing to show. We proceed via a proof by contradiction.

Suppose K is unbounded. Then K contains a subset $S := \{\vec{x}_1, \vec{x}_2, \cdots\}$ so that $\|\vec{x}_k\| \geq k$ for all $k \in \mathbb{N}$. Since S is an infinite subset of K, we have that that S has a limit point $\vec{z} \in K$. Theorem (2.3.4) implies $S \cap B_1(\vec{z})$ has infinitely many elements in it. Now let $m > \|\vec{z}\| + 1$. For such an m we have $\|\vec{x}_m\| > m > \|\vec{z}\| + 1$, and so $\vec{x}_m \notin B_1(\vec{z})$. Hence

$$\{\vec{x}_m, \vec{x}_{m+1}, \cdots\} \cap B_1(\vec{z}) = \emptyset.$$

Thus we see that $S \cap B_1(\vec{z}) \subseteq \{\vec{x}_1, \vec{x}_2, \cdots, \vec{x}_{m-1}\}$. This is a contradiction. Hence K must be bounded.

Suppose that K is not closed. Then K^c is not open. So there is a point $\vec{z} \in K^c$ so that for any $B_r(\vec{z})$, $B_r(\vec{z}) \cap K \neq \emptyset$. Thus we see that $\vec{z}$ is a limit point of K that is not in K. Thus for $r = \frac{1}{n}$, $n \in \mathbb{N}$ there is $\vec{x}_n \in K \cap B_{\frac{1}{n}}(\vec{z})$. Moreover, we may also require $\vec{x}_n \notin \{\vec{x}_1, \vec{x}_2, \cdots, \vec{x}_{n-1}\}$ for each n.

Let $S = \{\vec{x}_n : n \in \mathbb{N}\}$. Since S is an infinite subset of K, S has a limit point in $\vec{y} \in K$. Since $\vec{y}$ is a limit point of S, $S \cap B_{\frac{1}{n}}(\vec{y})$ contains

2.4. COMPACTNESS

infinitely many points for each $n \in \mathbb{N}$. Thus there is strictly increasing divergent sequence $\{k_i\}_{i=1}^{\infty}$ of natural numbers all larger than n so that we have $\vec{x}_{k_i} \in B_{\frac{1}{n}}(\vec{y})$ for all $i \in \mathbb{N}$. Hence for such an $\vec{x}_{k_i}$ we have $\|\vec{x}_{k_i} - \vec{y}\| \leq \frac{1}{n}$.

So
$$\|\vec{z} - \vec{y}\| = \|\vec{z} - \vec{x}_{k_i} + \vec{x}_{k_i} - \vec{y}\|$$
$$\leq \|\vec{z} - \vec{x}_{k_i}\| + \|\vec{x}_{k_i} - \vec{y}\|$$
$$\leq \frac{1}{k_i} + \frac{1}{n} \leq \frac{2}{n}.$$

Since $n \in \mathbb{N}$ is arbitrary and $\inf\{\frac{1}{n} : n \in \mathbb{N}\} = 0$ we see that $\|\vec{z} - \vec{y}\| = 0$, and this is a contradiction since $\vec{y} \in K$ and $\vec{z} \in K^c$. Hence K is closed.
□

Thus we see that theorems (2.4.5), (2.4.9) and (2.4.8) tell us that in $\mathbb{R}^n$ the following are equivalent:

1. K is compact.

2. K is closed and bounded.

3. Every infinite subset of K has a limit point in K.[3]

To conclude our section on compactness, there is another item to note which is commonly called the **Bolzano-Weierstrass theorem**.

Bolzano-Weierstrass Theorem 2.4.10 *Every bounded infinite subset of $\mathbb{R}^n$ has a limit point in $\mathbb{R}^n$.*

Proof: Since our set in question is bounded there is an n-dimensional cube centred at the origin that contains this set. Noting that such n-dimensional cubes are compact, we see that the result follows from the equivalent notions of compactness noted above.
□

---

[3]In general, we have that K is compact if and only if every net in K has a convergent subnet. See [8], section IV.3 for more details.

2.4.1 Homework Exercises

1. For $(0, 1] \subseteq \mathbb{R}$ construct an open cover that does not contain a finite subcover.

2. For $(0, 1) \subseteq \mathbb{R}$ construct an open cover that does not contain a finite subcover.

3. For $(0, \infty) \subseteq \mathbb{R}$ construct an open cover that does not contain a finite subcover.

4. If possible, construct a compact set $K \subseteq \mathbb{R}$ whose limit points form a countably infinite set.

5. Find a collection of compact subsets of $\mathbb{R}$ $\{K_i : i \in \mathbb{N}\}$ so that $\bigcup_{j=1}^{\infty} K_j$ is not compact.

6. For the collection in the previous exercise is $\bigcup_{j=1}^{N} K_j$ compact for each N?

7. Find a collection of compact subsets of $\mathbb{R}$ $\{K_i : i \in \mathbb{N}\}$ so that $\bigcup_{j=1}^{\infty} K_j$ is compact.

8. Show that a non-empty compact set $K \subseteq \mathbb{R}$ has $\inf K$, and $\sup K$ as limit points or as elements of K. Conclude that if either is a limit point, then it must also be an element of K.

2.5 Connectedness

A set S in a metric space (M, d) is said to be **disconnected** if $S = A \cup B$, where $A, B \subseteq M$ are non-empty and $A \cap \overline{B}, \overline{A} \cap B = \emptyset$. If S is not disconnected, it is said to be **connected**.

A observation to note in the setting of $\mathbb{R}$ is the following:

Theorem 2.5.1 *$S \subseteq \mathbb{R}$ is connected if and only if for any $a, b \in S$ with $a < b$ we have $[a, b] \subseteq S$.*

2.6. THE CANTOR SET

Proof: To see this note that if $a < b$ were elements of S and $x \in (a, b)$ was not an element of S, then

$$S = (S \cap (-\infty, x)) \cup (S \cap (x, \infty)).$$

Thus we see that S is disconnected.

To prove the converse, suppose S is disconnected. Then there are $A, B \subseteq \mathbb{R}$ so that $S = A \cup B$, and $A \cap \overline{B}, \overline{A} \cap B = \emptyset$. Pick $x \in A$ and $y \in B$, and without loss of generality assume $x < y$. Let $z := \sup(A \cap [x, y])$. Then $z \in \overline{A}$, and thus $z \notin B$. In particular, $x \leq z < y$.

If $z \notin A$, then $x < z < y$, and $z \notin S$. If $z \in A$, then $z \notin \overline{B}$. Thus there exists w so that $z < w < y$ and $w \notin B$. Then $x < w < y$ and $w \notin S$. □

2.6 The Cantor Set

2.6.1 Perfect Subsets of Euclidean Space

An observation that we will make is that a non-empty perfect[4] set in $\mathbb{R}^n$ must necessarily be uncountable. To see this we first observe that a perfect set $P \subseteq \mathbb{R}^n$ is infinite, since it has limit points. If P were to be countable, then we may write P in the form

$$P = \{\vec{x}_1, \vec{x}_2, \vec{x}_3, \cdots, \vec{x}_k, \cdots\}.$$

Let $N_1 = B_r(\vec{x}_1)$ be a neighbourhood of $\vec{x}_1$, with $r > 0$. Suppose N_k has been constructed so that $N_k \cap P \neq \emptyset$. Since every point of P is a limit point of P there is a neighbourhood N_{k+1} so that

1. $\overline{N_{k+1}} \subseteq N_k$,

2. $\vec{x}_k \notin \overline{N_{k+1}}$,

3. $N_{k+1} \cap P \neq \emptyset$.

---

[4]Recall that a set is perfect if it is closed and every point in it is a limit point for the set.

Notice that the collection $\overline{N_1}, \overline{N_2}, \overline{N_3}, \cdots$ is a collection of closed and bounded sets, and hence each $\overline{N_k}$ is compact. Let $K_j = \overline{N_j} \cap P$, then K_j is compact. Moreover $\vec{x}_j \notin K_{j+1}$. Thus we see that no point of P lies in every K_i and so $\bigcap_{j=1}^{\infty} K_j = \emptyset$. However, each K_j is non-empty and the collection $\{K_j : j \in \mathbb{N}\}$ satisfy

$$\cdots \subseteq K_{j+1} \subseteq K_j \subseteq \cdots \subseteq K_2 \subseteq K_1.$$

By theorem (2.4.3) we know $\bigcap_{j=1}^{\infty} K_j \neq \emptyset$, and this is a contradiction. Hence P must be uncountable.

2.6.2 The Construction of the Cantor Set

In this section we will construct the **Cantor set** which a perfect subset of the real numbers which had an empty interior.

Let $C_0 := [0,1] \subseteq \mathbb{R}$, and define $C_1 = C_0 \setminus (\frac{1}{3}, \frac{2}{3}) = [0, \frac{1}{3}] \cup [\frac{2}{3}, 1]$. Remove the open middle one thirds of these two intervals, and let C_2 be

$$C_2 := [0, \frac{1}{9}] \cup [\frac{2}{9}, \frac{3}{9}] \cup [\frac{6}{9}, \frac{7}{9}] \cup [\frac{8}{9}, 1].$$

Continuing in this way, we define a collection $\{C_n : n \in \mathbb{N}\}$ which satisfies

1. $\cdots \subset C_k \subset \cdots \subset C_3 \subset C_2 \subset C_1 \subset C_0$

2. C_k is a union of 2^k disjoint closed intervals, each of length $\frac{1}{3^k}$.

Then

$$C = \bigcap_{k=0}^{\infty} C_k$$

is called the **Cantor Set**. Clearly C is non-empty by theorem (2.4.3), and C is compact, since it is closed and bounded. Moreover, if C were to have a length[5], it must be zero. This is because C is contained in each C_j, and the length of C_j is $(\frac{2}{3})^j$. Hence the length of C must be less than or equal to $(\frac{2}{3})^j$ for any $j \in \mathbb{N}$. Thus it must have length zero.

---

[5]We will show later that C has measure zero. In particular, it is an example of an uncountable set of measure zero.

2.6. THE CANTOR SET

To see that C has an empty interior we note that any open interval of the form $(\frac{3k+1}{3^m}, \frac{3k+2}{3^m})$ intersected with C is empty, for any integers $k, m > 0$. Moreover, any open interval $(a, b) \subset \mathbb{R}$ contains such an open interval $(\frac{3k+1}{3^m}, \frac{3k+2}{3^m})$, whenever $\frac{1}{3^m} < \frac{b-a}{6}$.

To see that C is perfect, we will show that C has no isolated points. Let $x \in C$, and a, b any real numbers so that $a < x < b$. Let I_k be the interval of C_k that contains x. If k is large enough, then $I_k \subset (a, b)$. Let x_k be one of the endpoints of I_k, with the restriction that $x_k \neq x$. From the construction of C we see that $x_k \in C$. Thus x is a limit point of C, since it is a limit point of the collection $\{x_k\}$, and hence C is perfect.

2.6.3 Homework Exercises

1. Using **ternary expansions** for elements in the interval $[0, 1]$ – that is writing numbers in the interval $[0, 1]$ in the form $x = 0.x_1 x_2 x_3 \cdots$ where $x = \sum_{k=1}^{\infty} \frac{x_k}{3^k}$ and $x_k \in \{0, 1, 2\}$ – show that x is an element of the Cantor Set if and only if each $x_k \neq 1$.[6]

2. Use exercise one and binary expansions to produce a bijection between the Cantor set and $[0, 1]$.

---

[6]Up to examples like
$$\frac{1}{3} = 0.1000 \cdots = 0.0222 \cdots,$$
which are elements of the Cantor Set.

Chapter 3

Sequences and Series of Real Numbers

A **sequence of real numbers** is a function
$$f : \mathbb{N} \to \mathbb{R}.$$

Typically, we use a subscript notation for a sequence and we write
$$a_1 = f(1), a_2 = f(2), \cdots, a_n = f(n), \cdots,$$

and it is common to denote the sequence
$$a_1, a_2, a_3, \cdots$$

by
$$\{a_j\}_{j=1}^{\infty}.$$

3.1 Limits of Sequences

A sequence $\{a_j\}_{j=1}^{\infty}$ is said to have a limit L if given any $\varepsilon > 0$, there exists a $N \in \mathbb{N}$ so that for $k \geq N$ we have
$$L - \varepsilon < a_k < L + \varepsilon.$$

A sequence which has a limit is said to **converge**, and if a sequence does not have a limit, it is said to **diverge**. If a sequence $\{a_n\}_{n=1}^{\infty}$ has a limit L we often use the notation
$$\lim_{n \to \infty} a_n = L.$$

The **Fibonacci sequence** is the sequence f_j given by $f_1 = 1, f_2 = 1$, and in general for $n \geq 3$ $f_n = f_{n-1} + f_{n-2}$. Thus the Fibonacci sequence is

$$1, 1, 2, 3, 5, 8, 13, 21, 34, \cdots$$

This sequence actually diverges since $f_n \geq n - 1$.

Consider the sequence

$$1, \frac{1}{2}, \frac{1}{4}, \frac{1}{8}, \frac{1}{16}, \frac{1}{32}, \cdots,$$

then this sequence can be written as $a_n = \frac{1}{2^{n-1}}$. Here we have the limiting value of the sequence is 0. To see this, let $\varepsilon > 0$, and we may find a natural number N so that $n \geq N$ which implies $2^n > \frac{1}{\varepsilon}$.

Consider the sequence

$$-1, 1, -1, 1, -1, 1, -1, \cdots,$$

then this sequence may be written as

$$a_n = (-1)^n.$$

This sequence has no limit since the difference $|a_n - a_{n+1}| = 2$, and thus for $\varepsilon < 1$, and any $L > 0$, if $|a_n - L| < \varepsilon$ then $|a_{n+1} - L| > \varepsilon$.

We note that in a general metric space (M, d) we may define a sequence $\{a_n\}_{n=1}^\infty$ in a similar manner. Here $a_n = f(n)$ where $f : \mathbb{N} \to M$. We say that $\{a_n\}_{n=1}^\infty$ has a limit $L \in M$ if for any $\varepsilon > 0$, there is an $N \in \mathbb{N}$ so that for $n > N$ we have $d(a_n, L) < \varepsilon$.

A sequence $\{a_n\}_{n=1}^\infty$ is said to be **bounded** if there is a $R > 0$ and some fixed element x in the metric space so that $a_n \in B_R(x)$ for every n. In particular, for sequences of real numbers, we have $\{a_n\}_{n=1}^\infty$ is bounded if and only if there is a $B > 0$ so that $|a_n| \leq B$ for all naturals n.

In a metric space (M, d) we have the following results:

3.1. LIMITS OF SEQUENCES

Theorem 3.1.1 *If a sequence $\{a_n\}_{n=1}^{\infty}$ has a limit value, it must be unique.*

Proof: Suppose $\{a_n\}_{n=1}^{\infty}$ has two limit values L and M. Then for a given $\varepsilon > 0$, we may find $N \in \mathbb{N}$ so that for $n > N$ we have $d(a_n, L) < \frac{\varepsilon}{2}$. Also, we may find $K \in \mathbb{N}$ so that for $n > K$ we have $d(a_n, M) < \frac{\varepsilon}{2}$. Thus for $n > \max\{N, K\}$ we have

$$d(L, M) \leq d(L, a_n) + d(a_n, M) < \frac{\varepsilon}{2} + \frac{\varepsilon}{2} = \varepsilon.$$

Since $\varepsilon > 0$ was arbitrary, we have $d(L, M) = 0$,[1] and thus $L = M$.
□

Theorem 3.1.2 *Suppose x is a limit point of a set S in a metric space (M, d). Then there exists a sequence of distinct elements $\{x_n\}_{n=1}^{\infty}$, with $x_n \in S$ for all n, so that*

$$\lim_{n \to \infty} x_n = x.$$

Proof: Let $\varepsilon > 0$ be given. Since x is a limit point of S, for any open ball $B_r(x)$ we have $[B_r(x) \setminus \{x\}] \cap S$ has infinitely many elements in it. Since $\inf\{\frac{1}{n} : n \in \mathbb{N}\} = 0$, we may choose $N \in \mathbb{N}$ so that for $n > N$ we have $0 < \frac{1}{n} < \varepsilon$. We define our sequence as follows:

1. Let $x_1 \in [B_{\frac{1}{N+1}}(x) \setminus \{x\}] \cap S$.

2. Let $x_2 \in [B_{\frac{1}{N+2}}(x) \setminus \{x\}] \cap S$, $x_2 \neq x_1$.

3. Let $x_3 \in [B_{\frac{1}{N+3}}(x) \setminus \{x\}] \cap S$, $x_3 \notin \{x_1, x_2\}$.

4. In general, for $k \in \mathbb{N}$, let $x_k \in [B_{\frac{1}{N+k}}(x) \setminus \{x\}] \cap S$, with $x_k \notin \{x_1, x_2, \cdots, x_{k-1}\}$.

Then clearly

$$d(x_k, x) < \frac{1}{N+k} < \varepsilon, \quad \forall k \in \mathbb{N}$$

and thus

$$\lim_{n \to \infty} x_n = x.$$

□

---

[1] It is an exercise to show that in a metric space M: $x = y \iff \forall \varepsilon > 0$ we have $d(x, y) < \varepsilon$.

3.1.1 Homework Exercises

1. Let $a_n := \frac{1}{n}$, show that $\lim_{n\to\infty} a_n = 0$ using the definition of a limit.

2. Let $a_n := \frac{1}{n^2}$, show that $\lim_{n\to\infty} a_n = 0$ using the definition of a limit.

3. Let $a_n := \frac{1}{\sqrt{n}}$, show that $\lim_{n\to\infty} a_n = 0$ using the definition of a limit.

4. Let $a_n := c$ with c any fixed element of a metric space. Show that $\lim_{n\to\infty} a_n = c$ using the definition of a limit.

3.1.2 Properties of Limits

Theorem 3.1.3 *Suppose $\{a_n\}_{n=1}^{\infty}$ is a sequence in a metric space (M, d) so that*
$$\lim_{n\to\infty} a_n = a,$$
then $\{a_n\}_{n=1}^{\infty}$ is a bounded sequence.

Proof: Since $\lim_{n\to\infty} a_n = a$, we know that there is an $N \in \mathbb{N}$ so that for all $n > N$ we have $d(a_n, a) < 1$, that is, $a_n \in B_1(a)$ for $n > N$. We define R by
$$R = \max\{d(a, a_1), d(a, a_2), \cdots, d(a_N, a), 1\}.$$
Then for all $n \in \mathbb{N}$, we have
$$a_n \in B_R(a).$$
$\square$

The following theorem gives a list of limit properties, often called **limit laws**.

Theorem 3.1.4 *Let $\{a_n\}_{n=1}^{\infty}$ and $\{b_n\}_{n=1}^{\infty}$ be any two convergent sequences of real numbers with limit values a and b respectively. Then we have the following*

1.
$$\lim_{n\to\infty}(a_n + b_n) = a + b,$$

3.1. LIMITS OF SEQUENCES

2. For any constant c,
$$\lim_{n\to\infty} (ca_n) = ca,$$

3.
$$\lim_{n\to\infty} (a_n \cdot b_n) = a \cdot b,$$

4. If $b \neq 0$, we have
$$\lim_{n\to\infty} \frac{1}{b_n} = \frac{1}{b}.$$

Proof: Let $\varepsilon > 0$ be given.

1. We may find $N \in \mathbb{N}$ so that for $n > N$ we have $|a_n - a| < \frac{\varepsilon}{2}$ and we may find $M \in \mathbb{N}$ so that for $n > M$ we have $|b_n - b| < \frac{\varepsilon}{2}$. Then for $n > \max\{N, M\}$ we have

$$|(a_n + b_n) - (a + b)| = |(a_n - a) + (b_n - b)|$$
$$\leq |a_n - a| + |b_n - b|$$
$$< \frac{\varepsilon}{2} + \frac{\varepsilon}{2} = \varepsilon.$$

2. If $c = 0$ then $ca_n = 0$ for every n, and so by a homework exercise from the previous section we know $ca_n \to 0$. If $c \neq 0$ we may find $N \in \mathbb{N}$ so that for $n > N$ we have $|a_n - a| < \frac{\varepsilon}{|c|}$. For such $n > N$ we have

$$|ca_n - ca| = |c| \cdot |a_n - a|$$
$$< |c| \cdot \frac{\varepsilon}{|c|} = \varepsilon.$$

3. Since $\{a_n\}_{n=1}^{\infty}$ is convergent, there is an $A > 0$ so that $|a_n| < A$ for any $n \in \mathbb{N}$. We first assume that $b \neq 0$. We may find $N \in \mathbb{N}$ so that for $n > N$ we have $|a_n - a| < \frac{\varepsilon}{2|b|}$ and we also find $M \in \mathbb{N}$ so that for $n > M$ we have $|b_n - b| < \frac{\varepsilon}{2A}$. Thus, for $n > \max\{N, M\}$ we have

$$|a_n b_n - ab| = |a_n b_n - a_n b + a_n b - ab|$$
$$\leq |a_n b_n - a_n b| + |a_n b - ab|$$
$$= |a_n| \cdot |b_n - b| + |b| \cdot |a_n - a|$$

$$\leq A \cdot |b_n - b| + |b| \cdot |a_n - a|$$

$$\leq A \cdot \frac{\varepsilon}{2A} + |b| \cdot \frac{\varepsilon}{2|b|} = \varepsilon.$$

If $b = 0$, then as above we find A so that $|a_n| < A$ for all n, and thus $|a_n b_n| \leq A|b_n|$ for all n. We now choose $N \in \mathbb{N}$ so that for $n > N$ we have $|b_n| = |b_n - 0| < \frac{\varepsilon}{A}$. So for such n, we have $|a_n b_n| < A|b_n| < \varepsilon$.

4. Since b_n converges to $b \neq 0$ will be assumed, we will only consider n as well so that $|b_n| > \frac{|b|}{2} > 0$. This is possible since there is an $N \in \mathbb{N}$, so that for $n > N$ we have

$$|b_n - b| < \frac{|b|}{2}.$$

However, $||b_n| - |b|| \leq |b_n - b|$, and so for $n > N$ we have

$$-\frac{|b|}{2} < |b_n| - |b| < \frac{|b|}{2},$$

and thus $|b_n| > \frac{|b|}{2}$. Note as well that we have $\frac{1}{|b_n|} < \frac{2}{|b|}$. Now there is an $M \in \mathbb{N}$ so that for $n > M$ we have

$$|b_n - b| < \frac{b^2 \cdot \varepsilon}{2}.$$

Now let $n > \max\{N, M\}$, and so

$$\left|\frac{1}{b_n} - \frac{1}{b}\right| = \left|\frac{b - b_n}{b \cdot b_n}\right|$$

$$= \frac{|b_n - b|}{|b \cdot b_n|}$$

$$\leq \frac{2}{b^2}|b_n - b|$$

$$\leq \frac{2}{b^2} \cdot \frac{b^2 \cdot \varepsilon}{2} = \varepsilon.$$

□

3.1. LIMITS OF SEQUENCES

Corollary 3.1.5 *Let $c, k \in \mathbb{R}$ and let $\{a_n\}_{n=1}^{\infty}$ and $\{b_n\}_{n=1}^{\infty}$ be any two convergent sequences of real numbers with limit values a and b respectively, then we have*
$$\lim_{n \to \infty} (ca_n + kb_n) = ca + kb.$$
In addition, if $b \neq 0$, then
$$\lim_{n \to \infty} \frac{a_n}{b_n} = \frac{a}{b}.$$

Proof: To show $\lim_{n \to \infty}(ca_n + kb_n) = ca + kb$, we simply apply item two in theorem (3.1.4) to $\{ca_n\}_{n=1}^{\infty}$ and $\{kb_n\}_{n=1}^{\infty}$. We then apply item one from theorem (3.1.4) to these two sequences as well. To see $\lim_{n \to \infty} \frac{a_n}{b_n} = \frac{a}{b}$ we observe $\frac{a_n}{b_n} = a_n \cdot \frac{1}{b_n}$, and use items three and four in theorem (3.1.4).
□

Observe that statements about limits are statements about eventual behaviour. If we $\lim_{n \to \infty} b_n = b \neq 0$, then initially the terms in the sequence $\{b_n\}_{n=1}^{\infty}$ may be zero. However, for n sufficiently large, we showed above that $b_n \neq 0$. So when considering the sequence $\{\frac{1}{b_n}\}_{n=1}^{\infty}$, we may need to ignore the first so many terms because for small n, $b_n = 0$ is possible.

Also, in the future we may use the notation
$$a_n \to a \quad \text{as} \quad n \to \infty,$$
or simply
$$a_n \to a,$$
in place of
$$\lim_{n \to \infty} a_n = a.$$

3.1.3 Bounded Monotone Sequences

A sequence or real numbers $\{a_n\}_{n=1}^{\infty}$ is said to be

1. **Increasing** if $a_{n+1} \geq a_n$ for all n.

2. **Strictly increasing** if $a_{n+1} > a_n$ for all n.

3. **Decreasing** if $a_{n+1} \leq a_n$ for all n.

4. **Strictly decreasing** if $a_{n+1} < a_n$ for all n.

5. **Monotone** if it is increasing, strictly increasing, decreasing or strictly decreasing.

The following theorem if often called the **bounded monotone convergence theorem.**

Theorem 3.1.6 *If $\{a_n\}_{n=1}^{\infty}$ is a bounded monotone sequence of real numbers, then it is a convergent sequence.*

Proof: Without loss of generality, by otherwise considering the sequence $\{-a_n\}_{n=1}^{\infty}$, we may assume that the sequence is increasing or strictly increasing. Then $\sup\{a_n : n \in \mathbb{N}\}$ exists as a real number; let a denote this number.

Let $\varepsilon > 0$ be given. We know that there is an $N \in \mathbb{N}$ so that $a_N \in (a-\varepsilon, a]$, since otherwise $a - \varepsilon$ would be an upper bound for $\{a_n : n \in \mathbb{N}\}$. For $n > N$ we have $a_n \geq a_N$, and so
$$a - \varepsilon < a_n \leq a.$$
Hence $a_n \to a$.
□

Example: A sequence $\{a_n\}_{n=1}^{\infty}$ is given recursively by
$$a_1 = 1, \quad a_{n+1} = \sqrt{2 + a_n} \quad n > 1.$$
We claim that a_n is bounded by 2 and increasing, and we will find $\lim_{n\to\infty} a_n$.

$$a_1 = 1, \quad a_2 = \sqrt{2+1} = \sqrt{3}, \quad a_3 = \sqrt{2+\sqrt{3}}, \cdots$$

First, we'll show $a_n \leq 2$ for all n by mathematical induction. Clearly $a_1 = 1 \leq 2$. Now, if $a_n \leq 2$ we have
$$a_{n+1} = \sqrt{2 + a_n} \leq \sqrt{2+2} = \sqrt{4} = 2.$$
Thus, we get
$$a_n \leq 2 \quad \text{for all} \quad n.$$
Next, we'll show that $\{a_n\}_{n=1}^{\infty}$ is an increasing sequence. Suppose not, then there is some $n \in \mathbb{N}$ so that $a_1 \leq a_2 \leq \cdots \leq a_n$ but
$$a_{n+1} = \sqrt{2 + a_n} < a_n.$$

3.1. LIMITS OF SEQUENCES

Thus
$$2 + a_n \leq a_n^2,$$
and so
$$a_n^2 - a_n - 2 \geq 0.$$
However, the parabolic function
$$f(x) = x^2 - x - 2$$
is positive if and only if $x > 2$ or $x < -1$. This implies $a_n > 2$, which cannot happen. Thus a_n is increasing.

We may conclude that $\lim_{n \to \infty} a_n = a$ exists. Moreover, by taking the limit of both sides of $a_{n+1} = \sqrt{2 + a_n}$ we see that a must satisfy
$$a = \sqrt{2 + a},$$
and so we see $a = 2$.

3.1.4 Homework Exercises

1. For $\{x_n\}_{n=1}^\infty$ a positive sequence of real numbers, show
$$x_n \to x \quad \text{as} \quad n \to \infty \iff \sqrt{x_n} \to \sqrt{x} \quad \text{as} \quad n \to \infty.$$

2. For any polynomial function $f(x)$, show
$$x_n \to x \quad \text{as} \quad n \to \infty \Rightarrow f(x_n) \to f(x) \quad n \to \infty.$$

   Here $f(x)$ is a **polynomial** with real coefficients if $f(x) = a_n x^n + \cdots + a_2 x^2 + a_1 x + a_0$, for some $n \in \mathbb{N} \cup \{0\}$ and $a_i \in \mathbb{R}$ for $i = 0, 1, \cdots, n$.

3. Does the sequence $a_n = \sqrt{n+1} - \sqrt{n}$ converge? If so, show it converges and find its limit. If it diverges, please explain why it diverges.

4. For the sequence defined recursively by $s_1 = \sqrt{2}$ and $s_{n+1} = \sqrt{2 + \sqrt{s_n}}$ for $n \in \mathbb{N}$, show $s_n < L$, where $L^2 - \sqrt{L} - 2 = 0$, and $\{s_n\}_{n=1}^\infty$ converges. If possible, find its limit value.

5. Show that the sequence $a_n = (1+\frac{1}{n})^n$ is bounded by using the binomial theorem[2] to expand $(1+\frac{1}{n})^n$.

6. Does the sequence $a_n = \sqrt{n^2+n} - n$ converge? If so, show it converges and find its limit. If it diverges, please explain why it diverges.

7. Define a sequence a_n by
$$a_n = 1 + \frac{1}{2^2} + \frac{1}{3^2} + \cdots + \frac{1}{n^2}.$$
Show that this sequence converges.

8. Define a sequence a_n by
$$a_n = 1 + \frac{1}{2} + \frac{1}{2^2} + \cdots + \frac{1}{2^n}.$$
Show that this sequence converges.

9. For $c > 1$, show that the sequence $a_n = \sqrt[n]{c}$ converges to 1.

10. For $0 < c < 1$, show that the sequence $a_n = \sqrt[n]{c}$ converges to 1.

3.2 Subsequences

Let $\{a_n\}_{n=1}^\infty$ be a sequence of terms in a metric space (M, d), and let $\{k_i\}_{i=1}^\infty$ be a strictly increasing sequence of natural numbers. Then we call the composed sequence $\{a_{k_i}\}_{i=1}^\infty$ a **subsequence** of the sequence $\{a_n\}_{n=1}^\infty$. If the sequence defined by $b_i = a_{k_i}$ converges to some limit value b, its limit is called a **subsequential limit** of the sequence $\{a_n\}_{n=1}^\infty$.

The following theorem tells us of the existence of subsequences which converge to subsequential limits in the case where our sequence is bounded.

---

[2]The **Binomial Theorem** says that
$$(x+y)^k = \sum_{m=0}^{k} \binom{k}{m} x^m y^{k-m}.$$

3.2. SUBSEQUENCES

Theorem 3.2.1 *Let $\{a_n\}_{n=1}^{\infty}$ be a sequence contained in a compact subset of a metric space, then there is a subsequence of this sequence which converges to a subsequential limit. That is $\{a_n\}_{n=1}^{\infty}$ has a convergent subsequence.*

Proof: Consider the set $S = \{a_n : n \in \mathbb{N}\}$. Then S is a bounded set in (M, d). If S is a finite set, then clearly one of its elements, say s, have $a_n = s$ for infinitely many $n \in \mathbb{N}$. In this case, we take the subsequence to be the collection of a_n's for which $a_n = s$.

In the case where S has infinitely many elements, we know that S must have a limit point. This follows from theorem (2.4.5). Let b be such a limit point of S. Then we know that every neighbourhood of b contains infinitely many elements of S. Choose $k_1 \in \mathbb{N}$ so that $d(a_{k_1}, b) < 1$. Having chosen $k_1, k_2, \cdots, k_n$ choose $k_{n+1} > k_n$ so that $d(a_{k_{n+1}}, b) < \frac{1}{n+1}$. Then $\{a_{k_i}\}_{i=1}^{\infty}$ is the desired sequence.
□

Corollary 3.2.2 *Bounded sequences in Euclidean space $\mathbb{R}^n$ have a convergent subsequence.*

It should be noted here that if a sequence converges, then every subsequence converges to the same subsequential limit value, namely the limit of the sequence.

3.2.1 Homework Exercises

1. Find 2 different convergent subsequences of the following sequence which have different subsequential limit values:

$$1, -1, 1, -1, \cdots, (-1)^{n+1}, \cdots$$

2. Construct a sequence of real numbers with exactly three different subsequential limit values.

3. Construct a sequence of real numbers with exactly four different subsequential limit values.

4. Construct a sequence of real numbers with exactly five different subsequential limit values.

5. Construct a sequence of real numbers with exactly n, $n \in \mathbb{N}$ different subsequential limit values.

6. If possible construct a sequence of real numbers which has a countably infinite number of subsequential limit values which are different. If this is not possible, please explain why not.

7. If possible, construct a sequence which has subsequences converging to all the different rational values. If this is not possible, please explain why not. If it is possible, find the set of all subsequential limit values.

8. If possible construct a sequence of real numbers which has $\mathbb{N}$ as the set of subsequential limit values. If this is not possible, please explain why not.

3.3 Limit Superior and Limit Inferior

Theorem 3.3.1 *For a given sequence $\{a_n\}_{n=1}^{\infty}$ in a metric space (M, d) the collection L of subsequential limit values is a closed set.*

Proof: If L has only finitely many points, then clearly L is closed since all finite sets are closed because they have no limit points.

Let p be a limit point of L. We must show $p \in L$. In this case L has infinitely many points. We may choose k_1 so that $a_{k_1} \neq p$ and let $\varepsilon := d(a_{k_1}, p)$. Now suppose that an increasing collection $k_1, k_2, \cdots, k_n$ has been chosen. Since p is a limit point of L there is a $q \in L$ so that $d(q, p) < \frac{\varepsilon}{2^{n+1}}$. Since q is a subsequential limit value we may choose $k_{n+1} > k_n$ so that $d(a_{k_{n+1}}, q) < \frac{\varepsilon}{2^{n+1}}$. Thus

$$d(a_{k_{n+1}}, p) \leq d(a_{k_{n+1}}, q) + d(p, q) \leq 2\frac{\varepsilon}{2^{n+1}} = \frac{\varepsilon}{2^n}.$$

Thus we see $a_{k_n} \to p$ as $n \to \infty$.
□

In the context of $\mathbb{R}$, for a given sequence $\{a_n\}_{n=1}^{\infty}$ with a non-empty L we define the **limit supremum** and the **limit infimum** to be

$$\limsup_{n \to \infty} a_n := \sup L$$

3.3. LIMIT SUPERIOR AND LIMIT INFERIOR

and
$$\liminf_{n\to\infty} a_n := \inf L$$

respectively, where L is the set of all subsequential limit values of $\{a_n\}_{n=1}^{\infty}$, allowing $\pm\infty$ as values of subsequential limits. In the case where $\sup L = \pm\infty$, we let $\limsup_{n\to\infty} a_n = \pm\infty$. If $\{a_n : n \in \mathbb{N}\}$ is not bounded from above, we have $\sup L = \infty$ and so $\limsup_{n\to\infty} a_n = \infty$. In the case where $\inf L = \pm\infty$, we let $\liminf_{n\to\infty} a_n = \pm\infty$. If $\{a_n : n \in \mathbb{N}\}$ is not bounded from below, we have $\inf L = -\infty$ and so $\liminf_{n\to\infty} a_n = -\infty$. Note that if $\{a_n : n \in \mathbb{N}\}$ is bounded, we must have L contain a real number, and in such a case we must also have L is bounded.

Sometimes the limit supremum is called **limit superior** or even **limsup** for short. Likewise, the limit infimum may be called **limit inferior** or even **liminf** for short.

Since the collection of all subsequential limit values L is closed, if $\sup L \in \mathbb{R}$, then there is a subsequence which converges to this value. Likewise, if $\inf L \in \mathbb{R}$, then there is a subsequence which converges to this value.

3.3.1 Homework Exercises

1. Let $\{r_k\}_{k=1}^{\infty}$ be an enumeration of the rational numbers and let $x \in \mathbb{R}$. Show that there is a subsequence which converges[3] to x.

2. Let $\{r_n : n \in \mathbb{N}\}$ be an enumeration of the rational numbers. Consider the sequence $\{r_n\}_{n=1}^{\infty}$.

    (a) Show $\limsup_{n\to\infty} r_n = \infty$.
    (b) Show $\liminf_{n\to\infty} r_n = -\infty$.

3. Show that a sequence of real numbers $\{b_n\}_{n=1}^{\infty}$ converges to b if and only if $\limsup_{n\to\infty} b_n = \liminf_{n\to\infty} b_n = b$.

4. For a sequence of real numbers, prove the following: If $x < \liminf_{n\to\infty} a_n$, then there is an $N \in \mathbb{N}$ so that $n > N$ implies $a_n > x$.

---

[3] Thus one sees that the collection of all subsequential limit values L is the collection of all real numbers. In particular, one sees that the collection of all subsequential limit values could be uncountable.

5. For a sequence of real numbers, prove the following: If $y > \limsup_{n \to \infty} a_n$, then there is an $N \in \mathbb{N}$ so that $n > N$ implies $a_n < y$.

6. Let $\{a_k\}_{k=1}^{\infty}$ be a sequence of real numbers. Prove the following characterizations for lim sup and lim inf:

$$\limsup_{k \to \infty} a_k = \lim_{j \to \infty} (\sup\{a_k : k \geq j\}), \tag{3.1}$$

$$\liminf_{k \to \infty} a_k = \lim_{j \to \infty} (\inf\{a_k : k \geq j\}). \tag{3.2}$$

7. If $\{a_n\}_{n=1}^{\infty}$ and $\{b_n\}_{n=1}^{\infty}$ are sequences of real numbers which satisfy

$$a_n \leq b_n \quad \forall n \in \mathbb{N},$$

prove the following:

$$\liminf_{n \to \infty} a_n \leq \liminf_{n \to \infty} b_n$$

and

$$\limsup_{n \to \infty} a_n \leq \limsup_{n \to \infty} b_n.$$

8. Investigate the behaviour of the sequence $a_n = \sqrt[n]{n}$. Determine the collection of subsequential limit values.

9. For any real number r, show that there is a sequence of rational numbers which converges to r.

10. For any real number r, show that there is a sequence of irrational numbers which converges to r.

11. For any two sequences of real numbers $\{a_n\}_{n=1}^{\infty}$ and $\{b_m\}_{m=1}^{\infty}$ prove that

$$\liminf_{k \to \infty} a_k + \liminf_{k \to \infty} b_k \leq \liminf_{k \to \infty} (a_k + b_k)$$

$$\leq \limsup_{k \to \infty} (a_k + b_k) \leq \limsup_{k \to \infty} a_k + \limsup_{k \to \infty} b_k.$$

12. In the above exercise, show that if $\lim_{k \to \infty} a_k$ and $\lim_{k \to \infty} b_k$ both exist, then the inequalities are all equalities.

3.4 Cauchy Sequences

In a metric space (M, d) a sequence $\{a_n\}_{n=1}^{\infty}$ is called a **Cauchy sequence** if given any $\varepsilon > 0$, there is an $N \in \mathbb{N}$, so that

$$d(a_n, a_m) < \varepsilon \quad \forall m, n > N. \tag{3.3}$$

Clearly convergent sequences are Cauchy sequences, since if $a_n \to a$ as $n \to \infty$, then there is an $N \in \mathbb{N}$ so that for $n > N$ we have $d(a_n, a) < \frac{\varepsilon}{2}$. Thus for $n, m > N$ we have

$$d(a_n, a_m) \leq d(a_n, a) + d(a, a_m) < \frac{\varepsilon}{2} + \frac{\varepsilon}{2} = \varepsilon.$$

Let $\{a_n\}_{n=1}^{\infty}$ be a sequence contained in a compact subset K of a metric space (M, d). If $\{a_n\}_{n=1}^{\infty}$ is a Cauchy sequence, then $\{a_n\}_{n=1}^{\infty}$ converges.

To see this, we must first recall what we call the **diameter** of a set $S \subseteq M$, which is
$$\text{diam } S = \sup\{d(x, y) : x, y \in S\}. \tag{3.4}$$

Clearly any bounded subset of a metric space has finite diameter, and the diameter of a ball $B_r(x)$ is $2r$.

Lemma 3.4.1 *For any set S in a metric space (M, d), we have*

$$\text{diam } \overline{S} = \text{diam } S. \tag{3.5}$$

Proof: Since $S \subseteq \overline{S}$, we have $\{d(x, y) : x, y \in S\} \subseteq \{d(x, y) : x, y \in \overline{S}\}$, and thus
$$\text{diam} S \leq \text{diam } \overline{S}.$$
Let ε be any positive number, and let $x, y \in \overline{S}$, then there are points $w, z \in S$ so that $d(x, w) < \frac{\varepsilon}{2}$ and $d(y, z) < \frac{\varepsilon}{2}$, and so

$$d(x, y) \leq d(x, w) + d(w, z) + d(z, y)$$
$$\leq \varepsilon + d(w, z) \leq \varepsilon + \text{diam } S.$$

Since $x, y \in \overline{S}$ were arbitrary, we see that
$$\text{diam } \overline{S} \leq \text{diam } S + \varepsilon.$$

Since $\varepsilon > 0$ was arbitrary, we get

$$\text{diam } \overline{S} \leq \text{diam } S.$$

Hence we have the desired result.

□

Lemma 3.4.2 *Suppose that $K_1, K_2, K_3, \cdots$ is a collection of compact nested sets in a metric space (M, d) which satisfy*

1. $\cdots \subseteq K_{n+1} \subseteq K_n \subseteq \cdots \subseteq K_2 \subseteq K_1$

2. $\lim_{n \to \infty} \text{diam } K_n = 0$

Then $\bigcap_{n=1}^{\infty} K_n = \{x\}$ for some $x \in M$.

Proof: By theorem (2.4.3) we know $\bigcap_{n=1}^{\infty} K_n \neq \emptyset$. Let $K = \bigcap_{n=1}^{\infty} K_n$. If $x, y \in K$ are two distinct points, then $d(x, y) > 0$, and diam $K \geq d(x, y)$. Moreover, $x, y \in K_n$ for all n, and so we have diam $K_n \geq d(x, y)$ for all n. However, we know diam $K_n \to 0$ as $n \to \infty$. Thus, there is an $N \in \mathbb{N}$ so that for $m > N$ we have diam $K_m < \frac{d(x,y)}{2}$. Thus implies $d(x, y) \leq$ diam $K_m < \frac{d(x,y)}{2}$, which is impossible. Thus K must contain only one point.
□

Now, for our Cauchy sequence $\{a_n\}_{n=1}^{\infty}$ whose values are contained in a compact set K of M let $K_n := \{a_k : k > n\}$. Then the $\overline{K_n}$'s are compact since they are closed subsets of K. Moreover the $\overline{K_n}$'s satisfy the hypotheses of lemma (3.4.2), since for any $\varepsilon > 0$ there is an $N \in \mathbb{N}$ so that for $n, m > N$ we have $d(a_n, a_m) < \varepsilon$. Thus diam $\overline{K_j} < \varepsilon$ for any $j > N$. Hence $\lim_{n \to \infty}$ diam $\overline{K_n} = 0$. Therefore, $\bigcap_{n=1}^{\infty} \overline{K_n} = \{a\}$, for some $a \in M$.

Now let $\varepsilon > 0$ be given. Then there is an $N \in \mathbb{N}$ so that for $n > N$ we have diam $\overline{K_n} < \varepsilon$. Also, $a \in \overline{K_j}$ for all $j \in \mathbb{N}$. Hence for $n > N$ we have $d(a_n, a) < \varepsilon$ and thus $a_n \to a$ as $n \to \infty$.

Thus we have the following theorem:

Theorem 3.4.3 *Let $\{a_n\}_{n=1}^{\infty}$ be a sequence contained in a compact subset K of a metric space (M, d). If $\{a_n\}_{n=1}^{\infty}$ is a Cauchy sequence, then $\{a_n\}_{n=1}^{\infty}$ converges.*

3.4. CAUCHY SEQUENCES

In particular, we also have the so-called **Cauchy convergence criterion**:

Theorem 3.4.4 *If $\{\vec{a}_k\}_{k=1}^\infty$ is a Cauchy sequence in Euclidean space $\mathbb{R}^n$, then $\lim_{k\to\infty} \vec{a}_k$ exists.*

Proof: Let $\varepsilon > 0$ be given. Then there is an $N \in \mathbb{N}$ so that for $m, n > N$ we have
$$\|\vec{a}_m - \vec{a}_n\| < \varepsilon.$$
Thus for any $k \in \mathbb{N}$ we have $\|\vec{a}_k\|$ satisfies
$$\|\vec{a}_k\| \leq \max\{\|\vec{a}_1\|, \|\vec{a}_2\|, \cdots, \|\vec{a}_N\|, \|\vec{a}_{N+1}\| + \varepsilon\},$$
since for $k > N$ we have $\|\vec{a}_k\| = \|\vec{a}_k - \vec{a}_{N+1}\| + \|\vec{a}_{N+1}\| \leq \varepsilon + \|\vec{a}_{N+1}\|$. Hence $\{\vec{a}_j\}$ is a sequence that lies is a closed and bounded set in Euclidean space, and so theorem (3.4.3) gives us the desired result.
□

A metric space (M, d) for which every Cauchy sequence converges is called a **complete metric space**. The above theorem tells us that Euclidean space is complete, and any compact metric space[4] is complete.

3.4.1 Homework Exercises

1. Give an example of a bounded sequence of real numbers that is not a Cauchy sequence.

2. Give an example of a monotonic sequence of real numbers that is not a Cauchy sequence.

3. If $\{a_n\}_{n=1}^\infty$ is a Cauchy sequence of real numbers, use the definition of being a Cauchy sequence to show $\{a_n\}_{n=1}^\infty$ is bounded.

4. If $\{a_n\}_{n=1}^\infty$ and $\{b_n\}_{n=1}^\infty$ are Cauchy sequences of real numbers, prove that $\{a_n + b_n\}_{n=1}^\infty$ and $\{a_n \cdot b_n\}_{n=1}^\infty$ are Cauchy sequences.

5. Suppose that $\{a_n\}_{n=1}^\infty$ is a Cauchy sequence of natural numbers. Prove that $\{a_n\}_{n=1}^\infty$ is eventually a constant sequence.

---
[4]Here a compact metric space (M, d) is one where the set M is compact.

6. Suppose that for $r \in (0, 1)$ that a sequence $\{a_n\}_{n=1}^{\infty}$ in a metric space satisfies $d(x_n, x_{n+1}) \leq r^n$. Show that the sequence is a Cauchy sequence.

7. The sequence $a_n = \sqrt{n}$ satisfies $\lim_{n \to \infty} |a_{n+1} - a_n| = 0$. However, show that it is not a Cauchy sequence, and please explain why not.

3.5 Series

Given a sequence $\{a_n\}_{n=1}^{\infty}$ of real numbers, consider the sequence $\{S_n\}_{n=1}^{\infty}$ defined by

$$S_n = \sum_{k=1}^{n} a_k := a_1 + a_2 + \cdots + a_n.$$

$\{S_n\}_{n=1}^{\infty}$ is called the **sequence of partial sums** of the sequence $\{a_n\}_{n=1}^{\infty}$. A natural question to ask is does this sequence converge? If so, what is its limit value?

If the sequence $\{a_n\}_{n=1}^{\infty}$ has values which are non-negative, then clearly $\{S_n\}_{n=1}^{\infty}$ is monotone increasing. In such a case we see that $\lim_{n \to \infty} S_n$ exists if and only if $\{S_n\}_{n=1}^{\infty}$ is bounded.

If the sequence $\{S_n\}_{n=1}^{\infty}$ converges, we say that the **infinite series** or simply the **series**

$$\sum_{n=1}^{\infty} a_n = a_1 + a_2 + a_3 + \cdots + a_n + \cdots$$

converges and its value is $\lim_{n \to \infty} S_n$. Otherwise, we say that the series diverges.

Example: Consider the sequence $a_n = 1$, $n \in \mathbb{N}$. The partial sums for this sequence are

$$S_n = \underbrace{1 + 1 + 1 \cdots + 1}_{n \text{ times}} = \sum_{k=1}^{n} 1 = n.$$

3.5. SERIES

Thus $S_n \to \infty$ as $n \to \infty$, and hence the series

$$\sum_{n=1}^{\infty} 1 = 1 + 1 + 1 + 1 + \cdots$$

diverges.

Example: Consider the sequence $a_n = (-1)^n$, for $n \in \mathbb{N}$. This has partial sums

$$S_n = \begin{cases} 0 & \text{if } n \text{ is even,} \\ -1 & \text{if } n \text{ is odd.} \end{cases}$$

We see that $\{S_n\}_{n=1}^{\infty}$ does not have a limit, and thus

$$\sum_{n=1}^{\infty} (-1)^n = -1 + 1 - 1 + 1 - 1 + \cdots$$

diverges.

Example: Consider the sequence $a_n = 2^{-n+1}$, $n \in \mathbb{N}$. Here

$$S_n = \sum_{k=1}^{n} \frac{1}{2^{k-1}} = 1 + \frac{1}{2} + \cdots + \frac{1}{2^{n-1}}.$$

One can show that

$$S_n = \frac{1 - \frac{1}{2^n}}{1 - \frac{1}{2}} = 2 - \frac{1}{2^{n-1}},$$

and this clearly converges to 2 as $n \to \infty$. Hence $\sum_{k=1}^{\infty} \frac{1}{2^{n-1}}$ converges, and

$$\sum_{k=1}^{\infty} \frac{1}{2^{n-1}} = 2.$$

The Cauchy convergence criterion for a series is seen by examining the quantities $|S_n - S_m|$ for $m \leq n$, where $S_k = \sum_{j=1}^{k} a_j$. We see that

$$|S_n - S_m| = |\sum_{j=m+1}^{n} a_j|.$$

We know that $\lim_{n \to \infty} S_n$ exists if and only if $\{S_n\}_{n=1}^{\infty}$ is a Cauchy sequence. That is for any $\varepsilon > 0$, there is an $N \in \mathbb{N}$ so that for $n, m > N$ and $m < n$

we have $|\sum_{j=m+1}^{n} a_j| < \varepsilon$. Thus, if $\lim_{n\to\infty} S_n$ exists, by taking $n = m+1$, $m > N$ we see that $|a_m| < \varepsilon$. Hence we have the so-called **divergence theorem for series**, which is given below.

Divergence Theorem 3.5.1 *If $\sum_{n=1}^{\infty} a_n$ converges, then $\lim_{n\to\infty} a_n = 0$.*

Clearly the converse of this theorem is not true. To see this consider the so-called **harmonic series**

$$\sum_{n=1}^{\infty} \frac{1}{n}. \tag{3.6}$$

Example: The harmonic series $\sum_{n=1}^{\infty} \frac{1}{n}$ diverges.

Let $S_k := \sum_{j=1}^{k} \frac{1}{j}$. We will examine the subsequence $\{S_{2^n}\}_{n=1}^{\infty}$ of the sequence of partial sums $\{S_k\}_{k=1}^{\infty}$.

$$S_2 = 1 + \frac{1}{2},$$

$$S_{2^2} = S_4 = 1 + \frac{1}{2} + \frac{1}{3} + \frac{1}{4} > 1 + \frac{1}{2} + \frac{1}{4} + \frac{1}{4} = 1 + \frac{2}{2},$$

$$S_{2^3} = S_8 = S_4 + \frac{1}{5} + \frac{1}{6} + \frac{1}{7} + \frac{1}{8}$$

$$> S_4 + \frac{1}{8} + \frac{1}{8} + \frac{1}{8} + \frac{1}{8}$$

$$> 1 + \frac{2}{2} + \frac{1}{2} = 1 + \frac{3}{2},$$

$$S_{2^4} = S_{16} = S_8 + \frac{1}{9} + \cdots + \frac{1}{16}$$

$$> S_8 + \underbrace{\frac{1}{16} + \cdots + \frac{1}{16}}_{8 \text{ times}}$$

$$> 1 + \frac{3}{2} + \frac{1}{2} = 1 + \frac{4}{2}.$$

In general,

$$S_{2^n} > 1 + \frac{n}{2} \to \infty \quad \text{as} \quad n \to \infty.$$

Thus $\{S_{2^n}\}_{n=1}^{\infty}$ cannot converge to a real-valued limit, and so $\sum_{n=1}^{\infty} \frac{1}{n}$ diverges.

3.5.1 Geometric and Telescoping Series

Consider the following series

$$\sum_{k=1}^{\infty} \frac{1}{k(k+1)}.$$

Here

$$a_k = \frac{1}{k(k+1)},$$

which can be written as

$$a_k = \frac{1}{k} - \frac{1}{k+1}$$

by using a partial fractions decomposition.

To test for convergence, we examine the sequence of partial sums $\{S_n\}_{n=1}^{\infty}$.

$$S_1 = 1 - \frac{1}{2},$$

$$S_2 = 1 - \frac{1}{2} + \frac{1}{2} - \frac{1}{3} = 1 - \frac{1}{3},$$

$$S_3 = S_2 + \frac{1}{3} - \frac{1}{4} = 1 - \frac{1}{4}.$$

Continuing in this manner, we see that

$$S_n = 1 - \frac{1}{n+1}.$$

Thus

$$\lim_{n \to \infty} S_n = 1,$$

and so the series converges. Moreover,

$$\sum_{n=1}^{\infty} \frac{1}{n(n+1)} = 1.$$

Given a sequence $\{c_n\}_{n=1}^{\infty}$, a series of the form

$$\sum_{j=1}^{\infty}(c_j - c_{j+k}), \quad \text{or} \quad \sum_{j=1}^{\infty}(c_{j+k} - c_j),$$

for $k \in \mathbb{N}$, is called a **telescoping series**. In the case where we have $\sum_{j=1}^{\infty}(c_j - c_{j+k})$, the sequence of partial sums for $n > k$ satisfies

$$S_n = \sum_{j=1}^{n}(c_j - c_{j+k}) = c_1 + c_2 + \cdots + c_k - c_{n+1} - c_{n+2} - \cdots - c_{n+k},$$

and in the case where we have $\sum_{j=1}^{\infty}(c_{j+k} - c_j)$,

$$S_n = c_{n+1} + c_{n+2} + \cdots + c_{n+k} - c_1 - c_2 - \cdots - c_k.$$

In either case, S_n converges whenever $\lim_{n \to \infty} c_n =$ exists.

Now, fix any real numbers a and x, and consider the series

$$\sum_{n=0}^{\infty} ax^n = a + ax + ax^2 + \cdots.$$

Such a series is called a **geometric series**. For a geometric series, the sequence of partial sums are given by

$$S_n = a + ax + \cdots ax^n = a\frac{1 - x^{n+1}}{1 - x}.$$

Clearly this converges to $\frac{a}{1-x}$ if and only if $x \in (-1, 1)$. Moreover, the geometric series diverges for $x \in (-\infty, -1] \cup [1, \infty)$.

Example: We may use a geometric series to find the value of a repeating decimal expansion. For instance, consider

$$y = 0.1232323\cdots.$$

We write y as a geometric series

$$y = \frac{1}{10} + \frac{1}{1000}\sum_{n=0}^{\infty} 23 \cdot (\frac{1}{100})^n$$

$$= \frac{1}{10} + \frac{23}{990} = \frac{61}{495}.$$

Using this method we see that $x \in \mathbb{R}$ is rational if its decimal expansion terminates, that is eventually has all 0's, or it eventually has a repeating pattern. We note that the converse is true as well.

3.5.2 Series of Non-negative Terms

It was observed earlier that if a sequence $\{a_n\}_{n=1}^\infty$ has values which are non-negative, then clearly $\{S_n\}_{n=1}^\infty$ is monotone increasing. In such a case we see that $\lim_{n\to\infty} S_n$ exists if and only if $\{S_n\}_{n=1}^\infty$ is bounded.

The next result is often called the **comparison test** for series.

Theorem 3.5.2 *Suppose that $\sum_{k=1}^\infty a_k$, $\sum_{k=1}^\infty b_k$ are series with non-negative terms.*

1. *If $\sum_{k=1}^\infty b_k$ converges and $a_n \leq b_n \ \forall n$, then $\sum_{k=1}^\infty a_k$ also converges.*

2. *If $\sum_{k=1}^\infty b_k$ diverges and $a_n \geq b_n \ \forall n$, then $\sum_{k=1}^\infty a_k$ also diverges.*

Proof: Consider the sequences of partial sums

$$A_n = \sum_{k=1}^n a_k, \quad B_n = \sum_{k=1}^n b_n.$$

Since the terms a_n, b_n are non-negative, we have $\{A_n\}_{n=1}^\infty, \{B_n\}_{n=1}^\infty$ are monotone increasing sequences. Moreover, if $b_n \geq a_n$, we have $B_n \geq A_n$. Thus if $B_n \to B$, then B is the least upper bound for the sequence $\{B_n\}_{n=1}^\infty$. Thus $A_n \leq B$ as well, and so $\{A_n\}_{n=1}^\infty$ is monotone and bounded, hence convergent. If $a_n \geq b_n$ and $\{B_n\}_{n=1}^\infty$ is divergent, it must be that $A_n \geq B_n$ and $B_n \to \infty$ as $n \to \infty$. Therefore

$$A_n \to \infty \quad \text{as} \quad n \to \infty.$$

Thus $\{A_n\}_{n=1}^\infty$ diverges.
□

Example: Determine whether or not the series

$$\sum_{n=1}^\infty \frac{1}{2^n + 1}$$

converges.

Notice that $2^n + 1 > 2^n$, and thus $\frac{1}{2^n+1} < \frac{1}{2^n}$. Thus by the comparison test (3.5.2) the series

$$\sum_{k=1}^\infty \frac{1}{2^n + 1}$$

converges.

Example: Determine whether or not the series

$$\sum_{n=1}^{\infty} \frac{1}{n+1}$$

converges.

To make this determination we examine the leading terms in the numerator and the denominator. The numerator is 1 and so 1 is the leading term, and the leading term of the denominator is n. Thus we shall try to compare our series to a series of the form $\sum_{k=1}^{\infty} c\frac{1}{n}$, which is a divergent series. We need to choose our constant $c > 0$ so that

$$\frac{1}{n+1} \geq c\frac{1}{n}.$$

That is, we need

$$\frac{n}{n+1} \geq c \quad \forall n.$$

This is an increasing sequence, since $\frac{n}{n+1} = 1 - \frac{1}{n+1}$. Thus we can choose $c = \frac{1}{1+1} = \frac{1}{2}$, and for such a c we have

$$\frac{1}{n+1} \geq \frac{1}{2n}.$$

By the comparison test (3.5.2) we get that

$$\sum_{n=1}^{\infty} \frac{1}{n+1}$$

diverges.

The limit comparison test is another comparison test that is actually a bit more user-friendly and it says the following:

Theorem 3.5.3 *Suppose that*

$$\sum_{n=1}^{\infty} a_n, \quad \sum_{n=1}^{\infty} b_n$$

3.5. SERIES

are two series with positive terms. If

$$\lim_{n \to \infty} \frac{a_n}{b_n} = c \in (0, \infty),$$

then either both series converge or both diverge.

Proof: Let $\varepsilon > 0$ be given (but sufficiently small so that $\varepsilon < \frac{c}{2}$). Then there is an $N \in \mathbb{N}$ so that $\forall n \geq N$ we have

$$0 < c - \varepsilon \leq \frac{a_n}{b_n} \leq c + \varepsilon < \infty.$$

Thus

$$(c - \varepsilon)b_n \leq a_n \leq (c + \varepsilon)b_n.$$

By applying the comparison test (3.5.2) we get the desired result.
□.

The next result is often called the **Cauchy condensation test**, and it gives us another useful test for convergence.[5]

Theorem 3.5.4 *Suppose that $\{a_n\}_{n=1}^{\infty}$ is a decreasing sequence of non-negative real numbers, then $\sum_{n=0}^{\infty} a_n$ converges if and only if $\sum_{k=1}^{\infty} 2^k a_{2^k}$ converges.*

Proof: We define

$$S_n = \sum_{j=1}^{n} a_j = a_1 + a_2 + \cdots + a_n$$

and

$$T_k = \sum_{j=1}^{k} 2^j a_{2^j} = a_1 + 2a_2 + 4a_4 + \cdots + 2^k a_{2^k}.$$

We will now show that $S_{2^k} \leq T_k$. Since $\{a_n\}_{n=1}^{\infty}$ is decreasing, we have $a_2 + a_3 \leq 2a_2$, $a_4 + a_5 + a_6 + a_7 \leq 4a_4$, and in general $a_{2^k} + a_{2^k+1} + \cdots + a_{2^{k+1}-1} \leq 2^k a_{2^k}$. So we see $S_{2^k} \leq T_k$.

---

[5]At this point it will serve the purpose that the integral test often serves when one encounters the integral test for convergence of series in calculus.

We will also show $S_{2^k} \geq \frac{1}{2}(T_k - a_1)$. Observe that $a_1 \geq a_2$, $a_2 + a_3 \geq 2a_3 \geq 2a_4$, $a_4 + a_5 + a_6 + a_7 \geq 4a_7 \geq 4a_8$, and in general $a_{2^k} + a_{2^k+1} + \cdots + a_{2^{k+1}-1} \geq 2^k a_{2^{k+1}}$. So we see

$$S_{2^k-1} \geq a_2 + 2a_4 + \cdots + 2^{k-1}a_{2^k} \geq \frac{1}{2}(T_k - a_1).$$

Thus we see $\{S_n\}_{n=1}^\infty$ is bounded if and only if $\{T_n\}_{n=1}^\infty$ is bounded. By theorem (3.5.2) we see that one converges if and only if the other converges. □

p-Series

Let p be a positive rational number. We call a series of the form $\sum_{n=1}^\infty \frac{1}{n^p}$ a **p-series**. Earlier we showed that the harmonic series $\sum_{n=1}^\infty \frac{1}{n}$ diverges. Thus by theorem (3.5.2), for $p \leq 1$ the p-series diverges. This follows since the harmonic series diverges, and for $p < 1$ we have $n^p < n$ and so $\frac{1}{n} < \frac{1}{n^p}$.

We will now show that for $p > 1$ the p-series converges. Let $a_n = \frac{1}{n^p}$ and consider $2^k a_{2^k} = (\frac{1}{2^{p-1}})^k$. Using theorem (3.5.2) we get the p-series converges for $p > 1$, since $\sum_{k=1}^\infty (\frac{1}{2^{p-1}})^k$ is a convergent geometric series in that case.

3.5.3 The Root Test and the Ratio Test

Given a series $\sum_{k=1}^\infty a_k$, with $a_k \geq 0$, we consider the following test for convergence called the **root test**.

Theorem 3.5.5 *Consider $\alpha := \limsup_{n \to \infty} \sqrt[n]{a_n}$.*

- *$\alpha < 1$ implies $\sum_{k=1}^\infty a_k$ converges.*

- *$\alpha > 1$ implies $\sum_{k=1}^\infty a_k$ diverges.*

- *If $\alpha = 1$, then no information is given by this test.*

Proof: If $\alpha < 1$, then given any $\beta \in (\alpha, 1)$ we have that there is an $N \in \mathbb{N}$ so that $n \geq N$ we have

$$\sqrt[n]{a_n} < \beta.$$

Then for all such n's we have

$$a_n < \beta^n.$$

3.5. SERIES

The series $\sum_{k=n}^{\infty} \beta^k$ is a convergent geometric series, and thus $\sum_{k=n}^{\infty} a_k$ converges by the comparison test theorem (3.5.2).

If $\alpha > 1$ then there is a subsequence $\{a_{n_k}\}_{k=1}^{\infty}$ so that $\lim_{k\to\infty} a_{n_k} = \alpha$. So $\sqrt[n_k]{a_{n_k}} > 1$ for k large, and thus $a_{n_k} > 1$ for k large. Thus $a_n \not\to 0$ as $n \to \infty$, and thus by the divergence theorem (3.5.1) we get that $\sum_{n=1}^{\infty} a_n$ diverges.

To see that if $\alpha = 1$ then no conclusion can be drawn we consider two series with $\alpha = 1$, namely

$$\sum_{n=1}^{\infty} \frac{1}{n}, \quad \text{and} \quad \sum_{n=1}^{\infty} \frac{1}{n^2}.$$

The first series diverges, and the second converges.
□

Given a series $\sum_{k=1}^{\infty} a_k$ with $a_k > 0$ for each k, we consider the following test for convergence called the **ratio test**.

Theorem 3.5.6 *Consider* $\alpha := \limsup_{n\to\infty} \frac{a_{n+1}}{a_n}$.

- $\alpha < 1$ *implies* $\sum_{k=1}^{\infty} a_k$ *converges.*

- $\frac{a_{n+1}}{a_n} \geq 1$ *for* $n \geq N$ *implies* $\sum_{k=1}^{\infty} a_k$ *diverges.*

- *If* $\lim_{n\to\infty} \frac{a_{n+1}}{a_n} = 1$, *then no information is given by this test.*

Proof: If $\alpha < 1$, then given any $\beta \in (\alpha, 1)$ we have that there is an $N \in \mathbb{N}$ so that $n \geq N$ satisfies

$$\frac{a_{n+1}}{a_n} < \beta.$$

Then for such n's we have

$$a_{n+1} < a_n \beta.$$

In particular, we know

$$a_{N+1} < a_N \beta,$$
$$a_{N+2} < a_{N+1} \beta < a_N \beta^2,$$

etc., and so

$$a_{N+k} < a_N \beta^k.$$

The series $\sum_{n=1}^{\infty} a_N \beta^n$ is a convergent geometric series, and thus $\sum_{k=N+1}^{\infty} a_k$ converges by the comparison test (3.5.2).

If $\frac{a_{n+1}}{a_n} \geq 1$ for $n > N$, then a_n is eventually an increasing sequence., ans so $a_n \not\to 0$ as $n \to \infty$. So by the divergence theorem (3.5.1) we get that $\sum_{n=1}^{\infty} a_n$ diverges.

To see that if $\lim_{n \to \infty} \frac{a_{n+1}}{a_n} = 1$ then no conclusion can be drawn we consider two series with $\lim_{n \to \infty} \frac{a_{n+1}}{a_n} = 1$, namely

$$\sum_{n=1}^{\infty} \frac{1}{n}, \quad \text{and} \quad \sum_{n=1}^{\infty} \frac{1}{n^2}.$$

The first series diverges, and the second converges.
□

Example: Consider the series

$$\sum_{n=1}^{\infty} a_n = \frac{1}{2} + \frac{1}{3} + \frac{1}{2^2} + \frac{1}{3^2} + \frac{1}{2^3} + \frac{1}{3^3} + \frac{1}{2^4} + \frac{1}{3^4} + \cdots + \frac{1}{2^n} + \frac{1}{3^n} + \cdots$$

This series converges by the root test, but the ratio test is inconclusive, since

$$\liminf_{n \to \infty} \frac{a_{n+1}}{a_n} = 0,$$

$$\liminf_{n \to \infty} \sqrt[n]{a_n} = \frac{1}{\sqrt{3}},$$

$$\limsup_{n \to \infty} \sqrt[n]{a_n} = \frac{1}{\sqrt{2}},$$

and

$$\limsup_{n \to \infty} \frac{a_{n+1}}{a_n} = \infty.$$

In general, if the ratio test shows convergence then so will the root test. Moreover, whenever the root test is inconclusive, the ratio test will be as well. To see this we note the following theorem.

Theorem 3.5.7 *Let $\{c_n\}_{n=1}^{\infty}$ be a sequence of positive real numbers, then*

$$\liminf_{n \to \infty} \frac{c_{n+1}}{c_n} \leq \liminf_{n \to \infty} \sqrt[n]{c_n} \leq \limsup_{n \to \infty} \sqrt[n]{c_n} \leq \limsup_{n \to \infty} \frac{c_{n+1}}{c_n}. \qquad (3.7)$$

3.5. SERIES

Proof: Note the the middle inequality is always true, and thus we only need to establish the other two.

Let $\alpha = \limsup_{n \to \infty} \frac{c_{n+1}}{c_n}$. If $\alpha = \infty$ then there is nothing to show, so we consider the case $\alpha < \infty$. Let $\beta > \alpha$. Then there is an $N \in \mathbb{N}$ so that for $n \geq N$ we have $\frac{c_{n+1}}{c_n} < \beta$. Using this we may easily see that $c_{N+k+1} < c_{N+k}\beta$ for any k, and so $c_{N+m} < c_N \beta^m$. In other words,

$$c_n < c_N \beta^{-N} \beta^n \quad \forall n \geq N.$$

Taking n^{th} roots we see that

$$\sqrt[n]{c_n} < \sqrt[n]{c_N \beta^{-N}} \beta \quad \forall n \geq N.$$

Moreover, $\lim_{n \to \infty} \sqrt[n]{c_N \beta^{-N}} = 1$ and so

$$\limsup_{n \to \infty} \sqrt[n]{c_n} \leq \beta.$$

Since this is true for any $\beta > \alpha$ we get

$$\limsup_{n \to \infty} \sqrt[n]{c_n} \leq \alpha.$$

Let $\gamma = \liminf_{n \to \infty} \frac{c_{n+1}}{c_n}$. If $\gamma = -\infty$ then there is nothing to show, so we consider the case $\gamma > -\infty$. Let $\delta < \gamma$. Then there is an $N \in \mathbb{N}$ so that for $n \geq N$ we have $\frac{c_{n+1}}{c_n} > \delta$. Using this we may easily see that $c_{N+k+1} > c_{N+k}\delta$ for any k, and so $c_{N+m} > c_N \delta^m$. In other words,

$$c_n > c_N \delta^{-N} \delta^n \quad \forall n \geq N.$$

Taking n^{th} roots we see that

$$\sqrt[n]{c_n} > \sqrt[n]{c_N \delta^{-N}} \delta \quad \forall n \geq N.$$

Moreover, $\lim_{n \to \infty} \sqrt[n]{c_N \delta^{-N}} = 1$ and so

$$\liminf_{n \to \infty} \sqrt[n]{c_n} \geq \delta.$$

Since this is true for any $\delta < \gamma$ we get

$$\liminf_{n \to \infty} \sqrt[n]{c_n} \geq \gamma.$$

□

3.5.4 Homework Exercises

For a given non-negative integer n we define n-**factorial**, denoted $n!$ by: $0! = 1$, $1! = 1$, $2! = 2$, $3! = 3 \cdot 2 = 6$ and in general $n! = n \cdot (n-1)!$.

1. Use the ratio or root test to test $\sum_{k=1}^{\infty} \frac{1}{k^k}$ for convergence.

2. Use the ratio test to test the series below for convergence
$$\frac{1}{2} + 1 + \frac{1}{8} + \frac{1}{4} + \frac{1}{32} + \frac{1}{16} + \cdots.$$

3. Use the root test to test the series below for convergence
$$\frac{1}{2} + 1 + \frac{1}{8} + \frac{1}{4} + \frac{1}{32} + \frac{1}{16} + \cdots.$$

4. Use the ratio or root test to test $\sum_{k=1}^{\infty} \frac{2^k}{k!}$ for convergence.

5. For the Fibonacci sequence $\{f_n\}_{n=1}^{\infty}$ show that $\sum_{k=1}^{\infty} \frac{1}{f_k}$ converges.

6. For any $x \in (0,1)$ show that
$$\sum_{k=1}^{\infty} kx^k$$
converges, and use this to conclude $\lim_{n \to \infty} nx^n = 0$ whenever $x \in (0,1)$.

7. Suppose
$$\sum_{k=1}^{\infty} a_k$$
converges, and $a_k > 0$ for all k. Define $b_n = \frac{1}{n} \sum_{k=1}^{n} a_k$. Show
$$\sum_{k=1}^{\infty} b_k$$
diverges.

3.5. SERIES

8. Test the series
$$\sum_{n=1}^{\infty} \frac{1}{n^{1+\frac{1}{n}}}$$
for convergence.

9. Test the series
$$\sum_{n=1}^{\infty} \frac{n!}{n^n}$$
for convergence.

10. Test the series
$$\sum_{n=1}^{\infty} \frac{1^2 + 2^2 + \cdots + n^2}{n!}$$
for convergence.

11. Test the series
$$\sum_{k=1}^{\infty} \frac{\sqrt{k+1}}{\sqrt[3]{k^5 + k^3 + 1}}$$
for convergence.

12. Test the series
$$\sum_{n=1}^{\infty} \left(\frac{n}{n^2 + 1}\right)^n$$
for convergence.

13. Test the series
$$\sum_{n=1}^{\infty} \left(\frac{n}{n^2 + 1}\right)^{n^{1.0001}}$$
for convergence.

14. Test the series
$$\sum_{n=1}^{\infty} 2^{\frac{(-1)^n}{n}}$$
for convergence.

15. Test the series
$$\sum_{n=1}^{\infty} (\frac{5+(-1)^n}{2})^{-n}$$
for convergence.

16. Test the series
$$\sum_{n=1}^{\infty} (\frac{5+(-1)^n}{2})^n$$
for convergence.

3.5.5 The Number e

We define what is often called **Euler's number**, denoted by e, to be the number defined by

$$e := \sum_{n=0}^{\infty} \frac{1}{n!}. \tag{3.8}$$

By the ratio test we see that the series $\sum_{n=0}^{\infty} \frac{1}{n!}$ converges and so this definition of e is well defined. In this section we will show that there is another limit that equals e, namely $\lim_{n\to\infty}(1+\frac{1}{n})^n$, and we will also show that e is irrational.[6]

The sequence of partial sums

$$S_n = \sum_{k=0}^{n} \frac{1}{k!}$$

is an increasing sequence, and so we see that $e > S_n$ for any $n \in \mathbb{N}$. For instance $S_2 = 2.5$, $S_{15} = 2.718281828\cdots$. Hence we know $e > 2.718281828$. Moreover,

$$S_n = 1 + 1 + \frac{1}{1\cdot 2} + \frac{1}{1\cdot 2\cdot 3} + \frac{1}{1\cdot 2\cdot 3\cdot 4} + \cdots + \frac{1}{1\cdot 2\cdot 3\cdots(n-1)\cdot n}$$

$$< 1 + 1 + \frac{1}{2} + \frac{1}{2^2} + \frac{1}{2^3} + \cdots + \frac{1}{2^{n-1}} < 1 + \sum_{k=0}^{\infty} (\frac{1}{2})^k = 3.$$

---

[6]The symbol e was first used by the mathematician Leonard Euler in 1727. In 1737 Euler showed e is irrational, and in 1873 the mathematician Charles Hermite proved that e was transcendental.

3.5. SERIES

Hence $e \in (2.718281828, 3)$.

In fact the rate at which $S_n = \sum_{k=0}^{n} \frac{1}{k!}$ converges to e can be estimated, since

$$e - S_n = \sum_{k=n+1}^{\infty} \frac{1}{k!} = \frac{1}{(n+1)!} + \frac{1}{(n+2)!} + \frac{1}{(n+3)!} + \frac{1}{(n+4)!} \cdots$$

$$< \frac{1}{(n+1)!}(1 + \frac{1}{n+1} + \frac{1}{(n+1)^2} + \frac{1}{(n+1)^3} + \cdots)$$

$$= \frac{1}{(n+1)!} \cdot \frac{1}{1 - \frac{1}{n+1}} = \frac{1}{n! \cdot n}.$$

So we actually see that e actually lies in the interval $(S_{15}, S_{15} + \frac{1}{15! \cdot 15})$.

Theorem 3.5.8

$$e = \lim_{n \to \infty} (1 + \frac{1}{n})^n. \tag{3.9}$$

Proof: Define the sequence $\{T_n\}_{n=1}^{\infty}$ by $T_n = (1+\frac{1}{n})^n$. Using the binomial theorem we see that

$$T_n = (1 + \frac{1}{n})^n = \sum_{k=0}^{n} \binom{n}{k} \frac{1}{n^k}$$

$$= 1 + 1 + \frac{1}{2!}(1 - \frac{1}{n}) + \frac{1}{3!}(1 - \frac{1}{n})(1 - \frac{2}{n}) + \cdots + \frac{1}{n!} \prod_{j=1}^{n-1}(1 - \frac{j}{n})$$

$$< S_n.$$

Thus $T_n < S_n$ and so

$$\limsup_{n \to \infty} T_n \le e.$$

However, for $n > m$ we have

$$T_n > 1 + 1 + \frac{1}{2!}(1 - \frac{1}{n}) + \frac{1}{3!}(1 - \frac{1}{n})(1 - \frac{2}{n}) + \cdots + \frac{1}{m!} \prod_{j=1}^{m-1}(1 - \frac{j}{n}).$$

Keeping m fixed, we let $n \to \infty$ to obtain

$$\liminf_{n \to \infty} T_n \ge 1 + 1 + \frac{1}{2!} + \cdots + \frac{1}{m!} = S_m.$$

Letting $m \to \infty$ we obtain
$$e \leq \liminf_{n \to \infty} T_n.$$

Thus we get
$$\limsup_{n \to \infty} T_n \leq e \leq \liminf_{n \to \infty} T_n,$$
and so
$$e = \lim_{n \to \infty} T_n = \lim_{n \to \infty} (1 + \frac{1}{n})^n.$$

□

Theorem 3.5.9 *e is irrational.*

Proof: Suppose $e = \frac{m}{n}$, $m, n \in \mathbb{N}$. Then $n! \cdot e \in \mathbb{N}$. Moreover, for $S_n = \sum_{k=0}^{n} \frac{1}{k!}$ we have

$$n! \cdot S_n = n!(1 + 1 + \frac{1}{2!} + \cdots + \frac{1}{(n-1)!} + \frac{1}{n!}) \in \mathbb{N}.$$

So we see $n! \cdot (e - S_n)$ must be an integer. Recall that we showed $e - S_n < \frac{1}{n! \cdot n}$, and so
$$0 < n! \cdot (e - S_n) < \frac{1}{n} < 1.$$

This is a contradiction.
□

3.5.6 Alternating Series

An **alternating series** is an infinite series whose terms alternate between positive and negative. For example, the series

$$\sum_{n=1}^{\infty} (-1)^{n-1} = 1 - 1 + 1 - 1 + 1 - 1 + \cdots$$

and the series

$$\sum_{n=0}^{\infty} \frac{(-1)^n}{n} = 1 - \frac{1}{2} + \frac{1}{3} - \frac{1}{4} + \cdots$$

3.5. SERIES

are alternating series.

In general an alternating series is a series of the form

$$\sum_{n=1}^{\infty}(-1)^n b_n = -b_1 + b_2 - b_3 + \cdots, \qquad (3.10)$$

or

$$\sum_{n=1}^{\infty}(-1)^{n-1} b_n = b_1 - b_2 + b_3 - b_4 + \cdots, \qquad (3.11)$$

where for each $n \in \mathbb{N}$

$$b_n \geq 0.$$

Theorem 3.5.10 *Let $\{b_n\}_{n=1}^{\infty}$ be a sequence which satisfies $b_n \geq 0$, $b_{n+1} \leq b_n$ $\forall n$, and $\lim_{n \to \infty} b_n = 0$. Then the alternating series*

$$\sum_{n=1}^{\infty}(-1)^n b_n = -b_1 + b_2 - b_3 + \cdots$$

and

$$\sum_{n=1}^{\infty}(-1)^{n-1} b_n = b_1 - b_2 + b_3 - b_4 + \cdots$$

both converge.

Proof: Without loss of generality[7], we consider only

$$\sum_{n=1}^{\infty}(-1)^{n-1} b_n = b_1 - b_2 + b_3 - b_4 + \cdots$$

Here we have that the sequence of partial sums satisfies

$$S_2 = b_1 - b_2 \geq 0,$$

$$S_4 = S_2 + b_3 - b_4 \geq S_2,$$

$$S_6 = S_4 + b_5 - b_6 \geq S_4,$$

---
[7]Since the other case is simply the negative of this case.

and thus we see that
$$0 \leq S_2 \leq S_4 \leq S_6 \leq \cdots.$$
However, we also have
$$S_2 = b_1 - b_2 \leq b_1,$$
$$S_4 = b_1 - (b_2 - b_3) - b_4 \leq b_1,$$
$$S_6 = b_1 - (b_2 - b_3) - (b_4 - b_5) - b_6 \leq b_1,$$
and thus we see that
$$S_2, S_4, S_6, \cdots \leq b_1.$$
Thus the sequence $\{S_{2n}\}_{n=1}^{\infty}$ is monotone increasing and bounded from above, and hence is convergent. Let S be its limit, then
$$\lim_{n \to \infty} S_{2n} = S \leq b_1.$$
Observe as well that $S_3 = S_2 + b_3$, $S_5 = S_4 + b_5$, etc., and in general
$$S_{2n+1} = S_{2n} + b_{2n+1}.$$
By taking limits of both sides, since the right hand side converges to S, we see that the left hand side does as well, and so $\lim_{n \to \infty} S_{2n+1} = S$.
□

The alternating series test can be used to show that the **alternating harmonic series**
$$\sum_{n=1}^{\infty} \frac{(-1)^n}{n} = 1 - \frac{1}{2} + \frac{1}{3} - \frac{1}{4} + \cdots$$
is convergent. Since $b_n = \frac{1}{n}$, we see that this series converges to a limit value $S \leq 1$.

Alternating Series Limit Estimation and Rate of Convergence

If $\sum_{n=1}^{\infty}(-1)^{n-1}b_n$ is an alternating series with $b_n \geq 0$ decreasing to a limit value of zero, then we have that the limit $S = \sum_{n=1}^{\infty}(-1)^{n-1}b_n$ satisfies:
$$|S - S_n| \leq b_{n+1}.$$

3.5. SERIES

We know that the limit value S lies between any two consecutive partial sums, and thus
$$|S - S_n| \leq |S_{n+1} - S_n| = b_{n+1}.$$

Example: Find the sum of the series
$$\sum_{n=1}^{\infty} (-1)^{n-1} \frac{1}{n}$$
accurate to a margin of error of at most 0.1.

Recall that
$$S_n = 1 - \frac{1}{2} + \frac{1}{3} - \frac{1}{4} + \cdots + (-1)^{n-1} \frac{1}{n}$$
and the sum of the series S satisfies
$$|S - S_n| \leq \frac{1}{n+1}.$$

We need this to be less than 0.1. We can do this if $n = 10$, and so S_{10} will approximate S with an error less than $\frac{1}{10}$, where
$$S_{10} = 1 - \frac{1}{2} + \frac{1}{3} - \frac{1}{4} + \frac{1}{5} - \frac{1}{6} + \frac{1}{7}$$
$$- \frac{1}{8} + \frac{1}{9} - \frac{1}{10} = 0.645634921.$$

3.5.7 Absolute and Conditional Convergence

Given any series
$$\sum_{k=1}^{\infty} a_k = a_1 + a_2 + a_3 + a_4 + \cdots$$
we may create another series
$$\sum_{k=1}^{\infty} |a_k| = |a_1| + |a_2| + |a_3| + |a_4| + \cdots$$

We say that the series $\sum_{k=1}^{\infty} a_k$ is **absolutely convergent** if the series $\sum_{k=1}^{\infty} |a_k|$ converges. Clearly a convergent series with non-negative terms is always absolutely convergent. It is the series which may have negative terms such as an alternating series that we must consider here. Since the alternating harmonic series $\sum_{k=1}^{\infty}(-1)^{k+1}\frac{1}{k}$ converges by the alternating series test but the harmonic series $\sum_{k=1}^{\infty}\frac{1}{k}$ diverges, we see that not every convergent series is absolutely convergent. A series $\sum_{k=1}^{\infty} a_k$ which converges but the series $\sum_{k=1}^{\infty} |a_k|$ diverges is said to be **conditionally convergent**. Note that we often say that a series **converges absolutely** if it is an absolutely convergent series.

Example: Test the series for absolute or conditional convergence:

$$\sum_{n=1}^{\infty}(-1)^{n+1}\frac{1}{n^2}, \quad \sum_{n=1}^{\infty}(-1)^{n+1}\frac{1}{\sqrt{n}}.$$

By the alternating series test, both series converge. The series

$$\sum_{n=1}^{\infty}\frac{1}{n^2}, \quad \sum_{n=1}^{\infty}\frac{1}{\sqrt{n}}$$

are both p-series, and the first converges since $p = 2$ but the second diverges since $p = \frac{1}{2}$. Thus the series $\sum_{n=1}^{\infty}(-1)^{n+1}\frac{1}{n^2}$ converges absolutely and the series $\sum_{n=1}^{\infty}(-1)^{n+1}\frac{1}{\sqrt{n}}$ is converges conditionally.

Theorem 3.5.11 *If $\sum_{k=1}^{\infty} a_k$ is absolutely convergent, then it is convergent.*

Proof: Observe that

$$0 \leq a_n + |a_n| \leq 2|a_n|.$$

Since $\sum_{k=1}^{\infty} 2|a_k|$ converges, by the comparison test (3.5.2) we have that $\sum_{k=1}^{\infty}(a_k + |a_k|)$ converges, and so

$$\sum_{k=1}^{\infty} a_k = \sum_{k=1}^{\infty}[(a_k + |a_k|) - |a_k|]$$

$$\sum_{k=1}^{\infty}(a_k + |a_k|) - \sum_{k=1}^{\infty}|a_k|.$$

Thus we see that $\sum_{n=1}^{\infty} a_n$ converges.
□

3.5. SERIES

3.5.8 The Algebra of Series

Theorem 3.5.12 *Suppose $\sum_{k=0}^{\infty} a_k$ and $\sum_{k=0}^{\infty} b_k$ are convergent series, then for any $c, d \in \mathbb{R}$*

$$\sum_{k=0}^{\infty}(ca_k + db_k) = c\sum_{k=0}^{\infty} a_k + d\sum_{k=0}^{\infty} b_k.$$

Proof: To see this, let $A_n = \sum_{k=0}^{n} a_k$ and $B_n = \sum_{k=0}^{n} b_k$, then the sequence of partial sums for $\sum_{k=0}^{\infty}(ca_k + db_k)$ is $\{cA_n + dB_n\}_{n=0}^{\infty}$ and the result follows by properties of limits.
□

The theorem below says that the so-called **Cauchy product** converges if either of our series is absolutely convergent. Here the Cauchy product is $\sum_{k=0}^{\infty} c_k$ as given below. Note that if the series are not absolutely convergent, then the result may not hold, and to see this we take $a_k = b_k = (-1)^k \frac{1}{\sqrt{k+1}}$.

Theorem 3.5.13 *Suppose $\sum_{k=0}^{\infty} a_k$ and $\sum_{k=0}^{\infty} b_k$ are two convergent series and $\sum_{k=0}^{\infty} a_k$ is absolutely convergent, then for $c_k = \sum_{j=0}^{k} a_j b_{k-j}$ we have $\sum_{k=0}^{\infty} c_k$ converges to $\sum_{k=0}^{\infty} a_k \cdot \sum_{k=0}^{\infty} b_k$.*

Proof: Define the following quantities:

$$A = \sum_{k=0}^{\infty} a_k, \quad \alpha = \sum_{k=0}^{\infty} |a_k|, \quad B = \sum_{k=0}^{\infty} b_k,$$

$$A_n = \sum_{k=0}^{n} a_k, \quad B_n = \sum_{k=0}^{n} b_k, \quad \beta_n = B - B_n, \quad C_n = \sum_{k=0}^{n} c_k.$$

Since $B_n \to B$ as $n \to \infty$ we have $\beta_n \to 0$ as $n \to \infty$.

$$C_n = a_0 b_0 + (a_0 b_1 + a_1 b_0) + \cdots + (a_0 b_n + a_1 b_{n-1} + \cdots + a_n b_0)$$

$$= \sum_{k=0}^{n} a_k B_{n-k}$$

$$= \sum_{k=0}^{n} a_k (B - \beta_{n-k})$$

$$= A_n B - \sum_{k=0}^{n} a_k \beta_{n-k}.$$

Since $A_n B \to AB$ as $n \to \infty$ to show $C_n \to AB$ we see that it suffices to show $\sum_{k=0}^{n} a_{n-k} \beta_k \to 0$ as $n \to \infty$.

Let $\varepsilon > 0$ be given. Since $\lim_{n \to \infty} \beta_n = 0$, there is an $N \in \mathbb{N}$ so that for $n \geq N$ we have $|\beta_n| < \varepsilon$. Thus for $n > N$ we have

$$\left| \sum_{k=0}^{n} a_{n-k} \beta_k \right| \leq \left| \sum_{k=0}^{N} a_{n-k} \beta_k \right| + \left| \sum_{k=N+1}^{n} a_{n-k} \beta_k \right|$$

$$\leq \left| \sum_{k=0}^{N} a_{n-k} \beta_k \right| + \sum_{k=N+1}^{n} |a_{n-k}||\beta_k|$$

$$\leq \left| \sum_{k=0}^{N} a_{n-k} \beta_k \right| + \varepsilon \cdot \alpha$$

$$= |\beta_0 a_n + \beta_1 a_{n-1} + \cdots + \beta_N a_{n-N}| + \varepsilon \cdot \alpha.$$

Keeping N fixed, we let $n \to \infty$. Since $\sup\{|\beta_0|, |\beta_1|, \cdots, |\beta_N|\} < \infty$ and $a_n \to 0$ as $n \to \infty$ we know that the term $|\beta_0 a_n + \beta_1 a_{n-1} + \cdots + \beta_N a_{n-N}| \to 0$ as $n \to \infty$. Thus

$$\limsup_{n \to \infty} \left| \sum_{k=0}^{n} a_{n-k} \beta_k \right| \leq \varepsilon \cdot \alpha.$$

Since $\varepsilon > 0$ is arbitrary we get $\left| \sum_{k=0}^{n} a_{n-k} \beta_k \right| \to 0$ as $n \to \infty$. $\square$

3.5.9 Rearrangements of Series

Let
$$k_1, k_2, k_3, k_4, \cdots$$
be a sequence in which every positive integer appears once and only once. 1.e. $k(n) = k_n$ is a bijection whose domain and range are $\mathbb{N}$. Then the series

$$\sum_{n=1}^{\infty} a_{k_n}$$

3.5. SERIES

is called a **rearrangement** of the series

$$\sum_{j=1}^{\infty} a_j.$$

Suppose that

$$S_n = \sum_{j=1}^{n} a_j, \quad T_n = \sum_{i=1}^{n} a_{k_i},$$

then the two sequences of partial sums $\{S_n,\}_{n=1}^{\infty}$ and $\{T_n\}_{n=1}^{\infty}$ may be entirely different. So we are led to the problem of determining the conditions for which they both converge to the same limit S.

Example: Consider the alternating harmonic series

$$1 - \frac{1}{2} + \frac{1}{3} - \frac{1}{4} + \frac{1}{5} - \frac{1}{6} + \cdots$$

and one of its rearrangements

$$1 + \frac{1}{3} - \frac{1}{2} + \frac{1}{5} + \frac{1}{7} - \frac{1}{4} + \frac{1}{9} + \frac{1}{11} - \frac{1}{6} + \cdots.$$

The alternating harmonic series converges to a limit value S which satisfies

$$S \leq S_1, S_3, S_5, \cdots$$

In particular,

$$S < S_3 = 1 - \frac{1}{2} + \frac{1}{3} = \frac{5}{6}.$$

Note that for any positive k we have that

$$\frac{1}{4k-3} + \frac{1}{4k-1} - \frac{1}{2k} > 0,$$

and thus

$$1 + \frac{1}{3} - \frac{1}{2} > 0,$$

$$\frac{1}{5} + \frac{1}{7} - \frac{1}{4} > 0,$$

$$\frac{1}{9} + \frac{1}{11} - \frac{1}{6} > 0,$$

etc. Hence the rearrangement's partial sums satisfy

$$\frac{5}{6} = T_3 < T_6 < T_9 < \cdots.$$

So we see that if $\{T_n\}_{n=1}^{\infty}$ converges[8], it must converge to a limit greater than $\frac{5}{6}$. Thus we see that the alternating harmonic series and this particular rearrangement of the alternating harmonic series do not converge to the same limit value S.

The following result due to the mathematician Riemann shows that for any conditionally convergent series $\sum_{k=1}^{\infty} a_k$ and any real number r one can find a rearrangement $\sum_{n=1}^{\infty} a_{k_n}$ that will converge r.

Theorem 3.5.14 *Let $\{a_n\}_{n=1}^{\infty}$ be a sequence of real numbers for which $\sum_{k=1}^{\infty}$ converges, but $\sum_{k=1}^{\infty} |a_k|$ diverges. Let a, b satisfy*

$$-\infty \leq a \leq b \leq \infty,$$

Then there exists a rearrangement $\sum_{n=1}^{\infty} a_{k_n}$ of the series $\sum_{n=1}^{\infty} a_n$ for which its sequence of partial sums $T_n = \sum_{j=1}^{n} a_{k_j}$ satisfies

$$\liminf_{n \to \infty} T_n = a, \quad \limsup_{n \to \infty} T_n = b.$$

Proof: Let $u_n = \frac{|a_n|+a_n}{2}$ and $l_n = \frac{|a_n|-a_n}{2}$, then $|a_n| = u_n + l_n$, $a_n = u_n - l_n$, and $u_n, l_n \geq 0$ for all $n \in \mathbb{N}$.

We will now show that the series $\sum_{k=1}^{\infty} l_k$ and $\sum_{k=1}^{\infty} u_k$ diverge. Clearly if both converge then

$$\sum_{k=1}^{\infty} |a_k| = \sum_{k=1}^{\infty} (u_k + l_k) = \sum_{k=1}^{\infty} u_k + \sum_{k=1}^{\infty} l_k < \infty.$$

If $\sum_{k=1}^{\infty} u_k$ converges, then the comparison test (3.5.2) tells us that $\sum_{k=1}^{\infty} l_k$ also converges, since $0 \leq l_k \leq u_k$. However, if $\sum_{k=1}^{\infty} l_k$ converges we have

$$\sum_{k=1}^{\infty} |a_k| = \sum_{k=1}^{\infty} (a_k + 2l_k) = \sum_{k=1}^{\infty} a_k + 2\sum_{k=1}^{\infty} l_k$$

---

[8] It actually does converge.

3.5. SERIES

would hold, and in particular $\sum_{k=1}^{\infty} |a_k|$ would converge. Now if $\sum_{n=1}^{\infty} l_k$ converges, then since $u_n = a_n + l_n$ we would have

$$\sum_{n=1}^{\infty} u_n = \sum_{n=1}^{\infty} (a_n + l_n) = \sum_{n=1}^{\infty} a_n + \sum_{n=1}^{\infty} l_n,$$

and so $\sum_{n=1}^{\infty}$ would also converge.

Let P_k and N_k we two sequences which are defined as follows: Let

$$\{P_1, P_2, P_3, \cdots, P_n, \cdots\}$$

be the positive terms in the sequence $\{a_j\}_{j=1}^{\infty}$ in the order in which they occur, and let $\{N_1, N_2, N_3, \cdots\}$ be the absolute value of the negative terms in the sequence $\{a_j\}_{j=1}^{\infty}$ in the order in which they occur. Clearly both collections must be infinite, or $\sum_{k=1}^{\infty} |a_k|$ would converge. Moreover, it must also be that $\sum_{j=1}^{\infty} P_j$ and $\sum_{j=1}^{\infty} N_j$ both diverge, otherwise $\sum_{k=1}^{\infty} |a_k|$ would converge.

We will now construct sequences of natural numbers $\{r_k\}_{k=1}^{\infty}$ and $\{s_k\}_{k=1}^{\infty}$ and consider the sums

$$P_1 + \cdots + P_{r_1} - N_1 - \cdots - N_{s_1} + P_{r_1+1} + \cdots + P_{r_2} - N_{s_1+1} - \cdots - N_{s_2} + \cdots.$$

This defines a rearrangement of the series $\sum_{j=1}^{\infty} a_j$ and let $\sum_{k=1}^{\infty} a_{n_k}$ denote this rearrangement with $\{T_n\}_{n=1}^{\infty}$ being the rearrangement's sequence of partial sums. We will create the sequences $\{r_k\}_{k=1}^{\infty}$ and $\{s_k\}_{k=1}^{\infty}$ so that $\liminf_{n \to \infty} T_n = a$ and $\limsup_{n \to \infty} T_n = b$.

Now choose any sequences $\{\alpha_k\}_{k=1}^{\infty}$ and $\{\beta_k\}_{k=1}^{\infty}$ so that $\lim_{k \to \infty} \alpha_k = a$, $\lim_{k \to \infty} \beta_k = b$, $\alpha_k < \beta_k$ for all $k \in \mathbb{N}$, and $\beta_1 > 0$.

Let r_1, s_1 be the smallest such natural numbers so that

$$P_1 + \cdots + P_{r_1} > \beta_1,$$

and

$$P_1 + \cdots + P_{r_1} - N_1 - \cdots - N_{s_1} < \alpha_1.$$

Then choose $r_2 > r_1$ and $s_2 > s_1$ to be the smallest natural numbers so that

$$P_1 + \cdots + P_{r_1} - N_1 - \cdots - N_{s_1} + P_{r_1+1} + \cdots + P_{r_2} > \beta_2$$

and

$$P_1 + \cdots + P_{r_1} - N_1 - \cdots - N_{s_1} + P_{r_1+1} + \cdots + P_{r_2} - N_{s_1+1} - \cdots - N_{s_2} < \alpha_2.$$

We continue in this manner, and we note that it is possible to do so since $\sum_{j=1}^{\infty} P_j = \infty$ and $\sum_{j=1}^{\infty} N_j = \infty$. Let $\{T_{j_k}\}_{k=1}^{\infty}$ be the subsequence of partial sums of our rearrangement whose final term is P_{r_k} and let $\{T_{i_k}\}_{k=1}^{\infty}$ be the subsequence of of partial sums of our rearrangement whose final term is $-N_{s_k}$. Then $|T_{j_k} - \beta_k| \leq P_{r_k}$ and $|T_{i_k} - \alpha_k| \leq N_{s_k}$. Since $a_t \to 0$ as $t \to \infty$, it must be that $P_t, N_t \to 0$ as $t \to \infty$. Thus we see $T_{j_k} \to b$ and $T_{i_k} \to a$ as $k \to \infty$.

We also note that any other subsequence of $\{T_j\}_{j=1}^{\infty}$ with a subsequential limit must have its limit value in $[a, b]$. If $b < \infty$, let $c > b$. Let $N \in \mathbb{N}$ be such that for any $k > N$ we have $|\beta_k - b| < \frac{c-b}{3}$ and $P_{r_k} < \frac{c-b}{3}$. Thus we have for such k, $T_{j_k} < c - \frac{b-c}{3}$. Moreover, for any i so that $j_k \leq i \leq j_{k+1}$ we have $T_i \leq \max\{T_{j_k}, T_{j_{k+1}}\}$. Hence $T_i < c - \frac{c-b}{3}$. This implies $\limsup_{n \to \infty} T_n < c$. Since $c > b$ was arbitrary, we get $\limsup_{n \to \infty} T_n = b$. If $a > -\infty$, let $d < a$. Let $N \in \mathbb{N}$ be such that for any $k > N$ we have $|\alpha_k - a| < \frac{a-d}{3}$ and $N_{s_k} < \frac{a-d}{3}$. Thus we have for such k, $T_{i_k} > d + \frac{a-d}{3}$. Moreover, for any j so that $i_k \leq j \leq i_{k+1}$ we have $T_j \geq \sup\{T_{i_k}, T_{i_{k+1}}\}$. Hence $T_j > d + \frac{a-d}{3}$. This implies $\liminf_{n \to \infty} T_n > d$. Since $d < a$ was arbitrary, we get $\liminf_{n \to \infty} T_n = a$. □

The following theorem is due to the mathematician Dirichlet.

Theorem 3.5.15 *If $\sum_{n=1}^{\infty} a_n$ converges absolutely, then any rearrangement of $\sum_{n=1}^{\infty} a_n$ converges to the same limit.*

Proof: Let $S_n = \sum_{i=1}^{n} a_i$ and $T_n = \sum_{i=1}^{n} a_{k_i}$ where $\sum_{i=1}^{\infty} a_{k_i}$ is a rearrangement of $\sum_{i=1}^{\infty} a_i$, and $\sum_{i=1}^{\infty} a_i$ converges absolutely. Since $\sum_{j=1}^{\infty} |a_j|$ converges, we know that given any $\varepsilon > 0$ there is an $N \in \mathbb{N}$ so that for any $m \geq n \geq N$ we have

$$\sum_{j=n}^{m} |a_j| < \varepsilon.$$

Now choose p so that

$$\{1, 2, \cdots, N\} \subseteq \{k_1, k_2, \cdots k_p\}.$$

3.5. SERIES

Let $T_n = \sum_{j=1}^n a_{k_j}$ and $S_n = \sum_{j=1}^n a_n$. Then, if $n > p$ we have that $a_1, a_2, \cdots, a_N$ will cancel in the difference $S_n - T_n$ and so

$$|S_n - T_n| < \varepsilon.$$

Hence the sequence $\{T_n\}_{n=1}^\infty$ converges to the same limit as the sequence $\{S_n\}_{n=1}^\infty$.

□

3.5.10 Power Series

Given a sequence $\{a_n\}_{n=0}^\infty$ of real numbers we consider the series

$$\sum_{n=0}^\infty a_n x^n = a_0 + a_1 x + a_2 x^2 + a_3 x^3 + a_4 x^4 + \cdots \quad (3.12)$$

for any fixed x. Such a series is called a **power series** in x, and the sequence terms

$$a_0, a_1, a_2, a_3, a_4, a_5, \cdots$$

are the **coefficients** of the power series. In the case where the sequence of coefficients are constant we have that the power series is a geometric series in x. In such a case the series converges if and only if $|x| < 1$.

In general we desire to know the values of x for which the given power series $\sum_{n=0}^\infty a_n x^n$ converges. Such values of x are elements of the **interval of convergence**, noting that this set is always non-empty since a power series $\sum_{n=1}^\infty a_n a^n$ always converges for $x = 0$.

More generally an infinite series of the form

$$\sum_{n=1}^\infty a_n (x-c)^n$$

$$= a_0 + a_1(x-c) + a_2(x-c)^2 + a_3(x-c)^3 + \cdots \quad (3.13)$$

is called a **power series** in $x - c$, or a **power series centred at** $x = c$ and if $c = 0$ this is simply the case we considered above. Similarly, the values of x for which 3.13 converges is also called the interval of convergence of the power series. Note that a power series centred at $x = c$ always converges for

$x = c$, and so the interval of convergence is always non-empty. However the power series need not converge at any other x value. To see this consider $\sum_{n=0}^{\infty} n! x^n$, or $\sum_{n=0}^{\infty} n!(x-c)^n$.

Our next task is to determining the interval of convergence of a given power series (3.13). To determine this we will use our tests for absolute converges such as the root test or the ratio test.

For instance, the root test says that

$$\limsup_{n\to\infty} \sqrt[n]{|a_n(x-c)^n|} = \limsup_{n\to\infty} \sqrt[n]{|a_n|} \cdot |x-c| < 1$$

is sufficient for convergence, and for the values of x which satisfy

$$|x-c| < \frac{1}{\limsup_{n\to\infty} \sqrt[n]{|a_n|}}$$

we have absolute convergence of the power series. Also, the root test fails if there is equality in the above inequality and thus the power series may or may not converge if

$$|x-c| = \frac{1}{\limsup_{n\to\infty} \sqrt[n]{|a_n|}}.$$

Also, the power series diverges if

$$|x-c| > \frac{1}{\limsup_{n\to\infty} \sqrt[n]{|a_n|}}.$$

Similarly, we may apply the ratio test to test for absolute convergence of the power series. Using the ratio test we see that

$$\limsup_{n\to\infty} \frac{|a_{n+1}(x-c)^{n+1}|}{|a_n(x-c)^n|} = \limsup_{n\to\infty} \frac{|a_{n+1}|}{|a_n|} \cdot |x-c| < 1$$

is sufficient for convergence. So for the values of x which satisfy

$$|x-c| < \frac{1}{\limsup_{n\to\infty} \frac{|a_{n+1}|}{|a_n|}}$$

3.5. SERIES

we have absolute convergence of the power series. Moreover, the series diverges if $|x - c|\frac{|a_{n+1}|}{|a_n|} > 1$ for all n large.

Using either the ratio or root test we see that the interval of convergence has an interior of the form $(c - R, c + R)$, where $R = \infty$ or 0 is a possibility. The quantity R is often called the **radius of convergence**.

Example: Consider the power series

$$\sum_{n=0}^{\infty} \frac{x^n}{n!}.$$

Find its radius and interval of convergence.

Solution: Because of the appearance of a factorial we shall use the ratio test. This is an approach we will take any time a factorial appears in our a_n terms. Here $a_n = \frac{1}{n!}$ and since we require

$$\limsup_{n \to \infty} \frac{|a_{n+1}|}{|a_n|} \cdot |x| < 1,$$

using

$$\frac{a_{n+1}}{a_n} = \frac{\frac{1}{(n+1)!}}{\frac{1}{n!}} = \frac{n!}{(n+1)!} = \frac{1}{n+1}$$

we see that we actually require

$$\limsup_{n \to \infty} \frac{1}{n+1} \cdot |x| < 1.$$

However, the limit is 0 regardless of x's value and thus this inequality always holds. Thus the radius of convergence is infinite, and the interval of convergence is $\mathbb{R}$.

Hence the function

$$\operatorname{Exp}(x) := \sum_{n=0}^{\infty} \frac{x^n}{n!} \tag{3.14}$$

has a natural domain of $\mathbb{R}$.

Similarly, one can show that the power series

$$\mathrm{Sin}(x) = \sum_{k=0}^{\infty} \frac{(-1)^k}{(2k+1)!} x^{2k+1} \tag{3.15}$$

and

$$\mathrm{Cos}(x) = \sum_{k=0}^{\infty} \frac{(-1)^k}{(2k)!} x^{2k} \tag{3.16}$$

converge for all $x \in \mathbb{R}$.

3.5.11 Homework Exercises

1. Find the radius of convergence for

$$\sum_{n=1}^{\infty} \frac{n!}{n^n} x^n.$$

2. Let $p \in \mathbb{N}$ and $q > 0$, find the radius of convergence for

$$\sum_{n=1}^{\infty} \frac{(pn)!}{(n!)^q} x^n.$$

3. Prove that

$$\sum_{n=0}^{\infty} a_n x^n, \quad \sum_{n=0}^{\infty} n a_n x^{n-1}, \quad \text{and} \quad \sum_{n=0}^{\infty} \frac{a_n}{n+1} x^{n+1}$$

have the same radii of convergence.

4. For the power series defined by

$$f(x) = \sum_{n=1}^{\infty} \frac{1}{n} x^n$$

find the interval and radius of convergence. Then define the function

$$\mathrm{Log}(x) = -f(1-x), \tag{3.17}$$

and determine the domain of this function.

3.5. SERIES

5. Determine the interval and radius of convergence for the series

$$J_0(x) = \sum_{n=0}^{\infty} \frac{(-1)^n}{(n!)^2 2^{2n}} x^{2n}.$$

6. Determine the interval and radius of convergence for the series

$$J_1(x) = \sum_{n=0}^{\infty} \frac{(-1)^n}{2n!(n+1)!2^{2n}} x^{2n+1}.$$

Note: The functions $J_0(x)$ and $J_1(x)$ are often called **Bessel functions**.

7. For $\text{Sin}(x)$ and $\text{Cos}(x)$ as defined before show that

$$\text{Sin}(x) \cdot \text{Cos}(x) = \frac{1}{2}\text{Sin}(2x).$$

Hint: Use theorem (3.5.13).

Chapter 4

Limits and Continuity

In this chapter we shall consider functions $f : D \to R$, where $D \subseteq M$ and $R \subseteq M'$, and (M, d) and (M', d') are metric spaces. In this setting we will define and discuss the concepts of limit and continuity for such functions.

4.1 Limits of Functions

Let $a \in \overline{D} \subseteq M$ be a limit point for D and suppose $f : D \to R$. We say that the **limit** of f exists at a and equals $L \in M'$ if given any $\varepsilon > 0$, there is $\delta > 0$ so that

$$0 < d(x, a) < \delta \quad \Rightarrow \quad d'(f(x), L) < \varepsilon. \tag{4.1}$$

We often denote L^1 by

$$\lim_{x \to a} f(x) := L. \tag{4.2}$$

Moreover, we often write $f(x) \to L$ as $x \to a$ to denote $L = \lim_{x \to a} f(x)$. To see that this notation is non-ambiguous, we note that if the limit of a function exists at a point a, then it must be unique. To see this suppose L, Λ are two limit values, then given $\varepsilon > 0$, there is a $\delta_1 > 0$ so that

$$0 < d(x, a) < \delta_1 \quad \Rightarrow \quad d'(f(x), L) < \frac{\varepsilon}{2},$$

and there is a $\delta_2 > 0$ so that

$$0 < d(x, a) < \delta_2 \quad \Rightarrow \quad d'(f(x), \Lambda) < \frac{\varepsilon}{2}.$$

---
[1] Note that L must be a limit point of the range of f.

Then for $\delta = \min\{\delta_1, \delta_2\}$, if $0 < d(x, a) < \delta$, we have

$$d'(L, \Lambda) \leq d'(L, f(x)) + d'(f(x), \Lambda) \leq \frac{\varepsilon}{2} + \frac{\varepsilon}{2} = \varepsilon.$$

Since $\varepsilon > 0$ was arbitrary, we must have $d'(L, \Lambda) = 0$ and so $L = \Lambda$. Thus we have the following theorem.

Theorem 4.1.1 *Suppose that $f : D \to R$ has a limit L at $a \in \overline{D}$ and a limit Λ at $a \in \overline{D}$, then $L = \Lambda$.*

Example: For $a > 0$, we will now show

$$\lim_{x \to a} \sqrt{x} = \sqrt{a}.$$

Let $\varepsilon > 0$ be given. Consider $x \in (\frac{a}{2}, \frac{3a}{2})$. For such an x we have

$$|\sqrt{x} - \sqrt{a}| = \frac{|x - a|}{\sqrt{x} + \sqrt{a}} \leq \frac{|x - a|}{\sqrt{\frac{a}{2}} + \sqrt{a}}.$$

Define $\delta = \min\{\frac{a}{2}, (1 + \frac{1}{\sqrt{2}})\sqrt{a} \cdot \varepsilon\}$, then for $|x - a| < \delta$ we have $|\sqrt{x} - \sqrt{a}| < \varepsilon$.

4.1.1 Properties of Limits

The following are a list of limit properties, often called **limit laws**. Here we will assume that our domain D of our functions is contained in a metric space (M, d), and our codomain is in either Euclidean space, with the usual norm $\|\ \|$, or any normed linear space over $\mathbb{R}$. In the case where we are multiplying functions together or taking the reciprocal of a function, we assume our functions are real-valued.

Theorem 4.1.2 *Let $f : D \to R$ and $g : D \to R$ be any two functions with limit values $\lim_{x \to a} f(x) = L$ and $\lim_{x \to a} g(x) = M$ respectively. Then we have the following*

1.
$$\lim_{x \to a}(f(x) + g(x)) = L + M,$$

2. *For any constant c,*
$$\lim_{x \to a}(cf(x)) = cL,$$

4.1. LIMITS OF FUNCTIONS

3.
$$\lim_{x \to a} (f(x)g(x)) = L \cdot M,$$

4. If $L \neq 0$, we have
$$\lim_{x \to a} \frac{1}{f(x)} = \frac{1}{L}.$$

Proof: Let $\varepsilon > 0$ be given.

1. We may find $\delta_f > 0$ so that for $0 < d(x,a) < \delta_f$ we have $\|f(x) - L\| < \frac{\varepsilon}{2}$. We may also find $\delta_g > 0$ so that for $0 < d(x,a) < \delta_g$ we have $\|g(x) - M\| < \frac{\varepsilon}{2}$. Then for $\delta = \min\{\delta_f, \delta_g\}$ and $0 < d(x,a) < \delta$ we have

$$\|(f(x) + g(x)) - (L + M)\| \leq \|f(x) - L\| + \|g(x) - M\|$$

$$< \frac{\varepsilon}{2} + \frac{\varepsilon}{2} = \varepsilon.$$

2. If $c = 0$ then $cf(x) = 0$ and $cL = 0$. One can see that for all $\varepsilon > 0$ any $\delta > 0$ will satisfy $d(x,a) < \delta$ implies $0 = \|cf(x) - cL\| < \varepsilon$. In the case where $c \neq 0$ we may find $\delta > 0$ so that for $0 < d(x,a) < \delta$ we have $\|f(x) - L\| < \frac{\varepsilon}{|c|}$. For such $0 < d(x,a) < \delta$ we have

$$\|cf(x) - cL\| = |c| \cdot \|f(x) - L\|$$

$$< |c| \cdot \frac{\varepsilon}{|c|} = \varepsilon.$$

3. First suppose that $L = 0$ or $M = 0$. Without loss of generality let's suppose $L = 0$. Then for some $\delta_1 > 0$, for $0 < d(x,a) < \delta_1$ we have $|g(x) - M| < 1$, and so $|g(x)| < |M| + 1$. Moreover, for some $\delta_2 > 0$, for $0 < d(a,x) < \delta_2$ we have $|f(x)| = |f(x) - 0| < \frac{\varepsilon}{|M|+1}$. Thus for $\delta = \min\{\delta_1, \delta_2\}$, if $0 < d(x,a) < \delta$, then we have

$$|f(x)g(x)| < (|M| + 1)|f(x)| < (|M| + 1)\frac{\varepsilon}{|M| + 1} = \varepsilon.$$

Now, assume $L, M \neq 0$. Then there is an $\delta_1 > 0$ so that for $0 < d(x,a) < \delta_1$ we have $|f(x) - L| < \frac{|L|}{2}$. Thus $\frac{|L|}{2} < |f(x)| < \frac{3|L|}{2}$ for all such $0 < d(x,a) < \delta_1$.

Also, we may find $\delta_2 > 0$ so that for $0 < d(x, a) < \delta_2$ we have $|f(x) - L| < \frac{\varepsilon}{2|M|}$. We may also find $\delta_3 > 0$ so that for $0 < d(x, a) < \delta_3$ we have $|g(x) - M| < \frac{\varepsilon}{3|L|}$. Thus, for $\delta < \min\{\delta_1, \delta_2, \delta_3\}$ and $0 < d(x, a) < \delta$ we have

$$|f(x)g(x) - LM| = |f(x)g(x) - f(x)M + f(x)M - LM|$$
$$\leq |f(x)g(x) - f(x)M| + |f(x)M - LM|$$
$$= |f(x)| \cdot |g(x) - M| + |M| \cdot |f(x) - L|$$
$$\leq \frac{3|L|}{2} \cdot |g(x) - M| + |M| \cdot |f(x) - L|$$
$$\leq \frac{3|L|}{2} \cdot \frac{\varepsilon}{3|L|} + |M| \cdot \frac{\varepsilon}{2|M|} = \varepsilon.$$

4. Since $\lim_{x \to a} f(x) \neq 0$ is assumed, there is a $\delta_1 > 0$ so that $0 < d(x, a) < \delta_1$ implies $|f(x) - L| < \frac{|L|}{2}$, and so $|f(x)| > \frac{|L|}{2}$. Thus we have $\frac{1}{|f(x)|} < \frac{2}{|L|}$ whenever $0 < d(x, a) < \delta_1$. Also there is an $\delta_2 > 0$ so that for $0 < d(x, a) < \delta_2$ we have

$$|f(x) - L| < \frac{L^2 \cdot \varepsilon}{2}.$$

Now let $\delta = \min\{\delta_1, \delta_2\}$, and for $0 < d(x, a) < \delta$ we have

$$\left|\frac{1}{f(x)} - \frac{1}{L}\right| = \left|\frac{L - f(x)}{f(x) \cdot L}\right|$$
$$= \frac{|f(x) - L|}{|f(x) \cdot L|}$$
$$\leq \frac{2}{L^2}|f(x) - L|$$
$$\leq \frac{2}{L^2} \cdot \frac{L^2 \cdot \varepsilon}{2} = \varepsilon.$$

$\square$

Corollary 4.1.3 *Suppose $f, g : D \to \mathbb{R}$, $\lim_{x \to a} f(x) = L$, $\lim_{x \to a} g(x) = M$. Then for any $c, k \in \mathbb{R}$ we have*

$$\lim_{x \to a}(cf(x) + kg(x)) = cL + kM. \tag{4.3}$$

4.2. CONTINUOUS FUNCTIONS

Corollary 4.1.4 *Suppose $f(x)$ is a polynomial with real coefficients whose domain is $\mathbb{R}$. Then for any $a \in \mathbb{R}$*

$$\lim_{x \to a} f(x) = f(a).$$

Corollary 4.1.5 *Suppose $f(x)$ is a real-valued rational function, whose domain D is a subset of $\mathbb{R}$. Then for any $a \in D$*

$$\lim_{x \to a} f(x) = f(a).$$

4.1.2 Homework Exercises

1. Let $f : D \to \mathbb{R}$, with $D \subseteq \mathbb{R}$ and $a \in \overline{D}$ a limit point for D. Define

$$\lim_{x \to a^-} f(x) = L^- \qquad (4.4)$$

by: for all $\varepsilon > 0$, there exists $\delta > 0$ so that $|f(x) - L^-| < \varepsilon$ whenever $a - \delta < x < a$. We call L^- the **left-hand limit** of $f(x)$ as x approaches a. Similarly, we define

$$\lim_{x \to a^+} f(x) = L^+ \qquad (4.5)$$

by: for all $\varepsilon > 0$, there exists $\delta > 0$ so that $|f(x) - L^+| < \varepsilon$ whenever $a < x < a + \delta$. We call L^+ the **right-hand limit** of $f(x)$ as x approaches a. Prove the following

$$\lim_{x \to a} f(x) = L \iff \lim_{x \to a^-} f(x) = L = \lim_{x \to a^+} f(x). \qquad (4.6)$$

2. Show that $\lim_{x \to a} \sqrt[3]{x} = \sqrt[3]{a}$ for any $a \in \mathbb{R}$.

3. Show that $\lim_{x \to a} \sqrt[n]{x} = \sqrt[n]{a}$ for any $a \in \mathbb{R}$ if n is odd, and for any $a \in [0, \infty)$ if n is even.

4.2 Continuous Functions

A function $f : D \to R$ is said to be **continuous** at $a \in D$ if

$$\lim_{x \to a} f(x) = f(a). \qquad (4.7)$$

In corollaries (4.1.4) and (4.1.5) we see that real-valued polynomials and rational functions of a real variable are continuous at every a in their domains.

A function $f : D \to R$ is said to be a **continuous** function if it is continuous at every point $a \in D$. If f is continuous at all $x \in E$, where $E \subseteq D$, then we say that f is **continuous on** E.

The following theorem tells us that composition of functions preserves continuity.

Theorem 4.2.1 *Suppose (M, d), (M', d'), (M'', d'') are metric spaces. Let $D \subseteq M$, $D' \subseteq M'$, $D'' \subseteq M''$ and $f : D \to D'$, $g : D' \to D''$. Suppose $\lim_{x \to L} g(x) = g(L)$, and $\lim_{x \to a} f(x) = L$, then*

$$\lim_{x \to a} g(f(x)) = g(L) = g(\lim_{x \to a} f(x)).$$

Moreover, if $L = f(a)$ then $g \circ f$ is continuous at $a \in D$.

Proof: Let $\varepsilon > 0$ be given, then there exists a $\delta' > 0$ so that $d'(y, L) < \delta'$ implies $d''(g(y), g(L)) < \varepsilon$. Also, there is a $\delta > 0$ so that $0 < d(x, a) < \delta$ implies $d'(f(x), L) < \delta'$. Thus

$$0 < d(x, a) < \delta \quad \Rightarrow \quad d'(f(x), L) < \delta' \quad \Rightarrow \quad d''(g(f(x)), g(L)) < \varepsilon.$$

□

Some Examples

Example: Consider the function

$$g(x) = \begin{cases} 1 & x \in \mathbb{Q}, \\ 0 & x \in \mathbb{R} \setminus \mathbb{Q}, \end{cases} \quad (4.8)$$

then g is called the **characteristic function of the rational numbers** and is often denoted by $\chi_\mathbb{Q}$. Claim that g is not continuous at any real number x. To see this take $\varepsilon = \frac{1}{2}$. Let $\delta > 0$ be given. If $x \in \mathbb{Q}$ then there is an irrational number y in the interval $(x-\delta, x+\delta)$, and so $|f(x) - f(y)| = 1 > \frac{1}{2}$. Likewise, if x is irrational, then there is a rational number y in the interval $(x - \delta, x + \delta)$, and so $|f(x) - f(y)| = 1 > \frac{1}{2}$.

4.2. CONTINUOUS FUNCTIONS

Example: Consider the so-called **Thomae function**

$$f(x) = \begin{cases} 0 & x = 0, \\ 0 & x \in \mathbb{R} \setminus \mathbb{Q}, \\ \frac{1}{n} & x = \frac{m}{n} \text{ in reduced form, } n > 0. \end{cases} \quad (4.9)$$

We will now show that $f(x)$ is continuous at every irrational x and not continuous at any rational number x.

First we will show that f is not continuous at any rational number. Let $a = \frac{m}{n}$, with $n > 0$ and the greatest common divisor of m and n being 1. Let $\varepsilon = \frac{1}{2n}$. Then for any $\delta > 0$ there is an irrational number x in the interval $(\frac{m}{n} - \delta, \frac{m}{n} + \delta)$. Then $|f(x) - f(\frac{m}{n})| = \frac{1}{n} > \frac{1}{2n}$. Hence f is not continuous at $\frac{m}{n}$.

Now, suppose f is not continuous at some point in $\mathbb{R} \setminus \mathbb{Q}$. Then there is an $x \in \mathbb{R} \setminus \mathbb{Q}$ and a $\varepsilon > 0$ so that for all $\delta > 0$ we have that there is some y so that $|x - y| < \delta$ and $f(y) = |f(x) - f(y)| \geq \varepsilon$. Note that clearly such y's would have to be rational. Consider any strictly decreasing positive sequence $\{\delta_k\}_{k=1}^{\infty}$ with $\lim_{k \to \infty} \delta_k = 0$. Then for each δ_k, let $\frac{m_k}{n_k}$ be the corresponding y_k. Moreover, suppose $n_k > 0$ and m_k and n_k have a greatest common divisor of 1. Thus we see that $|x - \frac{m_k}{n_k}| < \delta_k$ and $\frac{1}{n_k} \geq \varepsilon$. Hence $1 \leq n_k \leq \frac{1}{\varepsilon}$ for all $k \in \mathbb{N}$. However, by the corollary (3.2.2) we know that there is a convergent subsequence of $\{n_k\}_{k=1}^{\infty}$, say $\{n_{j_k}\}_{k=1}^{\infty}$. This subsequence is Cauchy, and since it is a sequence of integers it must eventually be constant. So there is an $N \in \mathbb{N}$ so that for $k > N$ we have $n_{j_k} = n^*$. Thus $f(\frac{m_{j_k}}{n_{j_k}}) = \frac{1}{n^*} \geq \varepsilon$ for $k > N$. We also know that $\frac{m_{j_k}}{n_{j_k}} \to x$ as $k \to \infty$. Thus $\lim_{k \to \infty} \frac{m_{j_k}}{n^*} = x$, and so $\lim_{k \to \infty} m_{j_k} = n^* x \in \mathbb{R} \setminus \mathbb{Q}$. This is a contradiction. Hence we see that f is continuous at every irrational number x.

This example shows us that the denominators of any sequence of rational numbers which converge to an irrational number must become unbounded.

4.2.1 The Exponential and Logarithmic Functions

Suppose we wish to show the existence of a real-valued continuous function f defined on $\mathbb{R}$ which satisfies the following properties:

$$f(1) = e, \qquad f(x+y) = f(x) \cdot f(y) \quad \forall x, y \in \mathbb{R}. \tag{4.10}$$

We will show that a function which satisfies (4.10) must satisfy $f(r) = e^r$, where $r \in \mathbb{Q}$. Moreover, upon showing the existence and uniqueness of such a continuous function, we will see that $y = e^x$ is this function.

Claim 4.2.2 *If f is a function which satisfies (4.10), then $f(r) = e^r$ for $r \in \mathbb{Q}$.*

Proof:

1. $f(0) = 1$: To see this observe that $f(0) = f(0+0) = (f(0))^2$ and so $f(0)^2 = f(0)$. Hence $f(0) = 0$ or $f(0) = 1$. If $f(0) = 0$, then for any x we have $f(x) = f(x+0) = f(x) \cdot f(0) = 0$ and this contradicts $f(1) = e$. Thus $f(0) = 1$.

2. $f(-x) = \frac{1}{f(x)}$: To see this notice that

$$1 = f(0) = f(x + (-x)) = f(x) \cdot f(-x)$$

   and so the result follows. Moreover, this also says that $f(x) \neq 0$ for all $x \in \mathbb{R}$ since $f(x)$ has a multiplicative inverse.

3. $f(n) = e^n$ for $n \in \mathbb{N}$: We shall prove this by induction. Clearly $f(1) = e = e^1$. Now suppose for some $n \in \mathbb{N}$ that $f(n) = e^n$. Then

$$f(n+1) = f(n) \cdot f(1) = e^n \cdot e^1 = e^{n+1}.$$

4. $f(m) = e^m$ for any $m \in \mathbb{Z}$ follows as well, since $f(-n) = \frac{1}{f(n)} = \frac{1}{e^n} = e^{-n}$ for any $n \in \mathbb{N}$.

5. $f(\frac{1}{n}) = e^{\frac{1}{n}}$ for $n \in \mathbb{N}$: To see this, notice that

$$e = f(1) = f(\underbrace{\frac{1}{n} + \frac{1}{n} + \cdots + \frac{1}{n}}_{n \text{ times}}) = \underbrace{f(\frac{1}{n}) \cdot f(\frac{1}{n}) \cdots f(\frac{1}{n})}_{n \text{ times}} = f(\frac{1}{n})^n,$$

4.2. CONTINUOUS FUNCTIONS

and so
$$f(\frac{1}{n}) = e^{\frac{1}{n}}.$$

6. $f(\frac{m}{n}) = e^{\frac{m}{n}}$ for any $n, m \in \mathbb{N}$: To see this we observe
$$f(\frac{m}{n}) = f(\underbrace{\frac{1}{n} + \frac{1}{n} + \cdots + \frac{1}{n}}_{m \text{ times}}) = f(\frac{1}{n})^m = (e^{\frac{1}{n}})^m = e^{\frac{m}{n}}.$$

7. For $r \in \mathbb{Q}$, $f(r) = e^r$: If $r \geq 0$ we are done. Otherwise, for $r < 0$ we have
$$f(r) = f(-|r|) = \frac{1}{f(|r|)} = \frac{1}{e^{|r|}} = e^{-|r|} = e^r.$$

□

Claim 4.2.3 *If f, g are functions which satisfies (4.10) and $f(x) = g(x)$ for all x in a dense subset D of $\mathbb{R}$, then $f = g$ on $\mathbb{R}$.*

Proof: Let $x \in \mathbb{R}$, and let $\varepsilon > 0$ be given. Since f is continuous at x, there is a $\delta_f > 0$ so that
$$|x - y| < \delta_f \quad \Rightarrow \quad |f(x) - f(y)| < \frac{\varepsilon}{2}.$$
Moreover, since g is continuous at x, there is a $\delta_g > 0$ so that
$$|x - y| < \delta_g \quad \Rightarrow \quad |g(x) - g(y)| < \frac{\varepsilon}{2}.$$
Then for $\delta = \min\{\delta_f, \delta_g\}$ and $|x - y| < \delta$ with $y \in D$ we have
$$|f(x) - g(x)| = |f(x) - f(y) + f(y) - g(x)|$$
$$= |f(x) - f(y) + g(y) - g(x)|$$
$$\leq |f(x) - f(y)| + |g(x) - g(y)|$$
$$\leq \frac{\varepsilon}{2} + \frac{\varepsilon}{2} = \varepsilon.$$
Since $\varepsilon > 0$ was arbitrary, we have $f(x) = g(x)$.

□

Recall the function
$$\operatorname{Exp}(x) = \sum_{k=0}^{\infty} \frac{x^k}{k!}.$$

Earlier we showed that $\operatorname{Exp}(x)$ is defined for all $x \in \mathbb{R}$, and we will now show that this function is continuous on $\mathbb{R}$.

Theorem 4.2.4 *$\operatorname{Exp}(x)$ is continuous on $\mathbb{R}$.*

Proof:

$$|\operatorname{Exp}(x) - \operatorname{Exp}(y)| = |\sum_{k=0}^{\infty} \frac{x^k}{k!} - \sum_{k=0}^{\infty} \frac{y^k}{k!}|$$

$$= |\sum_{k=1}^{\infty} \frac{x^k - y^k}{k!}|$$

$$= |(x-y) \sum_{k=1}^{\infty} \frac{1}{k!} \sum_{j=0}^{k-1} x^{k-1-j} y^j|$$

$$\leq |x-y| \sum_{k=1}^{\infty} \frac{1}{k!} \sum_{j=0}^{k-1} |x|^{k-1-j} |y|^j$$

$$\leq |x-y| \sum_{k=1}^{\infty} \frac{1}{k!} k (\max\{|x|+1, |y|+1\})^{k-1}$$

$$= |x-y| \sum_{k=0}^{\infty} \frac{1}{k!} (\max\{|x|+1, |y|+1\})^k$$

$$= |x-y| \cdot \operatorname{Exp}(\max\{|x|+1, |y|+1\}).$$

We claim that $\operatorname{Exp}(x)$ is increasing on $(0, \infty)$. To see this, notice that for $k \in \mathbb{N}$ we have $x < y$ implies $x^k < y^k$, and so $\frac{x^k}{k!} < \frac{y^k}{k!}$. Thus $\sum_{k=0}^{\infty} \frac{x^k}{k!} < \sum_{k=0}^{\infty} \frac{y^k}{k!}$, and so $\operatorname{Exp}(x) < \operatorname{Exp}(y)$.

So if we assume $|x - y| < 1$, then

$$|\operatorname{Exp}(x) - \operatorname{Exp}(y)| < |x-y| \cdot \operatorname{Exp}(|x|+2).$$

Let $\varepsilon > 0$ be given and choose $\delta = \min\{1, \frac{\varepsilon}{\operatorname{Exp}(|x|+2)}\}$, then for $|x-y| < \delta$ we have $|\operatorname{Exp}(x) - \operatorname{Exp}(y)| < \varepsilon$.
□

4.2. CONTINUOUS FUNCTIONS

Theorem 4.2.5
$$\text{Exp}(x+y) = \text{Exp}(x) \cdot \text{Exp}(y) \quad \forall x, y \in \mathbb{R}$$

Proof: Since $\text{Exp}(x)$ converges absolutely on $\mathbb{R}$, by theorem (3.5.13) we may consider the Cauchy product of $\text{Exp}(x) = \sum_{k=0}^{\infty} \frac{x^k}{k!}$ and $\text{Exp}(y) = \sum_{k=0}^{\infty} \frac{y^k}{k!}$. By this theorem (3.5.13) we know that the Cauchy product converges to $\text{Exp}(x) \cdot \text{Exp}(y)$. However, the Cauchy product is

$$\sum_{k=0}^{\infty} c_k = \sum_{k=0}^{\infty} (\sum_{j=0}^{k} \frac{x^j}{j!} \frac{y^{k-j}}{(k-j)!})$$

$$= \sum_{k=0}^{\infty} \frac{1}{k!} (\sum_{j=0}^{k} \frac{k!}{j!(k-j)!} x^j y^{k-j})$$

which by the Binomial theorem equals

$$\sum_{k=0}^{\infty} \frac{(x+y)^k}{k!} = \text{Exp}(x+y).$$

□

Thus we see that $\text{Exp}(x)$ is a continuous function that satisfies (4.10), and $\text{Exp}(r) = e^r$ for all $r \in \mathbb{Q}$. So $\text{Exp}(x)$ is the unique such function with this property, and we define

$$e^x := \text{Exp}(x) \quad \forall x \in \mathbb{R}. \tag{4.11}$$

Theorem 4.2.6 $f(x) = \text{Exp}(x)$ *is a strictly increasing function on $\mathbb{R}$ whose range is $(0, \infty)$.*

Proof: For $0 < x < y$ we have already established $\text{Exp}(x) < \text{Exp}(y)$. Moreover, $\text{Exp}(0) = 1$ and for $x > 0$ we have $\text{Exp}(x) > 1$. Now for any $x < 0$ we have $\text{Exp}(x) = \frac{1}{\text{Exp}(-x)} < 1$, and so $\text{Exp}(x) < \text{Exp}(y)$ for all $y \geq 0$. Hence, it remains to show $x < y < 0$ has $\text{Exp}(x) < \text{Exp}(y)$. To see this note that $0 < -y < -x$ and so $\text{Exp}(-y) < \text{Exp}(-x)$, and thus $\frac{1}{\text{Exp}(y)} < \frac{1}{\text{Exp}(x)}$, which yields $\text{Exp}(x) < \text{Exp}(x)$. Hence $\text{Exp}(x)$ is strictly increasing on $\mathbb{R}$.

Clearly for $x > 0$ we have $x < \text{Exp}(x)$ and so $\sup\{\text{Exp}(x) : x \in \mathbb{R}\} = \infty$. Moreover, $\text{Exp}(-x) = \frac{1}{\text{Exp}(x)} < \frac{1}{x} \to 0$ as $x \to \infty$. Thus we see that the range of $\text{Exp}(x)$ is $(0, \infty)$.
$\square$

By the above theorem (4.2.6) we see that $f(x) = \text{Exp}(x)$ has an inverse function $f^{-1}(x)$ whose domain is $(0, \infty)$ and whose range is $\mathbb{R}$. We define the **logarithm** function to be this function and we denote it by $f^{-1}(x) = \log(x)$.

Theorem 4.2.7 *For $x, y > 0$ we have the following properties of the logarithm function:*

$$\log(xy) = \log(x) + \log(y), \quad \log\left(\frac{1}{x}\right) = -\log(x). \tag{4.12}$$

Proof: Let $u := \log(x)$ and $z := \log(y)$. Then $e^u = x$ and $e^z = y$, and so

$$xy = e^u e^z = e^{u+z}.$$

Thus we have
$$\log(xy) = u + z = \log(x) + \log(y).$$

Note that $\log(\frac{1}{x}) = y$ if and only if $e^y = \frac{1}{x}$. Thus $e^{-y} = \frac{1}{e^y} = x$, and so $\log(x) = -y = -\log(\frac{1}{x})$.
$\square$

Another important property of $\text{Exp}(x)$ is the following:

Theorem 4.2.8 *For any $r \in \mathbb{Q}$ and any $x \in \mathbb{R}$ we have*

$$(e^x)^r = (\text{Exp}(x))^r = \text{Exp}(x \cdot r) = e^{x \cdot r}. \tag{4.13}$$

Proof: For $n \in \mathbb{N}$ we have

$$(\text{Exp}(x))^n = \underbrace{\text{Exp}(x) \cdot \text{Exp}(x) \cdots \text{Exp}(x)}_{n \text{ times}}$$

$$= \text{Exp}(\underbrace{x + x + \cdots + x}_{n \text{ times}}) = \text{Exp}(n \cdot x).$$

For $n = 0$ we have
$$1 = (\text{Exp}(x))^0 = \text{Exp}(0) = \text{Exp}(0 \cdot x).$$

4.2. CONTINUOUS FUNCTIONS

For $n \in \mathbb{N}$ we also have
$$(\text{Exp}(x))^{-n} = \frac{1}{(\text{Exp}(x))^n} = \frac{1}{\text{Exp}(n \cdot x)} = \text{Exp}(-n \cdot x),$$
and
$$\text{Exp}(x) = \text{Exp}(n \cdot \frac{x}{n}) = (\text{Exp}(\frac{x}{n}))^n.$$
By taking n^{th} roots we get
$$(\text{Exp}(x))^{\frac{1}{n}} = \text{Exp}(\frac{x}{n}).$$
For $m, n \in \mathbb{N}$ we have
$$\text{Exp}(\frac{m}{n}x) = (\text{Exp}(mx))^{\frac{1}{n}} = ((\text{Exp}(x))^m)^{\frac{1}{n}} = (\text{Exp}(x))^{\frac{m}{n}}.$$
Also,
$$\text{Exp}(-\frac{m}{n}x) = \frac{1}{\text{Exp}(\frac{m}{n}x)} = \frac{1}{(\text{Exp}(x))^{\frac{m}{n}}} = (\text{Exp}(x))^{-\frac{m}{n}}.$$
□

Corollary 4.2.9 *For any $x > 0$ and any $r \in \mathbb{Q}$ we have*
$$\log(x^r) = r \cdot \log(x) \tag{4.14}$$
Proof: $\log(x) = z \iff x = e^z$. Now $x^r = (e^z)^r = e^{z \cdot r}$, and so $\log(x^r) = z \cdot r = r \cdot \log(x)$.
□

This allows us to define for any $a > 0$ and any $x \in \mathbb{R}$
$$a^x := e^{\log(a) \cdot x}. \tag{4.15}$$
Moreover, by properties of the exponential, we see
$$a^{x+y} = e^{\log(a) \cdot (x+y)} = e^{\log(a)x + \log(a)y} = e^{\log(a)x} e^{\log(a)y} = a^x a^y.$$
Moreover
$$a^{-x} = e^{\log(a)(-x)} = e^{-\log(a)x} = \frac{1}{e^{\log(a)x}} = \frac{1}{a^x}.$$
Notice that a^x is a continuous function, since it is defined as a composition of continuous functions. Using this, we also see that for any $x, y \in \mathbb{R}$, $(e^x)^y = e^{\log(e^x) \cdot y}$. However, e^x and $\log x$ are inverse functions of one another, and thus $\log(e^x) = x$. Hence, $(e^x)^y = e^{x \cdot y}$. Likewise, for $a > 0$, $(a^x)^y = (e^{\log a \cdot x})^y = e^{(\log a \cdot x) \cdot y} = e^{\log a \cdot (xy)} = a^{xy}$. So in corollary 4.2.9 we may replace $r \in \mathbb{Q}$ by $r \in \mathbb{R}$.

4.2.2 A Topological Characterization of Continuity

Let (M, d) and (M', d') be metric spaces, with $D \subseteq M$ and $R \subseteq M'$. For a function $f : D \to R$ we define

$$f^{-1}(y) = \{x \in D : f(x) = y\}. \tag{4.16}$$

Clearly f is injective if and only if the cardinality of $f^{-1}(y)$ is zero or one for all $y \in M'$.

More generally we define

$$f^{-1}(V) = \{x \in D : f(x) \in V\}. \tag{4.17}$$

We say that $U \subseteq D$ is **open in** D if $U = \mathcal{O} \cap D$, where $\mathcal{O}$ is an open set in M. Likewise, we say that $U \subseteq D$ is **closed in** D if $U = \mathcal{C} \cap D$, where $\mathcal{C}$ is an closed set in M.

Theorem 4.2.10 *Let $f : D \to R$. Then f is continuous on D if and only if $f^{-1}(V) \subseteq D$ is open in D for all open $V \subseteq M'$.*

Proof: Suppose f is continuous on D and let $V \subseteq M'$ be an open set with $x \in f^{-1}(V)$. Then $f(x) \in V$, and since V is open we see that there is an $\varepsilon > 0$ so that $B_\varepsilon(f(x)) \subseteq V$. Since f is continuous at x there is a $\delta > 0$ so that $f(y) \in B_\varepsilon(f(x))$ for all $y \in B_\delta(x) \cap D$. Thus $f^{-1}(V)$ is open in D.

Conversely, suppose $f^{-1}(V)$ is open in D for all open sets $V \subseteq M'$. Let $x \in D$ and $\varepsilon > 0$ be given. Then $f^{-1}(B_\varepsilon(f(x)))$ is open in D. Thus there exists $\delta > 0$ so that $f(y) \in B_\varepsilon(f(x))$ for all $y \in B_\delta(x) \cap D$. Hence f is continuous at x.
□

Observe that for $V \subseteq M'$

$$(f^{-1}(V))^c = f^{-1}(V^c). \tag{4.18}$$

To see this note that $x \in (f^{-1}(V))^c$ if and only if $x \notin f^{-1}(V)$. This holds if and only if $f(x) \notin V$, which holds if and only if $f(x) \in V^c$, and this holds if

4.2. CONTINUOUS FUNCTIONS

and only if $x \in f^{-1}(V^c)$.

Thus we have the following corollary.

Corollary 4.2.11 *Let $f : D \to R$. Then f is continuous of D if and only if $f^{-1}(C) \subseteq D$ is closed in D for all closed $C \subseteq M'$.*

4.2.3 Continuity and Compactness

Let $f : D \to R$ where $D \subseteq M$ and $R \subseteq M'$ with (M, d), (M', d') metric spaces. The following result shows that the image of a compact set under a continuous function is compact, and this gives us the very useful and important result from calculus that says that a continuous function on a closed and bounded interval attains its maximum and minimum on that interval.

Theorem 4.2.12 *Let $f : K \to M'$ be continuous, with K compact, $K \subseteq M$, and (M, d), (M', d') metric spaces, then $f(K) = \{f(x) : x \in K\}$ is compact in M'.*

Proof: Let $\{V_\alpha : \alpha \in \Delta\}$ be an open cover of $f(K)$. Then theorem (4.2.10) guarantees $f^{-1}(V_\alpha)$ is open in K for all $\alpha \in \Delta$. That is $f^{-1}(V_\alpha) = U_\alpha \cap K$, where U_α is open in M. Hence $\{U_\alpha : \alpha \in \Delta\}$ is an open cover for K, and so there is a finite subcover $\{U_{\alpha_1}, \cdots, U_{\alpha_m}\}$. Hence $K \subseteq U_{\alpha_1} \cup \cdots \cup U_{\alpha_m}$, and therefore $K = \bigcup_{k=1}^{m} U_{\alpha_k} \cap K = \bigcup_{k=1}^{m} f^{-1}(V_{\alpha_k})$. Since $f(f^{-1}(E)) \subseteq E$, we have $f(K) = \bigcup_{k=1}^{m} f(f^{-1}(V_{\alpha_k})) \subseteq \bigcup_{k=1}^{m} V_{\alpha_k}$, and so $\{V_{\alpha_1}, V_{\alpha_2}, \cdots, V_{\alpha_m}\}$ is a finite subcover of $f(K)$.
□

Note that the function $f(x) = \frac{1}{x}$ on $(0, 1]$ does not attain a maximum, and in fact is not even bounded. Thus the condition of the domain being closed is necessary, since in this case $(0, 1]$ is clearly bounded. Another example to consider is the function $f(x) = \frac{x^2}{x^2+1}$. Clearly $|f(x)| \leq 1$, but note that f does not attain its maximum on $\mathbb{R}$. Thus we see that bounded functions need not attain their maximum or minimum values if the domain is not compact, even if the domain is closed.

4.2.4 Continuity and Connectedness

Theorem 4.2.13 *Let $f : D \to M'$, $D \subset M$, with (M, D), (M', d') metric spaces. Suppose f is continuous and D is a connected set, then $f(D) = \{f(x) : x \in D\}$ is a connected subset of M'.*

Proof: Suppose $f(D)$ is disconnected, then there exists nonempty $A, B \subseteq M'$ so that $f(D) = A \cup B$ with $A \cap \overline{B}, \overline{A} \cap B = \emptyset$. Then $D = f^{-1}(A) \cup f^{-1}(B)$, with $f^{-1}(A), f^{-1}(B) \neq \emptyset$.

Since $A \subseteq \overline{A}$, we have $f^{-1}(A) \subseteq f^{-1}(\overline{A})$, and $f^{-1}(\overline{A})$ is closed in D, since f is continuous. Hence $\overline{f^{-1}(A)} \subseteq f^{-1}(\overline{A})$. Thus it follows that $f(\overline{f^{-1}(A)}) \subseteq \overline{A}$. Since $f(f^{-1}(B)) = B$ and $\overline{A} \cap B = \emptyset$ we have $\overline{f^{-1}(A)} \cap f^{-1}(B) = \emptyset$. Reversing the roles of A and B and running the same argument we see that $f^{-1}(A) \cap \overline{f^{-1}(B)} = \emptyset$. Thus we see that D is disconnected, which is a contradiction.
□

The following result is a corollary of the above theorem (4.2.13) and theorem (2.5.1). It is often called the **intermediate value theorem** in calculus.

Intermediate Value Theorem 4.2.14 *Suppose that $f : [a, b] \to \mathbb{R}$ is a continuous function and $f(a) \neq f(b)$, then for any L strictly[2] between $f(a)$ and $f(b)$ there exists $c \in (a, b)$ so that $f(c) = L$.*

Proof: By theorem (2.5.1) we know $[a, b]$ is connected. Thus by theorem (4.2.13) $f([a, b])$ is also connected. The result then follows from theorem (2.5.1).
□

4.2.5 Uniform Continuity

Let $D \subseteq M$ where (M, d) and (M', d') are metric spaces. A function $f : D \to M'$ is said to be **continuous on** D if

$$\forall \varepsilon > 0, \quad \forall x \in D, \quad \exists \delta > 0 \quad \text{so that:} \quad y \in B_\delta(x) \quad \Rightarrow \quad f(y) \in B_\varepsilon(f(x)). \tag{4.19}$$

---
[2] that is $f(a) < L < f(b)$, or $f(b) < L < f(a)$

4.2. CONTINUOUS FUNCTIONS

Here we note that the δ depends both on the ε and on the x.

Let $D \subseteq M$ where (M, d) and (M', d') are metric spaces. A function $f : D \to M'$ is said to be **uniformly continuous on** D if

$$\forall \varepsilon > 0, \quad \exists \delta > 0, \quad \forall x \in D \quad \text{so that:} \quad y \in B_\delta(x) \quad \Rightarrow \quad f(y) \in B_\varepsilon(f(x)). \tag{4.20}$$

Note that here the $\delta > 0$ only depends on the $\varepsilon > 0$, and the same δ works for each x. If $\delta(x, \varepsilon)$ is the δ associated to an $x \in D$ in (4.19), then f is uniformly continuous on D if and only if $\delta(\varepsilon) = \inf\{\delta(x, \varepsilon) : x \in D\} > 0$ for every $\varepsilon > 0$.

Clearly a uniformly continuous function is continuous. The following is a partial converse to this.

Theorem 4.2.15 *Suppose $f : K \to M'$, with $K \subseteq M$ a compact set, and (M, d), (M', d') are metric spaces. If f is continuous on K, then f is uniformly continuous on K.*

Proof: Assume f is continuous on K. Then given $\varepsilon > 0$, for each $x \in K$ there exists $\delta(x, \varepsilon) > 0$ so that

$$y \in B_{\delta(x,\varepsilon)}(x) \quad \Rightarrow \quad f(y) \in B_{\frac{1}{2}\varepsilon}(f(x)).$$

However, the collection

$$\{B_{\frac{1}{2}\delta(x,\varepsilon)}(x) : x \in K\}$$

is an open cover of K, and so there is a finite subcover

$$\{B_{\frac{1}{2}\delta(x_i,\varepsilon)} : i = 1, 2, \cdots n\}$$

for some collection $x_1, x_2, \cdots, x_n \in K$. Let

$$\delta(\varepsilon) := \frac{1}{4} \min\{\delta(x_i, \varepsilon) : i = 1, 2, \cdots, n\}.$$

Clearly this is positive, since the minimum of a finite set is taken on by one of its members, say $\delta(x_1, \varepsilon)$.

Now let $x \in K$ and suppose $y \in B_{\frac{1}{4}\delta(\varepsilon)}(x)$. Now $x \in B_{\frac{1}{2}\delta(x_i,\varepsilon)}(x_i)$ for some i. Since $\delta(\varepsilon) \leq \delta(x_i, \varepsilon)$ we have $d(y, x_i) \leq d(y, x) + d(x, x_i) < \delta(x_i, \varepsilon)$, and so $f(y) \in B_{\frac{1}{2}\varepsilon}(f(x_i))$. Likewise, $f(x) \in B_{\frac{1}{2}\varepsilon}(f(x_i))$. Thus

$$d'(f(x), f(y)) \leq d'(f(x), f(x_i)) + d'(f(y), f(x_i)) < \varepsilon.$$

□

4.2.6 Discontinuities

Let $f : D \to M'$ for $D \subseteq M$ with (M, d), (M', d') metric spaces. If f is not continuous at $x \in D$ then f is said to be **discontinuous** at x or f is said to have a **discontinuity** at $x \in D$. In the setting of $D \subseteq \mathbb{R}$ we classify discontinuities as one of two types. However before we do this we need to recall some definitions:

Define
$$\lim_{x \to a^-} f(x) = L^- \qquad (4.21)$$

by: for all $\varepsilon > 0$, there exists $\delta > 0$ so that $|f(x) - L^-| < \varepsilon$ whenever $a - \delta < x < a$. We call L^- the **left-hand limit** of $f(x)$ as x approaches a. Similarly, we define
$$\lim_{x \to a^+} f(x) = L^+ \qquad (4.22)$$

by: for all $\varepsilon > 0$, there exists $\delta > 0$ so that $|f(x) - L^+| < \varepsilon$ whenever $a < x < a + \delta$. We call L^+ the **right-hand limit** of $f(x)$ as x approaches a.

Let f have a discontinuity at $z \in D$. Then f is said to have a **simple discontinuity** or a **discontinuity of the first kind at** z if both $\lim_{x \to z^-} f(x)$ and $\lim_{x \to z^+} f(x)$ exist.

If f has a discontinuity at z which is not a discontinuity of the first kind, then either $\lim_{x \to z^-} f(x)$ or $\lim_{x \to z^+} f(x)$ does not exist. In such a case f is said to have a **discontinuity of the second kind** at z.

Note that $f = \chi_\mathbb{Q}$ (as in (4.8)) is an example of a function for which every point in $\mathbb{R}$ f has a discontinuity of the second kind. The function $g = x \cdot \chi_\mathbb{Q}$ is continuous at 0, and at every other point x, g has a discontinuity of the second kind.

4.2.7 Semicontinuity, lim sup and lim inf

Let $f : D \to \mathbb{R}$ with $D \subseteq \mathbb{R}^n$, and let $\vec{z} \in \overline{D}$ be a limit point of D. We define the **punctured disk** centred at $\vec{z}$ with radius $\delta > 0$ by

$$B'_\delta(\vec{z}) := B_\delta(\vec{z}) \setminus \{\vec{z}\}. \tag{4.23}$$

Also, we define

$$M_\delta(f, \vec{z}) := \sup\{f(\vec{x}) : \vec{x} \in B'_\delta(\vec{z}) \cap D\}, \quad m_\delta(f, \vec{z}) := \inf\{f(\vec{x}) : \vec{x} \in B'_\delta(\vec{z}) \cap D\}. \tag{4.24}$$

Note that as δ decreases to 0, $M_\delta(f, \vec{z})$ decreases and $m_\delta(f, \vec{z})$ increases. We define the **limit supremum** or lim sup of f at $\vec{z}$ by

$$\limsup_{\vec{x} \to \vec{z}, \vec{x} \in D} f(\vec{x}) = \lim_{\delta \to 0} M_\delta(f, \vec{z}), \tag{4.25}$$

and we define the **limit infimum** or lim inf of f at $\vec{z}$ by

$$\liminf_{\vec{x} \to \vec{z}, \vec{x} \in D} f(\vec{x}) = \lim_{\delta \to 0} m_\delta(f, \vec{z}). \tag{4.26}$$

Clearly both of these quantities exist at any limit point $\vec{z}$ of the domain of a function f, noting however $\pm \infty$ as values for these limits are indeed possible. It is an exercise to show that $\lim_{\vec{x} \to \vec{z}} f(\vec{x})$ exists if and only if both $\liminf_{\vec{x} \to \vec{z}, \vec{x} \in D} f(\vec{x}) = \limsup_{\vec{x} \to \vec{z}, \vec{x} \in D} f(\vec{x})$.

We say that f is **upper semicontinuous** or **usc** at $\vec{z}$ if

$$\limsup_{\vec{x} \to \vec{z}, \vec{x} \in D} f(\vec{x}) \leq f(\vec{z}). \tag{4.27}$$

Similarly, we say that f is **lower semicontinuous** or **lsc** at $\vec{z}$ if

$$\liminf_{\vec{x} \to \vec{z}, \vec{x} \in D} f(\vec{x}) \geq f(\vec{z}). \tag{4.28}$$

Note that $\chi_\mathbb{Q}$ is usc on the rational numbers and lsc on the irrationals. It is an exercise to show that f is continuous at $\vec{z}$ if and only if $|f(\vec{z})| < \infty$ and f is both usc and lsc at $\vec{z}$.

4.3 Monotone Functions

A real-valued function of a real variable $f : D \to \mathbb{R}$, $D \subseteq \mathbb{R}$ is said to be

1. **Increasing** if $f(y) \geq f(x)$, whenever $y > x$.

2. **Strictly increasing** if $f(y) > f(x)$, whenever $y > x$.

3. **Decreasing** if $f(y) \leq f(x)$, whenever $y > x$.

4. **Strictly decreasing** if $f(y) < f(x)$, whenever $y > x$.

5. **Monotone** if f is increasing, strictly increasing, decreasing or strictly decreasing.

6. **Strictly monotone** if f is strictly increasing or strictly decreasing.

Theorem 4.3.1 *Suppose $f : (a,b) \to \mathbb{R}$ is monotone. At any point $c \in (a,b)$, if f is increasing, then*

$$\sup_{x \in (a,c)} f(x) = \lim_{x \to c^-} f(x) \leq f(c) \leq \lim_{x \to c^+} f(x) = \inf_{x \in (c,b)} f(x), \qquad (4.29)$$

and if f is decreasing, then

$$\inf_{x \in (a,c)} f(x) = \lim_{x \to c^-} f(x) \geq f(c) \geq \lim_{x \to c^+} f(x) = \sup_{x \in (c,b)} f(x). \qquad (4.30)$$

Moreover, monotone functions cannot have discontinuities of the second kind.

Proof: Without loss of generality, we may suppose f is increasing, since in the case where f is decreasing we may consider $g = -f$ which is increasing and observe that $\lim_{x \to z^+} f(x) = -\lim_{x \to z^+} g(x)$ and $\lim_{x \to z^-} f(x) = -\lim_{x \to z^-} g(x)$. Also, $\inf\{f(x) : a < x < c\} = -\sup\{g(x) : a < x < c\}$, and $\sup\{f(x) : c < x < b\} = -\inf\{g(x) : c < x < b\}$.

Consider $x \in (a,b)$ and let $a < z < x$, then $f(z) \leq f(x)$ and so $s =: \sup\{f(z) : a < z < x\} \leq f(x)$. Let $\varepsilon > 0$ be given, then there is a $\delta > 0$ so that $s - \varepsilon < f(x - \delta) \leq s$, by definition of s. By monotonicity, $x - \delta \leq t \leq x$ implies $s - \varepsilon < f(x - \delta) \leq f(t) \leq s$, and thus we see that $\lim_{z \to x^-} f(z) = \sup\{f(z) : a < z < x\}$.

4.3. MONOTONE FUNCTIONS

Consider $x \in (a,b)$ and let $x < z < b$, then $f(x) \leq f(z)$ and so $i :=$ $\inf\{f(z) : x < z < b\} \geq f(x)$. Let $\varepsilon > 0$ be given, then there is a $\delta > 0$ so that $i \leq f(x+\delta) < i + \varepsilon$, by definition of i. By monotonicity, $x \leq t \leq x + \delta$ implies $i \leq f(t) \leq f(x+\delta) < i + \varepsilon$, and thus we see that $\lim_{z \to x^+} f(z) = \inf\{f(z) : x < z < b\}$.

$\square$

Theorem 4.3.2 *If $f : (a,b) \to \mathbb{R}$ is a monotone function, then the set of discontinuities of f is at most countable.*

Proof: As in the above theorem we suppose f is increasing, for otherwise we may consider $g = -f$. Let D be the set of discontinuities of f on (a,b). For each $z \in D$ we associate a rational number r_z so that $\lim_{x \to z^-} f(x) < r_z < \lim_{x \to z^+} f(x)$.

Claim that for $y < z$, $y, z \in D$ we have $r_y \neq r_z$. To see this we'll show that $\lim_{x \to y^+} f(x) \leq \lim_{x \to z^-} f(x)$. Observe that $\lim_{x \to y^+} f(x) = \inf\{f(x) : y < x < b\} = \inf\{f(x) : y < x < z\}$ and $\lim_{x \to z^-} f(x) = \sup\{f(x) : a < x < z\} = \sup\{f(x) : y < x < z\}$, and clearly $\inf\{f(x) : y < x < z\} \leq \sup\{f(x) : y < x < z\}$. Hence we see $r_y < \lim_{x \to y^+} f(x) \leq \lim_{x \to z^-} f(x) < r_z$. Thus there is a one to one correspondence between D and a subset of $\mathbb{Q}$, which is countable.

$\square$

Theorem 4.3.3 *Suppose that f is strictly monotone and continuous on $[a,b]$. Then $f^{-1} : f([a,b]) \to [a,b]$ is continuous and strictly monotone.*

Proof: Without loss of generality we consider the case where f is strictly increasing, since in the case where f is strictly decreasing we take $F = -f$. In such a case $F^{-1}(-x) = f^{-1}(x)$, and so we will get the result in this case as well.

Since f is strictly increasing, we have $f([a,b]) = [f(a), f(b)]$. By the intermediate value theorem (4.2.14) we know that for any $f(a) < L < f(b)$ there is a point $c \in (a,b)$ so that $f(c) = L$. Moreover, by monotonicity, c is unique. We define $g(L) = c$ for all such L's, and let $g(f(a)) = a, g(f(b)) = b$. Clearly g is f^{-1}, since $g(f(c)) = c$ for all $c \in [a,b]$. Now let $y_1 < y_2$ be points in $[f(a), f(b)]$. If $f(x_1) = y_1$ and $f(x_2) = y_2$, then we must have $x_1 < x_2$.

Hence g is strictly increasing on $[f(a), f(b)]$.

Suppose $z \in [f(a), f(b)]$ satisfies either $a = g(f(a)) < \lim_{y \to f(a)+} g(y)$, $\lim_{y \to z-} g(y) < \lim_{y \to z+} g(y)$ or $\lim_{y \to f(b)-} g(y) < g(f(b)) = b$. Then g cannot take on any values in $(a, \lim_{y \to f(a)+} g(y))$, $(\lim_{y \to z-} g(y), \lim_{y \to z+} g(y)) \setminus \{g(z)\}$ or $(\lim_{y \to f(b)-} g(y), b)$ respectively. This is a subset of $[a, b]$, and this is a contradiction since g takes on all values in $[a, b]$.
□

Theorem 4.3.4 $\log(x)$ *is continuous everywhere on* $(0, \infty)$.

Proof: Let $c > 0$. Since the range of e^x is $(0, \infty)$, and $y = e^x$ is strictly increasing, we see that there is an $a, b \in \mathbb{R}$, $a < b$, so that $c \in (e^a, e^b)$. By theorem (4.3.3) we see that $\log(x)$ is continuous on $[e^a, e^b]$, which contains c. Since c is arbitrary, we see that $\log(x)$ is continuous on all of $(0, \infty)$.
□

4.4 Infinite Limits and Limits at Infinity

In this section we extend the definition of limits $\lim_{x \to a} f(x) = L$ to the setting of either L or a being infinite. In all definitions below $d(\cdot, \cdot)$ represents a metric on a metric space M, which in most cases is $\mathbb{R}$ with the usual metric.

1. $a = +\infty$, L not infinite:
$$\lim_{x \to \infty} f(x) = L$$
   if and only if: Given $\varepsilon > 0$, there exists an $M > 0$ so that $x > M$ implies $d(f(x), L) < \varepsilon$.

2. $a = -\infty$, L not infinite:
$$\lim_{x \to -\infty} f(x) = L$$
   if and only if: Given $\varepsilon > 0$, there exists an $M > 0$ so that $x < -M$ implies $d(f(x), L) < \varepsilon$.

4.4. INFINITE LIMITS AND LIMITS AT INFINITY

3. $a = +\infty$, $L = +\infty$:
$$\lim_{x \to \infty} f(x) = \infty$$
if and only if: Given $N > 0$, there exists an $M > 0$ so that $x > M$ implies $f(x) > N$.

4. $a = +\infty$, $L = -\infty$:
$$\lim_{x \to \infty} f(x) = -\infty$$
if and only if: Given $N > 0$, there exists an $M > 0$ so that $x > M$ implies $f(x) < -N$.

5. $a = -\infty$, $L = +\infty$:
$$\lim_{x \to -\infty} f(x) = \infty$$
if and only if: Given $N > 0$, there exists an $M > 0$ so that $x < -M$ implies $f(x) > N$.

6. $a = -\infty$, $L = -\infty$:
$$\lim_{x \to -\infty} f(x) = -\infty$$
if and only if: Given $N > 0$, there exists an $M > 0$ so that $x < -M$ implies $f(x) < -N$.

7. a not infinite, $L = +\infty$:
$$\lim_{x \to a} f(x) = \infty$$
if and only if: Given $N > 0$, there is a $\delta > 0$ so that $d(x, a) < \delta$ implies $f(x) > N$.

8. a not infinite, $L = -\infty$:
$$\lim_{x \to a} f(x) = \infty$$
if and only if: Given $N > 0$, there is a $\delta > 0$ so that $d(x, a) < \delta$ implies $f(x) < -N$.

4.5 Homework Exercises

1. Show that the function $f(x)$ defined below is nowhere continuous but its absolute value is everywhere continuous, where
$$f(x) = \begin{cases} 1 & \text{if } x \in \mathbb{Q}, \\ -1 & \text{if } x \in \mathbb{R} \setminus \mathbb{Q}. \end{cases}$$

2. Show that the following function $f(x)$, as defined below, is continuous at $x = 0$, and everywhere else is not continuous, where
$$f(x) = \begin{cases} x & \text{if } x \in \mathbb{Q}, \\ -x & \text{if } x \in \mathbb{R} \setminus \mathbb{Q}. \end{cases}$$

3. Show that the following function as defined below is nowhere continuous, and is everywhere locally unbounded, where
$$f(x) = \begin{cases} n & \text{if } x \in \mathbb{Q} \text{ and } x = \frac{m}{n} \text{ is in reduced form}, n \in \mathbb{N}, m \in \mathbb{Z}, \\ 0 & \text{if } x \in \mathbb{R} \setminus \mathbb{Q}. \end{cases}$$

4. Let $f : D \to \mathbb{R}$ with $D \subseteq \mathbb{R}$ and $(-r, r) \subseteq D$ for some $r > 0$. Suppose $f(0) > 0$ and f is continuous at 0. Show that there is a $\delta > 0$ so that f is positive on $(-\delta, \delta)$.

5. Suppose $f : \mathbb{R} \to \mathbb{R}$ is continuous, and $f(x) \geq 0$ for all rational numbers x. Show $f(x) \geq 0$ for all real numbers x. Will this still hold if every $\geq$ is replaced by $>$? Please Explain.

6. Let $\{r_n : n \in \mathbb{N}\}$ be an enumeration of the rational numbers in $[0, 1]$, and define f on $[0, 1]$ by
$$f(x) = \sum_{r_n \leq x} 2^{-n}.$$
Show that f is discontinuous at all points in $[0, 1] \cap \mathbb{Q}$.

7. Suppose that $f : \mathbb{R} \to \mathbb{R}$ satisfies
$$f(x + y) = f(x) + f(y).$$
If f is continuous at some point x_0, show that there is a constant $c \in \mathbb{R}$ so that
$$f(x) = cx \quad \forall x \in \mathbb{R},$$
and conclude that f is continuous on all of $\mathbb{R}$.

4.5. HOMEWORK EXERCISES

8. $f : \mathbb{R} \to \mathbb{R}$ is said to be **Lipschitz continuous** if there is a $K > 0$ so that for any $x \neq y$ in $\mathbb{R}$

$$|f(x) - f(y)| < K|x - y|. \tag{4.31}$$

Show that f is uniformly continuous whenever f is Lipschitz continuous.

9. $f : \mathbb{R} \to \mathbb{R}$ is said to be **Hölder continuous** if there is a $K > 0$ and $\alpha \in (0, 1)$ so that for any $x \neq y$ in $\mathbb{R}$

$$|f(x) - f(y)| < K|x - y|^\alpha. \tag{4.32}$$

Show that f is uniformly continuous whenever f is Hölder continuous.

10. Let $E(x)$ be defined by

$$E(x) = \lim_{n \to \infty} (1 + \frac{x}{n})^n. \tag{4.33}$$

Clearly $E(1) = e$ and $E(0) = 1$, and so $E(x)$ exists for some $x \in \mathbb{R}$. Show that $E(x) = e^x$ for all $x \in \mathbb{R}$. *Hint: Use the Binomial theorem to expand* $(1 + \frac{x}{n})^n$.

11. Let f be a continuous real valued function defined on a closed subset of $\mathbb{R}$ space. Show that

$$Z := \{x : f(x) = 0\}$$

is closed.

12. Let $f : (a, b) \to \mathbb{R}$. We say that f is **convex** if for any $x \neq y$ in (a, b)

$$f(\lambda x + (1 - \lambda)y) < \lambda f(x) + (1 - \lambda)f(y) \quad \forall \lambda \in (0, 1). \tag{4.34}$$

(a) Show for $a < s < t < u < b$ we have

$$\frac{f(t) - f(s)}{t - s} < \frac{f(u) - f(s)}{u - s} < \frac{f(u) - f(t)}{u - t}.$$

(b) Show f is continuous on (a, b).

13. Let f be a real valued uniformly continuous function whose domain $D \subseteq \mathbb{R}$ is bounded. Prove that f is bounded on D.

14. Suppose $f : [0,1] \to [0,1]$ is continuous, show that $f(x) = x$ has at least one solution.

15. Let f be a real valued function whose domain is $(a,b) \subseteq \mathbb{R}$. We say that f has a **simple discontinuity** or a **discontinuity of the first kind** at $t \in (a,b)$ if
$$\lim_{x \to t^+} f(x) \neq \lim_{x \to t^-} f(x). \tag{4.35}$$
Show that the number of simple discontinuities is at most countably infinite. *Hint:* Let S be the set where $\lim_{x \to t^+} f(x) > \lim_{x \to t^-} f(x)$. For each $t \in S$ associate a triple (p, q, r) of rational numbers so that

(a) $\lim_{x \to t^-} f(x) < p < \lim_{x \to t^+} f(x)$
(b) $a < q < x < t$ implies $f(x) < p$
(c) $t < z < r < b$ implies $f(z) > p$.

Clearly the set of all such ordered triples is at most countable. Show that each triple is associated with at most one point in S. Then explain what to do for the other types of simple discontinuities of f.

16. Suppose $f : (a,b) \to \mathbb{R}$ is continuous and satisfies
$$f(\frac{x+y}{2}) < \frac{f(x) + f(y)}{2} \tag{4.36}$$
for all $x, y \in (a,b)$, $x \neq y$. Show that f is convex.

17. Define $f : [-\frac{1}{2}, \frac{1}{2}] \to \mathbb{R}$ by
$$f(x) = \begin{cases} 0 & x = 0, \\ f(|x|) & x < 0, \\ -2k(2k+1)(x - \frac{1}{2k}) & x \in (\frac{1}{2k+1}, \frac{1}{2k}], \quad k \in \mathbb{N}, \\ (2k+1)(2k+2)(x - \frac{1}{2k+2}) & x \in (\frac{1}{2k+2}, \frac{1}{2k+1}], \quad k \in \mathbb{N}. \end{cases}$$

(a) Show that $f(x)$ is continuous on $[\frac{1}{2}, 0) \cup (0, \frac{1}{2}]$.
(b) Show that $\lim_{x \to 0^-} f(x)$ and $\lim_{x \to 0^+} f(x)$ do not exist.

4.5. HOMEWORK EXERCISES

(c) Conclude that f is not continuous at $x = 0$. Moreover, it has a non-simple discontinuity or a discontinuity of the second kind at $x = 0$.

18. Suppose that f is a continuous function on an interval I and suppose that the range of f contains only finitely many points. Show that f is a constant function.

19. If $|f|$ is continuous, does f have to be continuous? Please explain. If this is not true, provide a counterexample.

20. Suppose $f : [0,1] \to \mathbb{R}$ is continuous and satisfies $f(0) = f(1)$. Prove that there exists $c \in [0, \frac{1}{2}]$ so that $f(c + \frac{1}{2}) = f(c)$.

21. Suppose $f, g : D \to \mathbb{R}$, with $D \subseteq \mathbb{R}$, and f, g are continuous. Show that $\{x \in D : f(x) > g(x)\}$ is open in D.

22. Show that $y = \sqrt{x}$ is uniformly continuous on $[0, \infty)$.

23. Show that $xe^x = 1$ has a real solution. Moreover, find an interval of length at most $\frac{1}{100}$ which contains a solution to this equation.

24. Let f be a bounded real-valued function whose domain $D \subseteq \mathbb{R}^n$. Define the **modulus of continuity of** f for each $\delta > 0$ by

$$w(\delta) = \sup\{|f(\vec{x}) - f(\vec{y})| : \|\vec{x} - \vec{y}\| < \delta\}. \tag{4.37}$$

Show that $w(\delta)$ decreases as δ decreases to 0. Moreover, show that f is uniformly continuous if and only if $w(\delta) \to 0$ as $\delta \to 0$.

25. Let f be usc and real-valued on a compact set $E \subseteq \mathbb{R}^n$. Show that f is bounded above on E and f attains its maximum on E.

26. Show that if f is usc on E, then $-f$ is lsc on E.

Chapter 5

Differentiation

5.1 Definitions and Basic Differentiability Results

Let f be a real-valued function of a real variable, and let D be its domain with $x_0 \in D$. Consider the limit

$$f'(x_0) := \lim_{x \to x_0} \frac{f(x) - f(x_0)}{x - x_0}. \tag{5.1}$$

If this limit exists, we say that f is **differentiable** at x_0 and has **derivative** $f'(x_0)$. Moreover, we say that f is **differentiable** if f is differentiable at all $x_0 \in D$. Sometimes the notation $\frac{df}{dx}(x_0)$ or even $\frac{d}{dx}f(x_0)$ is used in place of $f'(x_0)$. If we express x as $x_0 + h$ in (5.1), then we may rewrite $f'(x_0)$ as

$$f'(x_0) = \lim_{h \to 0} \frac{f(x_0 + h) - f(x_0)}{h}, \tag{5.2}$$

since $x_0 + h = x \to x_0$ if and only if $h \to 0$.

A notational convention to note is $\frac{d}{dx}(\ )$ means the derivative of $(\ )$, and often we use this notation when we have a formula for a function in the parentheses.

The following result tells us that differentiability implies continuity.

Theorem 5.1.1 *Suppose f is differentiable at x_0, then f is continuous at x_0.*

Proof: Note that

$$\lim_{x \to x_0} f(x) = f(x_0) \iff \lim_{x \to x_0} (f(x) - f(x_0)) = 0.$$

For $x \neq x_0$, we may write

$$f(x) - f(x_0) = \frac{f(x) - f(x_0)}{x - x_0} \cdot (x - x_0),$$

and so

$$\lim_{x \to x_0} (f(x) - f(x_0)) = \lim_{x \to x_0} \frac{f(x) - f(x_0)}{x - x_0} \cdot (x - x_0) = f'(x_0) \cdot 0 = 0.$$

□

The following result tells us that the operation of differentiation is a linear operation.

Theorem 5.1.2 *Suppose f and g are functions so that $f'(x_0)$ and $g'(x_0)$ both exist. Then for any real numbers c, k, the function $q(x) = cf(x) + kg(x)$ is differentiable at x_0 and*

$$q'(x_0) = cf'(x_0) + kg'(x_0).$$

Proof:

$$q'(x_0) = \lim_{x \to x_0} \frac{q(x) - q(x_0)}{x - x_0}$$

$$= \lim_{x \to x_0} \frac{cf(x) + kg(x) - cf(x_0) - kg(x_0)}{x - x_0}$$

$$= \lim_{x \to x_0} \{c \frac{f(x) - f(x_0)}{x - x_0} + k \frac{g(x) - g(x_0)}{x - x_0}\}$$

$$= c \lim_{x \to x_0} \frac{f(x) - f(x_0)}{x - x_0} + k \lim_{x \to x_0} \frac{g(x) - g(x_0)}{x - x_0}$$

$$= cf'(x_0) + kg'(x_0).$$

□

5.1.1 Derivatives of Some Functions

Theorem 5.1.3 *For any constant function f, f' is identically 0.*

Proof: Let $f(x) = c$, then $\frac{f(x+h)-f(x)}{h} = \frac{c-c}{h} = 0$, and thus $f'(x) = \lim_{h \to 0} \frac{f(x+h)-f(x)}{h} = 0$.
□

Theorem 5.1.4 *For any $n \in \mathbb{N}$,*

$$\frac{d}{dx}(x^n) = nx^{n-1}$$

for all values of $x \in \mathbb{R}$.

Proof: Here we will make use of the binomial theorem to expand $(x+h)^n$.

$$\lim_{h \to 0} \frac{(x+h)^n - x^n}{h} = \lim_{h \to 0} \frac{\sum_{k=0}^{n} \binom{n}{k} x^{n-k} h^k - x^n}{h}$$

$$= \lim_{h \to 0} \frac{nx^{n-1}h + \binom{n}{2}x^{n-2}h^2 + \cdots + nxh^{n-1} + h^n}{h}$$

$$= \lim_{h \to 0} \{nx^{n-1} + \binom{n}{2}x^{n-2}h + \cdots + nxh^{n-2} + h^{n-1}\}$$

$$= nx^{n-1}.$$

□

Theorem 5.1.5

$$\frac{d}{dx}e^x = e^x.$$

Proof: We will first show that $\lim_{h \to 0} \frac{e^h - 1}{h} = 1$. To see this observe that

$$e^h - 1 - h = \sum_{k=2}^{\infty} \frac{h^k}{k!}$$

$$= \frac{h^2}{2!} + \frac{h^3}{3!} + \cdots + \frac{h^n}{n!} + \cdots$$

Thus for $|h| < \frac{1}{2}$ we have

$$|e^h - 1 - h| \leq \frac{|h|^2}{2!} + \frac{|h|^3}{3!} + \cdots + \frac{|h|^n}{n!} + \cdots$$

$$\leq |h|^2 + |h|^3 + \cdots + |h|^n + \cdots$$

$$= h^2 \cdot \frac{1}{1 - |h|}.$$

Hence, we see that

$$\frac{|e^h - 1 - h|}{|h|} \leq |h| \cdot \frac{1}{1 - |h|} \leq 2|h|.$$

Thus we see that $\lim_{h \to 0} \frac{e^h - 1}{h} = 1$.

Now

$$\lim_{h \to 0} \frac{e^{x+h} - e^x}{h} = \lim_{h \to 0} e^x \cdot \frac{e^h - 1}{h}$$

$$= e^x \cdot \lim_{h \to 0} \frac{e^h - 1}{h} = e^x.$$

□

Examples of Functions which are Continuous at a Point but not Differentiable

Earlier we observed that differentiability implies continuity at a point. The examples given here show that the converse is not true. In fact, we will soon see that there exist functions which are continuous everywhere, but differentiable nowhere.

Consider the absolute value function $f(x) = |x|$. Since $f(x) = \sqrt{x^2}$ is the composition of continuous functions, it is also continuous. We will now show $f'(0)$ does not exist, even though f is continuous at $x = 0$. To see this observe

$$\frac{f(h) - f(0)}{h} = \frac{|h|}{h} = \begin{cases} 1 & \text{if } h > 0, \\ -1 & \text{if } h < 0. \end{cases}$$

Thus we see $\lim_{h \to 0^+} \frac{f(h) - f(0)}{h} = 1$ and $\lim_{h \to 0^-} \frac{f(h) - f(0)}{h} = -1$. Hence $f'(0)$ does not exist.

5.1. DEFINITIONS AND BASIC DIFFERENTIABILITY RESULTS

Now we consider the function $f(x) = \sqrt[3]{x}$. It was a homework exercise to show f is continuous on $\mathbb{R}$. Here

$$\frac{f(h) - f(0)}{h} = \frac{\sqrt[3]{h}}{h} = \frac{1}{(\sqrt[3]{h})^2}.$$

Moreover

$$\lim_{h \to 0} \frac{f(h) - f(0)}{h} = \lim_{h \to 0} \frac{1}{(\sqrt[3]{h})^2} = \infty.$$

Hence $f'(0)$ does not exist.

Also, we consider $f(x) = x^{\frac{2}{3}}$, which is continuous on $\mathbb{R}$. In this case,

$$\frac{f(h) - f(0)}{h} = \frac{h^{\frac{2}{3}}}{h} = \frac{1}{\sqrt[3]{h}},$$

and so

$$\lim_{h \to 0^+} \frac{f(h) - f(0)}{h} = \infty, \quad \lim_{h \to 0^-} \frac{f(h) - f(0)}{h} = -\infty.$$

Thus we see three examples of functions which are continuous at 0, but for which $f'(0)$ does not exist. In the first case $y = |x|$ has a corner point at $(0,0)$ on its graph, $y = \sqrt[3]{x}$ has a vertical tangent at $(0,0)$ on its graph, and $y = x^{\frac{2}{3}}$ has a **cusp** at $(0,0)$ on its graph.

An Example of a Function Continuous Everywhere but Differentiable Nowhere

Let $f(x) = |x|$ with domain $[-1, 1]$. Extend f to a function F which is defined everywhere, periodic with period 2, and $F|_{[-1,1]} = f$. Clearly F is continuous on $\mathbb{R}$ since $\lim_{x \to -1^+} f(x) = f(-1) = f(1) = \lim_{x \to 1^-} f(x)$ and f is continuous on $[-1, 1]$. Let $f_n(x) = \frac{3^n}{4^n} F(4^n x)$, for $n \in \mathbb{N} \cup \{0\}$. So f_n is periodic with period $2 \cdot 4^{-n}$, and has a maximum value of $\frac{3^n}{4^n}$. We define $g : \mathbb{R} \to \mathbb{R}$ by

$$g(x) := \sum_{n=0}^{\infty} f_n(x). \tag{5.3}$$

Clearly

$$\left|\sum_{n=0}^{\infty} f_n(x)\right| \leq \sum_{n=0}^{\infty} |f_n(x)| \leq \sum_{n=0}^{\infty} \frac{3^n}{4^n} = 4.$$

Hence g exists for all $x \in \mathbb{R}$. We will now show that g is continuous on $\mathbb{R}$. Observe that for any pair $x, y \in \mathbb{R}$

$$|g(x) - g(y)| = |\sum_{n=0}^{\infty}(f_n(x) - f_n(y))| \leq \sum_{n=0}^{\infty}|f_n(x) - f_n(y)|$$

Let $\varepsilon > 0$ be given, then

$$\sum_{n=N}^{\infty}|f_n(x) - f_n(y)| \leq \sum_{n=N}^{\infty} 2\frac{3^n}{4^n} = 8 \cdot \frac{3^N}{4^N}.$$

We may find $N \in \mathbb{N}$ so that $8(\frac{3}{4})^N < \frac{1}{2}\varepsilon$. Moreover, each of the functions $f_0, f_1, f_2, f_3, \cdots, f_{N-1}$ is continuous on $\mathbb{R}$, since each is obtained by a composition of a continuous function F with functions of the type $y = cx$, followed by a scalar multiplication. Let $x \in \mathbb{R}$ be fixed, then for each $i \in \{0, 1, 2, 3, \cdots, N-1\}$ there exists $\delta_i > 0$ so that $|x - y| \leq \delta_i$ implies $|f_i(x) - f_i(y)| < \frac{1}{2N}\varepsilon$. Let $\delta = \min\{\delta_0, \delta_1, \delta_2, \cdots, \delta_{N-1}\}$. Then for $|x - y| < \delta$ we have

$$|g(x) - g(y)| \leq \sum_{i=0}^{N-1}|f_i(x) - f_i(y)| + \sum_{i=N}^{\infty}|f_i(x) - f_i(y)|$$

$$\leq \frac{1}{2}\varepsilon + \frac{1}{2}\varepsilon = \varepsilon.$$

Hence we see that $g(x)$ is continuous on $\mathbb{R}$.

We will now show that g is nowhere differentiable. Fix $x_0 \in \mathbb{R}$. For each $n \in \mathbb{N} \cup \{0\}$ choose $h_n = \pm\frac{1}{2 \cdot 4^n}$ so that there are no integers between $4^n x_0$ and $4^n(x_0 + h_n)$. Thus $|F(4^n x_0 + 4^n h_n) - F(4^n x_0)| = \frac{1}{2}$.

Likewise, for $m \leq n$ we have $|F(4^m x_0 + 4^m h_n) - F(4^m x_0)| \leq |4^m h_n|$.

Moreover, for $m > n$ we observe that $|F(4^m x_0 + 4^m h_n) - F(4^m x_0)| = |F(4^m x_0 \pm \frac{1}{2}4^{m-n}) - F(4^m x_0)| = 0$, since F is periodic with a period of 2.

Hence we see that

$$|\frac{g(x_0 + h_n) - g(x_0)}{h_n}| = |\sum_{k=0}^{n} \frac{3^k}{4^k} \frac{F(4^k x_0 + 4^k h_n) - F(4^k x_0)}{h_n}|$$

5.1. DEFINITIONS AND BASIC DIFFERENTIABILITY RESULTS

$$\geq 3^n - \sum_{k=0}^{n-1} \frac{3^k}{4^k} \left| \frac{F(4^k x_0 + 4^k h_n) - F(4^k x_0)}{h_n} \right|$$

$$\geq 3^n - \sum_{k=0}^{n-1} 3^k$$

$$> \frac{1}{2} 3^n.$$

Since $h_n \to 0$ as $n \to \infty$, and $\left| \frac{g(x_0+h_n)-g(x_0)}{h_n} \right| \to \infty$ as $n \to \infty$ we see that $g'(x_0)$ does not exist. Since $x_0 \in \mathbb{R}$ is arbitrary we see that g is not differentiable anywhere.

5.1.2 Approximation by Linear Functions

A function $y = f(x)$ is differentiable at x_0 if and only if $\lim_{x \to x_0} \frac{f(x_0)-f(x)}{x-x_0}$ exists. We will now see that this requirement actually guarantees that $f(x)$ can be approximated by the linear function $y = f(x_0) + f'(x_0) \cdot (x - x_0)$ for x near x_0. Hence differentiability guarantees that the graph of $y = f(x)$ is locally flat near $(x_0, f(x_0))$.

To see that $y = f(x)$ can be approximated by the linear function $y = f(x_0) + f'(x_0) \cdot (x - x_0)$ we notice that $\lim_{x \to x_0} \frac{f(x)-f(x_0)}{x-x_0} = f'(x_0)$ exists if and only if

$$\lim_{x \to x_0} \frac{|f(x) - f(x_0) - f'(x_0) \cdot (x - x_0)|}{|x - x_0|} = 0. \tag{5.4}$$

So differentiability implies that $|f(x) - f(x_0) - f'(x_0) \cdot (x - x_0)| \to 0$ faster than $|x - x_0| \to 0$.

5.1.3 The Product and Quotient Rule

Theorem (The Product Rule) 5.1.6 *Let f and g be real valued functions of a real variable and suppose $f'(x_0)$ and $g'(x_0)$ exist. Then for the function $k(x) = f(x) \cdot g(x)$, $k'(x_0)$ exists and*

$$k'(x_0) = f'(x_0) \cdot g(x_0) + f(x_0) \cdot g'(x_0).$$

Proof:
$$\frac{k(x) - k(x_0)}{x - x_0} = \frac{f(x)g(x) - f(x_0)g(x_0)}{x - x_0}$$
$$= \frac{f(x)g(x) - f(x_0)g(x) + f(x_0)g(x) - f(x_0)g(x_0)}{x - x_0}$$
$$= g(x)\frac{f(x) - f(x_0)}{x - x_0} + f(x_0)\frac{g(x) - g(x_0)}{x - x_0}.$$

Taking limits, we get
$$k'(x_0) = \lim_{x \to x_0} g(x) \cdot \lim_{x \to x_0} \frac{f(x) - f(x_0)}{x - x_0} + f(x_0) \cdot \lim_{x \to x_0} \frac{g(x) - g(x_0)}{x - x_0}$$
$$= g(x_0) \cdot f'(x_0) + f(x_0) \cdot g'(x_0).$$

□

Theorem (The Quotient Rule) 5.1.7 *Let f and g be real valued functions of a real variable and suppose $f'(x_0)$ and $g'(x_0)$ exist, and $g(x_0) \neq 0$. Then for the function $k(x) = \frac{f(x)}{g(x)}$, $k'(x_0)$ exists and*
$$k'(x_0) = \frac{f'(x_0) \cdot g(x_0) - f(x_0) \cdot g'(x_0)}{(g(x_0))^2}$$

Proof: By continuity of g at x_0 we know that for $|x - x_0| < \delta$ we have $|g(x)| > \frac{1}{2}|g(x_0)|$. Now for such x values,
$$\frac{k(x) - k(x_0)}{x - x_0} = \frac{\frac{f(x)}{g(x)} - \frac{f(x_0)}{g(x_0)}}{x - x_0}$$
$$= \frac{f(x)g(x_0) - f(x_0)g(x)}{(x - x_0)g(x)g(x_0)}$$
$$= \frac{f(x)g(x_0) - f(x_0)g(x_0) + f(x_0)g(x_0) - f(x_0)g(x)}{(x - x_0)g(x)g(x_0)}$$
$$= \frac{g(x_0)\frac{f(x) - f(x_0)}{x - x_0} - f(x_0)\frac{g(x) - g(x_0)}{x - x_0}}{g(x)g(x_0)}.$$

Using the continuity of g at x_0, and taking limits we get
$$k'(x_0) = \frac{g(x_0)f'(x_0) - f(x_0)g'(x_0)}{(g(x_0))^2}.$$

□

5.1. DEFINITIONS AND BASIC DIFFERENTIABILITY RESULTS

Theorem 5.1.8 *For any $n \in \mathbb{Z}$ the function $f(x) = x^n$ is differentiable on its domain. Moreover,*
$$\frac{d}{dx}(x^n) = nx^{n-1}.$$

Proof: If $n \in \mathbb{N}$, then this follows from theorem (5.1.4). If $n = 0$, then $x^n = 1$, and the derivative of a constant function is 0, which agrees with nx^{n-1}. For $-n$ with $n \in \mathbb{N}$ we have

$$\frac{d}{dx}(x^{-n}) = \frac{d}{dx}(\frac{1}{x^n})$$
$$= \frac{\frac{d}{dx}(1) \cdot x^n - \frac{d}{dx}(x^n) \cdot 1}{(x^n)^2}$$
$$= \frac{-nx^{n-1}}{x^{2n}}$$
$$= -n \frac{1}{x^{n+1}}$$
$$= -n \cdot x^{-n-1}.$$

□

5.1.4 The Chain Rule

Theorem (The Chain Rule) 5.1.9 *Let f be defined in a neighbourhood of x_0, and suppose that $f'(x_0)$ exists. Let g be defined in a neighbourhood of $y_0 = f(x_0)$, and suppose $g'(y_0)$ exists. Then the function $h(x) = (g \circ f)(x)$ is differentiable at x_0, and*

$$h'(x_0) = g'(f(x_0)) \cdot f'(x_0).$$

Proof: Given $\varepsilon > 0$ we need to show that there exists a $\delta > 0$ so that $|x - x_0| < \delta$ implies

$$\frac{|g(f(x)) - g(f(x_0)) - g'(f(x_0)) \cdot f'(x_0) \cdot (x - x_0)|}{|x - x_0|} < \varepsilon.$$

We rewrite $g(f(x)) - g(f(x_0)) - g'(f(x_0)) \cdot f'(x_0) \cdot (x - x_0)$ as

$$g(f(x)) - g(f(x_0)) - g'(f(x_0)) \cdot (f(x) - f(x_0))$$

$$+g'(f(x_0)) \cdot (f(x) - f(x_0)) - g'(f(x_0)) \cdot f'(x_0) \cdot (x - x_0)$$

Now,
$$g'(f(x_0)) \cdot (f(x) - f(x_0)) - g'(f(x_0)) \cdot f'(x_0) \cdot (x - x_0)$$
$$= g'(f(x_0))[f(x) - f(x_0) - f'(x_0)(x - x_0)].$$

Assuming $g'(f(x_0)) \neq 0$, since $f'(x_0)$ exists, there's a $\delta_0 > 0$ so that

$$\frac{|f(x) - f(x_0) - f'(x_0)(x - x_0)|}{|x - x_0|} < \frac{\varepsilon}{2|g'(f(x_0))|},$$

whenever $|x - x_0| < \delta_0$. In the case where $g'(f(x_0)) = 0$, then

$$g'(f(x_0))(f(x) - f(x_0)) - g'(f(x_0))f'(x_0)(x - x_0) = 0.$$

Now we consider the second term. Let $y_0 = f(x_0)$. Since $g'(y_0)$ exists, there is a $\delta_1 > 0$ so that $|y - y_0| < \delta_1$ implies

$$\frac{|g(y) - g(y_0) - g'(y_0)(y - y_0)|}{|y - y_0|} < \frac{\varepsilon}{2(1 + |f'(x_0)|)}.$$

By continuity of $f(x)$ at x_0 there is a $\delta_2 > 0$ so that $|x - x_0| < \delta_2$ implies $|f(x) - f(x_0)| < \delta_1$. Moreover, we may choose $0 < \delta_3 < 1$ so that for $|x - x_0| < \delta_3$ we have

$$\frac{|f(x) - f(x_0) - f'(x_0)(x - x_0)|}{|x - x_0|} < 1,$$

and so
$$|f(x) - f(x_0)| < [1 + |f'(x_0)|]|x - x_0|.$$

Now for $\delta < \min\{\delta_0, \delta_2, \delta_3\}$ and $|x - x_0| < \delta$ we have

$$\frac{|g(f(x)) - g(f(x_0)) - g'(f(x_0)) \cdot f'(x_0) \cdot (x - x_0)|}{|x - x_0|} < \varepsilon.$$

□

5.1.5 The Trigonometric Functions

Recall[1] $\text{Sin}(x)$ and $\text{Cos}(x)$ as defined by

$$\text{Sin}(x) = \sum_{k=0}^{\infty} \frac{(-1)^k}{(2k+1)!} x^{2k+1}, \tag{5.5}$$

and

$$\text{Cos}(x) = \sum_{k=0}^{\infty} \frac{(-1)^k}{(2k)!} x^{2k}. \tag{5.6}$$

By the comparison test, we see that $|\text{Sin}(x)|, |\text{Cos}(x)| \leq e^{|x|}$, and both converge absolutely for all $x \in \mathbb{R}$. We shall now show $\text{Sin}(x)$ and $\text{Cos}(x)$ are continuous on $\mathbb{R}$.

Theorem 5.1.10 $\text{Sin}(x)$ and $\text{Cos}(x)$ are continuous on $\mathbb{R}$.

Proof: Let $x, y \in \mathbb{R}$. Without loss of generality, suppose $x-1 < y < x+1$. Then,

$$|\text{Sin}(x) - \text{Sin}(y)| \leq \sum_{k=0}^{\infty} \frac{1}{(2k+1)!} |x^{2k+1} - y^{2k+1}|$$

$$\leq \sum_{k=0}^{\infty} \frac{|x-y|}{(2k+1)!} \sum_{j=0}^{2k} |x|^{2k-j}|y|^j$$

$$\leq |x-y| \cdot \sum_{k=0}^{\infty} \frac{1}{(2k+1)!} (2k+1)(\max\{|x|+1, |y|+1\})^{2k}$$

$$= |x-y| \sum_{k=0}^{\infty} \frac{1}{(2k)!} (\max\{|x|+1, |y|+1\})^{2k}$$

$$\leq |x-y| \sum_{m=0}^{\infty} \frac{1}{m!} (\max\{|x|+1, |y|+1\})^m$$

$$= |x-y| \cdot e^{\max\{|x|+1, |y|+1\}}$$

$$\leq |x-y| \cdot e^{|x|+2}.$$

---
[1] For a similar approach see [11]. Also, see [9] for a functional equation approach to developing the trigonometric functions.

Thus we see that $\mathrm{Sin}(x)$ is continuous on $\mathbb{R}$.

Also,
$$|\mathrm{Cos}(x) - \mathrm{Cos}(y)| \leq \sum_{k=0}^{\infty} \frac{1}{(2k)!}|x^{2k} - y^{2k}|$$

$$\leq \sum_{k=1}^{\infty} \frac{|x-y|}{(2k)!} \sum_{j=0}^{2k-1} |x|^{2k-1-j}|y|^j$$

$$\leq |x-y| \sum_{k=1}^{\infty} \frac{1}{(2k-1)!}(\max\{|x|+1, |y|+1\})^{2k-1}$$

$$\leq |x-y| e^{\max\{|x|+1, |y|+1\}}$$

$$\leq |x-y| \cdot e^{|x|+2}.$$

Hence we see $\mathrm{Cos}(x)$ is also continuous on $\mathbb{R}$.
$\square$

Theorem 5.1.11 *For any real numbers x, y we have*

$$\mathrm{Sin}(x+y) = \mathrm{Sin}(x)\mathrm{Cos}(y) + \mathrm{Cos}(x)\mathrm{Sin}(y). \tag{5.7}$$

Proof: We will make use of theorem (3.5.13) to rewrite:

$$\mathrm{Sin}(x)\mathrm{Cos}(y) + \mathrm{Cos}(x)\mathrm{Sin}(y)$$

$$= \sum_{k=0}^{\infty} (-1)^k \frac{x^{2k+1}}{(2k+1)!} \sum_{k=0}^{\infty} (-1)^k \frac{y^{2k}}{(2k)!} + \sum_{k=0}^{\infty} (-1)^k \frac{y^{2k+1}}{(2k+1)!} \sum_{k=0}^{\infty} (-1)^k \frac{x^{2k}}{(2k)!}$$

$$= \sum_{k=0}^{\infty} \sum_{j=0}^{k} \frac{(-1)^j x^{2j+1}}{(2j+1)!} \frac{(-1)^{k-j} y^{2(k-j)}}{(2(k-j))!} + \sum_{k=0}^{\infty} \sum_{j=0}^{k} \frac{(-1)^j x^{2j}}{(2j)!} \frac{(-1)^{k-j} y^{2(k-j)+1}}{(2(k-j)+1)!}$$

$$= \sum_{k=0}^{\infty} \sum_{j=0}^{k} [\frac{(-1)^j x^{2j+1}}{(2j+1)!} \frac{(-1)^{k-j} y^{2(k-j)}}{(2(k-j))!} + \frac{(-1)^j x^{2j}}{(2j)!} \frac{(-1)^{k-j} y^{2(k-j)+1}}{(2(k-j)+1)!}]$$

$$= \sum_{k=0}^{\infty} \sum_{j=0}^{k} (-1)^k [\frac{x^{2j+1}}{(2j+1)!} \frac{y^{2(k-j)}}{(2(k-j))!} + \frac{x^{2j}}{(2j)!} \frac{y^{2(k-j)+1}}{(2(k-j)+1)!}]$$

5.1. DEFINITIONS AND BASIC DIFFERENTIABILITY RESULTS 179

$$= \sum_{k=0}^{\infty} \sum_{j=0}^{k} \frac{(-1)^k}{(2k+1)!} [\frac{(2k+1)!}{(2j+1)!(2(k-j))!} x^{2j+1} y^{2(k-j)}$$

$$+ \frac{(2k+1)!}{(2j)!(2(k-j)+1)!} x^{2j} y^{2(k-j)+1}]$$

$$= \sum_{k=0}^{\infty} \sum_{j=0}^{k} \frac{(-1)^k}{(2k+1)!} [\binom{2k+1}{2j+1} x^{2j+1} y^{2k+1-(2j+1)} + \binom{2k+1}{2j} x^{2j} y^{2k+1-2j}]$$

$$= \sum_{k=0}^{\infty} \frac{(-1)^k}{(2k+1)!} \sum_{i=0}^{2k+1} \binom{2k+1}{i} x^i y^{2k+1-i}$$

$$= \sum_{k=0}^{\infty} \frac{(-1)^k}{(2k+1)!} (x+y)^{2k+1}$$

$$= \text{Sin}(x+y).$$

□

Theorem 5.1.12 *For any real numbers x, y we have*

$$\text{Cos}(x+y) = \text{Cos}(x)\text{Cos}(y) - \text{Sin}(x)\text{Sin}(y). \tag{5.8}$$

Proof: We will make use of theorem (3.5.13) to rewrite:

$$\text{Cos}(x)\text{Cos}(y) - \text{Sin}(x)\text{Sin}(y)$$

$$= \sum_{k=0}^{\infty} (-1)^k \frac{x^{2k}}{(2k)!} \sum_{k=0}^{\infty} (-1)^k \frac{y^{2k}}{(2k)!} - \sum_{k=0}^{\infty} (-1)^k \frac{x^{2k+1}}{(2k+1)!} \sum_{k=0}^{\infty} (-1)^k \frac{y^{2k+1}}{(2k+1)!}$$

$$= \sum_{k=0}^{\infty} \sum_{j=0}^{k} \frac{(-1)^j x^{2j}}{(2j)!} \frac{(-1)^{k-j} y^{2(k-j)}}{(2(k-j))!} - \sum_{k=0}^{\infty} \sum_{j=0}^{k} \frac{(-1)^j x^{2j+1}}{(2j+1)!} \frac{(-1)^{k-j} y^{2(k-j)+1}}{(2(k-j)+1)!}$$

$$= \sum_{k=0}^{\infty} \frac{(-1)^k}{(2k)!} \sum_{j=0}^{k} \frac{(2k)!}{(2j)!(2(k-j))!} x^{2j} y^{2k-2j}$$

$$- \sum_{k=0}^{\infty} \frac{(-1)^k}{(2k+2)!} \sum_{j=0}^{k} \frac{(2k+2)!}{(2j+1)!(2(k-j)+1)!} x^{2j+1} y^{2(k-j)+1}$$

$$= \sum_{k=0}^{\infty} \frac{(-1)^k}{(2k)!} [\sum_{j=0}^{k} \binom{2k}{2j} x^{2j} y^{2k-2j}] + \sum_{k=1}^{\infty} \frac{(-1)^k}{(2k)!} [\sum_{j=0}^{k-1} \binom{2k}{2j+1} x^{2j+1} y^{2k-(2j+1)}]$$

$$= \sum_{k=0}^{\infty} \frac{(-1)^k}{(2k)!} \sum_{i=0}^{2k} \binom{2k}{i} x^i y^{2k-i}$$

$$= \sum_{k=0}^{\infty} \frac{(-1)^k}{(2k)!} (x+y)^{2k}$$

$$= \mathrm{Cos}(x+y)$$

□

Theorem 5.1.13 *For any real number x we have the so-called* **Pythagorean identity**

$$(\mathrm{Sin}(x))^2 + (\mathrm{Cos}(x))^2 = 1. \tag{5.9}$$

Proof: We will make use of theorem (3.5.13) to rewrite:

$$(\mathrm{Sin}(x))^2 + (\mathrm{Cos}(x))^2$$

$$= \sum_{k=0}^{\infty} (-1)^k \frac{x^{2k+1}}{(2k+1)!} \sum_{k=0}^{\infty} (-1)^k \frac{x^{2k+1}}{(2k+1)!} + \sum_{k=0}^{\infty} (-1)^k \frac{x^{2k}}{(2k)!} \sum_{k=0}^{\infty} (-1)^k \frac{x^{2k}}{(2k)!}$$

$$= \sum_{k=0}^{\infty} \frac{(-1)^j x^{2j+1}}{(2j+1)!} \frac{(-1)^{k-j} x^{2(k-j)+1}}{(2(k-j)+1)!} + \sum_{k=0}^{\infty} \sum_{j=0}^{k} \frac{(-1)^j x^{2j}}{(2j)!} \frac{(-1)^{k-j} x^{2(k-j)}}{(2(k-j))!}$$

$$= \sum_{k=0}^{\infty} (-1)^k \sum_{j=0}^{k} \frac{x^{2k+2}}{(2j+1)!(2(k-j)+1)!} + \sum_{k=0}^{\infty} (-1)^k \sum_{j=0}^{k} \frac{x^{2k}}{(2j)!(2(k-j))!}$$

$$= \sum_{k=0}^{\infty} \frac{(-1)^k}{(2(k+1))!} x^{2(k+1)} \sum_{j=0}^{k} \binom{2(k+1)}{2j+1} + \sum_{k=0}^{\infty} \frac{(-1)^k}{(2k)!} x^{2k} \sum_{j=0}^{k} \binom{2k}{2j}$$

$$= \sum_{k=0}^{\infty} \frac{(-1)^k}{(2(k+1))!} x^{2(k+1)} \sum_{j=0}^{k} \binom{2(k+1)}{2j+1} + 1 + \sum_{k=1}^{\infty} \frac{(-1)^k}{(2k)!} x^{2k} \sum_{j=0}^{k} \binom{2k}{2j}$$

$$= \sum_{k=0}^{\infty} \frac{(-1)^k}{(2(k+1))!} x^{2(k+1)} \sum_{j=0}^{k} \binom{2(k+1)}{2j+1} + 1 + \sum_{k=0}^{\infty} \frac{(-1)^{k+1}}{(2(k+1))!} x^{2(k+1)} \sum_{j=0}^{k+1} \binom{2(k+1)}{2j}$$

5.1. DEFINITIONS AND BASIC DIFFERENTIABILITY RESULTS

$$= 1 + \sum_{k=0}^{\infty} \frac{(-1)^k}{(2(k+1))!} x^{2(k+1)} [\sum_{j=0}^{k} \binom{2(k+1)}{2j+1} - \sum_{j=0}^{k+1} \binom{2(k+1)}{2j}]$$

$$= 1 - \sum_{k=0}^{\infty} \frac{(-1)^k}{(2(k+1))!} x^{2(k+1)} \sum_{i=0}^{2k+2} \binom{2(k+1)}{i}(-1)^i 1^{2(k+1)-i}$$

$$= 1 - \sum_{k=0}^{\infty} \frac{(-1)^k}{(2(k+1))!} x^{2(k+1)} (1-1)^{2(k+1)}$$

$$= 1$$

□

Theorem 5.1.14 $\mathrm{Sin}(x)$ and $\mathrm{Cos}(x)$ are differentiable on $\mathbb{R}$. Moreover,

$$\frac{d}{dx}(\mathrm{Sin}(x)) = \mathrm{Cos}(x), \quad \text{and} \quad \frac{d}{dx}(\mathrm{Cos}(x)) = -\mathrm{Sin}(x). \quad (5.10)$$

Proof: First we will show the following limit holds

$$\lim_{x \to 0} \frac{\mathrm{Sin}(x)}{x} = 1. \quad (5.11)$$

To see this, note that for $|x| < \frac{1}{2}$

$$|\mathrm{Sin}(x) - x| = \left| -\frac{x^3}{3!} + \frac{x^5}{5!} - \frac{x^7}{7!} + \cdots \right|$$

$$\leq \frac{|x|^3}{3!} + \frac{|x|^5}{5!} + \frac{|x|^7}{7!} + \cdots$$

$$\leq |x|^3 + |x|^5 + |x|^7 + \cdots$$

$$\leq |x|^3 + |x|^4 + |x|^5 + \cdots$$

$$= |x|^3 \cdot \frac{1}{1 - |x|}$$

$$\leq |x|^2 \cdot \frac{1}{1 - |x|}.$$

So, for $|x| < \frac{1}{2}$ we have

$$0 \leq \frac{|\mathrm{Sin}(x) - x|}{|x|} \leq \frac{|x|}{1 - |x|} < 2|x|.$$

Letting $x \to 0$ we get the desired result. We shall also show

$$\lim_{x \to 0} \frac{\cos(x) - 1}{x} = 0 \qquad (5.12)$$

To see this, note that

$$|\cos(x) - 1| = \left| -\frac{x^2}{2!} + \frac{x^4}{4!} - \frac{x^6}{6!} + \cdots \right|$$

$$\leq \frac{x^2}{2!} + \frac{x^4}{4!} + \frac{x^6}{6!} + \cdots$$

$$\leq x^2 + x^4 + x^6 + \cdots$$

$$\leq |x|^2 + |x|^3 + |x|^4 + |x|^5 + |x|^6 + \cdots$$

$$= |x|^2 \cdot \frac{1}{1 - |x|}.$$

Now for $|x| < \frac{1}{2}$ we have

$$0 \leq \frac{|\cos(x) - 1|}{|x|} \leq \frac{|x|}{1 - |x|} < 2|x|.$$

Letting $x \to 0$ we get the desired result. Now,

$$\frac{\sin(x + h) - \sin(x)}{h} = \frac{\sin(x)\cos(h) + \sin(h)\cos(x) - \sin(x)}{h}$$

$$= \cos(x) \cdot \frac{\sin(h)}{h} + \sin(x) \cdot \frac{\cos(h) - 1}{h}$$

This quantity tends to $1 \cdot \cos(x) + \sin(x) \cdot 0 = \cos(x)$ as $h \to 0$, and so $\frac{d}{dx}(\sin(x)) = \cos(x)$.

Likewise,

$$\frac{\cos(x + h) - \cos(x)}{h} = \frac{\cos(x)\cos(h) - \sin(x)\sin(h) - \cos(x)}{h}$$

$$= \cos(x) \cdot \frac{\cos(h) - 1}{h} - \sin(x) \cdot \frac{\sin(h)}{h}.$$

This quantity tends to $\cos(x) \cdot 0 - \sin(x) \cdot 1 = -\sin(x)$ as $h \to 0$, and so we see that $\frac{d}{dx}(\cos(x)) = -\sin(x)$.

5.1. DEFINITIONS AND BASIC DIFFERENTIABILITY RESULTS

□

At this point we switch our notation and define the **trigonometric functions sine** and **cosine**. They are denoted by $y = \sin(x)$ or $y = \sin x$, and $y = \cos(x)$ or $y = \cos x$ respectively, are are defined by

$$\sin x := \mathrm{Sin}(x), \quad \text{and} \quad \cos x := \mathrm{Cos}(x). \tag{5.13}$$

We also define the **tangent function**, denoted $y = \tan(x)$ or $y = \tan x$ by

$$\tan x = \frac{\sin x}{\cos x}. \tag{5.14}$$

From the above results and the quotient rule, on their respective domains we see

$$\frac{d}{dx}\sin x = \cos x, \quad \frac{d}{dx}\cos x = -\sin x, \quad \text{and} \quad \frac{d}{dx}\tan x = \frac{1}{(\cos x)^2}. \tag{5.15}$$

Also, from the theorems above we have

$$\sin(x+y) = \sin x \cos y + \sin y \cos x, \quad \cos(x+y) = \cos x \cos y - \sin x \sin y, \tag{5.16}$$

and we have the **Pythagorean identity**

$$(\sin x)^2 + (\cos x)^2 = 1, \tag{5.17}$$

which is commonly written as $\sin^2 x + \cos^2 x = 1$.

We will work with equations (5.16) and (5.17) to establish some other properties of the trigonometric functions. Moreover, there are other trigonometric functions that are commonly defined, namely **cotangent, secant** and **cosecant**. The notation used for these is $y = \cot(x)$ or $y = \cot x$, $y = \sec(x)$ or $y = \sec x$, and $y = \csc(x)$ or $y = \csc x$ respectively, and they are defined by

$$\cot x := \frac{\cos x}{\sin x}, \quad \sec x := \frac{1}{\cos x}, \quad \text{and} \quad \csc x := \frac{1}{\sin x}. \tag{5.18}$$

Theorem 5.1.15 *For any $x \in \mathbb{R}$, we have*

$$\sin(-x) = -\sin x, \quad \text{and} \quad \cos(-x) = \cos x. \tag{5.19}$$

Proof: By (5.16) we have

$$1 = \cos 0 = \cos(x - x) = \cos x \cos(-x) - \sin x \sin(-x),$$

$$0 = \sin 0 = \sin(x - x) = \sin x \cos(-x) + \sin(-x) \cos x.$$

We multiply the first equation by $\cos x$ and the second equation by $\sin x$, and add together to get

$$\cos x = (\cos^2 x + \sin^2 x) \cos(-x) = \cos(-x).$$

Then substitute this in the second equation to get

$$0 = \sin x \cos x + \sin(-x) \cos x,$$

and so $\sin(-x) = -\sin x$ or $\cos x = 0$. However, if $\cos x = 0$, then $\cos(-x)$ also equals zero, and thus by the Pythagorean identity we have $(\sin x)^2 = (\sin(-x))^2 = 1$. Also, using $\cos x = 0$ in the first equation above we have $1 = -\sin x \cdot \sin(-x)$. So we see that $\sin(-x) = -\sin x$.
□

Using (5.16) and theorem (5.1.15) we see

$$\sin(x + y) - \sin(x - y) = 2 \sin y \cos x$$

and

$$\cos(x + y) - \cos(x - y) = -2 \sin x \sin y.$$

Letting $\theta = x + y$ and $\phi = x - y$ we see

$$\sin \theta - \sin \phi = 2 \cos \frac{\theta + \phi}{2} \sin \frac{\theta - \phi}{2}, \qquad (5.20)$$

$$\cos \theta - \cos \phi = -2 \sin \frac{\theta + \phi}{2} \sin \frac{\theta - \phi}{2}. \qquad (5.21)$$

Theorem 5.1.16 $\cos x = 0$ *has a solution for some* $x > 0$.

Proof:[2] Let $c = \inf\{\cos x : x > 0\}$. Since $y = \cos x$ is continuous and $\cos 0 = 1$, if $\cos x \neq 0$ for all x, then $c \geq 0$. Let's first suppose $c > 0$.

---

[2] Note that this part was inspired by the treatment done in [11]

5.1. DEFINITIONS AND BASIC DIFFERENTIABILITY RESULTS

By (5.20), we see that for any $m \in \mathbb{N}$ and any $x \in \mathbb{R}$ we have

$$\sin((m+1)x) - \sin(mx) = 2\cos((m+\frac{1}{2})x)\sin(\frac{x}{2})$$

$$\geq 2c\sin(\frac{x}{2}).$$

Since $\lim_{x \to 0} \frac{\sin x}{x} = 1$, we know that for $x > 0$ sufficiently small $\sin \frac{x}{2} > 0$. Fix such an x value for which we see that

$$\sin((m+1)x) - \sin(mx) \geq 2c\sin(\frac{x}{2}) > 0.$$

Thus the sequence $\{\sin(mx)\}_{m=1}^{\infty}$ is strictly increasing for such an x, and

$$\sin(Nx) \geq 2(N-1)c\sin\frac{x}{2}.$$

This can be made larger than 1 by picking $N \in \mathbb{N}$ sufficiently large. This is a contradiction to $\sin^2\theta + \cos^2\theta = 1$, which implies that $-1 \leq \sin\theta \leq 1$ for all θ. Hence we see that if $c \geq 0$, then $c = 0$ is the only possibility.

It is still possible that $\cos x > 0$ for all $x > 0$. We must now rule this out. For any $\theta \in \mathbb{R}$ and for any $h > 0$ we have

$$\sin(\theta + h) - \sin\theta = 2\cos(\frac{2\theta + h}{2})\sin\frac{h}{2}$$

by (5.20).

As above, for $h > 0$ sufficiently small we have $\sin\frac{h}{2} > 0$, and so

$$\sin(\theta + h) - \sin\theta > 0.$$

Hence for $h > 0$ small, $\sin(\theta + h) > \sin\theta$, and so we see that $\sin x$ is always increasing. Since $\sin x > 0$ for $x > 0$, we have $\sin^2 x$ is always strictly increasing as well. Since $\cos^2 x = 1 - \sin^2 x$, we see that $\cos^2 x$ must always be decreasing for $x > 0$. Since $\cos x > 0$, $\cos x$ must always be decreasing for $x > 0$. Since $c = 0$, we know that there is a $z > 0$ so that for $x > z$ we have $\cos x < \frac{5}{13}$, and hence $\sin x = \sqrt{1 - \cos^2 x} > \frac{12}{13}$. Moreover, by (5.16) we have

$$\frac{12}{13} < \sin(2x) = 2\sin x \cos x \leq 2 \cdot 1 \cdot \frac{5}{13} = \frac{10}{13}.$$

This is a contradiction. Hence $\cos x = 0$ must have a solution for some $x > 0$. □

Let $\frac{p}{2} := \inf\{x > 0 : \cos x = 0\}$. By continuity of $\cos x$, we must have $\cos \frac{p}{2} = 0$ and $p > 0$.[3] Moreover, $\sin \frac{p}{2} \in \{\pm 1\}$. We may rule out $\sin \frac{p}{2} = -1$ as follows: Since $\frac{d}{dx} \sin x = \cos x > 0$, for any $0 \le x < \frac{p}{2}$ and any $h > 0$ we have
$$\frac{|\sin(x+h) - \sin x - h \cos x|}{h} \to 0, \quad \text{as} \quad h \to 0^+.$$
Thus, there is an $H > 0$ so that for $0 < h < H$ we have
$$\frac{|\sin(x+h) - \sin x - h \cos x|}{h} < \frac{\cos x}{2}.$$
Thus $\sin(x+h) > \sin x + \frac{h}{2} \cos x$. So for $x \in (0, \frac{p}{2})$, $y = \sin x$ is strictly increasing. Thus we conclude $\sin \frac{p}{2} = 1$.

Theorem 5.1.17 *The following identities hold for $\sin x$ and $\cos x$*

$$\sin(x + \frac{p}{2}) = \cos x, \quad \cos(x + \frac{p}{2}) = -\sin x, \tag{5.22}$$

$$\sin(x + p) = -\sin x, \quad \cos(x + p) = -\cos x, \tag{5.23}$$

$$\sin(x + 2p) = \sin x, \quad \cos(x + 2p) = \cos x. \tag{5.24}$$

Proof: We will now use (5.16) to establish these results.
$$\sin(x + \frac{p}{2}) = \sin x \cos \frac{p}{2} + \cos x \sin \frac{p}{2} = \cos x,$$
$$\cos(x + \frac{p}{2}) = \cos x \cos \frac{p}{2} - \sin x \sin \frac{p}{2} = -\sin x.$$
Moreover,
$$\sin(x + p) = \cos(x + \frac{p}{2}) = -\sin x,$$
$$\cos(x + p) = -\sin(x + \frac{p}{2}) = -\cos x.$$
Thus,
$$\sin(x + 2p) = -\sin(x + p) = -(-\sin x) = \sin x,$$

---
[3]Later we will be able to conclude that $p = \pi = 3.14159\cdots$.

5.1. DEFINITIONS AND BASIC DIFFERENTIABILITY RESULTS

$$\cos(x + 2p) = -\cos(x + p) = -(-\cos x) = \cos x.$$

◻

Theorem 5.1.18 *The number p equals π.*

Proof: To show this we will first show that the for any $\theta, \phi \in \mathbb{R}$ we have for the Euclidean distance $d(\cdot, \cdot)$

$$d((\cos(\theta + \phi), \sin(\theta + \phi)), (\cos \theta, \sin \theta)) = d((\cos \phi, \sin \phi), (1, 0)). \quad (5.25)$$

This holds since

$$\sqrt{(\cos(\theta + \phi) - \cos \theta)^2 + (\sin(\theta + \phi) - \sin \theta)^2}$$
$$= \sqrt{(\cos \theta \cos \phi - \sin \theta \sin \phi - \cos \theta)^2 + (\sin \theta \cos \phi + \sin \phi \cos \theta - \sin \theta)^2}$$
$$= \sqrt{[\cos \theta (\cos \phi - 1) - \sin \theta \sin \phi]^2 + [\sin \theta (\cos \phi - 1) + \sin \phi \cos \theta]^2}$$
$$= (\cos^2 \theta (\cos \phi - 1)^2 - 2 \cos \theta \sin \theta \sin \phi (\cos \phi - 1) + \sin^2 \theta \sin^2 \phi$$
$$+ \sin^2 \theta (\cos \phi - 1)^2 + 2 \cos \theta \sin \theta \sin \phi (\cos \phi - 1) + \cos^2 \theta \sin^2 \phi)^{\frac{1}{2}}$$
$$= \sqrt{(\cos \phi - 1)^2 + \sin^2 \phi}.$$

Partition the interval $[0, 2p]$ into n subintervals

$$[0, \frac{2p}{n}], [\frac{2p}{n}, 2\frac{2p}{n}], [2\frac{2p}{n}, 3\frac{2p}{n}], \cdots, [(n-1)\frac{2p}{n}, n\frac{2p}{n}].$$

Consider the points on the unit circle $\{(x, y) : x^2 + y^2 = 1\}$ which are of the form $(\cos \theta, \sin \theta)$ where $\theta \in \{0, \frac{2p}{n}, 2\frac{2p}{n}, \cdots, n\frac{2p}{n}\}$. For any two points of this type $(\cos(i-1)\frac{2p}{n}, \sin(i-1)\frac{2p}{n})$ and $(\cos i\frac{2p}{n}, \sin i\frac{2p}{n})$, for $i = 1, 2, \cdots, n$, we have that the distance between such points is the same as the distance between the points $(\cos(-\frac{p}{n}), \sin(-\frac{p}{n}))$ and $(\cos \frac{p}{n}, \sin \frac{p}{n})$, which is $2 \sin \frac{p}{n}$. Thus an approximation to the circumference of the unit circle is given by

$$L_n = \sum_{i=1}^{n} d((\cos(i-1)\frac{2p}{n}, \sin(i-1)\frac{2p}{n}), (\cos i\frac{2p}{n}, \sin i\frac{2p}{n}))$$

$$= \sum_{i=1}^{n} 2 \sin \frac{p}{n}$$

$$= 2n \sin \frac{p}{n}$$

$$= 2p \frac{\sin \frac{p}{n}}{\frac{p}{n}}.$$

Letting $n \to \infty$, $L_n \to 2\pi$. However, we also have

$$\lim_{n\to\infty} L_n = \lim_{n\to\infty} 2p \frac{\sin \frac{p}{n}}{\frac{p}{n}}$$

$$= 2p.$$

Thus we conclude $2p = 2\pi$, and so $p = \pi = 3.1415926535897 \cdots$.
□

Theorem 5.1.19 *If* $\sin(x+u) = \sin x$ *and* $\cos(x+u) = \cos x$, *then* $u = 2k\pi$ *for some* $k \in \mathbb{Z}$.

Proof: We will show this by establishing that there is no $u \in (0, 2\pi)$ which satisfies this property. By (5.16) we have

$$\sin(x+u) = \sin x \cos u + \cos x \sin u = \sin x,$$

$$\cos(x+u) = \cos x \cos u - \sin x \sin u = \cos x.$$

Multiplying the first equation by $\sin x$ and the second by $\cos x$ and adding we get $\cos u = 1$. Thus $\sin x + \cos x \sin u = \sin x$ and so $\cos x \sin u = 0$. Since this must hold for every x, we must have $\sin u = 0$.

We showed above that on the interval $(0, \frac{\pi}{2})$ $\sin x$ is strictly increasing. Since $\cos x$ and $\sin x$ are positive on this interval, and $\sin^2 x + \cos^2 x = 1$, we may conclude that $\cos x$ is strictly decreasing on this interval. From this fact, and from theorem (5.1.17), we may conclude the following:

1. On the interval $(\frac{\pi}{2}, \pi)$ we have $\cos x$ is negative and $\sin x$ is positive. Moreover, on this interval $\sin x$ is strictly decreasing and $\cos x$ is strictly decreasing.

2. On the interval $(\pi, \frac{3\pi}{2})$ we have $\cos x$ is negative and $\sin x$ is negative. Moreover, on this interval $\sin x$ is strictly decreasing and $\cos x$ is strictly increasing.

5.1. DEFINITIONS AND BASIC DIFFERENTIABILITY RESULTS 189

3. On the interval $(\frac{3\pi}{2}, 2\pi)$ we have $\cos x$ is positive and $\sin x$ is negative. Moreover, on this interval $\sin x$ is strictly increasing and $\cos x$ is strictly increasing.

4. $\cos 0 = \cos 2\pi = 1$, $\sin 0 = \sin 2\pi = 0$, $\cos \frac{\pi}{2} = 0$, $\sin \frac{\pi}{2} = 1$, $\cos \pi = -1$, $\sin \pi = 0$, $\cos \frac{3\pi}{2} = 0$, and $\sin \frac{3\pi}{2} = -1$.

From these observations we see that there is no solution $u \in (0, 2\pi)$. The result then follows.
□

Corollary 5.1.20 *The domain of $y = \tan x$ is $\{x : x \neq \frac{\pi}{2} + k\pi, \; k \in \mathbb{Z}\}$. On the domain of tangent, if $\tan(x + u) = \tan x$ for all x, then $u = k\pi$, for some $k \in \mathbb{Z}$.*

At this point we compute the values of the trigonometric functions for the values $x = \frac{\pi}{6}, \frac{\pi}{4}$ and $\frac{\pi}{3}$, and to do this we shall use equation (5.16).

$$0 = \cos\frac{\pi}{2} = \cos(\frac{\pi}{4} + \frac{\pi}{4}) = \cos^2\frac{\pi}{4} - \sin^2\frac{\pi}{4}.$$

Using this and $\sin^2\frac{\pi}{4} + \cos^2\frac{\pi}{4} = 1$ we see that $2\cos^2\frac{\pi}{4} = 1$. Hence

$$\cos\frac{\pi}{4} = \sin\frac{\pi}{4} = \frac{1}{\sqrt{2}}.$$

Also,

$$0 = \cos\frac{\pi}{2} = \cos(\frac{\pi}{3} + \frac{\pi}{6}) = \cos\frac{\pi}{3}\cos\frac{\pi}{6} - \sin\frac{\pi}{3}\sin\frac{\pi}{6}$$

$$= (\cos^2\frac{\pi}{6} - \sin^2\frac{\pi}{6})\cos\frac{\pi}{6} - 2\sin^2\frac{\pi}{6}\cos\frac{\pi}{6}$$

$$= \cos\frac{\pi}{6}(\cos^2\frac{\pi}{6} - 3\sin^2\frac{\pi}{6})$$

$$= \cos\frac{\pi}{6}(4\cos^2\frac{\pi}{6} - 3).$$

Since $\cos\frac{\pi}{6} \neq 0$ we must have $4\cos^2\frac{\pi}{6} - 3 = 0$, and so $\cos\frac{\pi}{6} = \frac{\sqrt{3}}{2}$. Using the Pythagorean identity we see that $\sin\frac{\pi}{6} = \frac{1}{2}$.

Also,

$$\sin\frac{\pi}{3} = 2\sin\frac{\pi}{6}\cos\frac{\pi}{6} = 2 \cdot \frac{\sqrt{3}}{2} \cdot \frac{1}{2} = \frac{\sqrt{3}}{2},$$

and so we see that $\cos \frac{\pi}{3} = \frac{1}{2}$.

We define the **inverse trigonometric functions** as follows: $y = \sin^{-1} x$ or $y = \arcsin x$ is the inverse function of $y = \sin x$ when restricted to $[-\frac{\pi}{2}, \frac{\pi}{2}]$, where $\sin x$ is strictly increasing. $y = \cos^{-1} x$ or $y = \arccos x$ is the inverse function of $y = \cos x$ when restricted to $[0, \pi]$, where $\cos x$ is strictly decreasing. $y = \tan^{-1} x$ or $y = \arctan x$ is the inverse function of $y = \tan x$ when restricted to $(-\frac{\pi}{2}, \frac{\pi}{2})$, where $\tan x$ is strictly increasing.

5.1.6 Homework Exercises

1. Show that
$$f(x) = \begin{cases} \sin \frac{1}{x} & x \neq 0, \\ 0 & x = 0 \end{cases}$$
is continuous everywhere except $x = 0$. Moreover, for $x \neq 0$, $f'(x)$ exists.

2. Show that
$$f(x) = \begin{cases} x^2 \sin \frac{1}{x} & x \neq 0, \\ 0 & x = 0 \end{cases}$$
is differentiable everywhere, but $f'(x)$ is discontinuous at $x = 0$.

3. Show that
$$f(x) = \begin{cases} x \sin \frac{1}{x} & x \neq 0, \\ 0 & x = 0 \end{cases}$$
is continuous everywhere and differentiable differentiable everywhere except $x = 0$.

4. Suppose that for some $\alpha > 1$, f satisfies
$$|f(x) - f(y)| \leq |x - y|^\alpha$$
for any $x, y \in \mathbb{R}$. Prove $f'(x)$ exists everywhere and $f'(x) \equiv 0$.

5. Suppose f satisfies
$$f(x + y) = f(x) f(y)$$
for all $x, y \in \mathbb{R}$ and $f'(0)$ exists, show that $f'(x)$ exists for all $x \in \mathbb{R}$. Moreover, establish that
$$f'(x) = f(x) f'(0).$$

6. Prove the following **half-angle formulas** for sine and cosine

$$\cos\frac{x}{2} = \pm\sqrt{\frac{1+\cos x}{2}}, \quad \sin\frac{x}{2} = \pm\sqrt{\frac{1-\cos x}{2}},$$

where the $\pm$ sign is chosen appropriately based on the value of $\frac{x}{2}$.

5.2 Extrema for Functions

Suppose $f : D \to \mathbb{R}$. We say that f has a **global** or **absolute maximum** at $c \in D$ if $f(c) = \sup_{x \in D} f(x)$. Similarly, we say that f has a **global** or **absolute minimum** at $c \in D$ if $f(c) = \inf_{x \in D} f(x)$. If f has either a global maximum or global minimum at $x = c$ we say that f has an **absolute extremum** at $x = c$.

We say that f has a **local** or **relative maximum** at $c \in D$ if there is an open $U \subseteq D$, with $c \in U$, so that $f|_U$ has an absolute maximum at c. Likewise, we say that f has a **local** or **relative minimum** at $c \in D$ if there is an open $U \subseteq D$, with $c \in U$, so that $f|_U$ has an absolute minimum at c. Likewise, if f has a local maximum or local minimum at $x = c$, f is said to have a **relative extremum** at $x = c$.

If $c \in D$ satisfies either $f'(c) = 0$ or $f'(c)$ does not exist, then $x = c$ is called a **critical value** of f.

The following theorem tells us that a relative extremum must occur at a critical value for f.

Critical Point Theorem 5.2.1 *If $f : D \to \mathbb{R}$ has a relative extremum at $c \in D$, then c is a critical value for f.*

Proof: Suppose $f'(c)$ exists. By otherwise considering $y = -f(x)$, we may assume f has a relative maximum at $x = c$. Since $f'(c)$ exists, we know $\lim_{h \to 0^-} \frac{f(c+h) - f(c)}{h}$ exists and equals $f'(c)$. However, for h sufficiently small and negative we have $f(c+h) \leq f(c)$, and so

$$\frac{f(c+h) - f(c)}{h} \geq 0.$$

Thus we see $f'(c) \geq 0$. Likewise, $\lim_{h\to 0^+} \frac{f(c+h)-f(c)}{h} = f'(c)$. So for h sufficiently small and positive, we have $f(c+h) \leq f(c)$. Thus $\frac{f(c+h)-f(c)}{h} \leq 0$, and so $f'(c) \leq 0$. Putting these together we see that $f'(c) = 0$.
□

Corollary 5.2.2 *If $f : D \to \mathbb{R}$ has a global extremum at $x = c$, then either $c \in \partial D$, or c is a critical value.*

Proof: If $c \notin \partial D$, then $c \in D^\circ$ since $\partial D = \overline{D} \setminus D^\circ$. If $c \in D^\circ$, then f has a relative extremum at c. Using theorem (5.2.1) we get the desired result.
□

5.3 The Mean Value Theorem

Mean Value Theorem 5.3.1 *Let $f : [a,b] \to \mathbb{R}$ be continuous[4] on $[a,b]$ and differentiable on (a,b), then there is a $c \in (a,b)$ so that*

$$f'(c) = \frac{f(b) - f(a)}{b - a}.$$

Proof: We will first prove the result in the special case where $f(a) = f(b)$. This special case is usually referred to as Rolle's theorem. Since f is continuous on $[a,b]$ we know by theorem (4.2.12) that $f([a,b])$ is compact, and thus f attains it maximum and minumum on $[a,b]$. Unless f is constant on $[a,b]$, in which case every $c \in (a,b)$ satisfies $f'(c) = 0$, at least one of the maximum or minimum values must be attained in (a,b). Let c be the location of an absolute extremum for f in (a,b). At this point f also has a relative extremum, and thus f has a critical value at c. Since $f'(c)$ exists we must have $f'(c) = 0$ must hold.

In the case where $f(a) \neq f(b)$ we consider the function

$$g(x) = f(x) - \frac{f(b) - f(a)}{b - a}(x - a).$$

---
[4]Continuity at a and b means $\lim_{x \to a^+} f(x) = f(a)$ and $\lim_{x \to b^-} f(x) = f(b)$.

5.3. THE MEAN VALUE THEOREM

Clearly $g(a) = f(a)$ and

$$g(b) = f(b) - \frac{f(b) - f(a)}{b-a}(b-a) = f(b) - (f(b) - f(a)) = f(a).$$

Moreover, g is continuous on $[a,b]$ since f is continuous and $y = \frac{f(b)-f(a)}{b-a}(x-a)$ is linear. Also, $g'(x) = f'(x) - \frac{f(b)-f(a)}{b-a}$ on (a,b). We see that there is a $c \in (a,b)$ so that $g'(c) = 0$. At such a c we have $0 = f'(c) - \frac{f(b)-f(a)}{b-a}$.
□

Corollary 5.3.2 *Suppose $f : [a,b] \to \mathbb{R}$ is continuous on $[a,b]$ and differentiable on (a,b). Suppose $f'(x) \equiv 0$ on (a,b), then $f(x)$ is a constant function on $[a,b]$.*

Proof: Let $a \le \alpha < \beta \le b$. By applying theorem (5.3.1) to f on $[\alpha, \beta]$ we conclude that there is a $c \in (\alpha, \beta)$ so that $f'(c) = \frac{f(\beta)-f(\alpha)}{\beta-\alpha}$. However $f'(c) = 0$, and so we get $f(\beta) = f(\alpha)$. Since $\alpha < \beta$ were arbitrary, we get the desired result.
□

Corollary 5.3.3 *Suppose $f : [a,b] \to \mathbb{R}$ is continuous on $[a,b]$ and differentiable on (a,b). If $f'(x) > 0$ on (a,b), then $f(x)$ is strictly increasing on $[a,b]$. If $f'(x) < 0$ on (a,b), then f is strictly decreasing on $[a,b]$.*

Proof: Consider $a \le \alpha < \beta \le b$. By the Mean Value Theorem (5.3.1) there is a $c \in (\alpha, \beta)$ so that $\frac{f(\beta)-f(\alpha)}{\beta-\alpha} = f'(c)$. If $f' > 0$ on (a,b), then $f'(c) > 0$, and so we may conclude that $f(\beta) > f(\alpha)$. If $f' < 0$, then we may conclude that $f'(c) < 0$, and so $f(\beta) < f(\alpha)$.
□

Corollary: Generalized Mean Value Theorem 5.3.4 *Suppose f and g are continuous on $[a,b]$ and differentiable on (a,b), with $g(b) \ne g(a)$. Then there is a $c \in (a,b)$ so that*

$$\frac{f(b) - f(a)}{g(b) - g(a)} = \frac{f'(c)}{g'(c)}.$$

Proof: We consider the function $k(x) = (f(b) - f(a))g(x) - (g(b) - g(a))f(x)$, which is continuous on $[a, b]$, since it is a linear combination of continuous functions. Moreover, $k(x)$ is differentiable on (a, b) and $k'(x) = (f(b) - f(a))g'(x) - (g(b) - g(a))f'(x)$. We then apply theorem (5.3.1) to get the existence of $c \in (a, b)$ so that $k'(c) = \frac{k(b)-k(a)}{b-a}$. However, $k(b) - k(a) = 0$, and so $k'(c) = 0$. This gives the desired result.
□

5.3.1 The Inverse Function Theorem

Consider an injective real-valued function of a real variable $f : D \to R$. If f is differentiable at x_0, then we know that f may be approximated by a linear function $y = f(x_0) + f'(x_0) \cdot (x - x_0)$ around x_0. In particular, we call this linear function the **tangent line** to the graph of $y = f(x)$ at the point $(x_0, f(x_0))$. The inverse function $y = f^{-1}(x)$'s graph is obtained from the graph of $y = f(x)$ by interchanging the x and y coordinates. Thus we suspect that there should be a tangent line to $y = f^{-1}(x)$ at the point $(f(x_0), f^{-1}(f(x_0))) = (f(x_0), x_0)$, and the slope of this line should be $\frac{1}{f'(x_0)}$. This is indeed true if we make certain assumptions about $f'(x)$. This result is often called the **inverse function theorem**.

The Inverse Function Theorem 5.3.5 *Let f be an injection which is be differentiable on an interval (a, b) with a continuous derivative on this interval. Suppose that $f'(x) \neq 0$ on any (a, b). Then for any $x_0 \in (a, b)$ the inverse function f^{-1} is differentiable at $y_0 = f(x_0)$ and*

$$(f^{-1})'(y_0) = \frac{1}{f'(x_0)}.$$

Proof: Let $[\alpha, \beta] \subseteq (a, b)$, and let $[A, B] := f([\alpha, \beta])$. We consider f^{-1} on $[A, B]$. Let $y = f(x)$ and $y_0 = f(x_0)$ be in $[A, B]$ and consider $\frac{f^{-1}(y)-f^{-1}(y_0)}{y-y_0}$. Since $f'(x_0)$ exists at $x_0 \in [\alpha, \beta]$, we have $\lim_{x \to x_0} \frac{f(x)-f(x_0)}{x-x_0} = f'(x_0) \neq 0$, and so

$$\lim_{x \to x_0} \frac{x - x_0}{f(x) - f(x_0)} = \frac{1}{f'(x_0)}.$$

Hence given $\varepsilon > 0$, there is a $\delta > 0$ so that $|x - x_0| < \delta$ implies

$$\left| \frac{x - x_0}{y - y_0} - \frac{1}{f'(x_0)} \right| < \varepsilon, \quad \text{where} \quad y = f(x), y_0 = f(x_0).$$

5.3. THE MEAN VALUE THEOREM

To prove the theorem, since $f^{-1}(y) = x$ and $f^{-1}(y_0) = x_0$, we need to show that there is an $\eta > 0$ so that $0 < |y - y_0| < \eta$ implies $0 < |x - x_0| < \delta$. This follows if we know that f^{-1} is continuous at y_0. This fact we shall now prove.

Since f' is continuous and never zero on $[\alpha, \beta]$, we know that $|f'(x)| > \varepsilon_1 > 0$ on $[\alpha, \beta]$. The Mean Value Theorem tells us that $\left|\frac{f(x) - f(x_0)}{x - x_0}\right| = |f'(x_1)| > \varepsilon_1$, where x_1 is strictly between x_0 and x. Thus we see that

$$|f(x) - f(x_0)| > \varepsilon_1 |x - x_0|.$$

For $|y - y_0| < \varepsilon_1 \cdot \delta$, we have

$$\varepsilon_1 |x - x_0| < |f(x) - f(x_0)| = |y - y_0| \leq \varepsilon_1 \cdot \delta,$$

and so $|x - x_0| < \delta$. Hence the result follows.
□

Theorem 5.3.6 $\log(x)$ is differentiable for all $x > 0$, and

$$\frac{d}{dx}\log(x) = \frac{1}{x}.$$

Proof: $y = \log(x)$ is the inverse function for $y = e^x$. Moreover, $\frac{d}{dx}(e^x) = e^x > 0$ for all $x \in \mathbb{R}$. Hence $\log(x)$ is differentiable on $(0, \infty)$ by theorem (5.3.5), and for $x = e^y$ we have $\frac{d}{dx}\log(x) = \frac{1}{e^y} = \frac{1}{x}$.
□

Theorem (The Power Rule) 5.3.7 *The function* $f : (0, \infty) \to \mathbb{R}$ *given by*

$$f(x) = x^n,$$

with $n \in \mathbb{R}$ *any real number, is differentiable and*

$$\frac{d}{dx}(x^n) = nx^{n-1}.$$

Proof: We may write $f(x)$ as

$$f(x) = x^n = e^{n\log(x)}.$$

Hence, using theorems (5.1.9) and (5.3.6) we see

$$\frac{d}{dx}(x^n) = \frac{d}{dx}(e^{n\log(x)})$$

$$= e^{n\log(x)} \cdot \frac{d}{dx}(n\log(x))$$

$$= x^n \cdot \frac{n}{x}$$

$$= nx^{n-1}.$$

□

Theorem 5.3.8 *The inverse trigonometric functions* $y = \arcsin x$, $y = \arccos x$ *and* $y = \arctan x$ *are differentiable on* $(-1,1)$, $(-1,1)$ *and* $\mathbb{R}$ *respectively. Moreover,*

$$\frac{d}{dx}\arcsin x = \frac{1}{\sqrt{1-x^2}}, \tag{5.26}$$

$$\frac{d}{dx}\arccos x = -\frac{1}{\sqrt{1-x^2}}, \tag{5.27}$$

and

$$\frac{d}{dx}\arctan x = \frac{1}{1+x^2}. \tag{5.28}$$

Proof: By theorem (5.3.5) we see that for $y = \arcsin x$ we have

$$\frac{d}{dx}\arcsin x = \frac{1}{\cos y} = \frac{1}{\sqrt{1-x^2}}.$$

For $y = \arccos x$, we have

$$\frac{d}{dx}\arccos x = \frac{1}{-\sin y} = -\frac{1}{\sqrt{1-x^2}}.$$

Also, for $y = \arctan x$ we have

$$\frac{d}{dx}\arctan x = \frac{1}{\frac{1}{\cos^2 y}} = \cos^2 y = \frac{1}{1+x^2}.$$

□

5.3.2 The Continuity of Derivatives and the Intermediate Value Property

Consider $f : D \to \mathbb{R}$, $D \subseteq \mathbb{R}$. If f is continuous on D then we say $f \in C^0(D)$, or simply f is C^0. If $f'(x)$ exists and is continuous on D we say $f \in C^1(D)$, or simply f is C^1.

If $f'(x)$ exists everywhere on D, it need not be continuous. To see this consider the function $f(x) = x^2 \sin \frac{1}{x}$ for $x \neq 0$, and $f(0) = 0$. One can show that $f'(x)$ exists everywhere, but $f'(x)$ is not continuous at 0. Moreover, the discontinuity at 0 is a discontinuity of the second kind. The following theorem tells us that for a function f, the derivative f' cannot have discontinuities of the first kind. This result is often called the **intermediate value theorem for derivatives**.

Intermediate Value Theorem for Derivatives 5.3.9 *Let f be differentiable on $[a,b]$. If $a \leq \alpha < \beta \leq b$, and $f'(\alpha) \neq f'(\beta)$, then for any γ whose value lies strictly between $f'(\alpha)$ and $f'(\beta)$, there is a $c \in (\alpha, \beta)$ so that $f'(c) = \gamma$.*

Proof: Without loss of generality we suppose $f'(\alpha) < f'(\beta)$, since otherwise we may consider $-f(x)$ instead. Consider the function $g(x) = f(x) - \gamma x$. Then $g'(x) = f'(x) - \gamma$, and thus $g'(\alpha) < 0$ and $g'(\beta) > 0$.

Since $g'(\alpha) < 0$, we may find a $\delta > 0$ so that for $x \in (\alpha, \alpha + \delta)$ we have $|\frac{g(x)-g(\alpha)}{x-\alpha} - g'(\alpha)| < \frac{1}{2}|g'(\alpha)|$. Thus for $x \in (\alpha, \alpha + \delta)$ we have $g(x) < g(\alpha)$. Similarly, there is a $\delta' > 0$ so that for $x \in (\beta - \delta', \beta)$ we have $|\frac{g(\beta)-g(x)}{\beta-x} - g'(\beta)| < \frac{1}{2}g'(\beta)$. Thus for $x \in (\beta - \delta', \beta)$ we have $g(x) < g(\beta)$. Thus we see that g attains its minimum for the interval $[\alpha, \beta]$ on the open interval (α, β). Let $x \in (\alpha, \beta)$ be such a point where the minimum of g is attained, then by theorem (5.2.2) we have $g'(x) = 0$, and so $f'(x) = \gamma$.
□

Corollary 5.3.10 *If f is differentiable on $[a,b]$, then f' cannot have any simple discontinuities on $[a,b]$.*

5.3.3 L'Hôpital's Rule

The next result is called **L'Hôpital's rule**. Is is very useful in computing limits of ratios of quantities which are either both going to 0 or $\pm\infty$.

Theorem (L'Hôpital's Rule) 5.3.11 *Suppose* $\lim_{x \to a} \frac{f'(x)}{g'(x)} = A$, *where* $A \in \mathbb{R}$. *If* $\lim_{x \to a} f(x) = \lim_{x \to a} g(x) = 0$, *or if* $\lim_{x \to a} g(x) = \pm\infty$, *then*

$$\lim_{x \to a} \frac{f(x)}{g(x)} = A.$$

Proof: Let $0 < \varepsilon < 1$ be given. Suppose that $\lim_{x \to a} f(x) = \lim_{x \to a} g(x) = 0$. Since $\lim_{x \to a} \frac{f'(x)}{g'(x)} = A$, there is a $\delta > 0$ so that $|x - a| < \delta$ implies $|\frac{f'(x)}{g'(x)} - A| < \varepsilon$. First we consider $a - \delta < x < a$, and let y satisfy $a - \delta < x < y < a$. By theorem (5.3.4) there is a $t \in (x, y)$ so that $\frac{f(y) - f(x)}{g(y) - g(x)} = \frac{f'(t)}{g'(t)}$. Thus

$$A - \varepsilon < \frac{f(y) - f(x)}{g(y) - g(x)} < A + \varepsilon.$$

Letting $y \to a^-$, we get

$$A - \varepsilon \leq \frac{f(x)}{g(x)} \leq A + \varepsilon.$$

Thus we see $\lim_{x \to a^-} \frac{f(x)}{g(x)} = A$.

Next, let $a < x < a + \delta$, and consider z so that $a < z < x < a + \delta$. Then there is a $t \in (z, x)$ so that $\frac{f(x) - f(z)}{g(x) - g(z)} = \frac{f'(t)}{g'(t)}$. Thus

$$A - \varepsilon < \frac{f(x) - f(z)}{g(x) - g(z)} < A + \varepsilon.$$

Letting $z \to a^+$ we get

$$A - \varepsilon \leq \frac{f(x)}{g(x)} \leq A + \varepsilon.$$

Thus we see $\lim_{x \to a^+} \frac{f(x)}{g(x)} = A$.

5.3. THE MEAN VALUE THEOREM

Now consider the case where $g(x) \to \infty$ or $-\infty$. Since $g \to \infty$ or $-\infty$ we may suppose $g(z) \neq 0$ and $A - \frac{\varepsilon}{2} < \frac{f'(z)}{g'(z)} < A + \frac{\varepsilon}{2}$ for $z \in (a - \delta, a + \delta)$.

First suppose $a - \delta < x < a$ and let y satisfy $a - \delta < x < y < a$. Then there is a $t \in (x, y)$ so that $\frac{f(y) - f(x)}{g(y) - g(x)} = \frac{f'(t)}{g'(t)}$. Multiplying both sides by $\frac{g(y) - g(x)}{g(y)}$ we get $\frac{f(y)}{g(y)} - \frac{f(x)}{g(y)} = \frac{f'(t)}{g'(t)} - \frac{f'(t)}{g'(t)} \cdot \frac{g(x)}{g(y)}$, or simply

$$\frac{f(y)}{g(y)} = \frac{f'(t)}{g'(t)} - \frac{f'(t)}{g'(t)} \cdot \frac{g(x)}{g(y)} + \frac{f(x)}{g(y)}.$$

Thus we see that

$$A - \frac{\varepsilon}{2} - [|A| + 1]|\frac{g(x)}{g(y)}| - |\frac{f(x)}{g(y)}| < \frac{f(y)}{g(y)} < A + \frac{\varepsilon}{2} + [|A| + 1]|\frac{g(x)}{g(y)}| + |\frac{f(x)}{g(y)}|.$$

We then choose $\delta_1 < \delta$ so that for $y \in (a - \delta_1, a)$ we have $[|A| + 1]\frac{|g(x)|}{|g(y)|} + \frac{|f(x)|}{|g(y)|} < \frac{\varepsilon}{2}$. Thus for $y \in (a - \delta_1, a)$ we have

$$A - \varepsilon < \frac{f(y)}{g(y)} < A + \varepsilon.$$

Now we suppose $a < x < a + \delta$ and let y satisfy $a < z < x < a + \delta$. Then there is a $t \in (z, x)$ so that $\frac{f(x) - f(z)}{g(x) - g(z)} = \frac{f'(t)}{g'(t)}$. Multiplying both sides by $\frac{g(z) - g(x)}{g(z)}$ we get

$$\frac{f(z)}{g(z)} = \frac{f'(t)}{g'(t)} - \frac{f'(t)}{g'(t)} \cdot \frac{g(x)}{g(z)} + \frac{f(x)}{g(z)}.$$

Thus we see that

$$A - \frac{\varepsilon}{2} - [|A| + 1]|\frac{g(x)}{g(z)}| - |\frac{f(x)}{g(z)}| < \frac{f(z)}{g(z)} < A + \frac{\varepsilon}{2} + [|A| + 1]|\frac{g(x)}{g(z)}| + |\frac{f(x)}{g(z)}|.$$

We then choose $\delta_1 < \delta$ so that for $z \in (a, a + \delta_1)$ we have $[|A| + 1]\frac{|g(x)|}{|g(z)|} + \frac{|f(x)|}{|g(z)|} < \frac{\varepsilon}{2}$. Thus for $z \in (a, a + \delta_1)$ we have

$$A - \varepsilon < \frac{f(z)}{g(z)} < A + \varepsilon.$$

□

5.3.4 Homework Exercises

1. Suppose that
$$g(x) = \begin{cases} x & 0 \leq x \leq 1, \\ x-1 & 1 < x \leq 2 \end{cases}$$
Explain why g can not be the derivative of a function f which is differentiable on $[0, 2]$.

2. Suppose that f' is a constant function on $\mathbb{R}$, prove that f is a linear function. That is, $f(x) = mx + b$ for some $m, b \in \mathbb{R}$.

3. Suppose that for some $\alpha > 1$, f satisfies
$$|f(x) - f(y)| \leq |x - y|^\alpha$$
for any $x, y \in \mathbb{R}$. Prove that f is a constant function.

4. Let f be a differentiable function on $[a, b]$ and suppose $f(a) = f(b) = 0$. Show that $f'(x) = 0$ has a solution for some x strictly between a and b.

5. If f is continuous on $[a, b]$ and differentiable on (a, b) whose one-sided derivatives $f'(b^-) = \lim_{h \to 0^-} \frac{f(b+h)-f(b)}{h}$ and $f'(a^+) = \lim_{h \to 0^+} \frac{f(a+h)-f(a)}{h}$ exist. Suppose that f attains its global maximum and minimum at a and b respectively. What can you say about the signs of $f'(a^+)$ and $f'(b^-)$?

6. Let $f(x) = \sqrt[k]{x}$, for $k > 1$. Use the inverse function theorem to find $f'(x)$.

7. Suppose $f : [a, b] \to \mathbb{R}$ is continuous on $[a, b]$ and differentiable on (a, b). Suppose $f'(x)$ is bounded on (a, b). Show that f is Lipschitz continuous on $[a, b]$.

8. Suppose $f'(x) = g'(x)$ for all $x \in \mathbb{R}$, show that $f(x) = g(x) + C$ for some constant C.

9. Explain why the function $f(x) = |x|$ does not violate theorem (5.3.10).

10. Show that
$$\lim_{n \to \infty} (1 + \frac{x}{n})^n = e^x$$
for all $x \in \mathbb{R}$.

11. Calculate
$$\lim_{x \to 0^+} x^x.$$

12. Let $n \in \mathbb{N}$, show that
$$\lim_{x \to \infty} \frac{x^n}{e^x} = 0.$$

13. Let $\alpha > 0$, show that
$$\lim_{x \to 0^+} \frac{x^\alpha}{\log(x)} = 0.$$

5.4 Higher Order Derivatives

If f is differentiable in a neighbourhood of $a \in \mathbb{R}$, we may wish to see if $f'(x)$ is differentiable at a. If $\lim_{h \to 0} \frac{f'(a+h)-f'(a)}{h}$ exists, we denote it by either $f''(a)$, $\frac{d^2 f}{dx^2}(a)$ or $\frac{d^2}{dx^2} f(a)$ and this quantity is called the **second derivative of f at a**. Likewise if $f''(x)$ exists in a neighbourhood of a, we may wish to see if $\lim_{h \to 0} \frac{f''(a+h)-f''(a)}{h}$ exists. If so, we call it the **third derivative of f at $x = a$**, and we shall denote it by either $f'''(a)$, $f^{(3)}(a)$, $\frac{d^3 f}{dx^3}(a)$, or $\frac{d^3}{dx^3} f(a)$. Likewise we may define the fourth derivative, etc. In general, the nth derivative of f is the derivative of the $(n-1)^{st}$ derivative and it is denoted by $f^{(n)}$, $\frac{d^n f}{dx^n}$ or $\frac{d^n}{dx^n} f$.

Suppose that f has n derivatives on a set D, and $f^{(n)}$ is continuous on D, then we say that $f \in C^n(D)$, or simply f is C^n in D. If f is in $C^n(D)$, and the nth derivative is Hölder continuous for some $\alpha \in (0,1)$, we say $f \in C^{n,\alpha}(D)$. Recall that a function g is **Hölder continuous** for some $\alpha \in (0,1)$ if there is a $C > 0$ so that $|g(y) - g(x)| < C|y-x|^\alpha$ for all $x, y \in D$.

5.4.1 Homework Exercises

1. Suppose that $\frac{df}{dx}$ exists everywhere and is an nth degree polynomial. Show that f is a polynomial of degree $n + 1$.

2. Suppose f is n times differentiable everywhere, and $f^{(n)}(x) \equiv 0$. Show that f is a polynomial of degree no more than $n - 1$.

5.5 Monotonicity, Concavity and Convexity

5.5.1 Some Monotonicity Results

Recall that corollary (5.3.3) guarantees that if $f' > 0$ on (a, b), then f is strictly increasing on $[a, b]$, and if $f' < 0$ on (a, b), then f is strictly decreasing on $[a, b]$. These both follow directly from the Mean Value Theorem (5.3.1). The following theorem gives us a partial converse.

Theorem 5.5.1 *Suppose f' exists on (a, b). If f is monotone increasing on (a, b), then $f' \geq 0$ on (a, b). If f is monotone decreasing on (a, b), then $f' \leq 0$ on (a, b).*

Proof: First let's suppose f is monotone increasing on (a, b). Let $x \in (a, b)$ and let $h > 0$ be sufficiently small so $x + h \in (a, b)$, then $\frac{f(x+h)-f(x)}{h} \geq 0$. Thus $f'(x) = \lim_{h \to 0^+} \frac{f(x+h)-f(x)}{h} \geq 0$. If f is monotone decreasing, let $x \in (a, b)$ and let $h > 0$ be sufficiently small so that $x + h \in (a, b)$, then $\frac{f(x+h)-f(x)}{h} \leq 0$, and so $f'(x) = \lim_{h \to 0^+} \frac{f(x+h)-f(x)}{h} \leq 0$.
□

Note 5.5.2 *If the assumption f is monotone in the above theorem is replaced by f is strictly monotone, then we still may only conclude that f' is either non-negative or non-positive. To see this note that $f(x) = x^3$ is strictly monotone increasing on all of $\mathbb{R}$, and $f'(0) = 0$.*

5.5.2 The First and Second Derivative Tests

Recall that the Critical Point Theorem (5.2.1) which guarantees that if f has a relative extremum at $a \in \mathbb{R}$, then a is a critical value. Thus the set of critical values contains the set of places where f may attains relative extrema. The following result, often called the **first derivative test**, tells us if a critical value is indeed a place where f attains a relative extremum.

Theorem (First Derivative Test) 5.5.3 *Suppose a is a critical value of f, and suppose f is differentiable on an open interval I containing a, except possibly at a. If $f' < 0$ for $x < a$, $x \in I$, and $f' > 0$ for $x > a$, $x \in I$, then f has a relative minimum at a. Likewise if $f' > 0$ for $x < a$, $x \in I$, and $f' < 0$ for $x > a$, $x \in I$, then f has a relative maximum at a.*

5.5. MONOTONICITY, CONCAVITY AND CONVEXITY

Proof: Let $(a - \delta, a + \delta) \subseteq I$, for some $\delta > 0$. Consider the case where $f' > 0$ on $(a - \delta, a)$ and $f' < 0$ on $(a, a + \delta)$. Then f is strictly increasing on $[a - \delta, a]$, and so for $x \in [a - \delta, a]$ we have $f(x) < f(a)$. Moreover, f is strictly decreasing on $[a, a + \delta]$, and so for $x \in [a, a + \delta]$ we have $f(x) < f(a)$. Hence we see that at a f has a relative maximum.

Consider now the case where $f' < 0$ on $(a - \delta, a)$ and $f' > 0$ on $(a, a + \delta)$. Then f is strictly decreasing on $[a - \delta, a]$, and so for $x \in [a - \delta, a]$ we have $f(x) > f(a)$. Moreover, f is strictly increasing on $[a, a + \delta]$, and so for $x \in [a, a + \delta]$ we have $f(x) > f(a)$. Hence we see that at a f has a relative minimum.
□

The next result is often called the **second derivative test**, and it also tells us if a critical value is a place where f attains a relative extremum. However, if may only be used if the critical value c is of the form $f'(c) = 0$.

Theorem (Second Derivative Test) 5.5.4 *Suppose $f'(c) = 0$, and $f''(c)$ exists and is non-zero. If $f''(c) > 0$, then f attains a relative minimum at c. If $f''(c) < 0$, then f attains a relative maximum at c.*

Proof: Since $f''(c) \neq 0$, there is a $\delta > 0$ so that for $x \in (c - \delta, c + \delta)$ we have
$$\left| \frac{f'(x) - f'(c)}{x - c} - f''(c) \right| < \frac{|f''(c)|}{2},$$
and thus for $x \in (c - \delta, c + \delta)$
$$f''(c) - \frac{|f''(c)|}{2} < \frac{f'(x) - f'(c)}{x - c} < f''(c) + \frac{|f''(c)|}{2}.$$

If $f''(c) > 0$, then on $(c - \delta, c + \delta)$ we have
$$\frac{f'(x) - f'(c)}{x - c} > \frac{f''(c)}{2} > 0.$$

So for $x \in (c, c+\delta)$ we have $f'(x) > f'(c) = 0$, and thus f is strictly increasing on $(c, c + \delta)$. Likewise, for $x \in (c - \delta, c)$ we have $f'(x) < f'(c) = 0$, and so f is strictly decreasing on $(c - \delta, c)$. Thus, by the first derivative test (5.5.3), we know f attains a relative minimum at c.

Now suppose that $f''(c) < 0$. Then on $(c - \delta, c + \delta)$ we have
$$\frac{f'(x) - f'(c)}{x - c} < -\frac{|f''(c)|}{2} < 0.$$
Thus on $(c - \delta, c)$ we have $f'(x) > f'(c) = 0$. So f is strictly increasing on $(c - \delta, c)$. Moreover, on $(c, c + \delta)$ we have $f'(x) < f'(c) = 0$, and so f is strictly decreasing on $(c, c + \delta)$. Thus, by the first derivative test (5.5.3), we know f attains a relative maximum at c.
□

5.5.3 Concavity and Convexity

Given a function $f : D \to \mathbb{R}$, with $D \subseteq \mathbb{R}$, we say that f is **convex** on an interval $I \subseteq D$ if for any $a, b \in I$ with $a < b$, f satisfies

$$f(x) < \frac{f(b) - f(a)}{b - a}(x - a) + f(a), \quad x \in (a, b). \tag{5.29}$$

This condition says that on the interval (a, b) the graph of f lies below the line segment joining the points $(a, f(a))$ and $(b, f(b))$. Thus we may also express condition (5.29) as

$$f(x) < \frac{f(b) - f(a)}{b - a}(x - b) + f(b), \quad x \in (a, b). \tag{5.30}$$

Note as well that we may rearrange the terms in the inequality (5.29) and obtain an equivalent inequality

$$\frac{f(x) - f(a)}{x - a} < \frac{f(b) - f(a)}{b - a}, \quad x \in (a, b). \tag{5.31}$$

We say that f is **concave** on an interval $I \subseteq D$ if for any $a, b \in I$ with $a < b$, f satisfies

$$f(x) > \frac{f(b) - f(a)}{b - a}(x - a) + f(a), \quad x \in (a, b). \tag{5.32}$$

This condition says that on the interval (a, b) the graph of f lies above the line segment joining the points $(a, f(a))$ and $(b, f(b))$. Thus we may also express condition (5.32) as

$$f(x) > \frac{f(b) - f(a)}{b - a}(x - b) + f(b), \quad x \in (a, b). \tag{5.33}$$

5.5. MONOTONICITY, CONCAVITY AND CONVEXITY

Note that we may rearrange the terms in the inequality (5.32) and obtain an equivalent inequality

$$\frac{f(x) - f(a)}{x - a} > \frac{f(b) - f(a)}{b - a}, \quad x \in (a, b). \tag{5.34}$$

At this point we would like to relate convexity and concavity to the first and second derivatives of a function, when they exist. Noting that f is convex if and only if $-f$ is concave, anything we say about convex functions can easily be changed into a statement about concave functions, and vice versa.

Theorem 5.5.5 *Let f be convex on an open interval I, and $a \in I$. If f is differentiable at a, then in a neighbourhood of a we have $f(x) > f(a) + f'(a)(x - a)$ for $x \neq a$. Moreover, if f is differentiable at $a, b \in I$, and $a < b$, then $f'(a) < f'(b)$.*

Proof: Let $h > 0$ be so that $a, a + h \in I$. Consider $\eta \in (0, h)$. Then using equation (5.31) with the interval $[a, a + h]$, for $a < a + \eta < a + h$ we have

$$\frac{f(a + \eta) - f(a)}{\eta} < \frac{f(a + h) - f(a)}{h}.$$

So the quantity $\frac{f(a+h)-f(a)}{h}$ decreases as $h \to 0^+$. Hence if $f'(a)$ exists, it satisfies

$$f'(a) < \frac{f(a + h) - f(a)}{h}, \quad \forall h > 0.$$

So for $x > a$ we see that

$$f(x) > f(a) + f'(a)(x - a).$$

Let $h < 0$ be so that $a, a + h \in I$, and let $\eta \in (h, 0)$. Using (5.30) on $[a + h, a]$ with $x = a + \eta$, we get $\frac{f(a+\eta)-f(a)}{\eta} > \frac{f(a+h)-f(a)}{h}$. Thus we see that $\frac{f(a+h)-f(a)}{h}$ increases as $h \to 0^-$. So for any $h < 0$ we have

$$f'(a) > \frac{f(a + h) - f(a)}{h}.$$

Hence, for $x < a$ we have

$$f(x) > f(a) + f'(a)(x - a).$$

Now suppose $a, b \in I$ satisfy $a < b$, where $f'(a)$, $f'(b)$ both exist. Convexity gives us

$$f(x) < \frac{f(b) - f(a)}{b - a}(x - a) + f(a) = \frac{f(b) - f(a)}{b - a}(x - b) + f(b).$$

Letting $x = a + h \in (a, b)$ we get $f(a + h) < \frac{f(b)-f(a)}{b-a}h + f(a)$, and so

$$\frac{f(a + h) - f(a)}{h} < \frac{f(b) - f(a)}{b - a}.$$

Letting $h \to 0^+$ we get

$$f'(a) < \frac{f(b) - f(a)}{b - a}.$$

Likewise, for $b + h \in (a, b)$ we have

$$f(b + h) < \frac{f(b) - f(a)}{b - a}h + f(b),$$

and so

$$\frac{f(b + h) - f(b)}{h} > \frac{f(b) - f(a)}{b - a}.$$

Letting $h \to 0^-$ we get

$$f'(b) > \frac{f(b) - f(a)}{b - a}.$$

□

Corollary 5.5.6 *Let f be concave on an open interval I, and $a \in I$. If f is differentiable at a, then in a neighborhood of a we have $f(x) < f(a) + f'(a)(x - a)$ for $x \neq a$. Moreover, if f is differentiable at $a, b \in I$, and $a < b$, then $f'(a) > f'(b)$.*

Proof: We apply theorem (5.5.5) to $-f$.
□

If f is twice differentiable in an interval I, then we have the following result.

Corollary 5.5.7 *Suppose f is convex on an open interval I where f'' exists, then $f''(x) \geq 0$ on I. If f is concave on I and f'' exists on I, then $f''(x) \leq 0$ on I.*

5.5. MONOTONICITY, CONCAVITY AND CONVEXITY

Proof: This theorem follows from the results of theorem (5.5.5) and corollary (5.5.6).
□

We will now provide a converse to corollary (5.5.7).

Lemma 5.5.8 *Suppose f' is strictly increasing on an interval I. If $a, b \in I$ with $a < b$ and $f(a) = f(b)$, then for any $x \in (a, b)$, $f(x) < f(a)$.*

Proof: Suppose $f(x) \geq f(a)$ for some $x \in (a, b)$. Then the maximum of f on $[a, b]$ occurs at some point $c \in (a, b)$, with $f(c) \geq f(a)$. Moreover, $f'(c) = 0$. However, by the mean value theorem (5.3.1) applied to $[a, c]$, we know there is a $d \in (a, c)$ so that $f'(d) = \frac{f(c)-f(a)}{c-a} \geq 0$, contradicting that f' is strictly increasing.
□

Theorem 5.5.9 *If f' is strictly increasing on I, then f is convex on I. If f' is strictly decreasing on I, then f is concave on I.*

Proof: We will show f' strictly increasing implies f is convex. To get the second result, f' decreasing implies f concave, we apply the first result to $-f$.

Let $a < b$, with $a, b \in I$. Define $g(x) = f(x) - \frac{f(b)-f(a)}{b-a}(x-a)$. Then $g'(x) = f'(x) - \frac{f(b)-f(a)}{b-a}$, and so g' is strictly increasing as well. Moreover, $g(a) = g(b) = f(a)$, so we may apply lemma (5.5.8) to get $g(x) < f(a)$ for all $x \in (a, b)$. That is, $f(x) < \frac{f(b)-f(a)}{b-a}(x-a) + f(a)$ for all $x \in (a, b)$, and so f is convex.
□

Theorem 5.5.10 *If f is differentiable on an interval I, and the graph of f lies above each tangent line to the graph except at the point of contact, then f is convex.*

Proof: Let $a, b \in I$ with $a < b$. Let $g(x) = f'(a)(x-a) + f(a)$ and $h(x) = f'(b)(x-b) + f(b)$. Since the graph of f lies above the graph of g we have $f(b) > f'(a)(b-a) + f(a)$, and so $\frac{f(b)-f(a)}{b-a} > f'(a)$. Since the graph of f lies above the graph of h we get $f(a) > f'(b)(a-b) + f(b)$, and so $\frac{f(b)-f(a)}{b-a} < f'(b)$. Putting these together, we get $f'(a) < f'(b)$. Thus f' is strictly increasing, and so f is convex on I.
□

Corollary 5.5.11 *If f is differentiable on an interval I, and the graph of f lies below each tangent line to the graph except at the point of contact, then f is concave.*

Proof: Apply theorem (5.5.10) to $-f$.

□

5.5.4 Homework Exercises

1. Let $p_1, p_2, \cdots, p_n > 0$ be such that $\sum_{i=1}^{n} p_i = 1$. Let $x_1, x_2, \cdots, x_n \in I$. Prove what is commonly called **Jensen's Inequality**: f convex on I implies
$$f\left(\sum_{i=1}^{n} p_i x_i\right) < \sum_{i=1}^{n} p_i f(x_i).$$

2. Suppose f is convex on (a, b), show that f is continuous on (a, b).

3. Use the definition of convex to show $f(x) = x^2$ and $g(x) = |x|$ are convex functions (assuming we allow $\leq$ instead of $<$ in the definition of convexity).

4. Suppose f, g are convex, show that $f + g$ is as well.

5. Suppose f is convex, and $c \neq 0$ is a constant. Show that cf is convex if $c > 0$, and concave if $c < 0$.

6. Suppose f is convex on $\mathbb{R}$. Show that the set of all points for which f is not differentiable is a countable set.

5.6 Taylor's Theorem

Let $f : D \to \mathbb{R}$, with $D \subseteq \mathbb{R}$. If f is n times differentiable at $a \in D$, we may define the nth degree **Taylor polynomial** centred at a by

$$T_n(x) := f(a) + f'(a)(x-a) + \frac{f''(a)}{2!}(x-a)^2 + \cdots + \frac{f^{(n)}(a)}{n!}(x-a)^n. \quad (5.35)$$

Note that $T_1(x)$ is the tangent line to the graph of the function $f(x)$ at a, and in general $T_n(a) = f(a)$ for all n. Hence we may expect that $T_n(x)$

5.6. TAYLOR'S THEOREM

approximates $f(x)$ for x near a.

If $f^{(n)}(a)$ exists for all $n \in \mathbb{N}$, we may define the **Taylor series** of f centred at a by

$$T(x) := \sum_{n=0}^{\infty} \frac{f^{(n)}(a)}{n!}(x-a)^n, \qquad (5.36)$$

where $f^{(0)}(a) = f(a)$. Here we may inquire whether or not this series converges for values of x other than a, and if so whether or not $T(x) = f(x)$. The following theorem, often called **Taylor's Theorem**, addresses this inquiry. We note that if $n = 1$ in Taylor's Theorem, then we simply have the mean value theorem (5.3.1).

Taylor's Theorem 5.6.1 *Suppose $f : [a,b] \to \mathbb{R}$, $n \in \mathbb{N}$, $f \in C^{n-1}([a,b])$, and $f^{(n)}$ exists on all of (a,b). Let $c \in [a,b]$, and let $T_{n-1}(x)$ be the $(n-1)^{st}$ Taylor polynomial centred at c. Then for any $x \in [a,b]$, $x \neq c$, there is a $\xi \in (a,b)$ whose value lies strictly between c and x so that*

$$f(x) = T_{n-1}(x) + \frac{f^{(n)}(\xi)}{n!}(x-c)^n.$$

Proof: Let M be defined by $M = \frac{f(x) - T_{n-1}(x)}{(x-c)^n}$, when $x \neq c$, and define

$$g(t) = f(t) - T_{n-1}(t) - M(t-c)^n, \quad t \in [a,b].$$

Clearly g is n times differentiable on (a,b), and $g^{(n)}(t) = f^{(n)}(t) - n!M$. $g(c) = 0$ and $g(x) = 0$, and so by the mean value theorem (5.3.1), there is an x_1 strictly between c and x so that $g'(x_1) = 0$. Also $g'(c) = 0$, so by the mean value theorem, there is an x_2 strictly between c and x_1 so that $g''(x_2) = 0$. Observing that $g''(c) = g^{(3)}(c) = \cdots = g^{(n-1)}(c) = 0$, and continuing in this manner, using the mean value theorem we may find $x_3, x_4, \cdots, x_{n-1}$ strictly between c and x so that $g^{(3)}(x_3) = 0$, $g^{(4)}(x_4) = 0$, $\cdots$, $g^{(n-1)}(x_{n-1}) = 0$. Finally, since $g^{(n-1)}(c) = 0 = g^{(n-1)}(x_{n-1})$, we use the mean value theorem to obtain ξ strictly between c and x_{n-1} so that $g^{(n)}(\xi) = 0$.
□

We note that the term $\frac{f^{(n)}(\xi)}{n!}(x-c)^n$ in the above theorem shall be called the **Taylor remainder term**, and shall be denoted by $R_{n-1}(x)$. Thus we see $f(x) = T_{n-1}(x) + R_{n-1}(x)$.

5.6.1 Some Examples

1. Consider the function $f(x) = \log x$. We shall now compute the Taylor polynomials and the Taylor series centred at 1. Now $f'(x) = \frac{1}{x}$, $f''(x) = -\frac{1}{x^2}$, $f^{(3)}(x) = \frac{2}{x^3}$, $\cdots$, $f^{(n)}(x) = \frac{(-1)^{n-1}(n-1)!}{x^n}$. So

$$T_n(x) = \sum_{k=1}^{n} \frac{(-1)^{k-1}}{k}(x-1)^k.$$

The Taylor remainder term is $R_n(x) = \frac{(-1)^n}{(n+1)!}\frac{1}{\xi^{n+1}}(x-1)^{n+1}$, where ξ lies between 1 and x. For $1 < x \leq 2$ we see that this term clearly goes to 0 as $n \to \infty$, and so on $[1,2]$ we have

$$\log x = \sum_{n=1}^{\infty} \frac{(-1)^{n-1}}{n}(x-1)^n.$$

We may then use this to approximate $\log 2$ as follows:

$$\log 2 = \sum_{k=1}^{n} \frac{(-1)^{k-1}}{k} + \frac{(-1)^{n+1}}{n+1}\frac{1}{\xi^{n+1}}.$$

Clearly the remainder term is less than $\frac{1}{n+1}$ in absolute value. Using that $T(2)$ is an alternating series, we also see that $\log 2$ lies between $P_n(2)$ and $P_{n+1}(2)$. Using $n = 200$, we see that

$$P_{200}(2) = 0.69065343 < \log 2 < 0.695628555 = P_{201}(2).$$

2. Consider the function

$$f(x) = \begin{cases} e^{-\frac{1}{x^2}} & x \neq 0, \\ 0 & x = 0. \end{cases} \tag{5.37}$$

In this case, $f'(0) = \lim_{h \to 0} \frac{e^{-\frac{1}{h^2}}}{h} = \lim_{x \to \pm\infty} \frac{x}{e^{x^2}} = 0$ by L'Hôpital's Rule. Similarly, one can show $f^{(n)}(0) = 0$. Thus the nth order Taylor polynomial for f centred at 0 is identically zero. Moreover, the same is true for the Taylor series centred at 0. However, for any $x \neq 0$ we see that $f(x) > 0$. Thus the Taylor series centred at 0 only agrees with $f(x)$ at $x = 0$, even though it converges for all x,

5.7. NEWTON'S METHOD

A function whose values agree with the Taylor series centred at c on a given interval I is said to be **analytic** about c on I. In the above example, we see that both functions are n times differential for every value of x, and any $n \in \mathbb{N}$. However, the first example involving $\log x$ has $\log x$ analytic about 1 on $[1, 2]$, whereas the second function $f(x)$ is not analytic about 0 anywhere other than $\{0\}$.

5.6.2 Homework Exercises

1. Compute the Taylor series for $f(x) = e^x$ centred at 0, and show that e^x is analytic about 0 everywhere.

2. Compute the Taylor series for $f(x) = \sin x$ centred at 0, and show that $\sin x$ is analytic about 0 everywhere.

3. Compute the Taylor series for $f(x) = \cos x$ centred at 0, and show that $\cos x$ is analytic about 0 everywhere.

4. Compute the Taylor series for $f(x) = \sin x$ centred at $\frac{\pi}{6}$, and show that $\sin x$ is analytic about $\frac{\pi}{6}$ everywhere.

5. Compute the Taylor series for $f(x) = \cos x$ centred at $\frac{\pi}{4}$, and show that $\cos x$ is analytic about $\frac{\pi}{4}$ everywhere.

6. Using both $n = 2$ and 3 use Taylor's Theorem to evaluate both $\sqrt{99}$ and $\sqrt{101}$, and estimate the error in the approximations by finding upper bounds for the Taylor remainder terms.

5.7 Newton's Method

In this section we will discuss numerical methods for finding roots of equations. The main method we will discuss is often called **Newton's method** or the **Newton-Raphson method**, named for the mathematician Sir Isaac Newton, who is credited with using a version of this method (circa 1669) to find solutions to polynomial equations. It was in 1740 that the mathematician Thomas Simpson described an iterative procedure that we now call Newton's method.

Before we discuss Newton's method in detail we shall consider another method of finding roots, which is often called the **bisection method**. This method uses the intermediate value theorem (4.2.14) to find solutions to equations $f(x) = 0$ on an interval $[a, b]$, when f is continuous and $f(a), f(b)$ have different signs. The following example illustrates this method.

Consider the equation
$$x^3 + x + 1 = 0.$$
We will use the bisection method on the interval $[-1, 1]$. Letting $f(x) = x^3 + x + 1$, we observe that f is continuous on all of $\mathbb{R}$, since it's a polynomial. Moreover, $f(-1) = -1$ and $f(1) = 3$, and so by the intermediate value theorem (4.2.14), we know there is an $\xi \in (-1, 1)$ so that $f(\xi) = 0$. Now $f(0) = 1$, and so we may conclude that there is a root in the interval $(-1, 0)$. Consider now $f(-\frac{1}{2}) = \frac{3}{8}$, and so by the intermediate value theorem there is a root in $(-1, -\frac{1}{2})$. $f(-\frac{3}{4}) = -\frac{11}{64}$, and so the intermediate value theorem tells us that there is a root in $(-\frac{3}{4}, -\frac{1}{2})$. We may continue in this manner to pinpoint the location of the root of $f(x) = 0$ with more accuracy.

Newton's method attempts to pinpoint roots of equations $f(x) = 0$ in an interval $[a, b]$ when f is differentiable on $[a, b]$. To show convergence of this method, we will actually require a bit more regularity, and this shall be discussed below. To use Newton's method we start with an initial guess $x_0 \in [a, b]$. We then compute the tangent line $y = f(x_0) + f'(x_0)(x - x_0)$ to the graph of $f(x)$ at x_0. The graphs of f and the tangent line are close if x is near x_0, and so if x_0 is close to a root of $f(x) = 0$, we may expect that the solution to
$$0 = f(x_0) + f'(x_0)(x - x_0)$$
is close a root of $f(x) = 0$. We denote the solution of $0 = f(x_0) + f'(x_0)(x - x_0)$ by x_1, and we see that
$$x_1 = x_0 - \frac{f(x_0)}{f'(x_0)},$$
assuming $f'(x_0) \neq 0$. In fact, we will usually assume $f'(x) \neq 0$ on $[a, b]$. Otherwise we must use a different method.

We then use x_1 as our initial guess, and compute the tangent line to f at

5.7. NEWTON'S METHOD

$(x_1, f(x_1))$. We then find its x-intercept, which we call x_2. This is given by

$$x_2 = x_1 - \frac{f(x_1)}{f'(x_1)}.$$

Continuing in this manner, upon having found $x_1, x_2, x_3, \cdots x_n$, we define x_{n+1} by

$$x_{n+1} = x_n - \frac{f(x_n)}{f'(x_n)}. \tag{5.38}$$

This procedure defines a sequence $\{x_n\}_{n=0}^{\infty}$. We desire that this sequence converges. If it does, then the limit value ξ should satisfy

$$\xi = \xi - \frac{f(\xi)}{\lim_{x \to \xi} f'(x)}.$$

Assuming $\lim_{x \to \xi} f'(x)$ exists and is not zero, we see that this will require that $f(\xi) = 0$.

Again we consider the equation $x^3 + x + 1 = 0$. In this case $f(x) = x^3 + x + 1$, and we use $[-1, 1]$ as our interval. Here we use $x_0 = 0$. Then x_{n+1} is given by

$$x_{n+1} = x_n - \frac{x_n^3 + x_n + 1}{3x_n^2 + 1}.$$

We see that $x_1 = -1$, $x_2 = -0.75$, $x_3 = -0.686046512$, $x_4 = -0.682339583$, $x_5 = -0.682327804, \cdots$. For this equation, Newton's method pinpoints the root much more quickly than the bisection method.

At this point we shall give sufficient conditions for Newton's method to converge, given any initial guess in $[a, b]$.

Theorem 5.7.1 *Let $f \in C'([a, b])$ with f'' existing on $[a, b]$. Suppose $f(a)$ and $f(b)$ are of different sign, and $|f'| > 0$ on $[a, b]$. Suppose as well that for some M, $0 < |f''| \leq M$ on $[a, b]$. Then for $x_0 \in [a, b]$ whose distance to the unique solution of $f(x) = 0$ is less than $\frac{2 \inf_{x \in [a,b]} |f'(x)|}{M}$, the sequence $\{x_n\}_{n=0}^{\infty}$ given by Newton's method has $x_n \to \xi \in [a, b]$, the unique solution of $f(x) = 0$ on $[a, b]$.*

Proof: Since f' is continuous and $|f'| > 0$ on $[a, b]$, we know f' is of constant sign on $[a, b]$, and there is a $\delta > 0$ so that $|f'| \geq \delta$ on $[a, b]$. Since

$f(a)$ and $f(b)$ are of different sign, there is a solution ξ to $f(x) = 0$. Moreover, since f is strictly monotone, we know this solution is unique.

Use Taylor's theorem to write

$$f(x) = f(x_n) + f'(x_n)(x - x_n) + \frac{1}{2}f''(c_n)(x - x_n)^2,$$

c_n strictly between x_n and x, for any $x \in [a,b]$ and any x_n obtained by Newton's method.

Using $x = \xi$ we see

$$0 = f(x_n) + f'(x_n)(\xi - x_n) + \frac{1}{2}f''(c_n)(\xi - x_n)^2.$$

Since $f' \neq 0$ on $[a,b]$, we may divide by $f'(x_n)$ to obtain

$$0 = \frac{f(x_n)}{f'(x_n)} + \xi - x_n + (\xi - x_n)^2 \frac{f''(c_n)}{2f'(x_n)}.$$

Using $\frac{f(x_n)}{f'(x_n)} = x_n - x_{n+1}$ we may rewrite this as

$$0 = \xi - x_{n+1} + (\xi - x_n)^2 \frac{f''(c_n)}{2f'(x_n)}$$

and so

$$|\xi - x_{n+1}| = (\xi - x_n)^2 |\frac{f''(c_n)}{2f'(x_n)}| \leq (\xi - x_n)^2 \frac{M}{2\delta}.$$

Now iterating this, we obtain

$$|\xi - x_n| \leq \frac{2\delta}{M}[\frac{M}{2\delta}|\xi - x_0|]^{2^n}.$$

If $|x_0 - \xi| < \frac{2\delta}{M}$, then clearly $x_n \to \xi$ as $n \to \infty$.
$\square$

5.7.1 Homework Exercises

1. Use Newton's method to approximate $\sqrt{a}$, for any $a > 0$. Do this by applying Newton's method for the equation

$$x^2 - a = 0.$$

5.7. NEWTON'S METHOD

Show that Newton's method yields
$$x_{n+1} = \frac{1}{2}\left(x_n + \frac{a}{x_n}\right).$$
Also, derive the following estimate for the error
$$\sqrt{a} - x_{n+1} = -\frac{1}{2x_n}(\sqrt{a} - x_n)^2.$$

2. Use Newton's method to find a root of $x^3 - 2x^2 - 5 = 0$ on $[1, 4]$.

3. Explain why Newton's method does not work for $x_0 = 1$ and the equation
$$x^3 - 3x + 6 = 0.$$

4. Show that
$$-2x^4 + 3x^2 + \frac{11}{8} = 0$$
has at least two real solutions. Then use $x_0 = 2$ to estimate the value of one of the roots using Newton's method. What happens if you use $x_0 = \frac{1}{2}$ instead? Please examine this situation.

Chapter 6

The Riemann Integral

In this chapter we shall discuss the Riemann integral of a function $f : I \to \mathbb{R}$ over the interval I. Upon definition of this quantity, the natural question to ask is what is the largest class of functions for which this quantity exists? This question was addressed by the mathematician Lebesgue who obtained what is commonly called Lebesgue's Integrability Criterion, which gives a necessary and sufficient condition for Riemann integrability. In this chapter we will first consider some sufficient conditions for the existence of the Riemann integral, and we will conclude this chapter with a discussion of Lebesgue's criterion.[1]

6.1 Riemann Sums and the Riemann Integral

Let $I = [a, b]$. A **partition** of I is a collection of points $x_0, x_1, x_2, \cdots, x_{n-1}, x_n$ so that
$$a = x_0 < x_1 < x_2 < \cdots < x_{n-1} < x_n = b.$$

A partition $\mathcal{P} = \{x_i\}_{i=0}^n$ of I gives rise to a collection of non-overlapping [2] subintervals
$$I_1 := [x_0, x_1], I_2 := [x_1, x_2], \cdots, I_j = [x_{j-1}, x_j], \cdots I_n = [x_{n-1}, x_n]$$
whose union is I.

---

[1] Which may be optional upon a first reading, or optional for an introductory course.
[2] except that they share endpoints, otherwise their interiors are pair-wise disjoint.

For a given partition $\mathcal{P}$ of I, we define the **norm** or **mesh** of $\mathcal{P}$ by

$$\|\mathcal{P}\| := \max\{x_1 - x_0, x_2 - x_1, \cdots, x_n - x_{n-1}\}. \tag{6.1}$$

Clearly $\|\mathcal{P}\|$ is the length of the largest subinterval I_j defined by the partition $\mathcal{P}$. Note that for a partition $\mathcal{P}$ we often use the notation

$$\Delta x_i := x_i - x_{i-1}, \quad i = 1, 2, \cdots, n.$$

For a given partition $\mathcal{P} = \{x_k\}_{k=0}^n$ of I and a collection of points $\{t_k\}_{k=1}^n$, $t_k \in I_k$, $k = 1, 2, \cdots, n$, the quantity

$$\sum_{k=1}^n f(t_k)\Delta x_k, \tag{6.2}$$

is often called a **Riemann sum** for f on I. The partition $\mathcal{P}$ along with a collection of points $\{t_k\}_{k=1}^n$ is often called a **marked partition** and shall be denoted by $\mathcal{P}^\dagger$. For convenience we shall use $S(f, [a,b], \mathcal{P}^\dagger)$ to represent $\sum_{k=1}^n f(t_k)\Delta x_k$.

If there exists a quantity $\int_a^b f(x)dx$ so that for any $\varepsilon > 0$, there is a $\delta > 0$ so that if $\mathcal{P}^\dagger$ is a marked partition of $[a,b]$ with $\|\mathcal{P}\| < \delta$, then

$$|S(f, [a,b], \mathcal{P}^\dagger) - \int_a^b f(x)dx| < \varepsilon, \tag{6.3}$$

we say that f is **Riemann integrable** on $[a,b]$. The quantity $\int_a^b f(x)dx$ is called the **Riemann integral** of f on $[a,b]$, or simply the integral of f on $[a,b]$.[3]

We often think of $\int_a^b f(x)dx$ as a limit of Riemann sums, even though this is not the type of limit we have considered before, since it is a limit in terms of the norm of the partition. However, we will often use the notation

$$\int_a^b f(x)dx = \lim_{\|\mathcal{P}\| \to 0} S(f, [a,b], \mathcal{P}^\dagger). \tag{6.4}$$

If $f : [a,b] \to [0, \infty)$, then we may see $S(f, [a,b], \mathcal{P}^\dagger)$ as an approximation for the area of the region bounded by $x = a$, $x = b$, $y = 0$ and $y = f(x)$, if such

---

[3]Sometimes we may use $\int_a^b f dx$ to represent $\int_a^b f(x)dx$.

6.1. RIEMANN SUMS AND THE RIEMANN INTEGRAL

an area exists. If $\int_a^b f(x)dx$ exists, then the area of this region is $\int_a^b f(x)dx$. However, we still have an issue a hand, and that is the uniqueness of the quantity $\int_a^b f(x)dx$.

Theorem 6.1.1 *If there is a a quantity that serves as $\int_a^b f(x)dx$, then it is unique.*

Proof: Suppose there are two quantities S and L that serve as limits of the Riemann sums. Then given any $\varepsilon > 0$ there is a $\delta_1 > 0$ and a $\delta_2 > 0$ so that for any marked partition $\mathcal{P}^\dagger$ that satisfies $\|\mathcal{P}\| < \delta_1$ we have

$$|L - S(f, [a, b], \mathcal{P}^\dagger)| < \frac{\varepsilon}{2},$$

and for $\|\mathcal{P}\| < \delta_2$ we have

$$|S - S(f, [a, b], \mathcal{P}^\dagger)| < \frac{\varepsilon}{2}.$$

Moreover, for $\|\mathcal{P}\|$ less than the minimum of δ_1 and δ_2, we have

$$|S - L| = |S - S(f, [a, b], \mathcal{P}^\dagger) + S(f, [a, b], \mathcal{P}^\dagger) - L|$$

$$\leq |S - S(f, [a, b], \mathcal{P}^\dagger)| + |S(f, [a, b], \mathcal{P}^\dagger) - L| < \frac{\varepsilon}{2} + \frac{\varepsilon}{2} = \varepsilon.$$

Since $\varepsilon > 0$ was arbitrary, we must have $S = L$.
□

6.1.1 The Existence of Riemann Integrable Functions

A natural question to ask at this point is: on an interval $[a, b]$ does there exist a Riemann integrable function on $[a, b]$? The answer is yes, and in lieu of the area discussion above we consider constant functions $f(x) = k$ on $[a, b]$.

Theorem 6.1.2 *If $f : [a, b] \to \mathbb{R}$ is a constant function, $f(x) = k$, then*

$$\int_a^b f(x)dx = \int_a^b k\,dx = k(b - a).$$

Proof: Let $\mathcal{P} = \{x_i\}_{i=0}^n$ be any partition of $[a,b]$, and $\{t_k\}_{k=1}^n$ be any collection of points so that $t_k \in [x_{k-1}, x_k]$, $k = 1, 2, \cdots, n$, which gives rise to a marked partition $\mathcal{P}^\dagger$. Then

$$S(f,[a,b],\mathcal{P}^\dagger) = \sum_{j=1}^n f(t_j)\Delta x_j = k\sum_{j=1}^n (x_{j-1} - x_j) = k(b-a).$$

□

Thus we see that the Riemann integral for a constant function $f(x) = k$, $k \geq 0$, on $[a,b]$ has its integral value equal to the area of the rectangle bounded by $x = a$, $x = b$, $y = 0$ and $f(x) = k$.

6.1.2 Linearity and Comparative Properties of the Integral

We now introduce a notational convention for Riemann integrability. If f is Riemann integrable on $[a,b]$, we shall use the notation $f \in \mathcal{R}([a,b])$, where $\mathcal{R}([a,b])$ is the collection of all Riemann integrable functions on $[a,b]$.

We will now discuss the linearity properties of the Riemann integral. They are given in the theorem below, along with a comparison result.

Theorem 6.1.3 *If $f, g \in \mathcal{R}([a,b])$, and $k \in \mathbb{R}$, then $f + g, kf \in \mathcal{R}([a,b])$. Moreover,*

$$\int_a^b f(x) + g(x)\,dx = \int_a^b f(x)\,dx + \int_a^b g(x)\,dx,$$

$$\int_a^b kf(x)\,dx = k\int_a^b f(x)\,dx.$$

Also, if $f(x) \leq g(x)$ on $[a,b]$, then

$$\int_a^b f(x)\,dx \leq \int_a^b g(x)\,dx.$$

Proof: Let $\varepsilon > 0$ be given. Then there exists a $\delta > 0$ so that for $\|\mathcal{P}\| < \delta$ and any $\{t_k\}_{k=1}^n$, $|\sum_{k=1}^n f(t_k)\Delta x_k - \int_a^b f(x)\,dx| < \frac{\varepsilon}{2}$ and $|\sum_{k=1}^n g(t_k)\Delta x_k -$

6.1. RIEMANN SUMS AND THE RIEMANN INTEGRAL

$\int_a^b g(x)dx| < \frac{\varepsilon}{2}$. Since $S(f+g, [a,b], \mathcal{P}^\dagger) = \sum_{k=1}^n (f(t_k) + g(t_k))\Delta x_k = \sum_{k=1}^n f(t_k)\Delta x_k + \sum_{k=1}^n g(t_k)\Delta x_k$, we have

$$|\sum_{k=1}^n (f(t_k) + g(t_k))\Delta x_k - \int_a^b f(x)dx - \int_a^b g(x)dx|$$

$$\leq |\sum_{k=1}^n f(t_k)\Delta x_k - \int_a^b f(x)dx| + |\sum_{k=1}^n g(t_k)\Delta x_k - \int_a^b g(x)dx|$$

$$< \frac{\varepsilon}{2} + \frac{\varepsilon}{2} = \varepsilon.$$

If $k = 0$, then $0 \cdot f(x) \equiv 0$, and $0 \cdot \int_a^b f(x)dx = 0$, and so by theorem (6.1.2), we get the desired result. Thus we consider $k \neq 0$. Choosing $\delta > 0$ so that for $\|\mathcal{P}\| < \delta$ and taking any collection $\{t_j\}_{j=1}^n$, $t_j \in [x_{j-1}, x_j]$, we have $|\sum_{j=1}^n f(t_j)\Delta x_j - \int_a^b f(x)dx| < \frac{\varepsilon}{|k|}$. Thus we have

$$|\sum_{j=1}^n kf(t_j)\Delta x_j - k\int_a^b f(x)dx| = |k||\sum_{j=1}^n f(t_j)\Delta x_j - \int_a^b f(x)dx| < |k| \cdot \frac{\varepsilon}{|k|} = \varepsilon.$$

Assume $f \leq g$ on $[a, b]$. For any partition $\mathcal{P}$ with $\|\mathcal{P}\| < \delta$, and for any $\{t_k\}_{k=1}^n$ which gives rise to the marked partition $\mathcal{P}^\dagger$, we have

$$|\sum_{k=1}^n f(t_k)\Delta x_k - \int_a^b f(x)dx| < \frac{\varepsilon}{2}, \quad |\sum_{k=1}^n g(t_k)\Delta x_k - \int_a^b g(x)dx| < \frac{\varepsilon}{2}.$$

Clearly $f(t_k) \leq g(t_k)$, and so $\sum_{k=1}^n f(t_k)\Delta x_k \leq \sum_{k=1}^n g(t_k)\Delta x_k$. Moreover,

$$\int_a^b f(x)dx - \frac{\varepsilon}{2} < \sum_{k=1}^n f(t_k)\Delta x_k \leq \sum_{k=1}^n g(t_k)\Delta x_k < \int_a^b g(x)dx + \frac{\varepsilon}{2}.$$

Thus

$$\int_a^b f(x)dx < \int_a^b g(x)dx + \varepsilon.$$

Since $\varepsilon > 0$ is arbitrary, we have

$$\int_a^b f(x)dx \leq \int_a^b g(x)dx.$$

□

Corollary 6.1.4 *If $f \geq 0$ on $[a, b]$, and $f \in \mathcal{R}([a, b])$, then*

$$\int_a^b f(x)dx \geq 0.$$

Corollary 6.1.5 *If f is bounded on $[a, b]$,[4] and $f \in \mathcal{R}([a, b])$, then*

$$\inf_{x \in [a,b]} f(x) \cdot (b-a) \leq \int_a^b f(x)dx \leq \sup_{x \in [a,b]} f(x) \cdot (b-a).$$

Corollary 6.1.6 *Suppose $f, |f| \in \mathcal{R}([a, b])$, then*

$$\left| \int_a^b f(x)dx \right| \leq \int_a^b |f(x)|dx.$$

6.1.3 A Cauchy Convergence Criterion for Integrability

As in the case of limits of sequences, we have a **Cauchy criterion** for convergence of Riemann sums to the Riemann integral.

Theorem (Cauchy Criterion for Integrability) 6.1.7 $f \in \mathcal{R}([a, b])$ *if and only if for any $\varepsilon > 0$, there is a $\delta > 0$ so that if $\mathcal{P} = \{x_i\}_{i=0}^n$ and $\mathcal{Q} = \{\xi_j\}_{j=0}^m$ are two partitions with $\|\mathcal{P}\|, \|\mathcal{Q}\| < \delta$ and $\{t_k\}_{k=1}^n$ and $\{\tau_j\}_{j=1}^m$ are any points with $t_k \in [x_{k-1}, x_k]$ for $k = 1, 2, \cdots, n$ and $\tau_j \in [\xi_{j-1}, \xi_j]$, for $j = 1, 2, \cdots, m$, which give rise to the marked partitions $\mathcal{P}^\dagger$ and $\mathcal{Q}^\dagger$ respectively, then we have*

$$|S(f, [a, b], \mathcal{P}^\dagger) - S(f, [a, b], \mathcal{Q}^\dagger)| < \varepsilon. \tag{6.5}$$

Proof: Suppose $f \in \mathcal{R}([a, b])$. Then for any $\varepsilon > 0$ there is a $\delta > 0$ so that if $\mathcal{P}$ is any partition satisfying $\|\mathcal{P}\| < \delta$, with $\{t_k\}_{k=1}^n$ satisfying $t_k \in [x_{k-1}, x_k]$, which gives rise to the marked partition $\mathcal{P}^\dagger$, we have

$$\left| S(f, [a, b], \mathcal{P}^\dagger) - \int_a^b f(x)dx \right| < \frac{\varepsilon}{2}.$$

---

[4] We shall later show that $f \in \mathcal{R}([a, b])$ implies f is bounded.

6.1. RIEMANN SUMS AND THE RIEMANN INTEGRAL

Thus for any two partitions $\mathcal{P} = \{x_k\}_{k=0}^n$, $\mathcal{Q} = \{\xi_k\}_{k=0}^m$ whose mesh is less than δ, and any corresponding collections of points $\{t_k\}_{k=1}^n$ and $\{\tau_k\}_{k=1}^m$ which give rise to the marked partitions $\mathcal{P}^\dagger$ and $\mathcal{Q}^\dagger$ respectively, then we have

$$|\sum_{k=1}^n f(t_k)\Delta x_k - \sum_{j=1}^m f(\tau_j)\Delta \xi_j|$$

$$\leq |\sum_{k=1}^n f(t_k)\Delta x_k - \int_a^b f(x)dx| + |\int_a^b f(x)dx - \sum_{j=1}^m f(\tau_j)\Delta \xi_j|$$

$$\leq \frac{\varepsilon}{2} + \frac{\varepsilon}{2} = \varepsilon.$$

Conversely, suppose f satisfies: for any $\varepsilon > 0$, there is a $\delta > 0$ so that if $\mathcal{P} = \{x_k\}_{k=0}^n$ and $\mathcal{Q} = \{\xi_j\}_{j=0}^m$ are two partitions with $\|\mathcal{P}\|, \|\mathcal{Q}\| < \delta$, and letting $\{t_k\}_{k=1}^n$ and $\{\tau_j\}_{j=1}^m$ be any collections of points with $t_k \in [x_{k-1}, x_k]$ for $k = 1, 2, \cdots, n$ and $\tau_j \in [\xi_{j-1}, \xi_j]$, for $j = 1, 2, \cdots, m$, which give rise to the marked partitions $\mathcal{P}^\dagger$ and $\mathcal{Q}^\dagger$, then we have

$$|S(f, [a,b], \mathcal{P}^\dagger) - S(f, [a,b], \mathcal{Q}^\dagger)| < \varepsilon.$$

For each natural number N, there is a $\delta_N > 0$ so that for any two marked partitions $\mathcal{P}^\dagger$ and $\mathcal{Q}^\dagger$ whose mesh is less that δ_N, we have

$$|S(f, [a,b], \mathcal{P}^\dagger) - S(f, [a,b], \mathcal{Q}^\dagger)| < \frac{1}{N}.$$

Moreover, we may assume that $\{\delta_N\}_{N=1}^\infty$ is a decreasing sequence, since otherwise we may use $\delta_N' = \min\{\delta_1, \delta_2, \cdots, \delta_N\}$ instead. For each N we fix a marked partition $\mathcal{P}_N^\dagger$, and for $M > N$ we have $\|\mathcal{P}_N\|, \|\mathcal{P}_M\| < \delta_N$. So

$$|S(f, [a,b], \mathcal{P}_N^\dagger) - S(f, [a,b], \mathcal{P}_M^\dagger)| < \frac{1}{N}.$$

Thus we see that

$$\{S(f, [a,b], \mathcal{P}_k^\dagger)\}_{k=1}^\infty$$

is a Cauchy sequence of real numbers, and hence it has a limit value I. Moreover, for any natural number k, and $r \geq k$ we have

$$||S(f, [a,b], \mathcal{P}_k^\dagger) - I| - |I - S(f, [a,b], \mathcal{P}_r^\dagger)|| \leq |S(f, [a,b], \mathcal{P}_k^\dagger) - S(f, [a,b], \mathcal{P}_r^\dagger)| \leq \frac{1}{k}.$$

By then letting $r \to \infty$ we get

$$|S(f, [a,b], \mathcal{P}_k^\dagger) - I| \leq \frac{1}{k}.$$

Let $s \in \mathbb{N}$ satisfy $s > \frac{2}{\varepsilon}$, where $\varepsilon > 0$ was any preassigned number. For any marked partition $\mathcal{Q}^\dagger$ satisfying $\|\mathcal{Q}\| < \delta_s$, we have

$$|S(f, [a,b], \mathcal{Q}^\dagger) - I|$$
$$\leq |S(f, [a,b], \mathcal{Q}^\dagger) - S(f, [a,b], \mathcal{P}_s^\dagger)|$$
$$+ |S(f, [a,b], \mathcal{P}_s^\dagger) - I|$$
$$\leq \frac{1}{s} + \frac{1}{s} < \varepsilon.$$

Thus we see that $I = \int_a^b f(x)dx$.
□

6.1.4 Some Examples

An Example of Bounded Function that is not Integrable

At this point we will consider an example of a function that is not Riemann integrable, namely $f = \chi_{[a,b] \cap \mathbb{Q}}$, which has value 1 on the rational points in $[a,b]$ and 0 at the irrational points in $[a,b]$. Let $\varepsilon = \frac{b-a}{2}$, and suppose $\delta > 0$ is given. Let $\mathcal{P} = \{x_k\}_{k=0}^n$ be any partition satisfying $\|\mathcal{P}\| < \delta$, and let $I_k = [x_{k-1}, x_k]$, $k = 1, 2, \cdots, n$, be the corresponding subintervals defined by the partition. We choose any collection $\{t_k\}_{k=1}^n$ of points satisfying $t_k \in I_k \cap \mathbb{Q}$, and any collection $\{\tau_k\}_{k=1}^n$ of points satisfying $\tau_k \in I_k \cap (\mathbb{R} \setminus \mathbb{Q})$. Then

$$\sum_{k=1}^n f(t_k)\Delta x_k = b - a, \quad \sum_{k=1}^n f(\tau_k)\Delta x_k = 0$$

and so

$$\left|\sum_{k=1}^n f(t_k)\Delta x_k - \sum_{k=1}^n f(\tau_k)\Delta x_k\right| = b - a > \frac{b-a}{2} = \varepsilon.$$

Since δ was arbitrary, we see that $f \notin \mathcal{R}([a,b])$.

6.1. RIEMANN SUMS AND THE RIEMANN INTEGRAL

An Example of a Highly Discontinuous Function that is Integrable

Consider the so-called Thomae function given in (4.9), which is

$$f(x) = \begin{cases} 0 & x = 0, \\ 0 & x \in \mathbb{R} \setminus \mathbb{Q}, \\ \frac{1}{n} & x = \frac{m}{n} \text{ in reduced form, } n > 0. \end{cases}$$

We claim that $\int_0^1 f(x)dx = 0$.

Let $\varepsilon > 0$ be given, and consider the set

$$D(\varepsilon) := \{x \in [0,1] : f(x) \geq \frac{\varepsilon}{2}\}.$$

This is a finite set since there are only finitely many natural numbers $n < \frac{2}{\varepsilon}$, and for each such natural number only finitely many $m < n$ that have no common factors. Let $N(\varepsilon)$ be the cardinality of $D(\varepsilon)$.

If $N(\varepsilon) \in \mathbb{N}$ let $\delta = \frac{\varepsilon}{4N(\varepsilon)}$, and let $\mathcal{P} = \{x_k\}_{k=0}^n$ be any partition of $[0,1]$ whose norm is less than δ. Let $\{t_k\}_{k=1}^n$ be any collection of points so that $t_k \in [x_{k-1}, x_k]$, $k = 1, 2, \cdots, n$. Then we separate the points in $\{t_k\}_{k=1}^n$ according to whether they are in $D(\varepsilon)$ or not. The collection of $t_k's$ which are in $D(\varepsilon)$ cannot exceed $2N(\varepsilon)$. Thus the sum of the lengths of the intervals which contain $t_k's$ in $D(\varepsilon)$ is less than $2N(\varepsilon)\delta = \frac{\varepsilon}{2}$. Moreover, $f \leq 1$ and so the portion of the Riemann sum $\sum_{k=1}^n f(t_k)\Delta x_k$ with $t_k \in D(\varepsilon)$ has value less than $\frac{\varepsilon}{2}$. The remaining portion of this Riemann sum has $f(t_k) \leq \frac{\varepsilon}{2}$, for $t_k \notin D(\varepsilon)$. So we see

$$0 \leq \sum_{k=1}^n f(t_k)\Delta x_k = \sum_{t_k \in D(\varepsilon)} f(t_k)\Delta x_k + \sum_{t_k \notin D(\varepsilon)} f(t_k)\Delta x_k$$

$$< \frac{\varepsilon}{2} + \frac{\varepsilon}{2} = \varepsilon.$$

If $N(\varepsilon) = 0$, then since $f \leq \frac{\varepsilon}{2}$ we clearly have

$$0 \leq \sum_{k=1}^n f(t_k)\Delta x_k \leq \frac{\varepsilon}{2} \sum_{k=1}^n \Delta x_k = \frac{\varepsilon}{2}$$

for any partition $\mathcal{P} = \{x_k\}_{k=0}^n$ of $[0,1]$ and any collection of marked points $\{t_k\}_{k=1}^n$.

Since $\varepsilon > 0$ is arbitrary, we see that $\int_0^1 f(x)dx$ exists and is equal to zero.

6.1.5 Further Properties of Riemann Integrable Functions

In this section we shall establish that Riemann integrable functions on an interval $[a,b]$ must be bounded. We will also establish a density result, which says that the class of **step functions** are dense in $\mathcal{R}([a,b])$, where a step function is a finite sum of characteristic functions of intervals. Recall that the characteristic function of a set A, denoted by χ_A, has value 1 at points $x \in A$, and is otherwise has value 0. By establishing a necessary and sufficient condition for Riemann integrability in terms approximation by step functions we will also establish that continuous functions are Riemann integrable.

Integrability Implies Boundedness

Theorem 6.1.8 *If $f \in \mathcal{R}([a,b])$, then f is bounded on $[a,b]$.*

Proof: Suppose there is an $f \in \mathcal{R}([a,b])$ which is not bounded on $[a,b]$ and let $I = \int_a^b f(x)dx$. Then there is a $\delta > 0$ so that for any marked partition $\mathcal{P}^\dagger$ satisfying $\|\mathcal{P}\| < \delta$ we have $|S(f,[a,b],\mathcal{P}^\dagger) - I| < 1$, and thus

$$|S(f,[a,b],\mathcal{P}^\dagger)| < |I| + 1.$$

Let $\mathcal{P}$ be any partition of norm less than δ. Since f is unbounded on $[a,b]$ there is a subinterval of $[a,b]$ defined by the partition $\mathcal{P}$ for which f is unbounded on this subinterval. Let $[x_{i-1}, x_i]$ be such a subinterval. Thus for any $M > 0$ there exists $x \in [x_{i-1}, x_i]$ so that $|f(x)| > M$.

We now pick t_k's to create a marked partition $\mathcal{P}^\dagger$. Let t_k, for $k \neq i$, be any points in the corresponding subintervals $[x_{k-1}, x_k]$ defined by the partition $\mathcal{P}$. Let $t_i \in [x_{i-1}, x_i]$ be chosen so that

$$|f(t_i)| > \frac{1 + |I| + |\sum_{k \neq i} f(t_k) \Delta x_k|}{x_i - x_{i-1}}.$$

6.1. RIEMANN SUMS AND THE RIEMANN INTEGRAL

Thus we see that

$$|S(f, [a,b], \mathcal{P}^\dagger)| = |\sum_k f(t_k)\Delta x_k| \geq |f(t_i)(x_i - x_{i-1})| - |\sum_{k \neq i} f(t_k)\Delta x_k| > |I| + 1,$$

which is a contradiction.
□

Clearly this converse of this theorem is not true, since earlier we showed $\chi_{\mathbb{Q} \cap [a,b]} \notin \mathcal{R}([a,b])$ for any $[a,b]$ with $a < b$.

Density of Step Functions in $\mathcal{R}([a,b])$

Step functions are members of a more general class of functions called **simple functions**, where simple functions are finite sums of characteristic functions of sets. In the context of the Lebesgue integral, simple functions which are sums of characteristic functions of measurable sets [5] play a role similar to that of step functions in the context of the Riemann integral.

We shall now show that the the family of step functions defined on $[a,b]$ is dense in the collection $\mathcal{R}([a,b])$.

Theorem 6.1.9 *Step functions are Riemann integrable. More precisely, if* $a = x_0 < x_1 < x_2 < \cdots < x_n = b$ *and* $c_1, c_2, \cdots, c_n \in \mathbb{R}$ *and*

$$f(x) = \sum_{k=1}^{n} c_k \chi_{I_k},$$

where $I_k = (x_{k-1}, x_k), (x_{k-1}, x_k], [x_{k-1}, x_k)$ *or* $[x_{k-1}, x_k]$, *then*

$$\int_a^b f(x)dx = \sum_{k=1}^{n} c_k(x_k - x_{k-1}).$$

Proof: We consider first the case of a step function of the form $\chi_{[c,d]}, \chi_{[c,d)}, \chi_{(c,d]}$ or $\chi_{(c,d)}$. We will show that for $[c,d] \subseteq [a,b]$ the integral of any of these step functions exists and equals $d - c$. Upon establishing this, the general

---
[5] with respect to the Lebesgue measure

result follows from linearity of the integral as stated in theorem (6.1.3).

Consider $g = \chi_{(c,d)}$. For any marked partition $\mathcal{P}^\dagger$ we have $S(g, [a,b], \mathcal{P}^\dagger)$ equals a sum of the form $\sum^*(\xi_k - \xi_{k-1})$, where $\{\xi_i\}_{i=0}^N$ is the partition of $[a,b]$, and the sum is only taken over the subintervals for which the marked points lie in $[c,d]$. Now choose $p, q \in \{1, 2, \cdots, N\}$ so that $\xi_{p-1} \leq c < \xi_p$ and $\xi_{q-1} < d \leq \xi_q$. Let $\{t_k\}_{k=1}^N$ be the marked points for our partition, then $t_j \in (c,d)$ if $p+1 \leq j \leq q-1$ and $t_j \notin (c,d)$ if $j < p$ or $j > q$.

Therefore,
$$\sum_{p+1 \leq j \leq q-1} (\xi_j - \xi_{j-1}) \leq S(g, [a,b], \mathcal{P}^\dagger) \leq \sum_{p \leq j \leq q} (\xi_j - \xi_{j-1}).$$

By choice of p, q we have
$$d - c \leq \xi_q - \xi_{p-1} < (q - p + 1)\|\mathcal{P}\|.$$

So if $\|\mathcal{P}\|$ is sufficiently small, we must have $p + 1 \leq q - 1$, in which case
$$\xi_{q-1} - \xi_p \leq S(g, [a,b], \mathcal{P}^\dagger) \leq \xi_q - \xi_{p-1}.$$

Therefore,
$$(\xi_{q-1} - d) - (\xi_p - c) \leq S(g, [a,b], \mathcal{P}^\dagger) - (d - c) \leq (\xi_q - d) - (\xi_{p-1} - c).$$

The quantities $|\xi_{q-1} - d|$, $|\xi_p - c|$, $|\xi_q - d|$, $|\xi_{p-1} - c|$ are all less than $\|\mathcal{P}\|$, and so
$$|S(g, [a,b], \mathcal{P}^\dagger) - (d - c)| < 2\|\mathcal{P}\|.$$

Since $\|\mathcal{P}\|$ can be chosen to be less than any chosen positive number, we see that $\int_a^b g\,dx = d - c$.

In the other cases where $g = \chi_{[c,d)}$, $\chi_{(c,d]}$ or $\chi_{[c,d]}$, minor modifications of the above argument also yield $\int_a^b g\,dx = d - c$.
□

Theorem 6.1.10 $f \in \mathcal{R}([a,b])$ if and only if for each $\varepsilon > 0$, there are step functions f_-, f_+ so that
$$f_-(x) \leq f(x) \leq f_+(x)$$

6.1. RIEMANN SUMS AND THE RIEMANN INTEGRAL

and
$$\int_a^b f_+(x) - f_-(x)dx < \varepsilon.$$

Proof: Let $\varepsilon > 0$ be given. Suppose there are step functions f_-, f_+ defined on $[a, b]$ so that
$$f_-(x) \leq f(x) \leq f_+(x)$$
on $[a, b]$, and
$$\int_a^b f_+(x) - f_-(x)dx < \frac{\varepsilon}{3}.$$

Since f_- and f_+ are Riemann integrable on $[a, b]$ we may find a common $\delta > 0$ so that if $\mathcal{P}$ is any partition of norm less than δ, then
$$|S(f_-, [a,b], \mathcal{P}^\dagger) - \int_a^b f_-(x)dx|, |S(f_+, [a,b], \mathcal{P}^\dagger) - \int_a^b f_+(x)dx| < \frac{\varepsilon}{3}.$$

For the same δ, consider any partition $\mathcal{P} = \{x_0, x_1, \cdots, x_n\}$ of $[a, b]$ of norm less than δ. Let $\{t_k\}_{k=1}^n$ be a collection of points so that $t_k \in [x_{k-1}, x_k]$ for $k = 1, 2, \cdots, n$, and let $\mathcal{P}^\dagger$ be the corresponding marked partition of $\mathcal{P}$ using $\{t_k\}_{k=1}^n$ as the marked points. Then clearly we have
$$\sum_{k=1}^n f_-(t_k)\Delta x_k \leq S(f, [a,b], \mathcal{P}^\dagger) \leq \sum_{k=1}^n f_+(t_k)\Delta x_k.$$

Since
$$|S(f_-, [a,b], \mathcal{P}^\dagger) - \int_a^b f_-(x)dx|, |S(f_+, [a,b], \mathcal{P}^\dagger) - \int_a^b f_+(x)dx| < \frac{\varepsilon}{3},$$
we have
$$\int_a^b f_-(x)dx - \frac{\varepsilon}{3} < S(f, [a,b], \mathcal{P}^\dagger) < \int_a^b f_+(x)dx + \frac{\varepsilon}{3}.$$

Since $0 \leq \int_a^b f_+(x) - f_-(x)dx < \frac{\varepsilon}{3}$, we have that $S(f, [a,b], \mathcal{P}^\dagger)$ lies in an interval $(\int_a^b f_-(x)dx - \frac{\varepsilon}{3}, \int_a^b f_+(x)dx + \frac{\varepsilon}{3})$ of length at most ε. Clearly if $\mathcal{Q}$ is another partition of $[a, b]$ of norm at most δ, then $S(g, [a,b], \mathcal{Q}^\dagger)$ lies in this interval as well. Thus we have
$$|S(f, [a,b], \mathcal{P}^\dagger) - S(f, [a,b], \mathcal{Q}^\dagger)| < \varepsilon$$

for any two marked partitions $\mathcal{P}^\dagger$, $\mathcal{Q}^\dagger$ of norm less than δ. By the Cauchy Criterion theorem (6.1.7) we see that $f \in \mathcal{R}([a,b])$.

Conversely, suppose $f \in \mathcal{R}([a,b])$ and let $\varepsilon > 0$ be given. By theorem (6.1.7) we know that there is a $\delta > 0$ so that if $\mathcal{P}$ and $\mathcal{Q}$ are two partitions of norm less than δ and $\mathcal{P}^\dagger$ and $\mathcal{Q}^\dagger$ are corresponding marked partitions, then

$$|S(f, [a,b], \mathcal{P}^\dagger) - S(f, [a,b], \mathcal{Q}^\dagger)| < \frac{\varepsilon}{2}.$$

In particular, for $\mathcal{P} = \{x_0, x_1, \cdots, x_n\}$ and for any $t_k, \tau_k \in [x_{k-1}, x_k]$, $k = 1, 2, \cdots, n$, we have

$$|\sum_{k=1}^{n} f(t_k)\Delta x_k - \sum_{k=}^{n} f(\tau_k)\Delta x_k| = |\sum_{k=1}^{n}(f(t_k) - f(\tau_k))\Delta x_k| < \frac{\varepsilon}{2}.$$

Since f is bounded on $[a,b]$ we may define

$$m_i = \inf_{x \in [x_{i-1}, x_i]} f(x), \quad M_i = \sup_{x \in [x_{i-1}, x_i]} f(x),$$

for $i = 1, 2, \cdots, n$. Define

$$f_-(x) = \begin{cases} m_i & x_{i-1} < x < x_i, \quad i = 1, 2, \cdots, n \\ \min\{m_1, m_2, \cdots, m_n\} & x = x_i, \quad i = 0, 1, 2, \cdots, n, \end{cases}$$

$$f_+(x) = \begin{cases} M_i & x_{i-1} < x < x_i, \quad i = 1, 2, \cdots, n \\ \max\{M_1, M_2, \cdots, M_n\} & x = x_i, \quad i = 0, 1, 2, \cdots, n. \end{cases}$$

Then we have
$$f_-(x) \leq f(x) \leq f_+(x).$$

Let $\eta > 0$ be any positive number and choose $t_k, \tau_k \in [x_{k-1}, x_k]$ for $k = 1, 2, \cdots, n$ so that

$$f(t_k) < m_k + \eta, \quad f(\tau_k) > M_k - \eta.$$

Since

$$|\sum_{k=1}^{n}(f(\tau_k) - f(t_k))\Delta x_k| < \frac{\varepsilon}{2}$$

6.1. RIEMANN SUMS AND THE RIEMANN INTEGRAL

holds, and

$$\sum_{k=1}^{n}(f(\tau_k) - f(t_k))\Delta x_k > \sum_{k=1}^{n}(M_k - m_k - 2\eta)\Delta x_k$$

$$= \int_a^b (f_+(x) - f_-(x))dx - 2\eta(b-a),$$

we have

$$\int_a^b (f_+(x) - f_-(x))dx < \frac{\varepsilon}{2} + 2\eta(b-a).$$

Upon choosing $\eta < \frac{\varepsilon}{4(b-a)}$, we get the desired result.
□

Integrability of Continuous Functions

The following result tells us $C^0([a,b]) \subseteq \mathcal{R}([a,b])$, where $C^0([a,b])$ is the class of continuous functions defined on $[a,b]$.

Theorem 6.1.11 *Suppose f is a continuous function on the interval $[a,b]$, then $\int_a^b f(x)dx$ exists.*

Proof: Assume f is continuous on $[a,b]$, with $b > a$. Then f is uniformly continuous on $[a,b]$. So given $\varepsilon > 0$, there is a $\delta > 0$ so that for $x, y \in [a,b]$

$$|x - y| < \delta \quad \text{implies} \quad |f(x) - f(y)| < \frac{\varepsilon}{b-a}.$$

Let $\mathcal{P}$ be any partition of norm less than δ. If $\{x_k\}_{k=0}^n$ are our partition points, then we define $f_-(x)$ and $f_+(x)$ by: $f_-(x) = \inf_{t \in [x_{k-1}, x_k]} f(t)$, $f_+(x) = \sup_{t \in [x_{k-1}, x_k]} f(t)$, for $x \in (x_{k-1}, x_k)$ and $k = 1, 2, \cdots n$; and $f_-(x_k) = f_+(x_k) = f(x_k)$ for $k = 0, 1, 2, \cdots, n$. We see that

$$f_-(x) \leq f(x) \leq f_+(x), \quad x \in [a,b],$$

$$0 \leq \int_a^b f_+(x) - f_-(x)dx = \sum_{k=1}^{n} [\sup_{x \in [x_{k-1}, x_k]} f(x) - \inf_{x \in [x_{k-1}, x_k]} f(x)]\Delta x_k$$

$$< \sum_{k=1}^{n} \frac{\varepsilon}{b-a} \Delta x_k = \varepsilon.$$

By theorem (6.1.10) we get the desired result.
□

6.1.6 Upper and Lower Sums and the Darboux Integral

Another approach to defining the Riemann integral is through what are commonly called **upper** and **lower sums**. For $f(x)$ on $[a,b]$ and a partition $\mathcal{P} = \{x_i\}_{i=0}^n$ of $[a,b]$, the upper sum is defined by

$$U(f,[a,b],\mathcal{P}) = \sum_{k=1}^n \sup_{x \in [x_{k-1},x_k]} f(x) \cdot \Delta x_k, \tag{6.6}$$

and the lower sum is defined by

$$L(f,[a,b],\mathcal{P}) = \sum_{k=1}^n \inf_{x \in [x_{k-1},x_k]} f(x) \cdot \Delta x_k. \tag{6.7}$$

If $\inf_\mathcal{P} U(f,[a,b],\mathcal{P}) = \sup_\mathcal{P} L(f,[a,b],\mathcal{P})$, with the inf and sup taken over all partitions $\mathcal{P}$ of $[a,b]$, we define $(D)\int_a^b f(x)dx$ to be this quantity, and this is often called the **Darboux integral**.

For a given partition $\mathcal{P} = \{x_j\}_{j=0}^n$ of $[a,b]$ we say that a partition $\mathcal{Q} = \{\xi_k\}_{k=0}^m$ of $[a,b]$ is a **refinement** of $\mathcal{P}$ if $\{x_i : i = 0,1,\cdots,n\} \subseteq \{\xi_k : k = 0,1,\cdots,m\}$. If we have two different partitions $\mathcal{P}_1$ and $\mathcal{P}_2$, we may create the **common refinement** of both partitions by taking $\mathcal{P}_1 \cup \mathcal{P}_2$, the collection of all partition points from both partitions.

Theorem 6.1.12 *Let $\mathcal{P}$ be a partition of $[a,b]$, $f : [a,b] \to \mathbb{R}$, and $\mathcal{Q}$ a refinement of $\mathcal{P}$. Then*

$$L(f,[a,b],\mathcal{P}) \leq L(f,[a,b],\mathcal{Q}),$$

and

$$U(f,[a,b],\mathcal{Q}) \leq U(f,[a,b],\mathcal{P}).$$

Proof: We will first prove this in the case where $\mathcal{Q}$ contains one more point than $\mathcal{P}$, i.e. as collections of points, $\mathcal{Q} = \mathcal{P} \cup \{x^*\}$. If $\mathcal{P} = \{x_i\}_{i=0}^n$, then $x_{k-1} < x^* < x_k$ for some $k \in \{1,2,\cdots,n\}$.

6.1. RIEMANN SUMS AND THE RIEMANN INTEGRAL

Let $m_i = \inf_{x\in[x_{i-1},x_i]} f(x)$, $M_i = \sup_{x\in[x_{i-1},x_i]} f(x)$, for $i = 1, 2, \cdots, n$. Let $m_k^* = \inf_{x\in[x_{k-1},x^*]} f(x)$, $m_k^{**} = \inf_{x\in[x^*,x_k]} f(x)$, $M_k^* = \sup_{x\in[x_{k-1},x^*]} f(x)$ and $M_k^{**} = \sup_{x\in[x^*,x_k]} f(x)$. Then clearly

$$m_k \leq m_k^*, m_k^{**} \quad \text{and} \quad M_k \geq M_k^*, M_k^{**}.$$

Hence we see

$$L(f, [a, b], \mathcal{Q}) - L(f, [a, b], \mathcal{P}) = m_k^*(x^* - x_{k-1}) + m_k^{**}(x_k - x^*) - m_k(x_k - x_{k-1})$$

$$(m_k^* - m_k)(x^* - x_{k-1}) + (m_k^{**} - m_k)(x_k - x^*) \geq 0$$

and

$$U(f, [a, b], \mathcal{P}) - U(f, [a, b], \mathcal{Q}) = M_k(x_k - x_{k-1}) - M_k^*(x^* - x_{k-1}) - M_k^{**}(x_k - x^*)$$

$$(M_k - M_k^*)(x^* - x_{k-1}) + (M_k - M_k^{**})(x_k - x^*) \geq 0.$$

In the case where $\mathcal{Q}$ has N points more than $\mathcal{P}$, we simply repeat this type of reasoning N times to get the desired result.
□

Theorem 6.1.13 *Let $f : [a, b] \to \mathbb{R}$, then*

$$\inf_{\mathcal{P}} U(f, [a, b], \mathcal{P}) \geq \sup_{\mathcal{P}} L(f, [a, b], \mathcal{P})$$

where the inf and sup are taken over all partitions of $[a, b]$.

Proof: For any partitions $\mathcal{P}_1$ and $\mathcal{P}_2$ of $[a, b]$, for any common refinement $\mathcal{P}$ of $\mathcal{P}_1$ and $\mathcal{P}_2$ we have

$$L(f, [a, b], \mathcal{P}_1) \leq L(f, [a, b], \mathcal{P}) \leq U(f, [a, b], \mathcal{P}) \leq U(f, [a, b], \mathcal{P}_2)$$

and so

$$L(f, [a, b], \mathcal{P}_1) \leq U(f, [a, b], \mathcal{P}_2)$$

for any two partitions of $[a, b]$. Keeping $\mathcal{P}_2$ fixed, upon taking the least upper bound of the quantities $L(f, [a, b], \mathcal{P}_1)$ over all possible partitions $\mathcal{P}_1$ of $[a, b]$ we get

$$\sup_{\mathcal{Q}} L(f, [a, b], \mathcal{Q}) \leq U(f, [a, b], \mathcal{P}_2).$$

Next, we take the greatest lower bound of the quantities $U(f, [a, b], \mathcal{P}_2)$ over all possible partitions $\mathcal{P}_2$ of $[a, b]$ to get

$$\sup_{\mathcal{P}} L(f, [a, b], \mathcal{P}) \leq \inf_{\mathcal{P}} U(f, [a, b], \mathcal{P}).$$

□

Theorem 6.1.14 $(D) \int_a^b f dx$ *exists if and only if for any $\varepsilon > 0$ there is a partition $\mathcal{P}$ so that*

$$U(f, [a, b], \mathcal{P}) - L(f, [a, b], \mathcal{P}) < \varepsilon.$$

Proof: Suppose $(D) \int_a^b f(x) dx$ exists and let $\varepsilon > 0$ be given. Then there partitions $\mathcal{P}_1$ and $\mathcal{P}_2$ of $[a, b]$ so that

$$0 \leq (D) \int_a^b f(x) dx - L(f, [a, b], \mathcal{P}_1) < \frac{\varepsilon}{2},$$

$$0 \leq U(f, [a, b], \mathcal{P}_2) - (D) \int_a^b f(x) dx < \frac{\varepsilon}{2}.$$

Let $\mathcal{P}$ be the common refinement of $\mathcal{P}_1$ and $\mathcal{P}_2$. By theorem (6.1.12) we have

$$U(f, [a, b], \mathcal{P}) \leq U(f, [a, b], \mathcal{P}_2) \leq (D) \int_a^b f(x) dx + \frac{\varepsilon}{2}$$

$$< L(f, [a, b], \mathcal{P}_1) + \varepsilon \leq L(f, [a, b], \mathcal{P}) + \varepsilon.$$

Clearly, this implies

$$0 \leq U(f, [a, b], \mathcal{P}) - L(f, [a, b], \mathcal{P}) < \varepsilon.$$

Conversely, if for any $\varepsilon > 0$ there is a partition $\mathcal{P}$ so that

$$U(f, [a, b], \mathcal{P}) - L(f, [a, b], \mathcal{P}) < \varepsilon$$

then

$$0 \leq \inf_{\mathcal{Q}} U(f, [a, b], \mathcal{Q}) - \sup_{\mathcal{Q}} L(f, [a, b], \mathcal{Q}) < \varepsilon.$$

since $\varepsilon > 0$ is arbitrary, we have $(D) \int_a^b f(x) dx$ exists.
□

6.1. RIEMANN SUMS AND THE RIEMANN INTEGRAL

Since $L(f, [a, b], \mathcal{P})$ and $U(f, [a, b], \mathcal{P})$ can be seen as integrals of step functions f_- and f_+ defined on $[a, b]$, namely $f_-(x) = \inf_{t \in [x_{k-1}, x_k]} f(t)$, $f_+(x) = \sup_{t \in [x_{k-1}, x_k]} f(t)$, for $x \in (x_{k-1}, x_k)$ and $k = 1, 2, \cdots n$; and $f_-(x_k) = f_+(x_k) = f(x_k)$ for $k = 0, 1, 2, \cdots, n$, then by theorem (6.1.10) we see that $\int_a^b f(x)dx$ and $(D)\int_a^b f(x)dx$ either both exist or neither exists.

Theorem 6.1.15 *If $f : [a, b] \to \mathbb{R}$, then either both $\int_a^b f(x)dx$ and $(D)\int_a^b f(x)dx$ exist or neither exists. Moreover, if both exist, then*

$$(D)\int_a^b f(x)dx = \int_a^b f(x)dx.$$

Proof: The first part has been established already. Supposing that both $(D)\int_a^b f(x)dx$ and $\int_a^b f(x)dx$ exist, given $\varepsilon > 0$ we first choose a partition $\mathcal{Q}$ of $[a, b]$ so that

$$U(f, [a, b], \mathcal{Q}) - L(f, [a, b], \mathcal{Q}) < \frac{\varepsilon}{3}.$$

We observe that for any refinement $\mathcal{P}$ of $\mathcal{Q}$ we also have

$$U(f, [a, b], \mathcal{P}) - L(f, [a, b], \mathcal{P}) < \frac{\varepsilon}{3}.$$

Let $\delta > 0$ be so that if $\mathcal{P}$ has norm less than δ we have

$$\left|\int_a^b f(x)dx - S(f, [a, b], \mathcal{P}^\dagger)\right| < \frac{\varepsilon}{3}.$$

Let $\mathcal{P} = \{x_0, x_1, \cdots, x_n\}$ be a refinement of $\mathcal{Q}$ which has norm less than δ. For each interval $[x_{k-1}, x_k]$, $k = 1, 2, \cdots, n$, pick $t_k \in [x_{k-1}, x_k]$ so that $f(t_k) > \sup_{x \in [x_{k-1}, x_k]} f(x) - \frac{\varepsilon}{3(b-a)}$. Let $\mathcal{P}^\dagger$ be the corresponding marked partition using the t_k's as our marked points. Then

$$\left|(D)\int_a^b f(x)dx - \int_a^b f(x)dx\right|$$

$$\leq \left|(D)\int_a^b f(x)dx - U(f, [a, b], \mathcal{P})\right| + |U(f, [a, b], \mathcal{P}) - S(f, [a, b], \mathcal{P}^\dagger)|$$

$$+ \left|S(f, [a, b], \mathcal{P}^\dagger) - \int_a^b f(x)dx\right|$$

$$< \frac{\varepsilon}{3} + \sum_{k=1}^{n} (\sup_{x \in [x_{k-1}, x_k]} f(x) - f(t_k))\Delta x_k + \frac{\varepsilon}{3}$$

$$< \frac{\varepsilon}{3} + \frac{\varepsilon}{3(b-a)} \sum_{k=1}^{n} \Delta x_k + \frac{\varepsilon}{3}$$

$$= \frac{\varepsilon}{3} + \frac{\varepsilon}{3} + \frac{\varepsilon}{3} = \varepsilon.$$

Since $\varepsilon > 0$ is arbitrary, we see that

$$(D) \int_a^b f(x)dx = \int_a^b f(x)dx.$$

□

The following result tells us that bounded monotonic functions are Riemann integrable.

Theorem 6.1.16 *Let $f : [a,b] \to \mathbb{R}$ be monotonic and bounded, then $\int_a^b f(x)dx$ exists.*

Proof: By otherwise considering $-f(x)$, we may assume $f(x)$ is monotone increasing. Let $\varepsilon > 0$ be given and choose $n \in \mathbb{N}$ so that $n > \frac{(f(b)-f(a))(b-a)}{\varepsilon}$. Consider the partition $\mathcal{P}$ given by $x_0 = a$, $x_1 = a + \frac{b-a}{n}$, $x_{j+1} = x_j + \frac{b-a}{n}$, for $j = 1, 2, \cdots, n-1$, then clearly $\|\mathcal{P}\| = \frac{b-a}{n}$. Since f is monotone increasing, we have $U(f, [a,b], \mathcal{P}) = \sum_{k=1}^{n} f(x_k) \cdot \frac{b-a}{n}$ and $L(f, [a,b], \mathcal{P}) = \sum_{k=1}^{n} f(x_{k-1}) \cdot \frac{b-a}{n}$ and so

$$0 \leq U(f, [a,b], \mathcal{P}) - L(f, [a,b], \mathcal{P})$$

$$= \sum_{k=1}^{n} (f(x_k) - f(x_{k-1})) \cdot \frac{b-a}{n}$$

$$= (f(b) - f(a)) \cdot \frac{b-a}{n} < \varepsilon.$$

□

We conclude this subsection with a result regarding composition and integrability, and a result regarding products and integrability.

Theorem 6.1.17 *Let $f \in \mathcal{R}([a,b])$ with $m \leq f \leq M$ on $[a,b]$. Let g be a continuous function of $[m, M]$, and consider $h(x) = g(f(x))$. Then $h \in \mathcal{R}([a,b])$.*

6.1. RIEMANN SUMS AND THE RIEMANN INTEGRAL

Proof: Let $\varepsilon > 0$. Since g is uniformly continuous on $[m, M]$ we have $\sup_{y \in [m,M]} |g(y)| < \infty$. There is a $\delta > 0$ which is smaller than the minimum of $\frac{\varepsilon}{2(b-a)}$ and $\frac{\varepsilon}{4K}$, where K is any chosen positive number larger than $\sup_{y \in [m,M]} |g(y)|$ so that $|g(y) - g(z)| < \frac{\varepsilon}{2(b-a)}$ whenever $|y - z| < \delta$ and $y, z \in [m, M]$.

By integrability of f, we know that there is a partition $\mathcal{P} = \{x_j\}_{j=0}^n$ so that
$$U(f, [a,b], \mathcal{P}) - L(f, [a,b], \mathcal{P}) < \delta^2.$$
For $k = 1, 2, \cdots, n$ define $m_k := \inf_{t \in [x_{k-1}, x_k]} f(t)$, $M_k := \sup_{t \in [x_{k-1}, x_k]} f(t)$, $m_k^* := \inf_{t \in [x_{k-1}, x_k]} h(t)$ and $M_k^* := \sup_{t \in [x_{k-1}, x_k]} h(t)$.

Partition the set $\{1, 2, \cdots, n\}$ into two sets A and B where $i \in A$ if $M_i - m_i < \delta$ and $i \in B$ if $M_i - m_i \geq \delta$.

For $i \in A$, by our choice of δ we have $M_i^* - m_i^* < \frac{\varepsilon}{2(b-a)}$. For $i \in B$ we certainly have $M_i^* - m_i^* < 2K$. Since $U(f, [a,b], \mathcal{P}) - L(f, [a,b], \mathcal{P}) < \delta^2$ we have
$$\delta \sum_{k \in B} \Delta x_k \leq \sum_{k \in B} (M_k - m_k) \Delta x_k < \delta^2,$$
and so $\sum_{k \in B} \Delta x_k < \delta$. It follows that

$$U(h, [a,b], \mathcal{P}) - L(h, [a,b], \mathcal{P}) = \sum_{k=1}^n (M_k^* - m_k^*) \Delta x_k$$

$$= \sum_{k \in A} (M_k^* - m_k^*) \Delta x_k + \sum_{k \in B} (M_k^* - m_k^*) \Delta x_k$$

$$\leq \frac{\varepsilon}{2(b-a)} \sum_{k \in A} \Delta x_k + 2K \sum_{k \in B} \Delta x_k$$

$$< \frac{\varepsilon}{2} + 2K \cdot \delta$$

$$\leq \frac{\varepsilon}{2} + 2K \frac{\varepsilon}{4K} = \varepsilon.$$

Since ε was arbitrary, we know $h \in \mathcal{R}([a,b])$.
□

Corollary 6.1.18 *Suppose $f, g \in \mathcal{R}([a,b])$, then $fg \in \mathcal{R}([a,b])$.*

Proof: Using $g(y) = y^2$ in the above theorem (6.1.17), we see that $f \in \mathcal{R}([a,b])$ implies $f^2 \in \mathcal{R}([a,b])$. Using this and

$$fg = \frac{1}{4}(f+g)^2 - \frac{1}{4}(f-g)^2$$

we get the desired result.
□

6.1.7 A Sufficient Condition for Integrability

Earlier we showed that if f is continuous on $[a, b]$ then $\int_a^b f(x)dx$ exists. At this point we shall prove a slightly more general result.

Theorem 6.1.19 *Suppose $f : [a, b] \to \mathbb{R}$ is bounded, and continuous at all points in $[a, b]$ except possibly finitely many points, then $\int_a^b f(x)dx$ exists.*

Proof: If f is continuous everywhere, then we are done. Thus we assume that f is continuous except at $c_1, c_2, \cdots, c_N \in [a, b]$.

Let $M := \sup_{x \in [a,b]} |f(x)|$, and let $\varepsilon > 0$ be given. Let $I_j := (c_j - \frac{\varepsilon}{8NM}, c_j + \frac{\varepsilon}{8NM})$, and $D = [a,b] \setminus \cup_{k=1}^n I_k$. D is compact, and thus f is uniformly continuous on D. Therefore there is a $\delta > 0$ so that for $x, y \in D$ so that $|x - y| < \delta$ we have $|f(x) - f(y)| < \frac{\varepsilon}{2(b-a)}$.

Let $\mathcal{P}_1 = \{x_j\}_{j=1}^n$ be a partition of D of norm less than δ. Extend $\mathcal{P}_1$ to a partition of $[a, b]$ in the obvious way. Then clearly $\mathcal{P}$ contains the collection $\{c_j \pm \frac{\varepsilon}{8NM} : j = 1, 2, \cdots, N\}$ as partition points, and between any two such points there are no other partition points. Let $\mathcal{J} \subseteq \{1, 2, \cdots, n\}$ be the collection of j's for which $[x_{j-1}, x_j] = [c_k - \frac{\varepsilon}{8NM}, c_k + \frac{\varepsilon}{8MN}]$ for some $k \in \{1, 2, \cdots, N\}$. Then $i \in \{1, 2, \cdots, n\} \setminus \mathcal{J}$ if and only if $[x_{i-1}, x_i] \subseteq D$.

Let $M_i := \sup_{x \in [x_{i-1}, x_i]} f(x)$ and $m_i := \inf_{x \in [x_{i-1}, x_i]} f(x)$ for $i = 1, 2, \cdots, n$. Then

$$U(f, [a,b], \mathcal{P}) - L(f, [a,b], \mathcal{P}) = \sum_{i=1}^n (M_i - m_i) \Delta x_i$$

6.1. RIEMANN SUMS AND THE RIEMANN INTEGRAL

$$= \sum_{i \in \mathcal{J}} (M_i - m_i)\Delta x_i + \sum_{i \notin \mathcal{J}} (M_i - m_i)\Delta x_i$$

$$< N \cdot 2M \cdot \frac{\varepsilon}{4NM} + \frac{\varepsilon}{2(b-a)} \sum_{j \notin \mathcal{J}} \Delta x_j$$

$$\leq \frac{\varepsilon}{2} + \frac{\varepsilon}{2(b-a)} \cdot (b-a) = \varepsilon.$$

Thus we see $f \in \mathcal{R}([a,b])$.

□

6.1.8 Additive Properties of the Riemann Integral

At this point we shall address questions of the type:

1. If $f \in \mathcal{R}([a,b])$, and $[c,d] \subseteq [a,b]$ is $f \in \mathcal{R}([c,d])$?

2. If $f \in \mathcal{R}([a,b])$ and $f \in \mathcal{R}([b,c])$ with $a \leq b \leq c$, is $f \in \mathcal{R}([a,c])$? Moreover, if so, what is the relationship between $\int_a^b f dx$, $\int_b^c f dx$ and $\int_a^c f dx$?

3. If $f \in \mathcal{R}([a,b])$ and $f \in \mathcal{R}([b,c])$, is $f \in \mathcal{R}([a,c])$? Moreover, if so, what is the relationship between $\int_a^b f dx$, $\int_b^c f dx$ and $\int_a^c f dx$?

4. Can we define $\int_a^a f dx$? If so, what should its value be?

Theorem 6.1.20 *Let $f : [a,c] \to \mathbb{R}$, and let $b \in (a,c)$, then $f \in \mathcal{R}([a,c])$ if and only if $f \in \mathcal{R}([a,b])$ and $f \in \mathcal{R}([b,c])$. In such a case*

$$\int_a^c f(x)dx = \int_a^b f(x)dx + \int_b^c f(x)dx.$$

Proof: Suppose for any $b \in (a,c)$ we have $f \in \mathcal{R}([a,b]) \cap \mathcal{R}([b,c])$. Then given any $\varepsilon > 0$ there is a $\delta > 0$ so that for any partition $\mathcal{P}$ of $[a,b]$ of norm less than δ and any partition $\mathcal{Q}$ of $[b,c]$ of norm less than δ we have

$$|S(f,[a,b],\mathcal{P}^\dagger) - \int_a^b f dx|, |S(f,[b,c],\mathcal{Q}^\dagger) - \int_b^c f dx| < \frac{\varepsilon}{3}.$$

We know that $\sup_{x \in [a,b]} |f(x)|$ and $\sup_{x \in [b,c]} |f(x)|$ are finite, and so $|f(x)|$ is bounded on $[a,c]$, and let M be any upper bound for $|f|$ on $[a,c]$.

Let $\delta' > 0$ be the minimum of δ and $\frac{\varepsilon}{6M}$. Let $\mathcal{T} = \{x_i\}_{i=0}^n$ be any partition of $[a,c]$ of norm less than δ', and let $\{t_k\}_{k=1}^n$ be any collection of points which give rise to a marked partition $\mathcal{T}^\dagger$. We shall now bound $|S(f,[a,c],\mathcal{T}^\dagger) - (\int_a^b f\,dx + \int_b^c f\,dx)|$.

Split $\mathcal{T}$ into two parts, $\mathcal{T}_1$ and $\mathcal{T}_2$, where $\mathcal{T}_1$ contains the partition points that are in $[a,b]$ and $\mathcal{T}_2$ contains the partition points that are in $[b,c]$. We consider whether c is a partition point or not.

Suppose $c = x_{k-1}$ for some k, then $\mathcal{T}_1 = \{x_i\}_{i=0}^{k-1}$ and $\mathcal{T}_2 = \{x_i\}_{i=k-1}^n$, which are partitions of $[a,b]$ and $[b,c]$ respectively. Let $\mathcal{T}_1^\dagger$ and $\mathcal{T}_2^\dagger$ be the corresponding marked partitions obtained by inheriting the marked points from $\mathcal{T}^\dagger$. Then we have

$$|S(f,[a,b],\mathcal{T}_1^\dagger) - \int_a^b f\,dx|, |S(f,[b,c],\mathcal{T}_2^\dagger) - \int_b^c f\,dx| < \frac{\varepsilon}{3}.$$

Since

$$S(f,[a,c],\mathcal{T}^\dagger) = S(f,[a,b],\mathcal{T}_1^\dagger) + S(f,[b,c],\mathcal{T}_2^\dagger),$$

we have

$$|S(f,[a,c],\mathcal{T}^\dagger) - (\int_a^b f\,dx + \int_b^c f\,dx)|$$

$$= |S(f,[a,b],\mathcal{T}_1^\dagger) + S(f,[b,c],\mathcal{T}_2^\dagger) - (\int_a^b f\,dx + \int_b^c f\,dx)|$$

$$\leq |S(f,[a,b],\mathcal{T}_1^\dagger) - \int_a^b f\,dx| + |S(f,[b,c],\mathcal{T}_2^\dagger) - \int_b^c f\,dx| < \frac{\varepsilon}{3} + \frac{\varepsilon}{3} < \varepsilon.$$

In the case where $c \in (x_{k-1}, x_k)$ for some k we create partitions of $[a,b]$ and $[b,c]$ by defining

$$\hat{\mathcal{T}}_1 := \mathcal{T}_1 \cup \{c\}, \hat{\mathcal{T}}_2 := \mathcal{T}_2 \cup \{c\}.$$

Create marked partitions of these partitions by taking $\{t_j\}_{j=1}^{k-1} \cup \{c\}$ for $\hat{\mathcal{T}}_1$ and $\{t_j\}_{j=k+1}^n \cup \{c\}$ for $\hat{\mathcal{T}}_2$.

Thus we see that

$$S(f,[a,c],\mathcal{T}^\dagger) - [S(f,[a,b],\hat{\mathcal{T}}_1^\dagger) + S(f,[b,c],\hat{\mathcal{T}}_2^\dagger)]$$

6.1. RIEMANN SUMS AND THE RIEMANN INTEGRAL

$$= f(t_k)(x_k - x_{k-1}) - f(c)(c - x_{k-1}) - f(c)(x_k - c)$$
$$= f(t_k)(x_k - x_{k-1}) - f(c)(x_k - x_{k-1})$$
$$= (f(t_k) - f(c))(x_k - x_{k-1}),$$

and so
$$|S(f, [a, c], \mathcal{T}^\dagger) - [S(f, [a, b], \hat{\mathcal{T}}_1^\dagger) + S(f, [b, c], \hat{\mathcal{T}}_2^\dagger)]|$$
$$\leq |f(t_k) - f(c)|(x_k - x_{k-1})$$
$$\leq 2M\delta' < \frac{\varepsilon}{3}.$$

Hence
$$|S(f, [a, c], \mathcal{T}^\dagger) - (\int_a^b f\,dx + \int_b^c f\,dx)|$$
$$= |S(f, [a, c], \mathcal{T}^\dagger) - S(f, [a, b], \hat{\mathcal{T}}_1^\dagger) - S(f, [b, c], \hat{\mathcal{T}}_2^\dagger)$$
$$+ S(f, [a, b], \hat{\mathcal{T}}_1^\dagger) + S(f, [b, c], \hat{\mathcal{T}}_2^\dagger) - (\int_a^b f\,dx + \int_b^c f\,dx)|$$
$$\leq |S(f, [a, c], \mathcal{T}^\dagger) - S(f, [a, b], \hat{\mathcal{T}}_1^\dagger) - S(f, [b, c], \hat{\mathcal{T}}_2^\dagger)|$$
$$+ |S(f, [a, b], \hat{\mathcal{T}}_1^\dagger) - \int_a^b f\,dx| + |S(f, [b, c], \hat{\mathcal{T}}_2^\dagger) - \int_b^c f\,dx|$$
$$< \frac{\varepsilon}{3} + \frac{\varepsilon}{3} + \frac{\varepsilon}{3} = \varepsilon.$$

Thus we see $f \in \mathcal{R}([a, c])$ and $\int_a^c f\,dx = \int_a^b f\,dx + \int_b^c f\,dx$, whenever $f \in \mathcal{R}([a, b]) \cap \mathcal{R}([b, c])$.

Next we assume that $f \in \mathcal{R}([a, c])$, and let $\varepsilon > 0$. By theorem (6.1.7) we know that there is a $\delta > 0$ so that if $\mathcal{P}$ is a partition of $[a, c]$ with norm less than δ, then $|S(f, [a, c], \mathcal{P}^\dagger) - \int_a^c f\,dx| < \varepsilon$.

Let $b \in (a, c)$, let $\mathcal{Q}$ and $\mathcal{T}$ be partitions of $[a, b]$ each of norm less than δ, and let $\mathcal{Q}^\dagger$ and $\mathcal{T}^\dagger$ be corresponding marked partitions. Take any partition $\mathcal{S}$ of $[b, c]$ of norm less than δ, and consider $\hat{\mathcal{Q}} := \mathcal{Q} \cup \mathcal{S}$, $\hat{\mathcal{T}} := \mathcal{T} \cup \mathcal{S}$. Take any collection of points which make $\mathcal{S}$ a marked partition of $[b, c]$ and define $\hat{\mathcal{Q}}^\dagger$ to be the marked partition that uses the marked points for $\mathcal{Q}$ and $\mathcal{S}$, and similarly we define $\hat{\mathcal{T}}^\dagger$ to be the marked partitions that uses the marked points for $\mathcal{T}$ and $\mathcal{S}$. Then clearly we have

$$|S(f,[a,c],\hat{\mathcal{Q}}^\dagger) - S(f,[a,c],\hat{\mathcal{T}}^\dagger)| = |S(f,[a,b],\mathcal{Q}^\dagger) - S(f,[a,b],\mathcal{T}^\dagger)| < \varepsilon.$$

Hence by theorem (6.1.7) we see that $f \in \mathcal{R}([a,b])$. A similar argument shows $f \in \mathcal{R}([b,c])$. Moreover, the argument used above in the first part of the proof actually shows that

$$\int_a^c f\,dx = \int_a^b f\,dx + \int_b^c f\,dx.$$

□

An application of the above theorem[6] actually yields the following corollary:

Corollary 6.1.21 *Let $f \in \mathcal{R}([a,b])$. Then for any subinterval $[\alpha,\beta] \subseteq [a,b]$ we have $f \in \mathcal{R}([\alpha,\beta])$.*

For $a < b$, if $\int_a^b f(x)dx$ exists we define

$$\int_b^a f(x)dx := -\int_a^b f(x)dx.$$

Also, for any a in the domain of a function f, we define $\int_a^a f(x)dx = 0$.[7] Using this notation, we have the following corollary:

Corollary 6.1.22 *Let $f \in \mathcal{R}([a,b])$, and let $\alpha, \beta, \gamma \in [a,b]$, then*

$$\int_\alpha^\gamma f(x)dx = \int_\alpha^\beta f(x)dx + \int_\beta^\gamma f(x)dx.$$

---

[6]possibly twice

[7]Assume $f \in \mathcal{R}([a,b])$ with $b > a$, and let $m := \inf_{x \in [a,b]} f(x)$, and $M := \sup_{x \in [a,b]} f(x)$. For any $c \in (a,b]$ we have

$$m(c-a) \leq \int_a^c f(x)dx \leq M(c-a).$$

Letting $c \to a^+$ we see that $\lim_{c \to a^+} \int_a^c f(x)dx = 0$. Thus we see that $\int_a^a f(x)dx = 0$ is necessary.

6.1.9 Homework Exercises

1. Show that $f(x) = x$ is in $\mathcal{R}([a,b])$.

2. Define $f(x)$ by
$$f(x) = \begin{cases} 0 & x \in \mathbb{Q}, \\ \frac{1}{x} & x \notin \mathbb{Q}. \end{cases}$$
Show that $f \notin \mathcal{R}([0,1])$.

3. Let $f(x) : [a,b] \to \mathbb{R}$, with $a < b$, suppose $f(x)$ is zero valued except at finitely many points, show that $\int_a^b f(x)dx$ exists and has value zero.

4. Suppose $g \in \mathcal{R}([a,b])$ and $f(x) = g(x)$ except possibly at a finite number of points. Show that $f \in \mathcal{R}([a,b])$ and $\int_a^b f(x)dx = \int_a^b g(x)dx$.

5. Let
$$f(x) = \begin{cases} 0 & x = 0, \\ 0 & x \in \mathbb{R} \setminus \mathbb{Q}, \\ n & x = \frac{m}{n} \text{ in reduced form, } n > 0 \end{cases}$$
Show that $f \notin \mathcal{R}([a,b])$ for any interval $[a,b]$ with $a < b$.

6. If $f : [a,b] \to \mathbb{R}$ has $f([a,b])$ a finite set, is f necessarily a step function? Please explain. Is f necessarily in $\mathcal{R}([a,b])$? Please explain.

7. If $f : [a,b] \to \mathbb{R}$ has $f([a,b])$ a finite set, is f necessarily a simple function? Please explain. Is f necessarily in $\mathcal{R}([a,b])$? Please explain.

8. Suppose f is continuous on $[a,b]$, and $f \geq 0$ on $[a,b]$. Suppose $\int_a^b f(x)dx = 0$, show $f \equiv 0$ on $[a,b]$.

9. Prove the so-called **mean value theorem for integrals**: If f is continuous on $[a,b]$, $a < b$, show that there is a $c \in [a,b]$ so that
$$f(c) = \frac{1}{b-a} \int_a^b f(x)dx.$$

10. Suppose that f is continuous on $[a,b]$ and $f \geq 0$ on $[a,b]$. Suppose there is a $c \in [a,b]$ so that $f(c) > 0$. Show that
$$\int_a^b f(x)dx > 0.$$

11. Suppose $f^2 \in \mathcal{R}([a,b])$. Does it follow that $f \in \mathcal{R}([a,b])$?

12. Suppose $f^3 \in \mathcal{R}([a,b])$. Does it follow that $f \in \mathcal{R}([a,b])$?

13. Suppose $f \in \mathcal{R}([a,b])$. Does it follow that $|f| \in \mathcal{R}([a,b])$?

14. Suppose $f \in \mathcal{R}([c,b])$ for all $[c,d] \subset [a,b]$, $c > a$. If $\lim_{c \to a^+} \int_c^b f(x)dx$ exists and is finite, then we define $\int_a^b f(x)dx$ to be this limit value. If $\int_a^b f(x)dx$ actually exists – that is $f \in \mathcal{R}([a,b])$ – show that it agrees with this limit as well.

15. Suppose $f \in \mathcal{R}([a,b])$ for all $b > a$. Define $\int_a^\infty f(x)dx = \lim_{b \to \infty} \int_a^b f(x)dx$ whenever this limit exists and is finite. Assume that $f(x) \geq 0$ and decreases monotonically with $\lim_{t \to \infty} f(t) = 0$. Prove that $\int_a^\infty f dx$ exists if and only if $\sum_{n \geq a, n \in \mathbb{N}} f(n)$ converges. This is the so-called **integral test**.

16. Let $f(x) : [-1,1] \to \mathbb{R}$ be defined by $f(0) = 0$ and $f(x) = \sin \frac{1}{x}$ for $x \neq 0$. Does $\int_{-1}^1 f(x)dx$ exist? Please explain.

17. Let $f(x) : [-1,1] \to \mathbb{R}$ be defined by $f(0) = 0$ and $f(x) = x \sin \frac{1}{x}$ for $x \neq 0$. Does $\int_{-1}^1 f(x)dx$ exist? Please explain.

6.2 The Fundamental Theorem of Calculus

Let $f \in \mathcal{R}([a,b])$. By corollary (6.1.21), for any $x \in [a,b]$ we may define

$$F(x) := \int_a^x f(t)dt. \tag{6.8}$$

In this section we shall discuss some properties of this function $F(x)$, which is often called an **antiderivative** of $f(x)$ in the case where f is continuous on $[a,b]$.

Theorem 6.2.1 *Let $f \in \mathcal{R}([a,b])$ and let $F(x) = \int_a^x f(t)dt$ for $x \in [a,b]$, then F is continuous on $[a,b]$.*

Proof: Since $f \in \mathcal{R}([a,b])$, we know f is bounded on $[a,b]$. Let M be any positive upper bound for $|f|$ on $[a,b]$.

6.2. THE FUNDAMENTAL THEOREM OF CALCULUS

Let $\varepsilon > 0$. Then for any pair $x, y \in [a, b]$ so that $x \leq y$, and $y - x \leq \frac{\varepsilon}{M}$, we have

$$|F(y) - F(x)| = |\int_a^y f(t)dx - \int_a^x f(t)dt|$$

$$= |\int_x^y f(t)dt|$$

$$\leq M(y - x)$$

$$< \varepsilon.$$

□

Theorem 6.2.2 *Let $f \in \mathcal{R}([a, b])$ and let $F(x) = \int_a^x f(t)dt$ for $x \in [a, b]$. If f is continuous at $x_0 \in (a, b)$, then F is differentiable at x_0. Moreover,*

$$F'(x_0) = f(x_0).$$

Proof: Let $\varepsilon > 0$ and suppose f is continuous at $x_0 \in (a, b)$. Then there is a $\delta > 0$ so that if $|x - x_0| < \delta$ we have $|f(x) - f(x_0)| < \varepsilon$. So for any $x \in (x_0 - \delta, x_0 + \delta) \cap [a, b]$ we have

$$|\frac{F(x) - F(x_0)}{x - x_0} - f(x_0)| = |\frac{1}{x - x_0} \int_{x_0}^x f(t)dt - f(x_0)|$$

$$= \frac{1}{|x - x_0|} |\int_{x_0}^x (f(t) - f(x))dt|$$

$$\leq \frac{1}{|x - x_0|} |\int_{x_0}^x |f(t) - f(x)|dt|$$

$$\leq \frac{1}{|x - x_0|} |\int_{x_0}^x \varepsilon dt| = \frac{1}{|x - x_0|} \cdot \varepsilon |x - x_0| = \varepsilon.$$

□

Fundamental Theorem of Calculus 6.2.3 *Let $f \in \mathcal{R}([a, b])$ and suppose there is a function $g : [a, b] \to \mathbb{R}$ which satisfies $g'(x) = f(x)$ for all $x \in [a, b]$, then*

$$\int_a^b f(x)dx = g(b) - g(a).$$

Proof: Let $\varepsilon > 0$ be given, then there exists a $\delta > 0$ so that for any partition $\mathcal{P} = \{x_i\}_{i=0}^{n}$ of $[a,b]$ with norm less than δ, and any marked partition $\mathcal{P}^\dagger$ obtained from $\mathcal{P}$, we have

$$|S(f,[a,b],\mathcal{P}^\dagger) - \int_a^b f(x)dx| < \varepsilon.$$

By the mean value theorem (5.3.1) we have there exists $t_k \in (x_{k-1}, x_k)$, $k = 1, 2, \cdots, n$, so that

$$g(x_k) - g(x_{k-1}) = f(t_k)\Delta x_k.$$

So we have

$$g(b) - g(a) = \sum_{k=1}^{n}[g(x_k) - g(x_{k-1})] = \sum_{k=1}^{n} f(t_k)\Delta x_k,$$

and thus

$$|g(b) - g(a) - \int_a^b f(x)dx| < \varepsilon.$$

Since $\varepsilon > 0$ is arbitrary, we get the theorem's claim.
□

Note that if f is continuous on $[a,b]$ and Riemann integrable on $[a - \eta, b + \eta]$ for some $\eta > 0$, then the function $F(x) = \int_a^x f(t)dt$ clearly satisfies $F'(x) = f(x)$ for all $x \in [a,b]$. In this case clearly $F(b) = F(b) - F(a) = \int_a^b f(t)dt$ holds. However, the fundamental theorem of calculus tells us that any function g which satisfies $g' = f$ on $[a,b]$ has $g(b) - g(a) = \int_a^b f(x)dx$. Such a function g is called an **antiderivative** of f. Moreover, by a corollary of the mean value theorem, we know that any two antiderivatives of f must differ by a constant. Note that is also what the fundamental theorem of calculus is also giving us[8].

Corollary – Integration by Parts 6.2.4 Let $f, g \in \mathcal{R}([a,b])$ and suppose that there are functions F, G so that $F'(x) = f(x)$ and $G'(x) = g(x)$ for all $x \in [a,b]$. Then

$$\int_a^b F(x)g(x)dx = F(b)G(b) - F(a)G(a) - \int_a^b f(x)G(x)dx.$$

---

[8]But observe that the fundamental theorem of calculus does not require that f is continuous.

6.2. THE FUNDAMENTAL THEOREM OF CALCULUS

Proof: Let $H = FG$. By the product rule, $\frac{dH}{dx} = \frac{dF}{dx}G + F\frac{dG}{dx}$. Since F, G are differentiable, they are continuous, and so $F, G \in \mathcal{R}([a,b])$. Also $\frac{dF}{dx} = f \in \mathcal{R}([a,b])$ and $\frac{dG}{dx} = g \in \mathcal{R}([a,b])$. Thus by applying corollary (6.1.18) we get $\frac{dH}{dx} \in \mathcal{R}([a,b])$. By the fundamental theorem of calculus (6.2.3) we get

$$\int_a^b \frac{dH}{dx} dx = H(b) - H(a)$$

and so

$$\int_a^b fG + Fg\, dx = F(b)G(b) - F(a)G(a),$$

which is the desired result.
☐

6.2.1 Integral Representation of Some Transcendental Functions

The Natural Logarithm

Earlier we defined $y = \log x$ to be the inverse function of $y = e^x$, and we showed via the inverse function theorem (5.3.5) that $\frac{d}{dx}\log x = \frac{1}{x}$. Using this and $\log 1 = 0$, we see that for $x > 0$ we have

$$\log x = \int_1^x \frac{1}{t} dt. \tag{6.9}$$

From this observation we may deduce some of the familiar properties of $\log x$. For $x, y > 0$ we have $\log(xy) = \int_1^{xy} \frac{1}{t} dt$, and thus

$$\frac{d}{dx}\log(xy) = \frac{d}{dx}\int_1^{xy} \frac{1}{t} dt = \frac{1}{xy}\frac{d}{dx}(xy) = \frac{1}{x}.$$

Hence we see that $\log(xy) = \log x + C$, for some $C \in \mathbb{R}$. Letting $x = 1$ we see that $C = \log y$, and so

$$\log(xy) = \log x + \log y.$$

Let $r \in \mathbb{R}$, $r \neq 0$ and $x > 0$. Consider $\log x^r = \int_1^{x^r} \frac{1}{t} dt$. Now[9]

$$\frac{d}{dx}\frac{1}{r}\log x^r = \frac{1}{r}\frac{d}{dx}\int_1^{x^r} \frac{1}{t} dt = \frac{1}{rx^r}\frac{d}{dx}x^r = \frac{1}{x}.$$

---

[9] $\frac{d}{dx}x^r = \frac{d}{dx}e^{r\log x} = e^{r\log x}\frac{d}{dx}r\log x = e^{r\log x}\frac{r}{x} = x^r \cdot \frac{r}{x} = rx^{r-1}$.

So $\frac{1}{r} \log x^r = \log x + C$ for some $C \in \mathbb{R}$. Letting $x = 1$ we see $C = 0$, and thus
$$\log x^r = r \log x.$$

The Inverse Trigonometric Functions

Recall that $\frac{d}{dx} \arcsin x = \frac{1}{\sqrt{1-x^2}}$ and $\frac{d}{dx} \arccos x = \frac{-1}{\sqrt{1-x^2}}$ for $x \in (-1, 1)$, and $\frac{d}{dx} \arctan x = \frac{1}{1+x^2}$ for $x \in \mathbb{R}$. Since $\arcsin 0 = 0$ and $\arctan 0 = 0$ we have the following integral representations of the inverse trigonometric functions

$$\arcsin x = \int_0^x \frac{1}{\sqrt{1-t^2}} dt, \quad x \in (-1, 1) \tag{6.10}$$

and

$$\arctan x = \int_0^x \frac{1}{1+t^2} dt, \quad x \in \mathbb{R}. \tag{6.11}$$

Moreover, since $\arcsin x + \arccos x = \frac{\pi}{2}$, we have

$$\arccos x = \frac{\pi}{2} - \int_0^x \frac{1}{\sqrt{1-t^2}} dt, \quad x \in (-1, 1). \tag{6.12}$$

6.2.2 Homework Exercises

1. Suppose $g : [a, b] \to \mathbb{R}$ is strictly increasing, g' exists and is continuous on $[a, b]$ and f is continuous on $[g(a), g(b)]$. Prove the following change of variables formula, often called **u-substitution** in calculus courses (if we let $u(x) = g(x)$ in our formula):

$$\int_{g(a)}^{g(b)} f(x) dx = \int_a^b f(g(x)) g'(x) dx. \tag{6.13}$$

2. In the above exercise, can g strictly increasing be replaced by g strictly monotone? If so, prove the equivalent result. If not, provide a counterexample.

3. In exercise one, can g strictly increasing be removed? If so, prove the equivalent result. If not, provide a counterexample.

6.3 Functions of Bounded Variation

Let $f : [a, b] \to \mathbb{R}$, and let $\mathcal{P} = \{x_i\}_{i=0}^n$ be a partition of $[a, b]$. Consider the quantity

$$V(f, [a, b], \mathcal{P}) := \sum_{k=1}^n |f(x_k) - f(x_{k-1})|.$$

We define the **total variation** of f on $[a, b]$, denoted by $V(f, [a, b])$, to be the quantity

$$V(f, [a, b]) = \sup_{\mathcal{P}} V(f, [a, b], \mathcal{P}), \tag{6.14}$$

where the sup is taken over all possible partitions $\mathcal{P}$ of $[a, b]$.

Clearly $V(f, [a, b])$ can be infinite. To see this consider $\chi_\mathbb{Q}$ on any interval $[a, b]$, with $b > a$. Since $V(\chi_\mathbb{Q}, \mathcal{P}) \geq n - 2$ for any partition $\{x_k\}_{k=0}^n$ of $[a, b]$ where the x_k's alternate between rational and irrational values, we see that $V(f, [a, b]) = \infty$.

If $V(f, [a, b])$ is finite, we say that f is of **bounded variation** on $[a, b]$. One may ask what functions are of bounded variation on a given interval $[a, b]$? Does continuity or differentiability guarantee a function is of bounded variation on an interval? In light of this question, consider the function $f(x) = x \sin \frac{1}{x}$ for $x \neq 0$, and $f(0) = 0$. One can show that f is continuous everywhere, and differentiable everywhere except $x = 0$. However on $[0, 1]$ f is not of bounded variation. To see this, for any $n \in \mathbb{N}$ take the points $\{0, \frac{1}{1.5\pi}, \frac{1}{2.5\pi}, \cdots, \frac{1}{(n+0.5)\pi}, \frac{1}{\pi}, \frac{1}{2\pi}, \cdots, \frac{1}{n\pi}, 1\}$ to define a partition $\mathcal{P}$ of $[0, 1]$. Then $V(f, [0, 1], \mathcal{P}) = \sin 1 + \sum_{k=1}^n \frac{2}{(k+0.5)\pi}$. By making n sufficiently large this can be made larger than any preassigned number. Hence $V(f, [0, 1]) = \infty$.

So we see that a function f being continuous on a closed interval and being differentiable at every point except one is not enough to insure that f is of bounded variation on a closed interval. However, the following theorem does guarantee that f is of bounded variation on $[a, b]$ if f' exists and is continuous on $[a, b]$.

Theorem 6.3.1 *If $f : [a, b] \to \mathbb{R}$ is continuously differentiable on $[a, b]$, then $V(f, [a, b])$ is finite.*

Proof: Since f' is continuous on $[a,b]$, there is an $M > 0$ so that $|f'(x)| \leq M$ on $[a,b]$. Let $\mathcal{P} = \{x_j\}_{j=0}^n$ be any partition of $[a,b]$. By the mean value theorem, for each $j = 1, 2, \cdots, n$ there is a $t_j \in (x_{j-1}, x_j)$ so that

$$|f(x_k) - f(x_{k-1})| = |f'(t_k)|(x_k - x_{k-1}),$$

and so we have

$$V(f, [a,b], \mathcal{P}) = \sum_{k=1}^n |f(x_k) - f(x_{k-1})|$$

$$= \sum_{k=1}^n |f'(t_k)|(x_k - x_{k-1})$$

$$\leq M \sum_{k=1}^n (x_k - x_{k-1}) = M(b-a).$$

□

Theorem 6.3.2 *f is of bounded variation on $[a,b]$ if and only if $-f$ is of bounded variation on $[a,b]$, and $V(f, [a,b]) = V(-f, [a,b])$. Also, f of bounded variation on $[a,b]$ implies $|f|$ is as well, and $V(|f|, [a,b]) \leq V(f, [a,b])$.*

Proof: Clearly $|f(x) - f(y)| = |f(y) - f(x)| = |-f(x) - (-f(y))|$, and so for any partition of $\mathcal{P}$ of $[a,b]$ we have $V(f, [a,b], \mathcal{P}) = V(-f, [a,b], \mathcal{P})$. Thus we see $V(f, [a,b]) = V(-f, [a,b])$. Likewise, $||f(x)| - |f(y)|| \leq |f(x) - f(y)|$, and so $V(|f|, [a,b], \mathcal{P}) \leq V(f, [a,b], \mathcal{P}) \leq V(f, [a,b])$. By taking the sup of $V(|f|, [a,b], \mathcal{P})$ over all possible partitions of $[a,b]$ we get the desired result. □

The next theorem tells us that bounded monotone functions on closed intervals are of bounded variation.

Theorem 6.3.3 *If $f : [a,b] \to \mathbb{R}$ is bounded and monotone on $[a,b]$, then $V(f, [a,b])$ is finite. Moreover, $V(f, [a,b]) = |f(b) - f(a)|$.*

Proof: Since $V(f, [a,b]) = V(-f, [a,b])$, without loss of generality, by otherwise replacing f by $-f$, we may assume f is increasing on $[a,b]$. Then for any partition $\mathcal{P} = \{x_j\}_{j=0}^n$ of $[a,b]$ we have

$$V(f, [a,b], \mathcal{P}) = \sum_{k=1}^n |f(x_{k-1}) - f(x_k)|$$

6.3. FUNCTIONS OF BOUNDED VARIATION

$$= \sum_{k=1}^{n}(f(x_k) - f(x_{k-1}))$$
$$= f(b) - f(a),$$

and so $V(f,[a,b]) = f(b) - f(a) = |f(b) - f(a)|$.
□

The following theorem tells us that functions of bounded variation must be bounded.

Theorem 6.3.4 *If $f : [a,b] \to \mathbb{R}$ is of bounded variation on $[a,b]$, then f is bounded on $[a,b]$.*

Proof: Let $x \in [a,b]$ and let $\mathcal{P} = \{a,x,b\}$. Then

$$||f(x)| - |f(a)|| \le |f(x) - f(a)| \le V(f,[a,b],\mathcal{P}) \le V(f,[a,b]),$$

and so

$$|f(x)| \le |f(a)| + V(f,[a,b]).$$

□

Corollary 6.3.5 *If $f,g : [a,b] \to \mathbb{R}$ are of bounded variation on $[a,b]$, then so is fg.*

Proof: For any $x,y \in [a,b]$ we have

$$|f(x)g(x) - f(y)g(y)| = |f(x)g(x) - f(x)g(y) + f(x)g(y) - f(y)g(y)|$$
$$\le |f(x)||g(x) - g(y)| + |g(y)||f(x) - f(y)|$$
$$\le \sup_{t \in [a,b]} |f(t)| \cdot |g(x) - g(y)| + \sup_{t \in [a,b]} |g(t)| \cdot |f(x) - f(y)|.$$

Then for any partition $\mathcal{P}$ of $[a,b]$ we have

$$V(fg,[a,b],\mathcal{P}) \le \sup_{t \in [a,b]} |g(t)| \cdot V(f,[a,b],\mathcal{P}) + \sup_{t \in [a,b]} |f(t)| \cdot V(g,[a,b],\mathcal{P})$$
$$\le \sup_{t \in [a,b]} |g(t)| \cdot V(f,[a,b]) + \sup_{t \in [a,b]} |f(t)| \cdot V(g,[a,b]),$$

and so

$$V(fg, [a,b], \mathcal{P}) \leq \sup_{t\in[a,b]} |g(t)| \cdot V(f, [a,b]) + \sup_{t\in[a,b]} |f(t)| \cdot V(g, [a,b]).$$

By taking the sup of $V(fg, [a,b], \mathcal{P})$ over all partitions of $[a,b]$, we get

$$V(fg, [a,b]) \leq \sup_{t\in[a,b]} |g(t)| \cdot V(f, [a,b]) + \sup_{t\in[a,b]} |f(t)| \cdot V(g, [a,b]).$$

□

The next theorem tells us that the collection of functions of bounded variation on an interval $[a,b]$ form a linear space.

Theorem 6.3.6 *If $f, g : [a,b] \to \mathbb{R}$ and $c \in \mathbb{R}$, and f and g are of bounded variation, then $f + g$ and cf are also of bounded variation.*

Proof: Let $\mathcal{P} = \{x_k\}_{k=0}^n$ be a partition of $[a,b]$. Then

$$V(cf, [a,b], \mathcal{P}) = \sum_{k=1}^n |cf(x_k) - cf(x_{k-1})|$$

$$= |c| \sum_{k=1}^n |f(x_k) - f(x_{k-1})|$$

$$= |c| V(f, [a,b], \mathcal{P}),$$

and thus

$$V(cf, [a,b]) = |c| V(f, [a,b]).$$

Also,

$$V(f+g, [a,b], \mathcal{P}) = \sum_{k=1}^n |f(x_k) + g(x_k) - f(x_{k-1}) - g(x_{k-1})|$$

$$\leq \sum_{k=1}^n |f(x_k) - f(x_{k-1})| + \sum_{k=1}^n |g(x_k) - g(x_{k-1})|$$

$$= V(f, [a,b], \mathcal{P}) + V(g, [a,b], \mathcal{P})$$

$$\leq V(f, [a,b]) + V(g, [a,b]),$$

6.3. FUNCTIONS OF BOUNDED VARIATION

and thus
$$V(f+g, [a,b], \mathcal{P}) \leq V(f, [a,b]) + V(g, [a,b]).$$

Upon taking the sup of the left hand side of this inequality over all partitions of $[a,b]$ we get
$$V(f+g, [a,b]) \leq V(f, [a,b]) + V(g, [a,b]).$$

$\square$

Let $\mathcal{P}$ be a partition of $[a,b]$ and $\mathcal{Q}$ a refinement of $\mathcal{P}$, we will now show $V(f, [a,b], \mathcal{Q}) \geq V(f, [a,b], \mathcal{P})$ for any $f : [a,b] \to \mathbb{R}$. Suppose $\mathcal{P} = \{x_k\}_{k=0}^n$ and $x^* \in (x_{j-1}, x_j)$ for some $j \in \{1, 2, \cdots, n\}$. We consider first the case where $\mathcal{Q} = \mathcal{P} \cup \{x^*\}$. Then for any $f : [a,b] \to \mathbb{R}$,

$$V(f, [a,b], \mathcal{Q}) - V(f, [a,b], \mathcal{P})$$
$$= |f(x_j) - f(x^*)| + |f(x^*) - f(x_{j-1})| - |f(x_j) - f(x_{j-1})| \geq 0$$

by the triangle inequality. The general case can be seen inductively by adding one point at a time to the previous partition.

The next theorem tells us that total variation increases when going from a subset of an interval to the interval itself. Also, the total variation satisfies an additive property.

Theorem 6.3.7 *Let $f : [a,b] \to \mathbb{R}$. Then for any $[c,d] \subseteq [a,b]$ we have*
$$V(f, [c,d]) \leq V(f, [a,b]).$$

Moreover, for any $e \in (a,b)$, we have
$$V(f, [a,e]) + V(f, [e,b]) = V(f, [a,b]).$$

Proof: Consider $[c,d] \subset [a,b]$ and let $\mathcal{P}$ be a partition of $[c,d]$. Extend $\mathcal{P}$ to be a partition of $[a,b]$ by adding a, b to $\mathcal{P}$, and let $\mathcal{Q}$ denote this partition. Then clearly
$$V(f, [a,b]) \geq V(f, [a,b], \mathcal{Q})$$
$$= V(f, [c,d], \mathcal{P}) + |f(a) - f(c)| + |f(d) - f(b)| \geq V(f, [c,d], \mathcal{P}).$$

Thus
$$V(f, [c,d], \mathcal{P}) \leq V(f, [a,b]),$$

and so by taking the sup over all partitions of $[c,d]$ we see
$$V(f,[c,d]) \leq V(f,[a,b]).$$

Let $e \in (a,b)$, and let $\mathcal{P}_1$ and $\mathcal{P}_2$ be partitions of $[a,e]$ and $[e,b]$ respectively. Consider $\mathcal{P} := \mathcal{P}_1 \cup \mathcal{P}_2$. Then for any $f : [a,b] \to \mathbb{R}$ we have
$$V(f,[a,e],\mathcal{P}_1) + V(f,[e,b],\mathcal{P}_2) = V(f,[a,b],\mathcal{P})$$
$$\leq V(f,[a,b]).$$

By taking the sup of the left hand side over all partitions $\mathcal{P}_1$ of $[a,e]$ and taking the sup over all all partitions $\mathcal{P}_2$ of $[e,b]$ we see that
$$V(f,[a,e]) + V(f,[e,b]) \leq V(f,[a,b]).$$

Conversely, suppose $\mathcal{P}$ is a partition of $[a,b]$. Consider $\mathcal{Q} := \mathcal{P} \cup \{e\}$. For any $f : [a,b] \to \mathbb{R}$ we have
$$V(f,[a,b],\mathcal{P}) \leq V(f,[a,b],\mathcal{Q}).$$

We may slit $\mathcal{Q}$ into two partitions $\mathcal{Q}_1$ of $[a,e]$ and $\mathcal{Q}_2$ of $[e,b]$, and
$$V(f,[a,b],\mathcal{Q}) = V(f,[a,e],\mathcal{Q}_1) + V(f,[e,b],\mathcal{Q}_2).$$

Thus we see
$$V(f,[a,b],\mathcal{P}) \leq V(f,[a,e],\mathcal{Q}_1) + V(f,[e,b],\mathcal{Q}_2) \leq V(f,[a,e]) + V(f,[e,b]).$$

Upon taking the sup of $V(f,[a,b],\mathcal{P})$ over all partitions of $[a,b]$, we see
$$V(f,[a,b]) \leq V(f,[a,e]) + V(f,[e,b]).$$

□

For $x \in \mathbb{R}$ define $x^+ = x$ if $x \geq 0$ and $x^+ = 0$ if $x < 0$, and define $x^- = 0$ if $x > 0$ and $x^- = -x$ if $x \leq 0$. We call x^+ the **positive part** of x, and we call x^- the **negative part** of x. Then for any $x \in \mathbb{R}$ we have
$$x = x^+ - x^-, \quad |x| = x^+ + x^-,$$

6.3. FUNCTIONS OF BOUNDED VARIATION

and so x^+ and x^- may be written as

$$x^+ = \frac{1}{2}(x + |x|), \quad x^- = \frac{1}{2}(|x| - x).$$

Thus

$$(x+y)^+ = \frac{1}{2}(|x+y| + x + y) \le \frac{1}{2}(|x| + |y| + x + y) = x^+ + y^+,$$

and

$$(x+y)^- = \frac{1}{2}(|x+y| - (x+y)) \le \frac{1}{2}(|x| + |y| - x - y) = x^- + y^-.$$

Let $\mathcal{P} = \{x_k\}_{k=0}^n$ be a partition of $[a,b]$ and $f : [a,b] \to \mathbb{R}$. We define

$$P(f, [a,b], \mathcal{P}) := \sum_{k=1}^n (f(x_k) - f(x_{k-1}))^+,$$

and

$$N(f, [a,b], \mathcal{P}) := \sum_{k=1}^n (f(x_k) - f(x_{k-1}))^-.$$

One can see that

$$P(f, [a,b], \mathcal{P}) + N(f, [a,b], \mathcal{P}) = V(f, [a,b], \mathcal{P}),$$

and

$$P(f, [a,b], \mathcal{P}) - N(f, [a,b], \mathcal{P}) = f(b) - f(a).$$

We also define

$$P(f, [a,b]) := \sup_{\mathcal{Q}} P(f, [a,b], \mathcal{Q}), \quad N(f, [a,b]) := \sup_{\mathcal{Q}} N(f, [a,b], \mathcal{Q}),$$

with the sup taken over all possible partitions of $[a,b]$. We call $P(f, [a,b])$ the **positive variation** of f on $[a,b]$ and $N(f, [a,b])$ the **negative variation** of f on $[a,b]$.

If $\mathcal{P}$ is a partition of $[a,b]$ and $\mathcal{Q}$ is a refinement of $\mathcal{P}$, then we have $P(f, [a,b], \mathcal{Q}) \ge P(f, [a,b], \mathcal{P})$ and $N(f, [a,b], \mathcal{Q}) \ge N(f, [a,b], \mathcal{P})$. We will show this inductively, by adding one point at a time to our partition.

If $\mathcal{P} = \{x_k\}_{k=0}^n$ and $x^* \in (x_{j-1}, x_j)$ for some $j \in \{1, 2, \cdots, n\}$, then for the partition $\mathcal{P} \cup \{x^*\}$ we have

$$P(f, [a,b], \mathcal{P} \cup \{x^*\}) - P(f, [a,b], \mathcal{P})$$
$$= [f(x_j) - f(x^*)]^+ + [f(x^*) - f(x_{j-1})]^+ - [f(x_j) - f(x_{j-1})]^+ \geq 0$$

and

$$N(f, [a,b], \mathcal{P} \cup \{x^*\}) - N(f, [a,b], \mathcal{P})$$
$$= [f(x_j) - f(x^*)]^- + [f(x^*) - f(x_{j-1})]^- - [f(x_j) - f(x_{j-1})]^- \geq 0.$$

Theorem 6.3.8

$$P(f, [a,b]) + N(f, [a,b]) = V(f, [a,b])$$

and

$$P(f, [a,b]) - N(f, [a,b]) = f(b) - f(a),$$

whenever any of $V(f, [a,b])$, $P(f, [a,b])$ or $N(f, [a,b])$ are finite, and in this case all are finite. Moreover, in such a case

$$P(f, [a,b]) = \frac{1}{2}[V(f, [a,b]) + f(b) - f(a)],$$

and

$$N(f, [a,b]) = \frac{1}{2}[V(f, [a,b]) - f(b) + f(a)].$$

Proof: For any $f : [a,b] \to \mathbb{R}$ and any partition $\mathcal{P}$ of $[a,b]$ we have

$$V(f, [a,b], \mathcal{P}) = P(f, [a,b], \mathcal{P}) + N(f, [a,b], \mathcal{P}) \leq P(f, [a,b]) + N(f, [a,b]),$$

and so

$$V(f, [a,b], \mathcal{P}) \leq P(f, [a,b]) + N(f, [a,b]).$$

Upon taking the sup of $V(f, [a,b], \mathcal{P})$ over all partitions of $[a,b]$ we see

$$V(f, [a,b]) \leq P(f, [a,b]) + N(f, [a,b]).$$

Since $P(f, [a,b], \mathcal{P})$ and $N(f, [a,b], \mathcal{P})$ differ by the constant $f(b) - f(a)$ for any fixed partition $\mathcal{P}$, we see that $P(f, [a,b])$ is finite if and only if $N(f, [a,b])$ is finite. In such a case, we see $V(f, [a,b])$ is finite as well.

6.3. FUNCTIONS OF BOUNDED VARIATION

Also,
$$P(f,[a,b],\mathcal{P}) + N(f,[a,b],\mathcal{P}) = V(f,[a,b],\mathcal{P}) \leq V(f,[a,b])$$
for any partition $\mathcal{P}$, and so
$$P(f,[a,b],\mathcal{P}) + N(f,[a,b],\mathcal{P}) \leq V(f,[a,b]).$$

If $V(f,[a,b])$ is finite, then $N(f,[a,b],\mathcal{P})$ and $P(f,[a,b],\mathcal{P})$ must both be bounded for any partition $\mathcal{P}$, and thus $P(f,[a,b])$ and $N(f,[a,b])$ must also be finite. Thus we shall assume that $V(f,[a,b])$, $P(f,[a,b])$ and $N(f,[a,b])$ are all finite.

Let $\varepsilon > 0$ be given. Choose $\mathcal{P}_1$ so that $P(f,[a,b],\mathcal{P}_1) > P(f,[a,b]) - \frac{\varepsilon}{2}$, and choose $\mathcal{P}_2$ so that $N(f,[a,b],\mathcal{P}_2) > N(f,[a,b]) - \frac{\varepsilon}{2}$. Let $\mathcal{P}$ be the common refinement $\mathcal{P}_1 \cup \mathcal{P}_2$ of $[a,b]$. Then $P(f,[a,b],\mathcal{P}) > P(f,[a,b]) - \frac{\varepsilon}{2}$ and $N(f,[a,b],\mathcal{P}) > N(f,[a,b]) - \frac{\varepsilon}{2}$. Thus we get
$$P(f,[a,b]) + N(f,[a,b]) - \varepsilon < P(f,[a,b],\mathcal{P}) + N(f,[a,b],\mathcal{P}) \leq V(f,[a,b]).$$

Hence
$$P(f,[a,b]) + N(f,[a,b]) < V(f,[a,b]) + \varepsilon.$$
Since $\varepsilon > 0$ is arbitrary, we have
$$P(f,[a,b]) + N(f,[a,b]) \leq V(f,[a,b]).$$

Since
$$P(f,[a,b],\mathcal{P}) = N(f,[a,b],\mathcal{P}) + f(b) - f(a) \leq N(f,[a,b]) + f(b) - f(a)$$
for any partition $\mathcal{P}$ of $[a,b]$, we see that $P(f,[a,b]) \leq f(b) - f(a) + N(f,[a,b])$, or equivalently
$$P(f,[a,b]) - N(f,[a,b]) \leq f(b) - f(a).$$

Since $P(f,[a,b],\mathcal{P}) - N(f,[a,b],\mathcal{P}) = f(b) - f(a)$, for every partition of $[a,b]$, given $\varepsilon > 0$, we can find a partition $\mathcal{P}_1$ of $[a,b]$ so that
$$P(f,[a,b],\mathcal{P}_1) > P(f,[a,b]) - \varepsilon.$$

Thus
$$N(f, [a, b], \mathcal{P}_1) + f(b) - f(a) = P(f, [a, b], \mathcal{P}_1) > P(f, [a, b]) - \varepsilon.$$

Hence there is a partition $\mathcal{P}_1$ so that
$$N(f, [a, b], \mathcal{P}_1) + f(b) - f(a) > P(f, [a, b]) - \varepsilon.$$

Thus we must have
$$N(f, [a, b]) + f(b) - f(a) \geq P(f, [a, b]) - \varepsilon.$$

Since $\varepsilon > 0$ was arbitrary, we have
$$N(f, [a, b]) + f(b) - f(a) \geq P(f, [a, b]),$$

or equivalently
$$P(f, [a, b]) - N(f, [a, b]) \leq f(b) - f(a).$$

Hence we have established
$$V(f, [a, b]) = P(f, [a, b]) + N(f, [a, b]),$$

and
$$f(b) - f(a) = P(f, [a, b]) - N(f, [a, b]).$$

So we have
$$P(f, [a, b]) = \frac{1}{2}[V(f, [a, b]) + f(b) - f(a)]$$

and
$$N(f, [a, b]) = \frac{1}{2}[V(f, [a, b]) - f(b) + f(a)].$$

□

Theorem 6.3.9 *Let $f : [a, b] \to \mathbb{R}$. Then for any $[c, d] \subseteq [a, b]$ we have*
$$P(f, [c, d]) \leq P(f, [a, b]), \quad N(f, [c, d]) \leq N(f, [a, b]).$$

6.3. FUNCTIONS OF BOUNDED VARIATION

Proof: Consider $[c,d] \subset [a,b]$ and let $\mathcal{P}$ be a partition of $[c,d]$. Extend $\mathcal{P}$ to be a partition of $[a,b]$ by adding a,b to $\mathcal{P}$, and let $\mathcal{Q}$ denote this partition. Then clearly

$$P(f,[a,b]) \geq P(f,[a,b],\mathcal{Q})$$
$$= P(f,[c,d],\mathcal{P}) + [f(a)-f(c)]^+ + [f(d)-f(b)]^+ \geq P(f,[c,d],\mathcal{P}).$$

Thus
$$P(f,[c,d],\mathcal{P}) \leq P(f,[a,b]),$$
and so by taking the sup over all partitions of $[c,d]$ we see
$$P(f,[c,d]) \leq P(f,[a,b]).$$

Also,
$$N(f,[a,b]) \geq N(f,[a,b],\mathcal{Q})$$
$$= N(f,[c,d],\mathcal{P}) + [f(a)-f(c)]^- + [f(d)-f(b)]^- \geq N(f,[c,d],\mathcal{P}).$$

Thus
$$N(f,[c,d],\mathcal{P}) \leq N(f,[a,b]),$$
and so by taking the sup over all partitions of $[c,d]$ we see
$$N(f,[c,d]) \leq N(f,[a,b]).$$

□

The following theorem gives us a characterization of functions of bounded variation on a closed interval.

Theorem 6.3.10 *f is of bounded variation on $[a,b]$ if and only if f is the difference of two bounded increasing functions on $[a,b]$.*

Proof: If $f = f_1 - f_2$, with f_1, f_2 increasing and bounded on $[a,b]$, then we have f is of bounded variation on $[a,b]$. This follows from theorems (6.3.3) and (6.3.6).

Conversely, suppose f is of bounded variation of $[a,b]$. Then by theorem (6.3.7) we know f is of bounded variation on $[a,x]$ for any $x \in (a,b]$. Moreover, by theorem (6.3.8) we may write $f(x)$ as $f(x) = P(f,[a,x]) + f(a) - N(f,[a,x])$, and we know by theorem (6.3.9) $P(f,[a,x])$ and $N(f,[a,x])$ increase as x increases.

□

6.3.1 Homework Exercises

1. Show that
$$f(x) = \begin{cases} x^2 \sin \frac{1}{x} & x \neq 0, \\ 0 & x = 0 \end{cases}$$
is of bounded variation on any compact interval $[a, b] \subset \mathbb{R}$.

2. Suppose f is Lipschitz continuous on $[a, b]$, and in particular
$$|f(x) - f(y)| \leq K|x - y|$$
on $[a, b]$ for some $K > 0$. Show that $V(f, [a, b]) < \infty$. In particular, show $V(f, [a, b]) \leq K(b - a)$.

3. Suppose f is a step function defined on $[a, b]$. Show $V(f, [a, b])$ is finite. In particular, can you find a formula for $V(f, [a, b])$ in terms of the values in the range of f?

4. Suppose f is continuously differentiable on $[a, b]$, show
$$V(f, [a, b]) \leq \int_a^b |f'(x)|dx.$$

5. Suppose f and g are of bounded variation on $[a, b]$ and $|g| \geq \varepsilon > 0$ on $[a, b]$, show that $V(\frac{f}{g}, [a, b]) < \infty$.

6. Show that every function of bounded variation on a closed interval $[a, b]$ has at most a countable number of discontinuities, and they all are of the first kind.

6.4 Rectifiable Curves and Arclength

In this section we define the **arclength** or the **length of a curve** for graphs of functions, and we will see that this can be defined if and only if our functions are of bounded variation.

Suppose $f : [a, b] \to \mathbb{R}$ and $\mathcal{P} = \{x_k\}_{k=0}^n$ is a partition of $[a, b]$. Define
$$\lambda(f, [a, b], \mathcal{P}) := \sum_{k=1}^n \sqrt{(x_k - x_{k-1})^2 + (f(x_k) - f(x_{k-1}))^2}.$$

6.4. RECTIFIABLE CURVES AND ARCLENGTH

This is the sum of lengths of line segments which joint two consecutive points in the collection

$$\{(a, f(a)), (x_1, f(x_1)), \cdots, (x_{n-1}, f(x_{n-1})), (b, f(b))\}.$$

If $\mathcal{Q} = \mathcal{P} \cup \{x^*\}$, where $x^* \in (x_{j-1}, x_j)$ for some $j \in \{1, 2, \cdots, n\}$, then by the triangle inequality we have

$$\lambda(f, [a,b], \mathcal{Q}) - \lambda(f, [a,b], \mathcal{P})$$
$$= \sqrt{(x_k - x^*)^2 + (f(x_k) - f(x^*))^2} + \sqrt{(x^* - x_{k-1})^2 + (f(x^*) - f(x_{k-1}))^2}$$
$$- \sqrt{(x_k - x_{k-1})^2 + (f(x_k) - f(x_{k-1}))^2} \geq 0.$$

By induction we see that for any refinement $\mathcal{Q}$ of $\mathcal{P}$ we have

$$\lambda(f, [a,b], \mathcal{Q}) \geq \lambda(f, [a,b], \mathcal{P}).$$

Define
$$\Lambda(f, [a,b]) := \sup_{\mathcal{P}} \lambda(f, [a,b], \mathcal{P}), \qquad (6.15)$$

where the sup is taken over all possible partitions of $[a, b]$.

If $\Lambda(f, [a, b]) < \infty$ we say that the graph of f on $[a, b]$ is a **rectifiable curve**. Moreover, in such a case we define the **length** or **arclength** of the graph of f on $[a, b]$ to be $\Lambda(f, [a, b])$.

Clearly if f is not continuous on all of $[a, b]$, but has only finitely many discontinuities, this length is actually larger than the sum of the lengths of the graph of f over all the intervals for which f is continuous. Hence if f is not continuous this may not coincide with our intuitive notion of length. However, if f is continuous this extends the notion of length of a curve to curves which are graphs of functions. Previously we only had a notion of length for linear or piecewise linear continuous functions. Moreover, one can see that for linear or continuous piecewise linear functions $\Lambda(f, [a, b])$ coincides with the length given based on sums of lengths of line segments given by the distance formula.

Our next result says that the graph of a function is rectifiable if and only if the function is of bounded variation, and this gives us a useful characterization of rectifiable curves.

Theorem 6.4.1 *Let $f : [a, b] \to \mathbb{R}$. Then the graph of f on $[a, b]$ is rectifiable if and only f is of bounded variation on $[a, b]$.*

Proof: Let $f : [a, b] \to \mathbb{R}$ and $\mathcal{P}$ be a partition of $[a, b]$. Observe that for any real numbers x, y

$$\max\{|x|, |y|\} \leq \sqrt{x^2 + y^2} \leq |x| + |y|.$$

Using this we see that

$$V(f, [a, b], \mathcal{P}) \leq \lambda(f, [a, b], \mathcal{P}) \leq V(f, [a, b], \mathcal{P}) + b - a.$$

Hence either both $V(f, [a, b])$ and $\Lambda(f, [a, b])$ are finite, or neither is.
□

If f is continuously differentiable on $[a, b]$, then f is of bounded variation. Moreover, we have a convenient formula for computing the arclength of the graph of f, namely

$$\Lambda(f, [a, b]) = \int_a^b \sqrt{1 + (f'(x))^2} dx. \tag{6.16}$$

To see this, let $\mathcal{P} = \{x_k\}_{k=0}^n$ be a partition of $[a, b]$. Then by the mean value theorem there is an $t_j \in (x_{j-1}, x_j)$ for each $j = 1, 2, \cdots, n$ so that

$$\lambda(f, [a, b], \mathcal{P}) = \sum_{k=1}^n \sqrt{(x_k - x_{k-1})^2 + (f(x_k) - f(x_{k-1}))^2}$$

$$= \sum_{k=1}^n \sqrt{1 + (\frac{f(x_k) - f(x_{k-1})}{x_k - x_{k-1}})^2} \cdot (x_k - x_{k-1})$$

$$= \sum_{k=1}^n \sqrt{1 + (f'(t_k))^2} \cdot (x_k - x_{k-1}).$$

Let $\mathcal{P}^\dagger$ be the marked partition obtained from $\mathcal{P}$ using the marking points $\{t_j\}_{j=1}^n$. We see that $\lambda(f, [a, b], \mathcal{P}) = S(\sqrt{1 + (f')^2}, [a, b], \mathcal{P}^\dagger)$.

Let $\varepsilon > 0$ be given. Then there is a $\delta > 0$ so that if $\mathcal{P}$ has norm less than δ, then $|S(\sqrt{1 + (f')^2}, [a, b], \mathcal{P}^\dagger) - \int_a^b \sqrt{1 + (f'(x))^2} dx| < \frac{\varepsilon}{2}$.

6.4. RECTIFIABLE CURVES AND ARCLENGTH

Let $\mathcal{P}$ be a partition of $[a,b]$ so that $\lambda(f,[a,b],\mathcal{P}) > \Lambda(f,[a,b]) - \frac{\varepsilon}{2}$. By otherwise adding points[10] to $\mathcal{P}$ we may assume $\|\mathcal{P}\| < \delta$. So in such a case

$$|\lambda(f,[a,b],\mathcal{P}) - \int_a^b \sqrt{1+(f'(x))^2}dx| < \frac{\varepsilon}{2}.$$

Thus we have $\lambda(f,[a,b],\mathcal{P})$ is in the intersection of the intervals

$$(\Lambda(f,[a,b])-\frac{\varepsilon}{2},\Lambda(f,[a,b])]\cap(\int_a^b \sqrt{1+(f'(x))^2}dx-\frac{\varepsilon}{2},\int_a^b \sqrt{1+(f'(x))^2}dx+\frac{\varepsilon}{2}).$$

This implies

$$|\int_a^b \sqrt{1+(f'(x))^2}dx - \Lambda(f,[a,b])| < \varepsilon.$$

Since $\varepsilon > 0$ was arbitrary, we see that

$$\Lambda(f,[a,b]) = \int_a^b \sqrt{1+(f'(x))^2}dx.$$

6.4.1 Homework Exercises

1. Calculate the arclength of the graph of the function $f(x) = \sqrt{1-x^2}$ for $x \in [0,r]$, for $r < 1$. To do this compare the integral formula for the arclength to that of an integral representation of one of the inverse trigonometric functions. What can you say as $r \to 1^-$? Please explain.

2. For a parametrized curve in $\mathbb{R}^n$ given by $\vec{\gamma}(t) = (\gamma_1(t), \gamma_2(t), \cdots, \gamma_n(t))$ for $t \in [a,b]$ and any partition $\mathcal{P} = \{t_k\}_{k=0}^m$ of $[a,b]$, consider

$$\lambda(\vec{\gamma},[a,b],\mathcal{P}) := \sum_{k=1}^m \|\vec{\gamma}(t_k) - \vec{\gamma}(t_{k-1})\|,$$

and let

$$\Lambda(\vec{\gamma},[a,b]) = \sup_{\mathcal{P}} \lambda(\vec{\gamma},[a,b],\mathcal{P}).$$

We say that $\vec{\gamma}$ is rectifiable if $\Lambda(\vec{\gamma},[a,b]) < \infty$. In such a case we call $\Lambda(\vec{\gamma},[a,b])$ the length of $\vec{\gamma}$ over $[a,b]$.

---

[10]Which only makes $\lambda(f,[a,b],\mathcal{P})$ larger or the same in value

(a) Show that $\vec{\gamma}$ is rectifiable if and only if each of $\gamma_1, \gamma_2, \cdots, \gamma_n$ is of bounded variation.

(b) If each of the $\gamma_i(t)$'s are continuously differentiable show that

$$\Lambda(\vec{\gamma}, [a, b]) = \int_a^b \|\frac{d}{dt}\vec{\gamma}(t)\| dt.$$

3. Find the length of the curve which is the graph of $f(x) = \frac{1}{3}(x^2 + 2)^{\frac{3}{2}}$ on $[0, 3]$.

4. Find the length of the curve which is the graph of $f(x) = \frac{1}{4}x^4 + \frac{1}{8x^2}$ on $[1, 2]$.

6.5 Lebesgue's Integrability Criterion

The following section is a discussion of Lebesgue's integrability criterion. It is an optional section, and so nothing else in this text directly requires this result. We start this section by first discussing sets of measure zero in $\mathbb{R}$ and the oscillation of functions on both sets and at points, which is a measure of the range of a function. Using these notions we will then discuss Lebesgue's criterion, which says that f is Riemann integrable if and only if the set of discontinuities of f has measure zero.

6.5.1 Sets of Measure Zero

Let $A \subseteq \mathbb{R}$, we say that A has **measure zero** if given any $\varepsilon > 0$ there exists an at most countable collection of open intervals $\{(a_i, b_i) : i \in \mathcal{I}\}$ so that

$$A \subseteq \bigcup_{i \in \mathcal{I}} (a_i, b_i), \quad \text{with} \quad \sum_{i \in \mathcal{I}} (b_i - a_i) < \varepsilon. \tag{6.17}$$

Clearly a set of measure zero cannot contain any open interval of the form (a, b) with $b > a$.

Sets with finitely many points have measure zero. If $A = \{c_k\}_{k=1}^n$, then for any $\varepsilon > 0$ we may take $\{(c_k - \frac{\varepsilon}{3n}, c_k + \frac{\varepsilon}{3n}) : k = 1, 2, \cdots, n\}$ as our collection of open intervals that cover $\{c_1, c_2, \cdots, c_n\}$. Moreover, the sum of the lengths of these intervals is $\frac{2}{3}\varepsilon$. In fact, countably infinite sets are of measure zero.

6.5. LEBESGUE'S INTEGRABILITY CRITERION

If $A = \{c_k : k \in \mathbb{N}\}$, then for any $\varepsilon > 0$ we may cover A with the collection $\{(c_k - \frac{\varepsilon}{2^{k+2}}, c_k + \frac{\varepsilon}{2^{k+2}}) : k \in \mathbb{N}\}$. Moreover, the sum of the lengths of these intervals is $\sum_{k=1}^{\infty} \frac{\varepsilon}{2^{k+1}} = \frac{\varepsilon}{2}$.

Thus we see that the set of rational numbers has measure zero, and hence there are dense sets in $\mathbb{R}$ which are of measure zero.

6.5.2 The Oscillation of a Function

Given a real-valued function f with a domain D, for any $A \subseteq D$ we define the **oscillation** of f on A to be

$$\text{Osc}(f, A) := \sup_{x \in A} f(x) - \inf_{x \in A} f(x). \tag{6.18}$$

Similarly, at a point $x_0 \in D$ we define the oscillation of f at x_0 to be

$$\text{Osc}(f, x_0) := \lim_{\varepsilon \to 0^+} \text{Osc}(f, (x_0 - \varepsilon, x_0 + \varepsilon) \cap D). \tag{6.19}$$

Note that this limit exists since $\text{Osc}(f, (x_0 - \varepsilon, x_0 + \varepsilon) \cap D)$ decreases as ε decreases.

We observe that f is continuous at x_0 if and only if $\text{Osc}(f, x_0) = 0$. Moreover, the set of discontinuities of f is precisely the set $\{x : \text{Osc}(f, x) > 0\}$, and this set can be written as $\cup_{\delta > 0} \{x : \text{Osc}(f, x) \geq \delta\}$ or $\cup_{n=1}^{\infty} \{x : \text{Osc}(f, x) \geq \frac{1}{n}\}$.

Theorem 6.5.1 *Let $f : D \to \mathbb{R}$. Then for any $\delta > 0$ the set $\{x : \text{Osc}(f, x) < \delta\}$ is open in D.*

Proof: Suppose $\text{Osc}(f, x_0) < \delta$. Then there is a $\varepsilon > 0$ so that $\text{Osc}(f, (x_0 - \varepsilon, x_0 + \varepsilon) \cap D) < \delta$. Thus at any $y \in (x_0 - \varepsilon, x_0 + \varepsilon) \cap D$ we have $\text{Osc}(f, y) < \delta$. So we see that $(x_0 - \varepsilon, x_0 + \varepsilon) \cap D \subseteq \{x : \text{Osc}(f, x) < \delta\}$.
□

Corollary 6.5.2 *Let $f : D \to \mathbb{R}$. Then for any $\delta > 0$ the set $\{x : \text{Osc}(f, x) \geq \delta\}$ is closed in D.*

6.5.3 Lebesgue's Criterion for Riemann Integrability

For notational convenience, for $f : D \to \mathbb{R}$ we define

$$\Omega(f, D, \delta) := \{x \in D : \operatorname{Osc}(f, x) \geq \delta)\}. \tag{6.20}$$

Lemma 6.5.3 *Suppose $f \in \mathcal{R}([a, b])$ and $\delta > 0$, then $\Omega(f, [a, b], \delta)$ has measure zero. In particular, for any $\varepsilon > 0$ there is a finite collection of open intervals whose sum of their lengths is less than ε and whose union covers $\Omega(f, [a, b], \delta)$.*

Proof: Let $f \in \mathcal{R}([a, b])$, let $\delta > 0$ and $\varepsilon > 0$, then there is a partition $\mathcal{P} = \{x_i\}_{i=0}^n$ of $[a, b]$ so that

$$U(f, [a, b], \mathcal{P}) - L(f, [a, b], \mathcal{P}) < \frac{\delta\varepsilon}{2}.$$

Consider the collection of open intervals $\{(x_i - \frac{\varepsilon}{8n}, x_i + \frac{\varepsilon}{8n})\}_{i=0}^n$ that covers the points of $\mathcal{P}$. Define $M := \{k \in \{1, 2, \cdots, n\} : (x_{k-1}, x_k) \cap \Omega(f, [a, b], \delta) \neq \emptyset\}$. Then $\{(x_{k-1}, x_k) : k \in M\}$ is a collection of open intervals which covers $\Omega(f, [a, b], \delta) \setminus \mathcal{P}$.

Using the usual notation

$$M_k := \sup_{x \in [x_{k-1}, x_k]} f(x) \quad \text{and} \quad m_k := \inf_{x \in [x_{k-1}, x_k]} f(x)$$

we see that $k \in M$ implies $M_k - m_k \geq \delta$. Moreover, $\sum_{k \in M} \Delta x_k$, which is the sum of the lengths of the intervals in $\{(x_{k-1}, x_k) : k \in M\}$, satisfies:

$$\frac{\delta\varepsilon}{2} > \sum_{k \in M}(M_k - m_k)\Delta x_k \geq \delta \sum_{k \in M} \Delta x_k,$$

and so $\sum_{k \in M} \Delta x_k < \frac{\varepsilon}{2}$.
□

Lemma 6.5.4 *Suppose $f : [a, b] \to \mathbb{R}$ is bounded, and let $\varepsilon > 0$. If $\operatorname{Osc}(f, x) < \varepsilon$ for all $x \in [a, b]$, then there is a $\delta > 0$ so that for any $[c, d] \subseteq [a, b]$ with $d - c < \delta$ we have $\operatorname{Osc}(f, [c, d]) < \varepsilon$.*

6.5. LEBESGUE'S INTEGRABILITY CRITERION

Proof: Suppose $\text{Osc}(f, x) < \varepsilon$ for every $x \in [a, b]$. Thus for each x there is a $\delta_x > 0$ so that $\text{Osc}(f, (x - \delta_x, x + \delta_x) \cap [a, b]) < \varepsilon$.

$\{(x - \frac{1}{2}\delta_x, x + \frac{1}{2}\delta_x) : a \leq x \leq b\}$ is an open cover for $[a, b]$, and thus there is a finite subcover $\{(x_k - \frac{1}{2}\delta_{x_k}, x_k + \frac{1}{2}\delta_{x_k}) : k = 1, 2, \cdots, n\}$. Let $\delta := \frac{1}{2} \min\{\delta_{x_1}, \delta_{x_2}, \cdots, \delta_{x_n}\}$.

Let $[c, d]$ be any subinterval of $[a, b]$ with $d - c < \delta$. Then there is a $x_j \in \{x_1, x_2, \cdots, x_n\}$ so that $[c, d] \cap (x_j - \frac{1}{2}\delta_{x_j}, x_j + \frac{1}{2}\delta_{x_j}) \neq \emptyset$. Since $[c, d]$ had length less than $\frac{1}{2}\delta_{x_j}$, we have $[c, d] \subseteq (x_j - \delta_{x_j}, x_j + \delta_{x_j})$. Thus,

$$\text{Osc}(f, [c, d]) \leq \text{Osc}(f, (x_j - \frac{1}{2}\delta_{x_j}, x_j + \frac{1}{2}\delta_{x_j}) \cap [a, b]) < \varepsilon.$$

□

Theorem – Lebesgue's Integrability Criterion 6.5.5 *Suppose that a function $f : [a, b] \to \mathbb{R}$ is a bounded function, then $f \in \mathcal{R}([a, b])$ if and only if the set of discontinuities of f on $[a, b]$ has measure zero.*

Proof: Let us first suppose $f \in \mathcal{R}([a, b])$. Let $\varepsilon > 0$, and define $\Omega := \bigcup_{n=1}^{\infty} \Omega(f, [a, b], \frac{1}{n})$, which is the set of discontinuities of f on $[a, b]$. By lemma (6.5.3) we know that each $\Omega(f, [a, b], \frac{1}{n})$ can be covered by a finite collection of open intervals $C_n = \{I_{n,1}, I_{n,2}, \cdots, I_{n,k_n}\}$ whose sum of lengths is less than $\frac{\varepsilon}{2^n}$. Consider $C = \bigcup_{n=1}^{\infty} C_n$. Then C is the countable union of finite collections of intervals, and thus is a countable union of intervals. Moreover, Ω can be covered by the intervals in C, and the sum of the lengths of all the intervals in C is less than $\sum_{k=1}^{\infty} \frac{\varepsilon}{2^k} = \varepsilon$. Hence we see that Ω has measure zero.

Conversely, suppose that the set of discontinuities of f in $[a, b]$ has measure zero. We shall assume f is not constant on $[a, b]$, since otherwise we are done. Let $\varepsilon > 0$, and let $M := \sup_{x \in [a,b]} f(x)$, and $m := \inf_{x \in [a,b]} f(x)$. Using the notation Ω for the set of discontinuities of f on $[a, b]$, we see that

$$\Omega = \bigcup_{\delta > 0} \Omega(f, [a, b], \delta).$$

Since Ω has measure zero, each $\Omega(f, [a, b], \delta)$ also has measure zero. Thus for $\delta = \frac{\varepsilon}{2(b-a)}$ we have that there is a countable collection of intervals $\{I_j : j \in \mathcal{I}\}$ so that $\Omega(f, [a, b], \delta) \subseteq \bigcup_{j \in \mathcal{I}} I_j$ and the sum of the lengths of the intervals in $\{I_j : j \in \mathcal{I}\}$ is less than $\frac{\varepsilon}{2(M-m)}$. Since $\Omega(f,[a,b],\delta)$ is compact, there is a finite subset of $\{I_j : j \in \mathcal{I}\}$, let's say $\{I_1, I_2, \cdots, I_N\}$, whose union covers $\Omega(f,[a,b],\delta)$.

By DeMorgan's law we see $[a,b] \setminus \bigcup_{j=1}^{N} I_j = \bigcap_{j=1}^{N}([a,b] \setminus I_j)$. If $x \in \bigcap_{j=1}^{N}([a,b] \setminus I_j)$ we have $x \notin \Omega(f,[a,b],\delta)$, and so $\mathrm{Osc}(f,x) < \delta$.

By lemma (6.5.4), for each $j \in \{1, 2, \cdots, N\}$ there is a δ_j so that $[a,b] \setminus I_j$ can be subdivided into a finite collection of closed intervals $\{C_{j,k}\}_{k=1}^{K_j}$, each of length less than δ_j, so that

$$\mathrm{Osc}(f, C_{j,k}) < \delta.$$

Define a partition $\mathcal{P}$ of $[a,b]$ that consists of $\{a,b\}$, the endpoints of of I_j, $j = 1, 2, \cdots, N$, and the endpoints of all the intervals $\{C_{j,k}\}_{k=1}^{K_j}$, for $j = 1, 2, \cdots, N$. Letting $\mathcal{P} = \{x_i\}_{i=0}^{n}$ and using the usual notation we have

$$U(f,[a,b],\mathcal{P}) - L(f,[a,b],\mathcal{P}) = \sum_{i=1}^{n}(M_i - m_i)\Delta x_i$$

$$= \sum_{i \in A}(M_i - m_i)\Delta x_i + \sum_{i \in B}(M_i - m_i)\Delta x_i,$$

where $A = \{i \in \{1, 2, \cdots, n\} : [x_{i-1}, x_i] \cap \bigcup_{j=1}^{N} I_j \neq \emptyset\}$, and $B = \{i \in \{1, 2, \cdots, n\} : [x_{i-1}, x_i] \cap \bigcup_{j=1}^{N} I_j = \emptyset\}$.

Now,

$$\sum_{i \in A}(M_i - m_i)\Delta x_i \leq (M - m)\sum_{i \in A}\Delta x_i$$

$$< (M - m) \cdot \frac{\varepsilon}{2(M - m)} = \frac{\varepsilon}{2}.$$

Also, $i \in B$ implies $[x_{i-1}, x_i] \subseteq \bigcap_{j=1}^{N}([a,b] \setminus I_j)$. Thus $\mathrm{Osc}(f,[x_{i-1}, x_i]) < \delta$, which implies $M_i - m_i < \delta$. Hence,

$$\sum_{i \in B}(M_i - m_i)\Delta x_i < \delta \sum_{i \in B}\Delta x_i$$

$$\leq \delta(b - a) = \frac{\varepsilon}{2}.$$

Therefore,
$$U(f, [a,b], \mathcal{P}) - L(f, [a,b], \mathcal{P}) < \varepsilon.$$

□

6.5.4 Homework Exercises

1. Show that the Cantor set has measure zero. Conclude that there are uncountable sets of measure zero.

2. Use Lebesgue's criterion to show that Thomae's function f is integrable on any $[a,b]$ where f is given by: $f(x) = 0$ if $x \notin \mathbb{Q}$, $f(0) = 0$ and $f(\frac{m}{n}) = \frac{1}{n}$ where $m, n \in \mathbb{Z}$, $n > 0$, and the greatest common divisor of m and n is one.

3. Use Lebesgue's criterion to show $\chi_\mathbb{Q}$ is not Riemann integrable on any non-trivial interval $[a,b]$.

4. Suppose $f, g \in \mathcal{R}([a,b])$ and $\int_a^b |f - g| dx = 0$, show the set of x's in $[a,b]$ for which $f(x) \neq g(x)$ has measure 0.

5. Suppose $f \in \mathcal{R}([a,b])$ and define $F(x) = \int_a^x f(t) dt$. Let S be the set of x values for which $F'(x) = f(x)$ on $[a,b]$. Show that $[a,b] \setminus S$ has measure 0.

Chapter 7

Sequences and Series of Functions

7.1 Point-wise Convergence

Suppose $\{f_n(x)\}_{n=1}^{\infty}$ is a family of real-valued functions defined on an interval $[a,b]$. We say that f_n **converge point-wise** to a function f on $[a,b]$ if

$$\lim_{n\to\infty} f_n(x) = f(x) \quad \forall x \in [a,b].$$

We may use the notation

$$f_n \to f \quad point-wise$$

to denote that the sequence of functions $\{f_n\}_{n=1}^{\infty}$ converges to f point-wise.

We may wish to know what properties are preserved under point-wise convergence. For instance, if the f_n's are all continuous, differentiable, or integrable is f necessarily continuous, differentiable or integrable? More precisely, if $f_n \to f$ point-wise, does $\frac{d}{dx}f_n \to \frac{d}{dx}f$ point-wise? or does $\int_a^b f_n(x)dx \to \int_a^b f(x)dx$? or does

$$\lim_{x\to x_0} \lim_{n\to\infty} f_n(x) = \lim_{n\to\infty} \lim_{x\to x_0} f_n(x)?$$

In general, the answer to all of these questions is no.

An Example of a Sequence of Continuous Functions That Converges Point-Wise to a Function With a Discontinuity

Consider $f_n(x) : [0, 1] \to \mathbb{R}$ defined by

$$f_n(x) = \begin{cases} -2^n x + 1 & 0 \le x \le \frac{1}{2^n}, \\ 0 & \frac{1}{2^n} < x \le 1. \end{cases}$$

In such a case each f_n is continuous, and $f_n \to f$ point-wise where

$$f = \begin{cases} 1 & x = 0, \\ 0 & 0 < x \le 1. \end{cases}$$

Clearly f is not continuous at 0, and thus

$$0 = \lim_{x \to 0} \lim_{n \to \infty} f_n(x) \ne \lim_{n \to \infty} \lim_{x \to 0} f_n(x) = 1.$$

An Example of a Sequence of Functions That Converges Point-wise Whose Sequence of Integrals Does Not Converge to the Integral of the Limit Function

Consider the sequence of functions $f_n : [0, 1] \to \mathbb{R}$ given by

$$f_n(x) = \begin{cases} 4n^2 x & 0 \le x \le \frac{1}{2n}, \\ 4n - 4n^2 x & \frac{1}{2n} < x \le \frac{1}{n}, \\ 0 & \frac{1}{n} < x \le 1. \end{cases}$$

Then $f_n \to f \equiv 0$ point-wise, and so $\int_0^1 f(x)dx = 0$. However $\int_0^1 f_n(x)dx = 1$. Thus

$$1 = \lim_{n \to \infty} \int_0^1 f_n(x)dx \ne \int_0^1 \lim_{n \to \infty} f_n(x)dx = \int_0^1 f(x)dx = 0.$$

An Example of a Sequence of Functions That Converges Point-Wise Whose Derivatives Do Not Converge to the Derivative of the Limit Function

Consider the sequence of functions $\{f_n\}_{n=1}^\infty$ with

$$f_n(x) = \frac{\sin nx}{\sqrt{n}}, \quad x \in \mathbb{R}.$$

7.1. POINT-WISE CONVERGENCE

Then $f_n \to f \equiv 0$ point-wise on any compact interval $[a, b]$. However

$$\frac{d}{dx} f_n(x) = \sqrt{n} \cos nx$$

does not converge to any real-valued function point-wise on any interval $[a, b]$. Moreover, $\frac{d}{dx} f = 0$ and thus $\frac{d}{dx} f_n(x)$ does not converge point-wise to $\frac{d}{dx} f$.

Thus we see that point-wise convergence in general does not preserve continuity, and the sequences of derivatives or integrals may not converge to the derivative or integral of the limit function. To preserve such properties or have such limits we will need a stronger notion of convergence that will allow us to interchange certain limit operations.

7.1.1 Homework Exercises

1. For $f_n : [0, 1] \to \mathbb{R}$ given by $f_n(x) = x^n$, find f so that $f_n \to f$ point-wise. Is f continuous? or differentiable? or integrable? Moreover, if so, does $\frac{d}{dx} f_n \to \frac{d}{dx} f$ point-wise? Is $\lim_{x \to x_0} \lim_{n \to \infty} f_n(x) = \lim_{n \to \infty} \lim_{x \to x_0} f_n(x)$ for any $x_0 \in [0, 1]$? Does $\int_0^1 f_n(x) dx \to \int_0^1 f(x) dx$?

2. For $f_n : \mathbb{R} \to \mathbb{R}$ given by $f_n(x) = \frac{\sin nx}{n}$, find f so that $f_n \to f$ point-wise. For any $[a, b]$ is f continuous? or differentiable? or integrable? Moreover, if so, does $\frac{d}{dx} f_n \to \frac{d}{dx} f$ point-wise? Is $\lim_{x \to x_0} \lim_{n \to \infty} f_n(x) = \lim_{n \to \infty} \lim_{x \to x_0} f_n(x)$ for any $x_0 \in [a, b]$? Does $\int_a^b f_n(x) dx \to \int_a^b f(x) dx$?

3. Define $f_n : [0, 1] \to \mathbb{R}$ by

$$f_n(x) = \begin{cases} 0 & x = 0, \\ 2n - 2n^2 x & 0 < x \leq \frac{1}{n}, \\ 0 & \frac{1}{n} < x \leq 1. \end{cases}$$

Show f_n converges point-wise on $[0, 1]$ and find the limit function f. Show that f_n and f are all integrable on $[0, 1]$ but $\lim_{n \to \infty} \int_0^1 f_n(x) dx \neq \int_0^1 f(x) dx$.

4. Define $f_n : [0, 1] \to \mathbb{R}$ by

$$f_n(x) = \begin{cases} 0 & x = 0, \\ 3n + 1 & 0 < x \leq \frac{1}{n}, \\ 0 & \frac{1}{n} < x \leq 1. \end{cases}$$

Show f_n converges point-wise on $[0, 1]$ and find the limit function f. Show that f_n and f are all integrable on $[0, 1]$ but $\lim_{n \to \infty} \int_0^1 f_n(x)dx \neq \int_0^1 f(x)dx$.

5. Show that $\sqrt[2n-1]{x} \to 1$ on $(0, 1]$ and $\sqrt[2n-1]{x} \to -1$ on $[-1, 0)$. Use this to show that $f_n(x) := x^{\frac{2n}{2n-1}}$ converges point-wise to $f(x) := |x|$ on $[-1, 1]$. Moreover, show that each f_n is differentiable on $[-1, 1]$, but f is not differentiable at 0, and so $\frac{d}{dx}f_n \not\to \frac{d}{dx}f$ on $[-1, 1]$.

7.2 Uniform Convergence

For a sequence of functions $\{f_n\}_{n=1}^{\infty}$ with $f_n : [a, b] \to \mathbb{R}$, we say that $\{f_n\}_{n=1}^{\infty}$ **converges uniformly** to a limit function $f : [a, b] \to \mathbb{R}$ if given $\varepsilon > 0$, there is an $N \in \mathbb{N}$ so that for any $x \in [a, b]$ and any $n > N$ we have

$$|f_n(x) - f(x)| < \varepsilon.$$

We may use the notation

$$f_n \to f \quad uniformly$$

to denote that the sequence of functions $\{f_n\}_{n=1}^{\infty}$ converges to f uniformly.

Roughly speaking, uniform convergence requires not only point-wise convergence but also convergence at the same rate for each x. Hence uniform convergence implies point-wise convergence, but the converse is not true.

We define the notation

$$\|f\|_{\sup,A} := \sup_{x \in A} |f(x)|. \tag{7.1}$$

Then $f_n \to f$ uniformly on a set A if and only if

$$\|f_n - f\|_{\sup,A} \to 0 \quad as \quad n \to \infty.$$

To see this, we first suppose that f_n converges uniformly to f on A. Thus, given $\varepsilon > 0$, there is an $N \in \mathbb{N}$ so that for $n > N$ we have $|f_n(x) - f(x)| < \frac{\varepsilon}{2}$ for every $x \in A$. Thus we must have $\|f_n - f\|_{\sup,A} = \sup_{x \in A} |f_n(x) - f(x)| \leq$

7.2. UNIFORM CONVERGENCE

$\frac{\varepsilon}{2} < \varepsilon$ for $n > N$. Conversely, suppose for $\varepsilon > 0$ there is an $N \in \mathbb{N}$ so that for $n > N$ we have $\|f_n - f\|_{\sup,A} < \varepsilon$. Thus we clearly have $|f_n(x) - f(x)| < \varepsilon$ for $n > N$ and all $x \in A$.

The following theorem gives a **Cauchy convergence criterion** for uniform convergence.

Theorem 7.2.1 *Let $\{f_n\}_{n=1}^{\infty}$ be a sequence of functions defined on A. Then $\{f\}_{n=1}^{\infty}$ converges uniformly on A if and only if for any $\varepsilon > 0$ there is an $N \in \mathbb{N}$ so that for $m, n > N$ we have $|f_n(x) - f_m(x)| < \varepsilon$ for any $x \in A$.*

Proof: Let $\varepsilon > 0$ and suppose $\{f_n\}_{n=1}^{\infty}$ converges uniformly on A to some function f. Then there is an $N \in \mathbb{N}$ so that for $n > N$ we have $|f_n(x) - f(x)| < \frac{\varepsilon}{2}$ for all $x \in A$. Thus for any $x \in A$ and any $n, m > N$ we have

$$|f_n(x) - f_m(x)| = |f_n(x) - f(x) + f(x) - f_m(x)|$$
$$\leq |f_n(x) - f(x)| + |f_m(x) - f(x)| < \frac{\varepsilon}{2} + \frac{\varepsilon}{2} = \varepsilon.$$

Conversely, suppose that there is an $N \in \mathbb{N}$ so that for $n, m > N$ we have $|f_n(x) - f_m(x)| < \frac{\varepsilon}{2}$ for all $x \in A$. Then for each $x \in A$ the sequence $\{f_k(x)\}_{k=1}^{\infty}$ is a Cauchy sequence of real numbers, and hence has a limit value which we denote by $f(x)$. Thus $\{f_k\}_{k=1}^{\infty}$ converges to the function f pointwise on A. We need to show that this convergence is also uniform.

For any $x \in A$ and any $m, n > N$ we have

$$||f_n(x) - f(x)| - |f_m(x) - f(x)|| \leq |f_n(x) - f_m(x)| < \frac{\varepsilon}{2}.$$

Letting $m \to \infty$ and keeping $n > N$ fixed, we have

$$|f_n(x) - f(x)| \leq \frac{\varepsilon}{2} < \varepsilon.$$

Thus we see that $\{f_k\}_{k=1}^{\infty}$ converges to f uniformly on A.
□

Theorem 7.2.2 *Suppose $\{f_n\}_{n=1}^{\infty}$ is a sequence of bounded functions on a set A that converges uniformly on A to a function f, then f is bounded. Moreover, if $\|f_n\|_{\sup,A} \leq M$ for all $n \in \mathbb{N}$, then $\|f\|_{\sup,A} \leq M$.*

Proof: Let $x \in A$ and $\varepsilon > 0$. Then there is an $N \in \mathbb{N}$ so that if $n > N$ we have $|f(x) - f_n(x)| < \varepsilon$. Thus

$$|f(x)| = |f(x) - f_n(x) + f_n(x)|$$
$$\leq |f(x) - f_n(x)| + |f_n(x)|$$
$$\leq \varepsilon + \|f_n\|_{\sup,A}.$$

If $\|f_n\|_{L^\infty(A)} \leq M$, then we have

$$\|f\|_{\sup,A} \leq M + \varepsilon.$$

Since $\varepsilon > 0$ is arbitrary, we get $\|f\|_{\sup,A} \leq M$ as well.
□

7.2.1 Uniform Convergence and Continuity

The following theorem tells us that uniform convergence allows us to interchange limit operations. In particular, this guarantees that the uniform limit of continuous functions is also continuous.

Theorem 7.2.3 *Suppose $\{f_n\}_{n=1}^\infty$ is a sequence of functions that converges uniformly to a function f on a set $D \subseteq \mathbb{R}$. Let $x_0 \in \overline{D}$, and suppose for each f_n that $\lim_{x \to x_0} f_n(x)$ exists and is finite, then $\lim_{x \to x_0} f(x)$ exists and is finite. Moreover,*

$$\lim_{n \to \infty} \lim_{x \to x_0} f_n(x) = \lim_{x \to x_0} \lim_{n \to \infty} f_n(x).$$

Proof: Let $L_n := \lim_{x \to x_0} f_n(x)$. We will first show $L = \lim_{n \to \infty} L_n$ exists.

Let $\varepsilon > 0$ be given. Then there is an $N \in \mathbb{N}$ so that for $n, m > N$, we have

$$|f_n(x) - f_m(x)| < \frac{\varepsilon}{3}$$

for every $x \in D$. Moreover, there is a $\delta_n > 0$ so that for $x \in D$ which satisfies $|x - x_0| < \delta_n$ we have

$$|f_n(x) - L_n| < \frac{\varepsilon}{3}.$$

Thus we have for $m, n > N$ and $|x - x_0| < \min\{\delta_n, \delta_m\}$ we have

$$|L_m - L_n| = |L_m - f_m(x) + f_m(x) - f_n(x) + f_n(x) - L_n|$$

7.2. UNIFORM CONVERGENCE

$$\leq |L_m - f_m(x)| + |f_m(x) - f_n(x)| + |f_n(x) - L_n|$$
$$< \frac{\varepsilon}{3} + \frac{\varepsilon}{3} + \frac{\varepsilon}{3} = \varepsilon.$$

Hence $\{L_n\}_{n=1}^{\infty}$ is a Cauchy sequence of real numbers, and thus converges to some limit value L. By letting $m \to \infty$ in $|L_n - L_m|$ we see that for $n > N$ we have $|L_n - L| \leq \varepsilon$.

There is a $M \in \mathbb{N}$ so that for $n > M$ we have $|L_n - L| < \frac{\varepsilon}{3}$ and $|f_n(x) - f(x)| < \frac{\varepsilon}{3}$ for any $x \in D$. Moreover, for $x \in D$ satisfying $|x - x_0| < \delta_n$ we have $|L_n - f_n(x)| < \frac{\varepsilon}{3}$. Thus for such an x and n we have

$$|f(x) - L|$$
$$= |f(x) - f_n(x) + f_n(x) - L_n + L_n - L|$$
$$\leq |f(x) - f_n(x)| + |f_n(x) - L_n| + |L_n - L|$$
$$< \frac{\varepsilon}{3} + \frac{\varepsilon}{3} + \frac{\varepsilon}{3} = \varepsilon.$$

□

Corollary 7.2.4 *Suppose $\{f_n\}_{n=1}^{\infty}$ is a sequence of functions that converges uniformly to a function f on a set $D \subseteq \mathbb{R}$. Suppose each f_n is continuous at some point $x_0 \in D$, then f is continuous at x_0.*

Hence if $\{f_n\}_{n=1}^{\infty}$ is a sequence of continuous functions on a set D that also converge uniformly to a function f on D, then f is continuous on D.

The following theorem due to Dini is a partial converse of corollary (7.2.4).

Dini's Uniform Convergence Theorem 7.2.5 *Suppose $\{f_n\}_{n=1}^{\infty}$ is a sequence of continuous functions that converges point-wise to a continuous functions f on a compact set K. Suppose that for all $x \in K$ the sequences $\{f_n(x)\}_{n=1}^{\infty}$ are all monotone increasing or all monotone decreasing. Then $f_n \to f$ uniformly on K.*

Proof: By otherwise replacing each of the f_n's by $-f_n$, we may assume that each sequence $\{f_n(x)\}_{n=1}^{\infty}$ is monotone decreasing.

We proceed by contradiction. Suppose the contrary, then the sequence $\{\|f_n - f\|_{\sup,K}\}_{n=1}^{\infty}$ does not converge to zero. Thus there is a $\varepsilon > 0$ so that

$\|f_n - f\|_{\sup,K} > \varepsilon$ for infinitely many $n \in \mathbb{N}$. By otherwise restricting to a subsequence of $\{f_n\}_{n=1}^\infty$, we may assume $\|f_n - f\|_{\sup,K} > \varepsilon$ for all $n \in \mathbb{N}$.

Since K is compact and each f_n is continuous, there is a sequence $\{x_k\}_{k=1}^\infty$ so that $\|f_k - f\|_{\sup,K} = f_k(x_k) - f(x_k)$. By the Bolzano-Weierstrass theorem (2.4.10) we know that there is a subsequence of $\{x_k\}_{k=1}^\infty$ that converges to some point $\xi \in K$. Let $\{x_{n_k}\}_{k=1}^\infty$ denote such a subsequence. Hence $f_{n_k}(x_{n_k}) - f(x_{n_k}) > \varepsilon$ for $k = 1, 2, \cdots$.

Let $N \in \mathbb{N}$. Then for $n_k > N$ we know $f_N(x_{n_k}) \geq f_{n_k}(x_{n_k})$, and so for all k so that $n_k > N$ we have $f_N(x_{n_k}) - f(x_{n_k}) > \varepsilon$. By letting $k \to \infty$ we get via the continuity of f_N and f that $f_N(\xi) - f(\xi) \geq \varepsilon$. Since $N \in \mathbb{N}$ is arbitrary, this is a contradiction to the fact that $f_n(\xi) \to f(\xi)$ as $n \to \infty$. $\square$

7.2.2 Uniform Convergence and Integration

The next theorem tells us that if $\{f_n\}_{n=1}^\infty$ is a sequence of integrable functions on $[a, b]$ which converge uniformly on $[a, b]$ to a function f, then $f \in \mathcal{R}([a, b])$ and the sequence of integrals $\{\int_a^b f_n(x) dx\}_{n=1}^\infty$ converges to $\int_a^b f(x) dx$. Then this allows us to interchange limits and integration when a sequence of integrable functions converges uniformly on a compact interval.

Theorem 7.2.6 *Suppose $\{f_n\}_{n=1}^\infty$ is a sequence of integrable functions on $[a, b]$ that converge uniformly on $[a, b]$ to a function f. If $f_n \in \mathcal{R}([a, b])$ for all n, then $f \in \mathcal{R}([a, b])$. Moreover, $\int_a^b f_n(x) dx \to \int_a^b f(x) dx$ as $n \to \infty$.*

Proof: Let $\varepsilon > 0$ be given. Then there is an $N \in \mathbb{N}$ so that for any $x \in [a, b]$ and $n, m > N$ we have $|f_n(x) - f_m(x)| < \frac{\varepsilon}{b-a}$, and thus $|\int_a^b (f_n(x) - f_m(x)) dx| < \varepsilon$. Hence $\{\int_a^b f_n(x) dx\}_{n=1}^\infty$ is a Cauchy sequence of real numbers, and thus has a limit value.

Since $f_n \in \mathcal{R}([a, b])$ for each n, there is a a partition $\mathcal{P}_n$ so that

$$U(f_n, [a, b], \mathcal{P}_n) - L(f_n, [a, b], \mathcal{P}_n) < \frac{\varepsilon}{3}.$$

7.2. UNIFORM CONVERGENCE

Let $M \in \mathbb{N}$ be so that for $n > M$ we have $|f_n(x) - f(x)| < \frac{\varepsilon}{3(b-a)}$ for any $x \in [a,b]$. Then for the partition $\mathcal{P}_n = \{x_j\}_{j=0}^{K}$ we have

$$U(f, [a,b], \mathcal{P}_n) - L(f, [a,b], \mathcal{P}_n) = \sum_{j=1}^{K} (\sup_{x \in [x_{j-1}, x_j]} f(x) - \inf_{x \in [x_{j-1}, x_j]} f(x))\Delta x_j$$

$$= |\sum_{j=1}^{K} (\sup_{x \in [x_{j-1}, x_j]} f(x) - \sup_{x \in [x_{j-1}, x_j]} f_n(x) + \sup_{x \in [x_{j-1}, x_j]} f_n(x)$$

$$- \inf_{x \in [x_{j-1}, x_j]} f_n(x) + \inf_{x \in [x_{j-1}, x_j]} f_n(x) - \inf_{x \in [x_{j-1}, x_j]} f(x))\Delta x_j|$$

$$\leq \sum_{j=1}^{K} |\sup_{x \in [x_{j-1}, x_j]} f_n(x) - \sup_{x \in [x_{j-1}, x_j]} f(x)|\Delta x_j +$$

$$U(f_n, [a,b], \mathcal{P}_n) - L(f_n, [a,b], \mathcal{P}_n) + \sum_{j=1}^{K} |\inf_{x \in [x_{j-1}, x_j]} f_n(x) - \inf_{x \in [x_{j-1}, x_j]} f(x)|\Delta x_j$$

$$\leq \sum_{j=1}^{K} \frac{\varepsilon}{3(b-a)}\Delta x_j + \frac{\varepsilon}{3} + \sum_{j=1}^{K} \frac{\varepsilon}{3(b-a)}\Delta x_j$$

$$= \frac{\varepsilon}{3} + \frac{\varepsilon}{3} + \frac{\varepsilon}{3} = \varepsilon.$$

Thus we see $f \in \mathcal{R}([a,b])$. To see that $\int_a^b f_n(x)dx \to \int_a^b f(x)dx$ as $n \to \infty$ we observe that for $n > N$ we have

$$|\int_a^b f_n(x)dx - \int_a^b f(x)dx|$$

$$= |\int_a^b (f_n(x) - f(x))dx| < \varepsilon.$$

□

7.2.3 Uniform Convergence and Differentiation

Ideally we would like a result that says if $\{f_n\}_{n=1}^{\infty}$ is a sequence of differentiable functions on a set D that converges uniformly on D to some function f, then f is differentiable, and $f_n' \to f'$ on D point-wise. To see that this is not true in general, consider the function $f_n(x) = \frac{\sin nx}{\sqrt{n}}$ on any interval interval I. Here $f_n \to 0$ uniformly on I. However, $f_n'(x) = \sqrt{n} \cos nx$ does not converge even point-wise since $f_n' \to \infty$ at all x values. However in light of this we do have the following theorem.

Theorem 7.2.7 *Suppose $\{f_n\}_{n=1}^\infty$ is a sequence of differentiable functions on $[a,b]$ with the sequence of derivatives $\{f_n'(x)\}_{n=1}^\infty$ converging uniformly on $[a,b]$. Suppose for some $x_0 \in [a,b]$ that the sequence $\{f_n(x_0)\}_{n=1}^\infty$ converges, then the sequence of functions $\{f_n\}_{n=1}^\infty$ converges uniformly on $[a,b]$ to some function f which on $[a,b]$ satisfies*

$$f'(x) = \lim_{n \to \infty} f_n'(x).$$

Proof: Let $\varepsilon > 0$ be given and let $N \in \mathbb{N}$ be such that for $n, m > N$ we have $|f_n(x_0) - f_m(x_0)| < \frac{\varepsilon}{2}$ and $|f_n'(x) - f_m'(x)| < \frac{\varepsilon}{2(b-a)}$ for any $x \in [a,b]$. By the mean value theorem (5.3.1), for any $z > y$ in $[a,b]$ and $n, m > N$ we have

$$|f_n(z) - f_m(z) - f_n(y) + f_m(y)| = |f_n'(t) - f_m'(t)||z - y|$$

for some $t \in (y, z)$. Hence for $n, m > N$ we have

$$|f_n(z) - f_m(z) - f_n(y) + f_m(y)| < \frac{\varepsilon}{2(b-a)} \cdot |z - y| \le \frac{\varepsilon}{2}.$$

Thus for any $x \in [a,b]$ and $n, m > N$ we have

$$|f_n(x) - f_m(x)| = |f_n(x) - f_m(x) - f_n(x_0) + f_m(x_0) + f_n(x_0) - f_m(x_0)|$$
$$\le |f_n(x) - f_m(x) - f_n(x_0) + f_m(x_0)| + |f_n(x_0) - f_m(x_0)|$$
$$< \frac{\varepsilon}{2} + \frac{\varepsilon}{2} = \varepsilon.$$

Thus we see $\{f_n\}_{n=1}^\infty$ converges uniformly on $[a,b]$ to some function f and let g be the limit function of the sequence of functions $\{f_n'\}_{n=1}^\infty$ on $[a,b]$.

We will now show $g = f'$ on $[a,b]$, that is

$$\lim_{n \to \infty} \frac{d}{dx} f_n(x) = \frac{d}{dx} \lim_{n \to \infty} f_n(x)$$

on $[a,b]$. Fix $x \in [a,b]$ and consider the the quantities $h_n(t) := \frac{f_n(t) - f_n(x)}{t - x}$ for $t \in [a,b]$, $t \ne x$, $n \in \mathbb{N}$, and $h(t) := \frac{f(t) - f(x)}{t - x}$.

7.2. UNIFORM CONVERGENCE

Then for each $n \in \mathbb{N}$ we have $\lim_{t \to x} h_n(t) = f'_n(x)$. As above, we see that

$$|h_n(t) - h_m(t)| \leq \frac{\varepsilon}{2(b-a)}$$

for any $n, m > N$. Thus for any $t \in [a, b]$, $t \neq x$, we have that $\{h_n(t)\}_{n=1}^{\infty}$ converges uniformly to $h(t)$. By theorem (7.2.3), on $[a, b] \setminus \{x\}$ we have

$$\lim_{t \to x} \lim_{n \to \infty} h_n(t) = \lim_{n \to \infty} \lim_{t \to x} h_n(t).$$

That is

$$\lim_{t \to x} h(t) = \lim_{n \to \infty} f'_n(t).$$

This tells us that $f'(x) = \lim_{n \to \infty} f'_n(x) = g(x)$.
□

Now, if $\{f'_n\}_{n=1}^{\infty}$ are all continuous on $[a, b]$, then we may use the fundamental theorem of calculus (6.2.3) to simplify the proof of theorem (7.2.7). To see this observe that for any $x \in [a, b]$ we have

$$\int_{x_0}^{x} f'_n(t) dt = f_n(x) - f_n(x_0)$$

for every $n \in \mathbb{N}$. Let g be the limit function for the sequence $\{f'_n\}_{n=1}^{\infty}$. By theorem (7.2.6) we have $\lim_{n \to \infty} (f_n(x) - f_n(x_0))$ exists and equals $\int_{x_0}^{x} g(t) dt$. Thus we see that $\lim_{n \to \infty} f_n(x)$ exists, and let $f(x)$ be its limit value.[1] Thus we have

$$f(x) - f(x_0) = \int_{x_0}^{x} g(t) dt,$$

and so by the fundamental theorem of calculus (6.2.3) we have $f'(x) = g(x)$.

7.2.4 Homework Exercises

1. For $f_n(x) := x + \frac{1}{n}$, show that $f_n(x) \to f(x) := x$ point-wise on $\mathbb{R}$. Also, show that this convergence is uniform.

2. In the exercise above, does f_n^2 converge point-wise? If so does it also converge uniformly?

---
[1] Although we cannot conclude this convergence is uniform with this proof.

3. Show that the sequence of functions given by $f_n(x) = x^2 e^{-nx}$ converges uniformly on $[0, \infty)$.

4. Show that if $f_n \to f$ and $g_n \to g$ uniformly of A, then on A we have $f_n + g_n \to f + g$ uniformly, and $cf_n \to cf$ uniformly for any constant c.

5. Suppose $f_n \to f$ and $g_n \to g$ uniformly on A, and f_n and g_n are bounded on A for each n. Show that $f_n g_n \to fg$ uniformly on A.

6. Suppose $f_n \to f$ uniformly on A, and $\|f_n\|_{\sup,A} \leq M$ for all n. Suppose g is continuous on $[-M, M]$. Show $g \circ f_n \to g \circ f$ uniformly on A.

7. Consider the sequence of functions
$$\sqrt{x}, \sqrt{x + \sqrt{x}}, \sqrt{x + \sqrt{x + \sqrt{x}}}, \cdots.$$
Show that on $[0, \infty)$ this sequence converges point-wise, and find the limit function f. Is this convergence also uniform on $[0, \infty)$?

8. Suppose $f_n : [a, b] \to \mathbb{R}$ for $n = 1, 2, \cdots$. Suppose f_n is an increasing function for each n. If $f_n \to f$ point-wise, then f is increasing as well. If f_n is strictly increasing for all n, is f necessarily strictly increasing? If so, prove it. If not, provide a counterexample.

9. Find a sequence of Riemann integrable functions on an interval $[a, b]$ that converge point-wise to a function that is not Riemann integrable.

10. Find a uniformly convergent sequence of functions on $(0, 1)$ for which the sequence of derivative functions does not converge.

11. Suppose $\{f_n\}_{n=1}^\infty$ is a sequence of bounded functions on a set A that converges uniformly on A to a function f. Show $\lim_{n \to \infty} \|f_n\|_{\sup,A} = \|f\|_{\sup,A}$.

12. Find a sequence of functions that is discontinuous everywhere but converge uniformly to a differentiable function.

7.3 Series of Functions

Given a sequence of functions $f_n : D \to \mathbb{R}$, $n = 1, 2, \cdots$, we define the sequence of functions $S_n : D \to \mathbb{R}$, $n = 1, 2, \cdots$, given by

$$S_n(x) := \sum_{k=1}^{n} f_k(x).$$

If $\{S_n\}_{n=1}^{\infty}$ converges point-wise on D we say that the **series** $\sum_{k=1}^{\infty} f_k$ **converges point-wise** on D. Likewise, if $\{S_n\}_{n=1}^{\infty}$ converges uniformly on D we say that the **series** $\sum_{k=1}^{\infty} f_k$ **converges uniformly** on D. Also, if $\sum_{n=1}^{\infty} |f_n(x)|$ converges on D, we say that the **series** $\sum_{n=1}^{\infty} f_n(x)$ **converges absolutely** on D.

The following theorem is often called the **Weierstrass M-test**. This theorem gives us a sufficient condition for a series of functions to converge uniformly.

Theorem – Weierstrass M-Test 7.3.1 *Let $\{f_n\}_{n=1}^{\infty}$ be a sequence of functions defined on a set D and suppose $\|f_n\|_{\sup,D} \leq M_n$, $M_n \geq 0$, for $n = 1, 2, \cdots$. Then $\sum_{n=1}^{\infty} f_n(x)$ converges absolutely and uniformly on D whenever $\sum_{n=1}^{\infty} M_n$ converges.*

Proof: We know that $\sum_{j=1}^{\infty} f_j(x)$ converges uniformly on D if and only if given any $\varepsilon > 0$ there is an $N \in \mathbb{N}$ so that if $n \geq m > N$, then we have $|\sum_{k=m}^{n} f_k(x)| < \varepsilon$ for all $x \in D$.

Let $\varepsilon > 0$ be given. If $\sum_{k=1}^{\infty} M_k$ converges, then there is an $N \in \mathbb{N}$ so that for $n \geq m > N$ we have $|\sum_{k=m}^{n} M_k| < \varepsilon$. Thus for any $x \in D$ we have

$$\left|\sum_{k=m}^{n} f_k(x)\right| \leq \sum_{k=m}^{n} \|f_k\|_{\sup,D} \leq \sum_{k=m}^{n} M_k = \left|\sum_{k=m}^{n} M_k\right| < \varepsilon.$$

□

Suppose $\{a_n\}_{n=0}^{\infty}$ and $\{b_n\}_{n=1}^{\infty}$ are sequences so that $\sum_{k=0}^{\infty} a_k$ and $\sum_{k=1}^{\infty} b_k$ converge absolutely, then the function $f(x)$ defined by

$$f(x) := a_0 + \sum_{n=1}^{\infty} (a_n \cos nx + b_n \sin nx) \tag{7.2}$$

converges absolutely and uniformly on $\mathbb{R}$. A function f as defined by (7.2) is called a **Fourier series**.

7.3.1 Uniform Convergence of Power Series

Theorem 7.3.2 *Consider a power series $\sum_{n=0}^{\infty} a_n(x-c)^n$ which converges absolutely in an interval $(c-r, c+r)$ for some $r > 0$, then for any $\rho < r$ we have $\sum_{n=0}^{\infty} a_n(x-c)^n$ converges absolutely and uniformly on $[c-\rho, c+\rho]$.*

Proof: Suppose the power series $\sum_{k=0}^{\infty} a_k(x-c)^k$ converges absolutely on $(c-r, c+r)$. Then for any $\rho < r$ we may pick $\xi \in (c-r, c+r) \setminus [c-\rho, c+\rho]$ for which $\sum_{k=0}^{\infty} a_k(\xi-c)^k$ converges absolutely. Let's fix such a ξ.

Since $\sum_{k=0}^{\infty} a_k(\xi-c)^k$ converges, we know that $\lim_{j \to \infty} a_j(\xi-c)^k = 0$. Hence there is a $B \in (0, \infty)$ for which $|a_n(\xi-c)^k| \leq B$ for all $k \in \mathbb{N} \cup \{0\}$.

For any x in $[c-\rho, c+\rho]$ we have

$$|x-c| \leq \rho < |\xi-c|.$$

Moreover, for any such x

$$|a_k(x-c)^k| = |a_k(\xi-c)^k| \cdot |\frac{x-c}{\xi-c}|^k \leq B \cdot |\frac{\rho}{\xi-c}|^k$$

for all $k \in \mathbb{N}$. Now $\sum_{k=0}^{\infty} B \cdot |\frac{\rho}{\xi-c}|^k$ is a convergent geometric series since $|\frac{\rho}{\xi-c}| < 1$. Thus by theorem (7.3.1) we have that $\sum_{k=0}^{\infty} a_k(x-c)^k$ converges absolutely and uniformly on $[c-\rho, c+\rho]$.
□

The next lemma will aid in allowing us to show that for a power series $f(x) = \sum_{j=0}^{\infty} a_j(x-c)^j$, it's derivative $f'(x)$, its antiderivative $\int_a^x f(t)dt$ and the power series $f(x)$ have the same radius of convergence.

Lemma 7.3.3 *Consider a power series $\sum_{n=0}^{\infty} a_n(x-c)^n$ with a radius of convergence $R \in [0, \infty) \cup \{\infty\}$. Then the power series $\sum_{n=1}^{\infty} na_n(x-c)^{n-1}$ and $\sum_{n=0}^{\infty} \frac{a_n}{n+1}(x-c)^{n+1}$ each have the same radius of convergence R.*

Proof: The radius of convergence of the power series $\sum_{n=0}^{\infty} a_n(x-c)^n$ is given by $\frac{1}{\limsup_{n \to \infty} \sqrt[n]{|a_n|}}$. The radius of convergence of $\sum_{n=1}^{\infty} na_n(x-c)^{n-1} = \sum_{j=0}^{\infty}(j+1)a_{j+1}(x-c)^j$ is given by $\frac{1}{\limsup_{j \to \infty} \sqrt[j]{(j+1)|a_{j+1}|}}$, and the radius of convergence of the series $\sum_{n=0}^{\infty} \frac{a_n}{n+1}(x-c)^{n+1} = \sum_{j=1}^{\infty} \frac{a_{j-1}}{j}(x-c)^j$ is given by

7.3. SERIES OF FUNCTIONS

$$\frac{1}{\limsup_{j\to\infty} \sqrt[j]{\frac{|a_{j-1}|}{j}}}.$$

We will now show that $\limsup_{n\to\infty} \sqrt[n]{|a_n|} = \limsup_{n\to\infty} \sqrt[n]{(n+1)|a_{n+1}|}$. Note first that $\lim_{n\to\infty} \sqrt[n]{n+1} = 1$. An easy way to see this is to consider the sequence given by $\log \sqrt[n]{n+1} = \frac{1}{n}\log(n+1)$, $n = 1, 2, \cdots$. By L'Hôpital's rule we see that this sequence has a limit value of 0. Now $\sqrt[n]{n+1} = e^{\log \sqrt[n]{n+1}}$, and by continuity of e^x we see that $\sqrt[n]{n+1} \to e^0 = 1$ as $n \to \infty$.

Let $\{n_k\}_{k=1}^\infty$ be a subsequence of $\mathbb{N}$ so that $\sqrt[n_k]{|a_{n_k}|} \to \limsup_{n\to\infty} \sqrt[n]{|a_n|}$ as $k \to \infty$. For convenience we let $\alpha := \limsup_{n\to\infty} \sqrt[n]{|a_n|}$.

We consider the sequence $b_k = \sqrt[n_k-1]{|a_{n_k}|}$, for $k = 1, 2, \cdots$. Now

$$\log b_k = (1 + \frac{1}{n_k - 1})\log \sqrt[n_k]{|a_{n_k}|} \to \log \alpha$$

as $k \to \infty$, and thus by continuity[2] of e^x we see that $b_k \to \alpha$ as $k \to \infty$. Since $\sqrt[n_k-1]{n_k} \to 1$ as $k \to \infty$ we see that $\sqrt[n_k-1]{n_k} \sqrt[n_k-1]{|a_{n_k}|} \to \alpha$ as $k \to \infty$. Thus we see that

$$\limsup_{n\to\infty} \sqrt[n]{|a_n|} \leq \limsup_{n\to\infty} \sqrt[n]{(n+1)|a_{n+1}|}.$$

Now let $\{m_k\}_{k=1}^\infty$ be an increasing sequence of natural numbers so that

$$\sqrt[m_k]{(m_k+1)|a_{m_k+1}|} \to \limsup_{n\to\infty} \sqrt[n]{(n+1)|a_{n+1}|}$$

as $k \to \infty$. For simplicity let

$$\beta := \limsup_{n\to\infty} \sqrt[n]{(n+1)|a_{n+1}|}.$$

Since $\sqrt[m_k]{m_k+1} \to 1$ as $k \to \infty$ we know that $\sqrt[m_k]{|a_{m_k+1}|} \to \beta$ as $k \to \infty$. We claim that $\sqrt[m_k+1]{|a_{m_k+1}|} \to \beta$ as $k \to \infty$. To see this consider $\log \sqrt[m_k+1]{|a_{m_k+1}|} = \frac{m_k}{m_k+1}\log \sqrt[m_k]{|a_{m_k+1}|}$ which converges to $\log \beta$ as $k \to \infty$. Hence we get that $\sqrt[m_k+1]{|a_{m_k+1}|} \to e^{\log \beta} = \beta$ as $k \to \infty$. This allows us to conclude that

$$\limsup_{n\to\infty} \sqrt[n]{(n+1)|a_{n+1}|} \leq \lim_{n\to\infty} \sqrt[n]{|a_n|}.$$

---

[2]where $e^{-\infty}$ is 0 in the sense of limits

Thus we see that $\sum_{n=1}^{\infty} na_n(x-c)^{n-1}$ and $\sum_{n=0}^{\infty} a_n(x-c)^n$ have the same radius of convergence.

Next, consider $\sum_{n=0}^{\infty} \frac{a_n}{n+1}(x-c)^{n+1} = \sum_{j=1}^{\infty} \frac{a_{j-1}}{j}(x-c)^j$. By the argument given above this power series and the power series $\sum_{j=1}^{\infty} j \cdot \frac{a_{j-1}}{j}(x-c)^{j-1} = \sum_{j=1}^{\infty} a_{j-1}(x-c)^{j-1} = \sum_{n=0}^{\infty} a_n(x-c)^n$ have the same radius of convergence.
□

Theorem 7.3.4 *Consider a power series $f(x) = \sum_{n=0}^{\infty} a_n(x-c)^n$ with a radius of convergence $R \in (0, \infty) \cup \{\infty\}$. Then on $(c-R, c+R)$, $f'(x)$ exists and it is given by $f'(x) = \sum_{n=1}^{\infty} na_n(x-c)^{n-1}$. Moreover, for any $a \in (c-R, c+R)$ the function given by $g(x) = \int_a^x f(t)dt$ exists and has a power series representation on $(c-R, c+R)$ given by $g(x) = C + \sum_{n=0}^{\infty} \frac{a_n}{n+1}(x-c)^{n+1}$ for some $C \in \mathbb{R}$.*

Proof: By Lemma (7.3.3) we know that the power series given by

$$\sum_{n=1}^{\infty} na_n(x-c)^{n-1} \quad \text{and} \quad \sum_{n=0}^{\infty} \frac{a_n}{n+1}(x-c)^{n+1}$$

all have radii of convergence R, and on any compact subset on $(c-R, c+R)$ both these power series and the power series $\sum_{n=0}^{\infty} a_n(x-c)^n$ converge absolutely and uniformly.

Let $r < R$ be any positive real number. The sequence of partial sums $S_n = \sum_{k=0}^{n} a_k(x-c)^k$ is differentiable and integrable on $[c-r, c+r]$ and here $S_n'(x) = \sum_{k=1}^{n} ka_k(x-c)^{k-1}$ and $\int_a^x S_n(t)dt = \sum_{k=0}^{n} \frac{a_k}{k+1}(x-c)^{k+1} - \sum_{k=0}^{n} \frac{a_k}{k+1}(a-c)^{n+1}$, where $a \in (c-R, c+R)$ is fixed.

By theorems (7.2.6) and (7.2.7) we know that $f(t)$ is integrable and $\{\int_a^x S_n(t)dt\}_{n=1}^{\infty}$ converges uniformly on $[c-r, c+r]$ to $\int_a^x \lim_{n\to\infty} S_n(t)dt = \int_a^x f(t)dt$. Moreover, since $\{S_n'\}_{n=1}^{\infty}$ and $\{S_n\}_{n=1}^{\infty}$ converge uniformly on $[c-r, c+r]$, we get $f(x)$ is differentiable and it's derivative is given by $\lim_{n\to\infty} S_n'(x)$.

Since

$$\int_a^x S_n dt \to \sum_{n=0}^{\infty} \frac{a_n}{n+1}(x-c)^{n+1} - \sum_{n=0}^{\infty} \frac{a_n}{n+1}(a-c)^{n+1}$$

uniformly and
$$S'_n \to \sum_{n=1}^{\infty} n a_n (x-c)^{n-1}$$
uniformly on $[c-r, c+r]$, we get the result.
□

7.3.2 Homework Exercises

1. Suppose $f(x) = \sum_{n=0}^{\infty} a_n (x-c)^n$ converges absolutely on $(c-R, c+R)$ for some $R > 0$. Show that on any compact subset of $(a-R, a+R)$ all derivatives $f', f'', \cdots$ exist and find power series representations of these derivatives that converge uniformly on compact subsets of $(c-R, c+R)$.

2. If $f(x) = \sum_{n=0}^{\infty} a_n(x-c)^n$ converges absolutely on $(c-R, c+R)$ for some $R > 0$, show that for each $n \in \mathbb{N}$ that $a_n = \frac{1}{n!} f^{(n)}(c)$ and $a_0 = f(c)$.

3. Find a power series representation for $\int_0^x e^{-t^2} dt$.

4. For $\alpha \in \mathbb{R}$ and $x > -1$ find a power series representation for $f(x) = (1+x)^\alpha$ centred at 0. Hint: To do this observe that $f(0) = 1$ and f satisfies $(1+x) f'(x) = \alpha f(x)$.

7.4 Equicontinuity

By theorems (2.4.5) and (2.4.10) we know that a compact subset K of Euclidean space has the property that every infinite set $S \subseteq K$ has a limit point in K. In this section we seek a condition that will allow us to make a similar claim about sequences or families of functions. In particular, we seek a condition for sequences or families of functions that allow us to conclude the existence of a subsequence which converges uniformly.

7.4.1 Properties of Convergent Sequences of Functions

Consider the collection of continuous functions on a closed interval $[a, b]$, often denoted by $C([a, b])$ or $C^0([a, b])$. Does this family of functions have the property that every sequence of functions in this family has a convergent subsequence?[3] In this section we will seek a condition that guarantees the existence of a subsequence of a sequence in $C([a,b])$ that converges uniformly on $[a, b]$. We will look for this condition by examining both uniformly convergent sequences and sequences that do not have convergent subsequences.

Consider the sequences $f_n(x) = n$, for $n = 1, 2, \cdots$ on any closed interval $[a, b]$. The sequence $f_n(x) = n$ does not have any convergent subsequences since $f_n(x) \to \infty$ for any $x \in [a, b]$. Thus we see that a uniform boundedness assumption may be needed to insure convergence.

For a family of functions $\mathcal{F}$ on a set D, we say that $\mathcal{F}$ is **uniformly bounded** if there is an $M \geq 0$ so that for each $f \in \mathcal{F}$ we have $\|f\|_{\sup,D} \leq M$.

Next, consider the sequence of functions given by $f_n(x) = \sin nx$, $n = 1, 2, \cdots$ on any interval $[a, b]$. This sequence is uniformly bounded since $|\sin nx| \leq 1$. However, there is no uniformly convergent subsequence of this sequence.[4] Thus we see boundedness, in particular uniform boundedness is not enough to insure the existence of a uniformly convergent subsequence.

Consider the sequence of functions given by $f_n(x) = \frac{x^2}{x^2+(1-nx)^2}$, $n = 1, 2, \cdots$ on $[0, 1]$. Clearly $\|f_n\|_{\sup,[0,1]} \leq 1$ for all $n \in \mathbb{N}$. Also, $f_n(x) \to 0$ point-wise as $n \to \infty$ for each $x \in [0, 1]$. Since $f_n(\frac{1}{n}) = 1$ we cannot have any subsequence of this collection converging uniformly on $[0, 1]$. Hence we see that even if a sequence converges point-wise and is uniformly bounded, then there still may not be a subsequence that converges uniformly.

Suppose that $\{f_n\}_{n=1}^{\infty}$ is a sequence of bounded functions on a set K that converges uniformly to a function f on K. Then by theorem (7.2.2) we know the limit function f is also bounded, and let M be so that $\|f\|_{\sup,K} \leq M$. Let $\varepsilon > 0$, then there is an $N \in \mathbb{N}$ so that for all $n \geq N$ we have $\|f_n - f\|_{\sup,K} \leq \varepsilon$. Thus for all such n's we have

$$|f_n(x)| \leq |f_n(x) - f(x)| + |f(x)| \leq \varepsilon + M.$$

Thus we see that for all $k \in \mathbb{N}$ we have

$$\|f_k\|_{\sup,K} \leq \max\{M + \varepsilon, \|f_1\|_{\sup,K}, \|f_2\|_{\sup,K}, \cdots, \|f_{N-1}\|_{\sup,K}\}.$$

---

[3] Where the convergence here is either point-wise or uniformly on $[a, b]$.
[4] We leave it an exercise to show this.

7.4. EQUICONTINUITY

Hence any uniformly convergent sequence of bounded functions $\{f_j\}_{j=1}^{\infty}$ is uniformly bounded.

Also, we note that if a sequence of bounded functions converges point-wise, our sequence may not be uniformly bounded. To see this consider the sequence of functions on $[0,1]$ given by $f_n(x) = n^2 x(1-x^2)^n$, $n = 1, 2, \cdots$. This sequence converges point-wise to the function $f \equiv 0$, however this convergence is not uniform on $[0, 1]$. Moreover, this family of bounded functions $\{f_n : n \in \mathbb{N}\}$ is not uniformly bounded.

Clearly uniform boundedness is necessary but not sufficient for the existence of a uniformly convergent subsequence of a given sequence of functions. However, there is another property that in conjunction with uniform boundedness will insure that a sequence of continuous functions on a compact set will have a uniformly convergent subsequence, and this is what is referred to as **uniform equicontinuity**.

A family of functions $\mathcal{F}$ defined on a set D is said to be **uniformly equicontinuous** if given any $\varepsilon > 0$, there is a $\delta > 0$ do that for $x, y \in D$ which satisfy $|x - y| < \delta$ we have $|f(x) - f(y)| < \varepsilon$ for any $f \in \mathcal{F}$. One can check that our sequence $\{\sin nx\}_{n=1}^{\infty}$ is not uniformly equicontinuous on any interval $I \subseteq \mathbb{R}$.

If $\{f_n\}_{n=1}^{\infty}$ is a uniformly convergent sequence of continuous functions on a compact set K, then the sequence of functions is uniformly equicontinuous on K. To see this, suppose $\varepsilon > 0$ is given. Since f_n converges uniformly to some function f on K, we know that there is an $N \in \mathbb{N}$ so that for $n, m \geq N$ we have $|f_n(x) - f_m(x)| < \frac{\varepsilon}{3}$ for any $x \in K$. Since the f_n's are all uniformly continuous on K, there are positive numbers $\delta_1, \delta_2, \cdots$ so that for $x, y \in K$, if $|x - y| < \delta_i$, then $|f_i(x) - f_i(y)| < \varepsilon$ for $i = 1, 2, \cdots$. Let $\delta = \min\{\delta_1, \delta_2, \cdots, \delta_N\}$. Then for any $n > N$ and $|x - y| < \delta$ we have

$$|f_n(x) - f_n(y)| \leq |f_n(x) - f_N(x)| + |f_N(x) - f_N(y)| + |f_N(y) - f_n(y)|$$

$$< \frac{\varepsilon}{3} + \frac{\varepsilon}{3} + \frac{\varepsilon}{3} = \varepsilon.$$

Thus we see that the family $\{f_n : n \in \mathbb{N}\}$ is uniformly equicontinuous on K.

Hence we see that on a compact set if a continuous sequence of functions converges uniformly, then the sequence must be uniformly bounded and uniformly equicontinuous.

7.4.2 The Ascoli-Arzelà Theorem and Compactness

The following theorem of Ascoli and Arzelà gives a sufficient condition for the existence of a uniformly convergent subsequence in terms of uniform boundedness and uniform equicontinuity.

The Ascoli-Arzelà Theorem 7.4.1 *Let K be a compact set in euclidean space[5] $\mathbb{R}^n$ and suppose $\{f_n\}_{n=1}^{\infty}$ is a sequence of continuous functions on K. Suppose for each $x \in K$ we have the sequence of numbers $\{f_n(x)\}_{n=1}^{\infty}$ is bounded. Suppose as well $\{f_n\}_{n=1}^{\infty}$ is uniformly equicontinuous. Then $\{f_n\}_{n=1}^{\infty}$ is uniformly bounded on K. Moreover, $\{f_n\}_{n=1}^{\infty}$ contains a uniformly convergent subsequence.*

Proof: Let $\varepsilon > 0$ be given. By the uniform equicontinuity of $\{f_n\}_{n=1}^{\infty}$ there is a $\delta > 0$ so that if $|x - y| < \delta$ we have $|f_n(x) - f_n(y)| < \varepsilon$ for all $n \in \mathbb{N}$.

The collection of $\{B_\delta(x) : x \in K\}$, where $B_\delta(x) = \{y : |y - x| < \delta\}$, is an open cover of K. Thus there is a finite subcover of K given by

$$\{B_\delta(x_1), B_\delta(x_2), \cdots, B_\delta(x_m)\}.$$

For each x_i, $i = 1, 2, \cdots, m$, there is an M_i so that $|f_n(x_i)| \leq M_i$ for all $n \in \mathbb{N}$. Let $M := \max\{M_1, M_2, \cdots, M_m\}$. Then for any $x \in K$ there is a $x_j \in \{x_1, x_2, \cdots, x_m\}$ so that $x \in B_\delta(x_j)$. Thus for any $n \in \mathbb{N}$ we have

$$|f_n(x)| = |f_n(x) - f_n(x_j) + f_n(x_j)|$$

$$\leq |f_n(x) - f_n(x_j)| + |f_n(x_j)| \leq M + \varepsilon.$$

Thus we see that $\{f_n\}_{n=1}^{\infty}$ is uniformly bounded on K.

Let $\mathcal{Q}$ be a countable dense subset of K. For instance, we may take $\mathbb{Q}^n \cap K$ as this countable dense subset. Let $\{x_i\}_{i=1}^{\infty}$ be an enumeration of $\mathcal{Q}$.

---

[5] Or a separable metric space

7.4. EQUICONTINUITY

Consider the family of sequences $\{f_n(x_m)\}_{n=1}^\infty$, $m = 1, 2, \cdots$.

Since the sequence $\{f_n(x_1)\}_{n=1}^\infty$ is bounded, it has a convergent subsequence which we denote by $\{f_{1,k}(x_1)\}_{k=1}^\infty$. Then consider the sequence $\{f_{1,k}(x_2)\}_{k=1}^\infty$. Then this sequence also has a convergent subsequence, which we denote by $\{f_{2,k}(x_2)\}_{k=1}^\infty$.

Continuing in this manner, if $\{f_{m-1,k}(x_{m-1})\}_{k=1}^\infty$ is a convergent subsequence of $\{f_n(x_{m-1})\}_{n=1}^\infty$, we consider the sequence $\{f_{m-1,k}(x_m)\}_{k=1}^\infty$, and let $\{f_{m,k}(x_m)\}_{k=1}^\infty$ denote a convergent subsequence of the sequence $\{f_{m-1,k}(x_m)\}_{k=1}^\infty$.

Consider the array of functions

$$\begin{array}{cccccc} f_{1,1} & f_{1,2} & f_{1,3} & f_{1,4}, & \cdots \\ f_{2,1} & f_{2,2} & f_{2,3} & f_{2,4}, & \cdots \\ f_{3,1} & f_{3,2} & f_{3,3} & f_{3,4}, & \cdots \\ f_{4,1} & f_{4,2} & f_{4,3} & f_{4,4}, & \cdots \\ \vdots & \vdots & \vdots & \vdots & \ddots \end{array}$$

Then each sequence $\{f_{j,k}\}_{k=1}^\infty$ converges at the points $x_1, x_2, \cdots, x_j$. Moreover, $\{f_{j,k}\}_{k=1}^\infty$ is a subsequence of $\{f_{j-1,k}\}_{k=1}^\infty$.

Consider the subsequence of $\{f_n\}_{n=1}^\infty$ given by $\{f_{j,j}\}_{j=1}^\infty$. Then this sequence of functions converges for all elements of $\mathcal{Q}$ and let $\{f_{n_k}\}_{k=1}^\infty$ denote this subsequence. We shall now show that $\{f_{n_k}\}_{k=1}^\infty$ converges uniformly on K.

Let $\varepsilon > 0$ be given, and let $\delta > 0$ be so that $|f_n(x) - f_n(y)| < \frac{\varepsilon}{3}$ for each $n \in \mathbb{N}$, whenever $|x - y| < \delta$. Also, let $y_1, y_2, \cdots, y_r$ be so that $\{B_\delta(y_1), B_\delta(y_2), \cdots, B_\delta(y_r)\}$ is a finite subcover of the open cover $\{B_\delta(y) : y \in \mathcal{Q}\}$ of K.

Since $\{f_{n_k}(x)\}_{k=1}^\infty$ converges for each $x \in \mathcal{Q}$, we may find an $N \in \mathbb{N}$ so that if $i, j \geq N$

$$|f_{n_i}(y_s) - f_{n_j}(y_s)| < \frac{\varepsilon}{3}$$

for $s = 1, 2, \cdots, r$.

Let $x \in K$, then there is a y_s for some $s \in \{1, 2, \cdots, r\}$ so that $x \in B_\delta(y_s)$, and thus $|f_{n_k}(x) - f_{n_k}(y_s)| < \frac{\varepsilon}{3}$ for every $k \in \mathbb{N}$. Thus if $i, j \geq N$ we have

$$|f_{n_i}(x) - f_{n_j}(x)| = |f_{n_i}(x) - f_{n_i}(y_s) + f_{n_i}(y_s) - f_{n_j}(y_s) + f_{n_j}(y_s) - f_{n_j}(x)|$$

$$\leq |f_{n_i}(x) - f_{n_i}(y_s)| + |f_{n_i}(y_s) - f_{n_j}(y_s)| + |f_{n_j}(y_s) - f_{n_j}(x)|$$

$$< \frac{\varepsilon}{3} + \frac{\varepsilon}{3} + \frac{\varepsilon}{3} = \varepsilon.$$

□

7.4.3 Homework Exercises

1. Let $f : \mathbb{R} \to \mathbb{R}$ be continuous and define $f_n(x) := f(nx)$. Suppose $\{f_n\}_{n=1}^\infty$ is uniformly equicontinuous on $[0, 1]$. What can you say about f?

2. Suppose $\{f_n\}_{n=1}^\infty$ is a uniformly equicontinuous sequence of functions defined on some compact set K and suppose $f_n \to f$ point-wise on K. Show that $f_n \to f$ uniformly on K.

3. For a given interval $[a, b]$ and $\alpha \in (0, 1]$ we define $\|f\|_{C^{0,\alpha}([a,b])} = \sup_{x \in [a,b]} |f(x)| + \sup_{x \neq y; x, y \in [a,b]} \frac{|f(x) - f(y)|}{|x - y|^\alpha}$. f is said to be Hölder continuous on $[a, b]$ if $\|f\|_{C^{0,\alpha}([a,b])} < \infty$. Suppose $\{f_n\}_{n=1}^\infty$ is a sequence of Hölder continuous functions for which $\exists M < \infty$ so that $\forall n, \|f_n\|_{C^{0,\alpha}([a,b])} < M$. Show that the sequence is both uniformly equicontinuous and uniformly bounded on $[a, b]$. Conclude that there must be a subsequence which converges uniformly on $[a, b]$.

4. Show that $f_n(x) = \sin nx$, $n = 1, 2, \cdots$ is not a family of uniformly equicontinuous functions on any interval $[a, b]$, $b > a$.

5. Give an example of a sequence that is uniformly equicontinuous, but not uniformly bounded.

6. Give an example of a uniformly bounded and uniformly equicontinuous sequence of functions on $\mathbb{R}$ that does not have a uniformly convergent subsequence.

7.5 Uniform Approximation by Polynomials

The primary goal of this section is to discuss a result of Weierstrass which says that the set of polynomials is dense in the set of continuous functions on a compact interval $[a, b]$ with respect to the $\|\cdot\|_{\sup,[a,b]}$ norm. More precisely, on any interval $[a, b]$ given a continuous function f on $[a, b]$ and any $\varepsilon > 0$ there is a polynomial P so that $\|f - P\|_{\sup,[a,b]} < \varepsilon$. This result is often called the **Weierstrass approximation theorem**.

Approximation by polynomials is of interest to mathematicians since many properties of polynomials can be addressed algebraically, the value of a polynomial at any point can be determined by arithmetic operations, and this is not the case for all continuous functions.

Classically there are two methods of approximation by polynomials that may initially come to mind, approximation using Taylor polynomials, and by the method of Lagrange interpolation. Taylor's theorem gives us a means of approximating a given function by a polynomial of degree n around a point in the domain of a function, but it requires that the given function be n times differentiable to approximate by a polynomial of degree at most n. Thus it requires much regularity. For control of the difference between the polynomial and the given function we may use Taylor's remainder theorem, but of course this requires control of the $(n + 1)^{st}$ derivatives of the given function. To insure that the sequence of approximating Taylor polynomials converges point-wise to the given function we need that the function has derivatives of all orders. Even if so, the interval of convergence may be small.

The method of Lagrange interpolation finds a polynomial that fits given data, namely a collection of n points which we desire the polynomial to contain on its graph. This method may be used to approximate a given function. If the function is sufficiently regular then we may control the $\|\cdot\|_{\sup,[a,b]}$ norm of the difference of the given function and the Lagrange interpolation polynomial. However, if the function is simply continuous we may not have any control on how close the polynomial is to the given function. Moreover, simply adding more points may not decrease the $\|\cdot\|_{\sup,[a,b]}$ norm of the difference of the given function and the approximating polynomial.

7.5.1 The Weierstrass Approximation Theorem

The Weierstrass Approximation Theorem 7.5.1 *Let $f : [a, b] \to \mathbb{R}$ be a continuous function. Then for any $\varepsilon > 0$, there exists a polynomial P so that $\|f - P\|_{\sup, [a,b]} < \varepsilon$.*

Proof: Without loss of generality we assume $a = 0$ and $b = 1$, since if we have the result for $[0, 1]$, then for any continuous $f : [a, b] \to \mathbb{R}$ we define $g : [0, 1] \to \mathbb{R}$ by $g(x) = f((b - a)x + a)$. If $P(x)$ is a polynomial that approximates $g(x)$, then since $f(x) = g(\frac{x-a}{b-a})$ we have that $P(\frac{x-a}{b-a})$ is a polynomial that approximates f.

Moreover, we may assume $0 = f(0) = f(1)$ since otherwise we may consider $g(x) := f(x) - f(0) - x[f(1) - f(0)]$. Because $f(x) = g(x) + f(0) + x[f(1) - f(0)]$, if p is a polynomial that approximates g, then $q(x) := p(x) + f(0) + x[f(1) - f(0)]$ is a polynomial that approximates f and $|g - p| = |f - q|$.

We extend f to all of $\mathbb{R}$ by defining f to be 0 on $\mathbb{R} \setminus [0, 1]$. Thus we have f is uniformly continuous on all of $\mathbb{R}$.

Define $Q_n(x) := c_n(1 - x^2)^n$, where $c_n := \frac{1}{\int_{-1}^{1}(1-x^2)^n dx}$.

Consider the function $g(x) = (1 - x^2)^n - (1 - nx^2)$, then $g(0) = 0$ and $g'(x) = 2nx[1 - (1 - x^2)^{n-1}] > 0$ on $(0, 1)$. Thus we have

$$(1 - x^2)^n \geq 1 - nx^2$$

on $[0, 1]$. Hence,

$$\int_{-1}^{1}(1 - x^2)^n dx = 2\int_{0}^{1}(1 - x^2)^n dx$$

$$\geq 2\int_{0}^{\frac{1}{\sqrt{n}}}(1 - x^2)^n dx$$

7.5. UNIFORM APPROXIMATION BY POLYNOMIALS

$$\geq 2\int_0^{\frac{1}{\sqrt{n}}} 1 - nx^2 dx$$

$$= \frac{4}{3\sqrt{n}} > \frac{1}{\sqrt{n}}.$$

From this we see that $c_n < \sqrt{n}$.

For $0 < \delta < 1$, on $[-1, -\delta] \cup [\delta, 1]$ we have

$$c_n(1-x^2)^n \leq \sqrt{n}(1-x^2)^n \leq \sqrt{n}(1-\delta^2)^n,$$

which by L'Hôpital's rule goes to zero as $n \to \infty$. Hence $Q_n \to 0$ uniformly on $[-1, -\delta] \cup [\delta, 1]$.

Define

$$P_n(x) := \int_{-1}^{1} f(x+t)Q_n(t)dt$$

for each $x \in [0, 1]$. Since $f \equiv 0$ on $\mathbb{R} \setminus [0, 1]$, we have

$$P_n(x) = \int_{-x}^{1-x} f(x+t)Q_n(t)dt.$$

By (6.13) we see that

$$P_n(x) = \int_0^1 f(s)Q_n(s-x)ds.$$

Since $f(s)Q_n(s-x)$ can be written as a polynomial in powers of x with coefficients that are functions of s, upon integrating in s we see that P_n is a polynomial in x.

Let $\varepsilon > 0$ be given and let $M = \sup_{x \in [0,1]} |f(x)|$. Choose $\delta > 0$ so that $|x - z| < \delta$ implies $|f(x) - f(z)| < \frac{\varepsilon}{2}$. By L'Hôpital's rule we may choose $n \in \mathbb{N}$ so that $4M\sqrt{n}(1-\delta^2)^2 < \frac{\varepsilon}{2}$. Then,

$$|P_n(x) - f(x)| = |\int_{-1}^{1} [f(x+t) - f(x)]Q_n(t)dt|$$

$$\leq \int_{-1}^{1} |f(x+t) - f(x)|Q_n(t)dt$$

$$\leq 2M \int_{-1}^{-\delta} Q_n(t)dt + \frac{\varepsilon}{2} \int_{-\delta}^{\delta} Q_n(t)dt + 2M \int_{\delta}^{1} Q_n(t)dt$$

$$\leq 4M \int_{\delta}^{1} Q_n(t)dt + \frac{\varepsilon}{2}$$

$$\leq 4M\sqrt{n}(1-\delta^2)^n + \frac{\varepsilon}{2}$$

$$< \varepsilon.$$

□

Chapter 8

Functions of Several Real Variables

In this chapter we will discuss calculus in Euclidean space $\mathbb{R}^n$. To begin with, we will revisit topology in the setting of $\mathbb{R}^n$ when endowed with a norm $\|\|$, and then we will then discuss differential calculus in $\mathbb{R}^n$ with respect to this topology.

8.1 The Algebra and Topology of $\mathbb{R}^n$

Recall that $\mathbb{R}^n = \{(x_1, x_2, \cdots, x_n) : x_i \in \mathbb{R}, 1 \leq i \leq n\}$ may be endowed with a binary operation $+ : \mathbb{R}^n \times \mathbb{R}^n \to \mathbb{R}^n$ called vector addition and an operation $\cdot : \mathbb{R} \times \mathbb{R}^n \cdot \mathbb{R}^n$ called scalar multiplication, where

$$(x_1, x_2, \cdots, x_n) + (y_1, y_2, \cdots, y_n) := (x_1 + y_1, x_2 + y_2, \cdots, x_n + y_n)$$

and

$$c \cdot (x_1, x_2, \cdots, x_n) := (cx_1, cx_2, \cdots, cx_n).$$

Sets such as $\mathbb{R}^n$ which are closed under the operations of vector addition $+$ and scalar multiplication $\cdot$ endowed with some other properties are typically called vector spaces.

More specifically, a set S closed under operations $+, \cdot$ called vector addition and scalar multiplication is called a **vector space** over a field $\mathbb{F}$ if for all $\vec{x}, \vec{y}, \vec{z} \in S$ and any $c, k \in \mathbb{F}$ we have

$$\vec{x} + \vec{y} = \vec{y} + \vec{x},$$

$$(\vec{x}+\vec{y})+\vec{z} = \vec{x}+(\vec{y}+\vec{z}),$$
$$(ck)\cdot\vec{x} = c\cdot(k\cdot\vec{x}),$$
$$c\cdot(\vec{x}+\vec{y}) = (c\cdot\vec{x})+(c\cdot\vec{y}),$$
$$(c+k)\cdot\vec{x} = (c\cdot\vec{x})+(k\cdot\vec{x}),$$
$$1_\mathbb{F}\cdot\vec{x} = \vec{x}, \quad 0_{\vec{x}}\cdot\vec{x} = \vec{0},$$

where $\vec{0} \in S$ satisfies $\vec{0} + \vec{x} = \vec{x}$ for any $\vec{x} \in S$. Moreover, for any $\vec{x} \in S$ there is a unique element $-\vec{x} \in S$ so that $\vec{x}+(-\vec{x}) = \vec{0}$, and $-\vec{x} = (-1_\mathbb{F})\cdot\vec{x}$. Note that we shall observe the usual order of operations convention for vector addition and scalar multiplication, i.e. $(c\cdot\vec{x})+(k\cdot\vec{y})$ shall be written $c\cdot\vec{x}+k\cdot\vec{y}$.

We may endow $\mathbb{R}^n$ with an inner product, denoted by either $\cdot$ or $<\ ,\ >$, where $<\ ,\ >: \mathbb{R}^n \times \mathbb{R}^n \to \mathbb{R}$, and

$$(x_1, x_2, \cdots, x_n)\cdot(y_1, y_2, \cdots, y_n)$$
$$=<(x_1, x_2, \cdots, x_n),(y_1, y_2, \cdots, y_n)>:= \sum_{j=1}^n x_j y_j.$$

This inner product endows $\mathbb{R}^n$ with a norm $\|\ \| : \mathbb{R}^n \to [0, \infty)$ defined by

$$\|\vec{x}\| := \sqrt{<\vec{x},\vec{x}>}.$$

Moreover, this inner product on $\mathbb{R}^n$ defines a geometry and a topology on $\mathbb{R}^n$, where the distance between two vectors $\vec{x}$ and $\vec{y}$ is defined to be $\|\vec{y}-\vec{x}\|$, where $\vec{y}-\vec{x} := \vec{y}+(-1)\cdot\vec{x}$ or $\vec{y}+(-\vec{x})$ and the angle $\theta \in [0,\pi)$ between two vectors is defined by $\cos\theta = \frac{\vec{y}\cdot\vec{x}}{\|\vec{y}\|\|\vec{x}\|}$.

8.1.1 $\mathbb{R}^n$ as a Normed Linear Space

Recall that on $\mathbb{R}^n$ we have a norm[1] $\|\ \|$ defined by $\|\vec{x}\| = \sqrt{\vec{x}\cdot\vec{x}}$. However, there are other norms we may define on $\mathbb{R}^n$, and some examples are

$$\|\vec{x}\|_1 = \|(x_1, x_2, \cdots, x_n)\|_1 := \sum_{j=1}^n |x_j|,$$

---

[1]That is, $\|\vec{x}\| \geq 0$ and equals zero if and only if $\vec{x} = \vec{0}$. For any scalar c and vector $\vec{x}$ we have $\|c\cdot\vec{x}\| = |c|\cdot\|\vec{x}\|$. Moreover, for any vectors $\vec{x}$ and $\vec{y}$ we have the so-called triangle inequality $\|\vec{x}+\vec{y}\| \leq \|\vec{x}\| + \|\vec{y}\|$.

8.1. THE ALGEBRA AND TOPOLOGY OF $\mathbb{R}^N$

and more generally for $1 \leq p < \infty$

$$\|\vec{x}\|_p = \|(x_1, x_2, \cdots, x_n)\|_p := \left(\sum_{k=1}^{n} |x_k|^p\right)^{\frac{1}{p}}.$$

Also, we define

$$\|\vec{x}\|_\infty = \|(x_1, x_2, \cdots, x_n)\|_\infty := \max\{|x_1|, |x_2|, \cdots, |x_n|\}.$$

We leave it as an exercise to show these are norms on $\mathbb{R}^n$, and observe that $\|\vec{x}\|$ as defined is $\|\vec{x}\|_2$.

Basis and Dimension

Let V be a vector space over some field $\mathbb{F}$. A set of vectors $\{\vec{x}_1, \vec{x}_2, \cdots, \vec{x}_m\} \subseteq V$ is said to be **linearly independent** if for scalars $c_1, c_2, \ldots, c_m$

$$c_1 \cdot \vec{x}_1 + c_2 \cdot \vec{x}_2 + \cdots + c_m \cdot \vec{x}_m = \vec{0}$$

we must have $c_1 = c_2 = \cdots = c_m = 0$. Otherwise, we say that $\{\vec{x}_1, \cdots, \vec{x}_m\}$ is **linearly dependent**.[2] In a general collection of vectors $\{\vec{x}_1, \vec{x}_2, \cdots, \vec{x}_m\}$, if $\vec{0}$ is in this collection, then the collection is linearly dependent. To see this, suppose $\vec{x}_1 = \vec{0}$ and then

$$1 \cdot \vec{0} + 0 \cdot \vec{x}_2 + \cdots + 0 \cdot \vec{x}_m = \vec{0}.$$

Also, for any non-zero vector $\vec{v}$, the collection $\{\vec{v}\}$ is linearly independent, since $c \cdot \vec{v} = \vec{0}$ implies $c = 0$.

For a given collection of vectors $S := \{\vec{y}_1, \vec{y}_2, \cdots, \vec{y}_m\}$, the **span of S**, denoted by span S, is the collection

$$\{c_1 \cdot \vec{y}_1 + c_2 \cdot \vec{y}_2 + \cdots + c_m \cdot \vec{y}_m : c_1, c_2, \cdots, c_m \in \mathbb{F}\}.$$

If a collection $\{\vec{x}_1, \vec{x}_2, \cdots, \vec{x}_m\}$ spans V, then for any $\vec{y} \in V$, we have $\{\vec{y}, \vec{x}_1, \vec{x}_2, \cdots, \vec{x}_m\}$ also spans V. If $\{\vec{x}_1, \vec{x}_2, \cdots, \vec{x}_m\}$ spans V, and $\vec{x}_j = c_1\vec{x}_1 + \cdots + c_{j-1}\vec{x}_{j-1} + c_{j+1}\vec{x}_{j+1} + \cdots + c_m\vec{x}_m$, then $\{\vec{x}_1, \vec{x}_2, \cdots, \vec{x}_{j-1}, \vec{x}_{j+1}, \cdots, \vec{x}_m\}$

---

[2] If a set of vectors is linearly dependent they we may write one of the vectors in the collection as a linear combination of all the other vectors.

spans V. If $\vec{0} \in \{\vec{x}_1, \vec{x}_2, \cdots, \vec{x}_m\}$ spans V, then $\{\vec{x}_1, \vec{x}_2, \cdots, \vec{x}_m\} \setminus \{\vec{0}\}$ spans V.

A **basis** of a vector space is a linearly independent collection of vectors whose span is the entire vector space. If a vector space contains a linearly independent set of n vectors but there is no set of $n+1$ vectors that is linearly independent, then we say that n is the **dimension** of the vector space, and we denote it by $\dim V$. To see that this definition of the dimension of V is well defined, suppose that we have a basis $\{\vec{x}_1, \vec{x}_2, \cdots, \vec{x}_m\}$ of V. Then for any non-zero $\vec{v} \in V$, we have that

$$\vec{v} = c_1 \cdot \vec{x}_1 + c_2 \cdot \vec{x}_2 + \cdots + c_m \cdot \vec{x}_m$$

for some choice of $c_1, c_2, \cdots, c_m$, all not zero. Thus we see that the collection $\{\vec{v}, \vec{x}_1, \vec{x}_2, \cdots, \vec{x}_m\}$ is linearly dependent, since

$$\vec{0} = c_1 \cdot \vec{x}_1 + c_2 \cdot \vec{x}_2 + \cdots + c_m \cdot \vec{x}_m + (-1) \cdot \vec{v}.$$

Hence we see that if a basis has m elements in it, then any set which contains the elements of the basis and additional elements of the vector space will span S, but will be linearly dependent.

Suppose that $n = \dim V$ and $\{\vec{x}_1, \vec{x}_2, \cdots, \vec{x}_n\}$ is a linearly independent set in V. Any vector in the vector space may be written as a sum of scalar multiples of the vectors in this set, i.e. as a **linear combination** of these vectors. To see this suppose $\vec{y}$ is any other vector, and

$$c_1 \vec{x}_1 + c_2 \vec{x}_2 + \cdots + c_n \vec{x}_n + c_{n+1} \vec{y} = \vec{0}.$$

If $c_{n+1} = 0$, then

$$c_1 \vec{x}_1 + c_2 \vec{x}_2 + \cdots + c_n \vec{x}_n = \vec{0},$$

and so all the c_i's are zero. However, this implies that the set $\{\vec{x}_1, \vec{x}_2, \cdots, \vec{x}_n, \vec{y}\}$ is linearly independent, and this can not be so. Thus $c_{n+1} \neq 0$ and so we may write $\vec{y} = \sum_{i=1}^{n} \frac{-c_i}{c_{n+1}} \vec{x}_i$. So in such a case $\{\vec{x}_1, \vec{x}_2, \cdots, \vec{x}_n\}$ is a basis for V.

Theorem 8.1.1 *If S is a vector space that is spanned by a set of m vectors, then $\dim S$, the dimension of S, is less than or equal to m.*

8.1. THE ALGEBRA AND TOPOLOGY OF $\mathbb{R}^N$

Proof: If this were not the case, then there is a vector space S that has a collection of linearly independent vectors $\{\vec{y}_1, \vec{y}_2, \cdots, \vec{y}_{m+1}\}$ with S spanned by a set m vectors $\{\vec{x}_1, \vec{x}_2, \cdots, \vec{x}_m\}$.

Let $0 \leq j < m$, and suppose that S is spanned by the collection

$$C_j := \{\vec{y}_1, \vec{y}_2, \cdots, \vec{y}_j, \vec{x}_{k_1}, \cdots, \vec{x}_{k_{m-j}}\}.$$

Since S is spanned by C_j, $\vec{y}_{j+1}$ lies in the span of C_j. Hence there are scalars

$$a_1, a_2, \cdots, a_{j+1}, b_1, b_2, \cdots, b_{m-j}$$

with $a_{j+1} = 1$ so that

$$\sum_{i=1}^{j+1} a_i \vec{y}_i + \sum_{i=1}^{m-j} b_i \vec{x}_{k_i} = \vec{0}.$$

If all the b_i's were zero, the linear independence of the collection $\{\vec{y}_1, \cdots, \vec{y}_{m+1}\}$ would force all the a_i's to be zero, and this is not so. It thus follows that some $\vec{x}_{k_i} \in C_j$ is a linear combination of the elements of $C_{j+1} := (C_j \cup \{\vec{y}_{j+1}\}) \setminus \{\vec{x}_{k_i}\}$.

Continuing in this manner, we can inductively construct $C_{j+1}, \cdots, C_m$, collections whose span is all of S and each C_k contains $\vec{y}_1, \cdots, \vec{y}_k$. However, $\{\vec{y}_1, \cdots, \vec{y}_{m+1}\}$ is a linearly independent collection, and thus we can not have $\vec{y}_{m+1}$ in the span of C_m, which gives us the desired contradiction.
□

Thus we see that the cardinality of any basis is greater than or equal to the dimension of a vector space.

Theorem 8.1.2 *Suppose S is a vector space of dimension n. Then a set of n vectors in S spans S if and only if the set of n vectors is linearly independent.*

Proof: Let $C := \{\vec{x}_1, \vec{x}_2, \cdots, \vec{x}_n\}$ be a collection of n vectors in S. If $\vec{y} \in S$ in any other vector in S, then $\{\vec{x}_1, \vec{x}_2, \cdots, \vec{x}_n, \vec{y}\}$ is linearly dependent. If C is linearly independent, then the argument given above shows that $\vec{y} \in$ span C.

Suppose C is linearly dependent, then there is a vector $\vec{x}_j \in C$ so that $\vec{x}_j \in \text{span}\,(C \setminus \{\vec{x}_j\})$. This says that span $C = \text{span}\,(C \setminus \{\vec{x}_j\})$. Hence C can not span S, since if C did span S we would have by theorem (8.1.1) that the dimension of S would be less than or equal to $n-1$, and this is not true by assumption.
□

Theorem 8.1.3 *Suppose S is an n dimensional vector space, then S has a basis, and any basis consists of n vectors. Moreover, if $C := \{\vec{x}_1, \vec{x}_2, \cdots, \vec{x}_k\}$ with $1 \leq k \leq n$ is a linearly independent set in S, then there is a basis of S that contains the vectors in C.*

Proof: If dim $S = n$, then S contains a linearly independent set T with n vectors. By theorem (8.1.2) we know that this set T is a basis of S.

Let C be as in the statement of this theorem, and let $\{\vec{b}_1, \vec{b}_2, \cdots, \vec{b}_n\}$ be a basis of S. Then the set $B := \{\vec{x}_1, \vec{x}_2, \cdots, \vec{x}_k, \vec{b}_1, \vec{b}_2, \cdots, \vec{b}_n\}$ is linearly dependent, and clearly spans S. Thus there are scalars $a_1, a_2, \cdots, a_k, c_1, c_2, \cdots c_n$ not all zero so that

$$\sum_{i=1}^{k} a_i \vec{x}_i + \sum_{j=1}^{n} c_j \vec{b}_j = 0.$$

If all the c_j's are zero, then $\sum_{i=1}^{k} a_i \vec{x}_i = 0$, and by linear independence of the $\vec{x}_i$'s, we see that all the a_i's are zero. Thus at least one of the b_j's is not zero. By otherwise reordering and relabeling the $\vec{b}_j$'s we may assume $c_n \neq 0$ and so $\vec{b}_n$ may be written as a linear combination of the remaining elements in B. So span $B = \text{span}\,(B \setminus \{\vec{b}_n\})$. We see that we may continue this process for a total of k times until we have a set of n vectors that spans S which contains all the $\vec{y}_i$'s. Then by theorem (8.1.2) we see that this is a basis of S.
□

Hence we see that $\mathbb{R}^n$ has dimension n, and the basis for $\mathbb{R}^n$ given by $\vec{e}_1 = (1, 0, \cdots, 0)$, $\vec{e}_2 = (0, 1, 0, \cdots, 0)$, $\cdots$, $\vec{e}_n = (0, \cdots, 0, 1)$ is called the **canonical basis** for $\mathbb{R}^n$. Using the notation $\vec{x} = \sum_{j=1}^{n} x_j \vec{e}_j$ we call the collection $\{x_1, x_2, \cdots, x_n\}$ the **coordinates** of $\vec{x}$ with respect to the canonical basis. [3]Moreover, we see $\vec{x} = (x_1, x_2, \cdots, x_n)$.

---

[3]Note that the order of the scalars in the collection $\{x_1, x_2, \cdots, x_n\}$ matters when they

8.1. THE ALGEBRA AND TOPOLOGY OF $\mathbb{R}^N$

The Topology of $\mathbb{R}^n$ as a Normed Linear Space

A vector space V over $\mathbb{R}$, or possibly the complex numbers $\mathbb{C}$, is called a **normed linear space** if there is a function $\|\cdot\| : V \to [0, \infty)$ so that for all $\vec{x}, \vec{y} \in V$ and scalars c we have

1. $\|\vec{x}\| \geq 0$ and equals zero if and only if $\vec{x} = \vec{0}$.

2. $\|c \cdot \vec{x}\| = |c| \|\vec{x}\|$

3. $\|\vec{x} + \vec{y}\| \leq \|\vec{x}\| + \|\vec{y}\|$.

We will now show that all norms on $\mathbb{R}^n$ define the same topology.

Lemma 8.1.4 *Let $\{\vec{x}_1, \cdots, \vec{x}_m\}$ be a linearly independent set of vectors in a normed linear space of any dimension which is a vector space over $\mathbb{R}$, then there is a positive number k so that for any choice of scalars $c_1, c_2, \cdots, c_m$ we have*

$$\|c_1 \cdot \vec{x}_1 + c_2 \cdot \vec{x}_2 + \cdots + c_m \cdot \vec{x}_m\| \geq k(|c_1| + |c_2| + \cdots + |c_m|) \tag{8.1}$$

Proof: Let $r := |c_1| + |c_2| + \cdots + |c_n|$. If all the c_i's are zero, then clearly (8.1) holds for any positive number k, thus we may assume $r > 0$. Let $d_i = \frac{c_i}{r}$ for $i = 1, 2, \cdots, m$. Then (8.1) is equivalent to $\|\sum_{k=1}^{m} d_k \vec{x}_k\| \geq k$ for some $k > 0$, where $\sum_{k=1}^{m} |d_k| = 1$. Suppose that this does not hold. Then there is a sequence of vectors $\{\vec{y}_k\}_{k=1}^{\infty}$, $\vec{y}_k = \sum_{t=1}^{m} d_t^k \vec{x}_t$, with $\sum_{t=1}^{m} |d_t^k| = 1$ and $\|\vec{y}_m\| < \frac{1}{m}$ for $m \in \mathbb{N}$.

Since $\sum_{t=1}^{m} |d_t^k| = 1$ for all $k \in \mathbb{N}$ we have that $b_t^k \in [-1, 1]$ for all k and t. Hence the the collection of vectors given by $(d_1^k, d_2^k, \cdots, d_m^k) \in \mathbb{R}^m$ is bounded. Thus by the corollary (3.2.2) we know that $\{(d_1^k, d_2^k, \cdots, d_m^k)\}_{k=1}^{\infty}$ has a convergent subsequence which we denote by $\{(d_1^{j_k}, d_2^{j_k}, \cdots, d_m^{j_k})\}_{k=1}^{\infty}$, and let $\vec{\delta} = (\delta_1, \delta_2, \cdots, \delta_m)$ denote the limit of this subsequence. Thus the collection $\{\vec{y}_{j_k}\}_{k=1}^{\infty}$ converges with respect to the norm $\|\|$ in our vector space to the vector $\vec{y} = \sum_{j=1}^{m} \delta_j \cdot \vec{x}_j$. Since $\sum_{j=1}^{m} |d_j^k| = 1$ we must have $\sum_{j=1}^{m} |\delta_j| = 1$, so not all of the δ_j's can be zero. Since $\{\vec{x}_1, \vec{x}_2, \cdots, \vec{x}_m\}$ is linearly independent we cannot have $\sum_{j=1}^{m} \delta_k \cdot \vec{x}_j = \vec{0}$. Since $\vec{y}_{j_k} \to \vec{y}$ as

---

are the coordinates of $\vec{x}$ with respect to the canonical basis. Later we will introduce a special notation to represent the coordinates of a vector with respect to a fixed basis.

$k \to \infty$ we must have $\|\vec{y}_{j_k}\| \to \|\vec{y}\|$ as $k \to \infty$. However $\|\vec{y}_{j_k}\| \to 0$ and so we must have $\vec{y} = \vec{0}$, and this is a contradiction.
□

For a vector space S with two norms $\|\ \|$ and $\|\ \|_1$, $\|\ \|$ is said to be **equivalent** to $\|\ \|_1$ if there is a constant $k \geq 1$ so that for any $\vec{x} \in S$

$$\frac{1}{k}\|\vec{x}\|_1 \leq \|\vec{x}\| \leq k\|\vec{x}\|_1. \tag{8.2}$$

Note that this condition implies

$$\frac{1}{k}\|\vec{x}\| \leq \|\vec{x}\|_1 \leq k\|\vec{x}\|$$

as well. Hence if $\|\ \|$ is equivalent to $\|\ \|_1$, then $\|\ \|_1$ is equivalent to $\|\ \|$. Thus we usually say that $\|\ \|$ and $\|\ \|_1$ are **equivalent norms**.

Note that equivalent norms $\|\ \|$ and $\|\ \|_1$ define the same metric topology, since a set is open with respect to $\|\ \|$ if and only if it is open with respect to $\|\ \|_1$. This holds since a set is open if and only if it is the union of open balls. Clearly a union of open balls is open. However, an open set is always the union of open balls since each point is an interior point. Thus an open set $\mathcal{O}$ has the property that $x \in \mathcal{O}$ implies there is an $r(x) > 0$ so that $B_{r(x)}(x) \subseteq \mathcal{O}$. Hence $\bigcup_{x \in S} B_{r(x)}(x) \subseteq S$. Since the other inclusion is obvious, we see that $S = \bigcup_{x \in S} B_{r(x)}(x)$.

We conclude this section with the following theorem, which says that on finite dimensional metric spaces all norms are equivalent.

Theorem 8.1.5 *On a finite dimensional vector space over $\mathbb{R}$ all norms are equivalent.*

Proof: Let S be a finite dimensional vector space of dimension n, and let $\|\ \|$ and $\|\ \|_1$ be two norms on S. Let $\{\vec{x}_1, \vec{x}_2, \cdots, \vec{x}_n\}$ be a basis for S and let $\vec{x} \in S$. Then $\vec{x} = \sum_{k=1}^{n} a_k \cdot \vec{x}_k$, and by Lemma (8.1.4) we know that there is a $k > 0$ so that

$$\|\vec{x}\| \geq k \sum_{j=1}^{n} |a_j|.$$

8.1. THE ALGEBRA AND TOPOLOGY OF $\mathbb{R}^N$

Now the triangle inequality gives us

$$\|\vec{x}\|_1 \leq \sum_{j=1}^{n} |a_j| \|\vec{x}_j\|_1$$

$$\leq \max\{\|\vec{x}_1\|_1, \|\vec{x}_2\|_1, \cdots, \|\vec{x}_n\|_1\} \sum_{j=1}^{n} |a_j|$$

$$\leq \frac{\max\{\|\vec{x}_1\|_1, \|\vec{x}_2\|_1, \cdots, \|\vec{x}_n\|_1\}}{k} \|\vec{x}\|.$$

by reversing the roles of $\|\|$ and $\|\|_1$ in this argument just given we get the desired result.
□

8.1.2 Homework Exercises

1. Show that $\|\|_1$ is a norm on $\mathbb{R}^n$ where $\|(x_1, x_2, \cdots, x_n)\|_1 = \sum_{j=1}^{n} |x_j|$.

2. Show that $\|\|_p$ is a norm on $\mathbb{R}^n$ for any $p \geq 1$, where $\|(x_1, x_2, \cdots, x_n)\|_p = (\sum_{j=1}^{n} |x_j|^p)^{\frac{1}{p}}$.

3. Show that $\|\|_\infty$ is a norm on $\mathbb{R}^n$ where $\|(x_1, x_2, \cdots, x_n)\|_\infty = \max\{|x_1|, |x_2|, \cdots, |x_n|\}$.

4. Find a $K \geq 1$ so that for any $\vec{x} \in \mathbb{R}^n$ we have

$$\frac{1}{K}\|\vec{x}\| \leq \|\vec{x}\|_1 \leq K\|\vec{x}\|.$$

5. Find a $K \geq 1$ so that for any $\vec{x} \in \mathbb{R}^n$ we have

$$\frac{1}{K}\|\vec{x}\| \leq \|\vec{x}\|_p \leq K\|\vec{x}\|,$$

where $p \in [1, \infty)$ is fixed.

6. Find a $K \geq 1$ so that for any $\vec{x} \in \mathbb{R}^n$ we have

$$\frac{1}{K}\|\vec{x}\| \leq \|\vec{x}\|_\infty \leq K\|\vec{x}\|.$$

7. Find a $K \geq 1$ so that for any $\vec{x} \in \mathbb{R}^n$ we have
$$\frac{1}{K}\|\vec{x}\|_1 \leq \|\vec{x}\| \leq K\|\vec{x}\|_1.$$

8. Find a $K \geq 1$ so that for any $\vec{x} \in \mathbb{R}^n$ we have
$$\frac{1}{K}\|\vec{x}\|_p \leq \|\vec{x}\| \leq K\|\vec{x}\|_p,$$
where $p \in [1, \infty)$ is fixed.

9. Find a $K \geq 1$ so that for any $\vec{x} \in \mathbb{R}^n$ we have
$$\frac{1}{K}\|\vec{x}\|_\infty \leq \|\vec{x}\| \leq K\|\vec{x}\|_\infty.$$

10. Show that the space of all $m \times n$ matrices over a field $\mathbb{F}$ is a vector space of dimension mn, where here $+, \cdot$ are as usually defined.

11. Let V be the vector space over $\mathbb{R}$ given by $\{a_0 + a_1 x + a_2 x^2 : a_0, a_1, a_2 \in \mathbb{R}\}$. Let $c \in \mathbb{R}$, and show that $\{1, x+c, (x+c)^2\}$ is a basis of V.

8.2 Linear Transformations and Affine Functions

8.2.1 Linear Transformations on Vector Spaces

Let S, T be vector spaces over the same scalar field $\mathbb{F}$, and $L: S \to T$. We say that L is a **linear transformation** or a **linear mapping** if for any $\vec{x}, \vec{y} \in S$ and scalar $c \in \mathbb{F}$ we have

$$L(\vec{x} + \vec{y}) = L(\vec{x}) + L(\vec{y}), \quad L(c \cdot \vec{x}) = c \cdot L(\vec{x}). \tag{8.3}$$

Note that we often write $L\vec{x}$ instead of $L(\vec{x})$ whenever L is a linear transformation.

Since $L\vec{x} = L(\vec{x} + \vec{0}) = L\vec{x} + L\vec{0}$ we see that $L\vec{0}$ must be $\vec{0} \in T$. The subset of S which L sends to $\vec{0} \in T$ is usually called the **kernel** of L, and often denoted by Ker L. Clearly this set is non-empty since $\vec{0} \in S$ is always

8.2. LINEAR TRANSFORMATIONS AND AFFINE FUNCTIONS

in Ker L. From this we see that a linear transformation L is injective if and only if Ker $L = \{\vec{0}\}$, since $L\vec{x} = L\vec{y}$ if and only if $\vec{x} - \vec{y} \in$ Ker L.

By linearity (8.3) we see that the values of L on a basis of S completely determine the linear mapping L.

If L is a linear transformation from S to itself we say that L is a **linear operator** on S. We let $\mathcal{L}(S)$ denote the collection of linear operators on S, and more generally we let $\mathcal{L}(S,T)$ denote the collection of linear transformations from S to T. The collection $\mathcal{L}(S,T)$ is closed under $+$ and $\cdot$ as defined in T. For $a, c, k \in \mathbb{F}$ and $\vec{x}, \vec{y} \in S$, for $L, M \in \mathcal{L}(S,T)$ we define

$$(cL + kM)(\vec{x}) := cL\vec{x} + kM\vec{x}.$$

Since L, M are linear transformations, we have

$$(cL + kM)(\vec{x} + \vec{y}) = cL(\vec{x} + \vec{y}) + kM(\vec{x} + \vec{y})$$

$$= cL\vec{x} + cL\vec{y} + kM\vec{x} + kM\vec{y} = (cL + kM)\vec{x} + (cL + kM)\vec{y},$$

and

$$(cL + kM)(a\vec{x}) = cL(a\vec{x}) + kM(a\vec{x})$$

$$= acL\vec{x} + akM\vec{x} = a(cL + kM)\vec{x}.$$

Moreover, $\mathcal{L}(S)$ is closed under composition, since

$$L(M(a\vec{x} + b\vec{y})) = L(aM\vec{x} + bM\vec{y})$$

$$= aL(M\vec{x}) + bL(M\vec{y}) = a(LM)\vec{x} + b(LM)\vec{y}.$$

This same argument also shows that for $M \in \mathcal{L}(S,T)$ and $L \in \mathcal{L}(T,U)$ we have $LM = L \circ M$ is in $\mathcal{L}(S,U)$.

When our vector spaces[4] S and T are normed linear spaces we may define a **norm of a linear transformation** $L \in \mathcal{L}(S,T)$ by

$$\|L\|_{\mathcal{L}(S,T)} := \sup\{\|L\vec{x}\|_T : \|\vec{x}\|_S = 1\}. \tag{8.4}$$

---

[4]which are vector spaces over $\mathbb{R}$ or $\mathbb{C}$

Then for any $\vec{y} \in S$, we may write $\vec{y} = \|\vec{y}\|_S (\frac{1}{\|\vec{y}\|_S}\vec{y})$ and so

$$\|L\vec{y}\|_T = \|\vec{y}\|_S \|L(\frac{1}{\|\vec{y}\|_S}\vec{y})\|_T \leq \|L\|_{\mathcal{L}(S,T)} \|\vec{y}\|_S.$$

For $L, M \in \mathcal{L}(S,T)$ with S, T normed linear spaces, we have

$$\|L+M\|_{\mathcal{L}(S,T)} \leq \|L\|_{\mathcal{L}(S,T)} + \|M\|_{\mathcal{L}(S,T)}, \quad \|cL\|_{\mathcal{L}(S,T)} = |c|\|L\|_{\mathcal{L}(S,T)}. \quad (8.5)$$

To see this we note that

$$\|(L+M)\vec{x}\|_T = \|L\vec{x} + M\vec{x}\|_T$$
$$\leq \|L\vec{x}\|_T + \|M\vec{x}\|_T$$
$$\leq \|L\|_{\mathcal{L}(S,T)}\|\vec{x}\|_S + \|M\|_{\mathcal{L}(S,T)}\|\vec{x}\|_S.$$

Suppose that $\|\vec{x}\|_S = 1$. Then taking the supremum of $\|(L+M)\vec{x}\|_T$ over the collection of $\vec{x} \in S$ of norm one we get $\|L+M\|_{\mathcal{L}(S,T)} \leq \|L\|_{\mathcal{L}(S,T)} + \|M\|_{\mathcal{L}(S,T)}$.

Also,
$$\|cL\vec{x}\|_T = |c|\|L\vec{x}\|_T \leq |c|\|L\|_{\mathcal{L}(S,T)}\|\vec{x}\|_S,$$

and thus by taking the supremum over all $\vec{x} \in S$ of norm one we see that $\|cL\|_{\mathcal{L}(S,T)} \leq |c|\|L\|_{\mathcal{L}(S,T)}$. By then using this with $L = \frac{1}{c}(cL)$ we see that $\|L\|_{\mathcal{L}(S,T)} \leq \frac{1}{|c|}\|cL\|_{\mathcal{L}(S,T)}$, and so we see that $\|cL\|_{\mathcal{L}(S,T)} = |c|\|L\|_{\mathcal{L}(S,T)}$.

If $L \in \mathcal{L}(S,T), M \in \mathcal{L}(T,U)$ with S,T,U normed linear spaces, then we have

$$\|ML\|_{\mathcal{L}(S,U)} \leq \|M\|_{\mathcal{L}(T,U)} \|L\|_{\mathcal{L}(S,T)} \quad (8.6)$$

To see this note that

$$\|ML\vec{x}\|_U \leq \|M\|_{\mathcal{L}(T,U)} \|L\vec{x}\|_T \leq \|M\|_{\mathcal{L}(T,U)} \|L\|_{\mathcal{L}(S,T)} \|\vec{x}\|_S.$$

By taking the supremum of the left hand side over all $\vec{x}$ of norm one we get the desired result.

The next theorem tells us that linear transformations from $\mathbb{R}^n$ to $\mathbb{R}^m$ always have bounded norms linear transformations in $\mathcal{L}(\mathbb{R}^n, \mathbb{R}^m)$ and thus are uniformly continuous mappings.

8.2. LINEAR TRANSFORMATIONS AND AFFINE FUNCTIONS

Theorem 8.2.1 *Let $L \in \mathcal{L}(\mathbb{R}^n, \mathbb{R}^m)$, then $\|L\|_{\mathcal{L}(\mathbb{R}^n, \mathbb{R}^m)} < \infty$. Moreover, L is a uniformly continuous function as well.*

Proof: Let $\{\vec{e}_1, \cdots, \vec{e}_n\}$ be the canonical basis of $\mathbb{R}^n$, and let $\vec{x} \in \mathbb{R}^n$ be any vector of norm one. If $\vec{x} = \sum_{k=1}^n c_k \vec{e}_k$, then $|c_k| \leq 1$ for all k. Thus

$$\|L\vec{x}\|_{\mathbb{R}^m} = \|L \sum_{k=1}^n c_k \vec{e}_k\|_{\mathbb{R}^m}$$

$$= \|\sum_{k=1}^n c_k L\vec{e}_k\|_{\mathbb{R}^m}$$

$$\leq \sum_{k=1}^n |c_k| \|L\vec{e}_k\|_{\mathbb{R}^m}$$

$$\leq \sum_{k=1}^n \|L\vec{e}_k\|_{\mathbb{R}^m}.$$

Taking the supremum over the collection of $\vec{x}$'s of norm one we see $\|L\|_{\mathcal{L}(\mathbb{R}^n, \mathbb{R}^m)} \leq \sum_{k=1}^n \|L\vec{e}_k\|_{\mathbb{R}^m}$.

Since $\|L\vec{x} - L\vec{y}\|_{\mathbb{R}^m} = \|L(\vec{x} - \vec{y})\|_{\mathbb{R}^m} \leq \|L\|_{\mathcal{L}(\mathbb{R}^n, \mathbb{R}^m)} \|\vec{x} - \vec{y}\|_{\mathbb{R}^n}$ we see that L is uniformly continuous.
□

Matrix Representations of Linear Transformations

Let S, T be vector spaces over a field $\mathbb{F}$ and let $L : S \to T$ be a linear transformation. Let $B = \{\vec{x}_1, \vec{x}_2, \cdots, \vec{x}_n\}$ be an ordered basis of S and $C = \{\vec{y}_1, \vec{y}_2, \cdots, \vec{y}_m\}$ an ordered basis of T. We let $[\vec{x}]_B$ denote the **coordinates** of a vector with respect to the basis B. That is, for $\vec{x} \in S$ we define $[\vec{x}]_B = (a_1, a_2, \cdots, a_n)$ is the coordinates of $\vec{x}$ with respect to the ordered basis B if

$$\vec{x} = \sum_{j=1}^n a_j \vec{x}_j.$$

We know that the values of L on the elements of a basis of S completely determine L. In particular, if $L\vec{x}_j = \vec{v}_j$ for $j = 1, 2, \cdots, n$, we may represent

$\vec{v}_j$ as $\vec{v}_j = \sum_{i=1}^{m} L_{ij}\vec{y}_i$. Thus for $\vec{x} = \sum_{j=1}^{n} a_j\vec{x}_j$ we have

$$L\vec{x} = L\sum_{j=1}^{n} a_j\vec{x}_j$$

$$= \sum_{j=1}^{n} a_j L\vec{x}_j$$

$$= \sum_{j=1}^{n} a_j (\sum_{i=1}^{m} L_{ij}\vec{y}_i)$$

$$= \sum_{i=1}^{m} (\sum_{j=1}^{n} L_{ij} a_j)\vec{y}_i.$$

If we let $[L]_{B,C}$ denote the matrix $\{L_{ij}\}_{1 \le i \le m, 1 \le j \le n}$, then

$$[L]_{B,C}[\vec{x}]_B = [L\vec{x}]_C,$$

since

$$[L]_{B,C}[\vec{x}]_B = \begin{bmatrix} L_{11} & L_{12} & \cdots & L_{1n} \\ L_{21} & L_{22} & \cdots & L_{2n} \\ \vdots & \vdots & \ddots & \vdots \\ L_{i1} & L_{i2} & \cdots & L_{in} \\ \vdots & \vdots & \ddots & \vdots \\ L_{m1} & L_{m2} & \cdots & L_{mn} \end{bmatrix} \begin{bmatrix} a_1 \\ a_2 \\ \vdots \\ a_j \\ \vdots \\ a_n \end{bmatrix} = \begin{bmatrix} \sum_{j=1}^{n} L_{1j} a_j \\ \sum_{j=1}^{n} L_{2j} a_j \\ \vdots \\ \sum_{j=1}^{n} L_{ij} a_j \\ \vdots \\ \sum_{j=1}^{n} L_{mj} a_j \end{bmatrix}.$$

Note that for $L \in \mathcal{L}(S,T), M \in \mathcal{L}(T,U)$ with ordered bases A, B, C of the vector spaces S, T and U respectively, then

$$[ML]_{A,C} = [M]_{B,C}[L]_{A,B}.$$

In the case where our vector spaces are $\mathbb{R}^n$ and $\mathbb{R}^m$ with their standard canonical ordered bases A and B, for $L \in \mathcal{L}(\mathbb{R}^n, \mathbb{R}^m)$, for $[\vec{x}]_A = (a_1, \cdots, a_n)$ by Cauchy-Schwarz we have

$$\|L\vec{x}\|^2 = \sum_{i=1}^{m} (\sum_{j=1}^{n} L_{ij} a_j)^2$$

8.2. LINEAR TRANSFORMATIONS AND AFFINE FUNCTIONS

$$\leq \sum_{i=1}^{m}(\sum_{j=1}^{n} L_{ij}^2 \cdot \sum_{j=1}^{n} a_j^2)$$

$$= \sum_{i,j} L_{ij}^2 \|\vec{x}\|^2.$$

Thus we see

$$\|L\|_{\mathcal{L}(\mathbb{R}^n,\mathbb{R}^m)} \leq \sqrt{\sum_{i,j} L_{ij}^2}.$$

8.2.2 Homework Exercises

1. Let V be the space of $n \times 1$ matrices over a field $\mathbb{F}$, and W the space of $m \times 1$ matrices over $\mathbb{F}$. Let A be a fixed $m \times n$ matrix whose coefficients are in $\mathbb{F}$. Define $T(\vec{x}) = A\vec{x}$, $T : V \to W$. Prove that $T\vec{x} = \vec{0}$ for all $\vec{x}$ if and only if all the coefficients in A are $0_\mathbb{F}$.

2. Let V, W, T be as in the above exercise. Show that if $m < n$, then the kernel[5] of T is non-trivial.

3. Let V be the vector space over $\mathbb{R}$ given by $\{a_0 + a_1 x + a_2 x^2 : a_0, a_1, a_2 \in \mathbb{R}\}$. Let $c \in \mathbb{R}$. Earlier, it was an exercise to show that $B := \{1, x + c, (x+c)^2\}$ is a basis of V. For the polynomial $f(x) := a_0 + a_1 x + a_2 x^2$ find $[f(x)]_B$.

4. Suppose $L : V \to W$ is a linear transformation between vector spaces V and W that is one-to-one and onto, i.e. bijective. Show that $L^{-1} : W \to V$ is also a linear transformation.

8.2.3 Affine Functions

Let V, W be vector spaces over a field $\mathbb{F}$, $\vec{b} \in W$ and let $L : V \to W$ be a linear transformation. A function $F : V \to W$ of the form

$$F(\vec{x}) = L(\vec{x}) + \vec{b} \tag{8.7}$$

is called an **affine function**.

---

[5] The kernel is the subset of the domain that maps to zero in the range. Since zero will map to zero, one must show that the kernel contains more than just zero in the domain.

For $F : V \to W$ an affine function with respect to any ordered bases B, C of V and W respectively, we may represent F in the form

$$[F(\vec{x})]_C = [L]_{B,C}[\vec{x}]_B + [\vec{b}]_C.$$

In the setting of $F : \mathbb{R}^n \to \mathbb{R}^m$ using the usual representation of vectors, we see that affine functions are functions of the form

$$F(\vec{x}) = A\vec{x} + \vec{b},$$

where A is an $m \times n$ real matrix, and $\vec{b}$ is a fixed vector. In the case where $n = m = 1$ we call affine functions linear functions.[6]

8.3 Differentiation

Recall that for functions $f : \mathbb{R} \to \mathbb{R}$, the derivative of f at a point x_0 in f's domain is given by $f'(x_0) = \lim_{h \to 0} \frac{f(x_0+h)-f(x)}{h}$, or equivalently $f'(x_0) = \lim_{x \to x_0} \frac{f(x)-f(x_0)}{x-x_0}$. We may rewrite this condition and observe that $f'(x_0)$ exists if and only if

$$0 = \lim_{x \to x_0} \{\frac{f(x) - f(x_0)}{x - x_0} - f'(x_0)\}$$

$$= \lim_{x \to x_0} \frac{f(x) - f(x_0) - f'(x_0) \cdot (x - x_0)}{x - x_0},$$

which holds if and only if

$$0 = \lim_{x \to x_0} \frac{|f(x) - f(x_0) - f'(x_0) \cdot (x - x_0)|}{|x - x_0|}.$$

Observe that $y = f(x_0) + f'(x_0) \cdot (x - x_0)$ is the so-called tangent line to the graph of $y = f(x)$ at the point $(x_0, f(x_0))$. This motivates us to define a derivative of a function in any complete[7] normed linear space, or **Banach space**, in a similar manner.

Let V, W be any Banach spaces, $f : V \to W$, with norms $\|\cdot\|_V$ and $\|\cdot\|_W$ defined on V and W respectively. Let $\vec{x}_0 \in \text{domain}(f) \subseteq V$. We say that f

---

[6] Often this term **linear** is used in the more general setting as well.
[7] Cauchy sequences converge to limits in the space.

8.3. DIFFERENTIATION

is differentiable at $\vec{x}_0$ if there exists a linear transformation $L : V \to W$ so that for the function $g(\vec{x}) := L(\vec{x} - \vec{x}_0) + f(\vec{x}_0)$ we have

$$0 = \lim_{\vec{x} \to \vec{x}_0} \frac{\|f(\vec{x}) - f(\vec{x}_0) - L(\vec{x} - \vec{x}_0)\|_W}{\|\vec{x} - \vec{x}_0\|_V}. \tag{8.8}$$

We may think of $g(\vec{x})$ as the supporting hyperplane to the graph of f in $V \times W$. Thus we see that a function $f : V \to W$ is differentiable at a point $\vec{x}_0$ in its domain if it can be approximated locally (with respect to the $\|\cdot\|_W$ norm) by an affine function. Note that our notion of differentiation depends on the norms on V and W. However, in Euclidean case, since all norms are topologically equivalent, if $f : \mathbb{R}^n \to \mathbb{R}^m$ is differentiable at a point with respect to any two norms of $\mathbb{R}^n$ and $\mathbb{R}^m$, then it is differentiable at that point with respect to all norms.

The linear function L shall also be called the **derivative** of f at $\vec{x}_0$. The following tells us that the derivative is unique, hence the use of the definite article the is appropriate here. Moreover, due to its uniqueness, we will commonly use the notation Df for the derivative of f instead of L in all future discussions.

Theorem 8.3.1 *Let L be as in (8.8), with V, W Banach spaces, $f : V \to W$ and $\vec{x}_0 \in \mathrm{domain}(f)$. Then the derivative L of f at $\vec{x}_0$ is uniquely defined.*

Proof: Suppose that L_1, L_2 are two linear functionals that act as the derivative of f at $\vec{x}_0$. Then we see that

$$0 \le \frac{\|L_1(\vec{x} - \vec{x}_0) - L_2(\vec{x} - \vec{x}_0)\|_W}{\|\vec{x} - \vec{x}_0\|_V}$$

$$= \frac{\|L_1(\vec{x} - \vec{x}_0) + f(\vec{x}_0) - f(\vec{x}) + f(\vec{x}) - f(\vec{x}_0) - L_2(\vec{x} - \vec{x}_0)\|_W}{\|\vec{x} - \vec{x}_0\|_V}$$

$$\le \frac{\|L_1(\vec{x} - \vec{x}_0) + f(\vec{x}_0) - f(\vec{x})\|_W}{\|\vec{x} - \vec{x}_0\|_V} + \frac{\|f(\vec{x}) - f(\vec{x}_0) - L_2(\vec{x} - \vec{x}_0)\|_W}{\|\vec{x} - \vec{x}_0\|_V} \to 0$$

as $\vec{x} \to \vec{x}_0$ in V.

Thus we see that for any $\varepsilon > 0$ there is a $\delta > 0$ so that

$$\left\|(L_1 - L_2)\left(\frac{\vec{x} - \vec{x}_0}{\|\vec{x} - \vec{x}_0\|_V}\right)\right\|_W = \frac{\|(L_1 - L_2)(\vec{x} - \vec{x}_0)\|_W}{\|\vec{x} - \vec{x}_0\|_V} < \varepsilon$$

whenever $\|\vec{x} - \vec{x}_0\|_V < \delta$.

Now for any $\vec{y} \in V$ with norm one, there is an $\vec{x} \in V$ with $\|\vec{x} - \vec{x}_0\|_V < \delta$ and $\vec{y} = \frac{\vec{x} - \vec{x}_0}{\|\vec{x} - \vec{x}_0\|_V}$. Thus for any $\vec{y} \in V$ of norm one we have

$$\|(L_1 - L_2)(\vec{y})\|_W < \varepsilon.$$

Hence we see

$$\|L_1 - L_2\|_{\mathcal{L}(V,W)} < \varepsilon.$$

Since $\varepsilon > 0$ is arbitrary, we see that $\|L_1 - L_2\|_{\mathcal{L}(V,W)} = 0$. Hence we have $L_1 = L_2$ [8]

$\square$

Another important result says that if f is differentiable at $\vec{x}_0$, then it is continuous at $\vec{x}_0$, and this is given by the following theorem.

Theorem 8.3.2 *Let $f : V \to W$ be differentiable at $\vec{x}_0$, then f is continuous at $\vec{x}_0$.*

Proof: Let $\varepsilon > 0$ be given. By equation (8.8) there is an $\eta > 0$ so that if $\|\vec{x} - \vec{x}_0\|_V < \eta$ then

$$\|f(\vec{x}) - f(\vec{x}_0) - Df(\vec{x} - \vec{x}_0)\|_W \leq \varepsilon \cdot \|\vec{x} - \vec{x}_0\|_V.$$

So we have

$$\|f(\vec{x}) - f(\vec{x}_0)\|_W \leq (\varepsilon + \|Df\|_{\mathcal{L}(V,W)})\|\vec{x} - \vec{x}_0\|_V.$$

The right hand side can be made less that ε if $\|\vec{x} - \vec{x}_0\|_V < \min\{\eta, \frac{\varepsilon}{\varepsilon + \|Df\|_{\mathcal{L}(V,W)}}\}$.
$\square$

We say that $f : V \to W$ is **differentiable** if (8.8) holds for all $\vec{x} \in V$. Moreover, for any scalar λ (in $\mathbb{R}$, $\mathbb{C}$ or any complete field) and any functions f, g that are differentiable at a point $\vec{x}_0$ we have that λf and $f + g$ are also differentiable at $\vec{x}_0$.

---

[8]It is an exercise to show $\|L\|_{\mathcal{L}(V,W)} = 0$ if and only if $L = 0$.

8.3. DIFFERENTIATION

8.3.1 The Chain Rule

The following theorem is the chain rule in the setting of Banach spaces.

Theorem 8.3.3 *Let U, V, W be Banach spaces with domain$(f) \subseteq U$, $f :$ domain$(f) \to V$, domain$(g) \subseteq V$, $g :$ domain$(g) \to W$. Let $\vec{x}_0 \in$ domain(f), $f(\vec{x}_0) \in$ domain(g) and f,g differentiable at $\vec{x}_0$ and $f(\vec{x}_0)$ respectively, which we assume are interior points in the domains of f and g. If $g \circ f$ is defined in an open neighbourhood of $\vec{x}_0$, then $g \circ f$ is differentiable at $\vec{x}_0$ and if Df, Dg are the derivatives of f and g respectively, then*

$$D(g \circ f)(\vec{x}_0) = Dg(f(\vec{x}_0)) \circ Df(\vec{x}_0).$$

Proof: By theorem (8.3.2) we know f is continuous at $\vec{x}_0$. Thus there are open sets $\Omega \subseteq$ domain(f) and $\Omega' \subseteq$ domain(g) which contain $\vec{x}_0$ and $f(\vec{x}_0)$ respectively and $f(\Omega) \subseteq \Omega'$. The function $g \circ f$ is then defined on Ω. Now, for $\vec{x} \in \Omega$ we have

$$\|g(f(\vec{x})) - g(f(\vec{x}_0)) - Dg(f(\vec{x}_0)) \circ Df(\vec{x}_0)(\vec{x} - \vec{x}_0)\|_W$$
$$\leq \|g(f(\vec{x})) - g(f(\vec{x}_0)) - Dg(f(\vec{x}_0))(f(\vec{x}) - f(\vec{x}_0))\|_W$$
$$+ \|Dg(f(\vec{x}_0))(f(\vec{x}) - f(\vec{x}_0)) - Dg(f(\vec{x}_0)) \circ Df(\vec{x}_0)(\vec{x} - \vec{x}_0)\|_W.$$

Let $\varepsilon > 0$, then there is a $\sigma > 0$ so that for $\|\vec{y} - f(\vec{x}_0)\|_V < \sigma$ we have

$$\|g(\vec{y}) - g(f(\vec{x}_0)) - Dg(f(\vec{x}_0))(\vec{y} - f(\vec{x}_0))\|_W < \frac{\varepsilon \|\vec{y} - f(\vec{x}_0)\|_V}{2(1 + \|Df(\vec{x}_0)\|_{\mathcal{L}(U,V)})},$$

since g is differentiable at $f(\vec{x}_0)$. Moreover, there is a $\eta > 0$ so that for $\|\vec{x} - \vec{x}_0\|_U < \eta$ we have

$$\|f(\vec{x}) - f(\vec{x}_0) - Df(\vec{x}_0)(\vec{x} - \vec{x}_0)\|_V < \frac{\varepsilon}{2\|Dg(f(\vec{x}_0))\|_{\mathcal{L}(V,W)}} \|\vec{x} - \vec{x}_0\|_U$$

since f is differentiable at $\vec{x}_0$. By continuity of f at $\vec{x}_0$ there is a $\gamma > 0$ so that for $\|\vec{x} - \vec{x}_0\|_U \leq \gamma$ we have

$$\|f(\vec{x}) - f(\vec{x}_0)\|_V < [1 + \|Df(\vec{x}_0)\|_{\mathcal{L}(U,V)}]\|\vec{x} - \vec{x}_0\|_U,$$

and so this can be made less than σ if $\|\vec{x} - \vec{x}_0\|_U < \frac{\sigma}{1+\|Df(\vec{x}_0)\|_{\mathcal{L}(U,V)}}$. Thus for $\|\vec{x} - \vec{x}_0\|_U < \min\{\gamma, \eta, \frac{\sigma}{1+\|Df(\vec{x}_0)\|_{\mathcal{L}(U,V)}}\}$ we have

$$\|g(f(\vec{x})) - g(f(\vec{x}_0)) - Dg(f(\vec{x}_0)) \circ Df(\vec{x}_0)(\vec{x} - \vec{x}_0)\|_W < \varepsilon \|\vec{x} - \vec{x}_0\|_U.$$

□

8.3.2 The Gradient and Directional Derivatives

Let $\vec{u}$ be a unit vector[9] in a Banach space and $\vec{x}_0$ an element in the interior of the domain of a real-valued function f. We define the **directional derivative of f at $\vec{x}_0$ in the direction of $\vec{u}$** to be

$$\nabla_{\vec{u}} f(\vec{x}_0) := \lim_{t \to 0} \frac{f(\vec{x}_0 + t\vec{u}) - f(\vec{x}_0)}{t}, \tag{8.9}$$

whenever the limits exists. If we let $g(t) = f(\vec{x}_0 + t\vec{u})$, then we see that $g'(0) = \lim_{t \to 0} \frac{g(t) - g(0)}{t} = \lim_{t \to 0} \frac{f(\vec{x}_0 + t\vec{u}) - f(\vec{x}_0)}{t} = \nabla_{\vec{u}} f(\vec{x}_0)$. By the chain rule, we see that $\nabla_{\vec{u}} f(\vec{x}_0) = Df(\vec{x}_0) \vec{u}$.

Partial Derivatives

Let $f : \mathbb{R}^n \to \mathbb{R}$ and let $\vec{u} = \vec{e}_i$ be one of the canonical basis elements in $\mathbb{R}^n$. We say that $\nabla_{\vec{e}_i} f$ is the ith **partial derivative** of f, and denote it by

$$\frac{\partial f}{\partial x_i} := \nabla_{\vec{e}_i} f. \tag{8.10}$$

Other common notations for the partial derivative are $\frac{\partial}{\partial x_i} f$, f_{x_i}, $D_{x_i} f$ and $D_i f$ and we may use any of these notations in future discussions.

The Gradient

In the same setting – that is domain$(f) \subseteq \mathbb{R}^n$ and $f :$ domain$(f) \to \mathbb{R}$ – the derivative of f, Df, as a linear operator is often expressed using the so-called **gradient** of f. A common notation for the gradient of f is ∇f. In fact, the gradient can be thought of as the dual operator to Df since $Df(\vec{x}_0) \vec{z} = <\nabla f(\vec{x}_0), \vec{z}> = \nabla f(\vec{x}_0) \cdot \vec{z}$. Moreover, we see that $\frac{\partial f}{\partial x_i} = \nabla f \cdot \vec{e}_i$, and directional derivatives can be expressed in terms of the gradient as $\nabla_{\vec{u}} f = \nabla f \cdot \vec{u}$.

However, we note that f differentiable clearly implies f has all partial derivatives exist and more generally, all directional derivatives exist. However, if all partial derivatives exist, f may not be differentiable. We note that in the setting of a **Hilbert space**, i.e. a Banach space whose norm is given by an inner product, if there is a quantity ∇f so that $\nabla_{\vec{u}} f = <\nabla f, \vec{u}>$ we call ∇f the gradient of f.

---

[9]The norm of the vector is one.

8.3. DIFFERENTIATION

Higher Order Partial Derivatives

We may calculate the partial derivative of a partial derivative. For instance $\frac{\partial^2 f}{\partial x_i \partial x_j} := \frac{\partial}{\partial x_i}(\frac{\partial f}{\partial x_j})$ and this may also be represented by $D_{x_i x_j} f$ $D_{ij} f$, $f_{x_j x_i}$, or $\frac{\partial^2}{\partial x_i \partial x_j} f$. In general $D_{ij} f \neq D_{ji} f$. However, the following theorem of Clairaut gives a sufficient condition for $D_{ij} f = D_{ji} f$ to hold.

Theorem 8.3.4 *Suppose $f : D \to \mathbb{R}$, where $D \subseteq \mathbb{R}^n$. Let $\vec{x}_0 \in D$ and suppose that on $B_r(\vec{x}_0)$ we have $f_{x_i x_j}$ and $f_{x_j x_i}$ both exist and are continuous, then $f_{x_i x_j}(\vec{x}_0) = f_{x_j x_i}(\vec{x}_0)$.*

Proof: Consider the quantity $\triangle(h, k)$ defined by

$$\triangle(h, k) = \frac{[f(\vec{x}_0 + h\vec{e}_i + k\vec{e}_j) - f(\vec{x}_0 + h\vec{e}_i)] - [f(\vec{x}_0 + k\vec{e}_j) - f(\vec{x}_0)]}{hk}.$$

Notice that

$$\lim_{k \to 0}(\lim_{h \to 0} \triangle(h, k)) = \lim_{k \to 0} \frac{f_{x_i}(\vec{x}_0 + k\vec{e}_j) - f_{x_i}(\vec{x}_0)}{k} = f_{x_i x_j}(\vec{x}_0)$$

and

$$\lim_{h \to 0}(\lim_{k \to 0} \triangle(h, k)) = \lim_{k \to 0} \frac{f_{x_j}(\vec{x}_0 + h\vec{e}_i) - f_{x_j}(\vec{x}_0)}{h} = f_{x_j x_i}(\vec{x}_0).$$

Let $G(\vec{x}) = f(\vec{x} + h\vec{e}_i) - f(\vec{x})$, so

$$\triangle(h, k) = \frac{1}{h} \frac{G(\vec{x}_0 + k\vec{e}_j) - G(\vec{x}_0)}{k},$$

and by the Mean Value Theorem this may be written as

$$= \frac{1}{h} G_{x_j}(\vec{x}_0 + \theta_1 k\vec{e}_j) = \frac{1}{h}\{f_{x_j}(\vec{x}_0 + h\vec{e}_i + \theta_1 k\vec{e}_j) - f_{x_j}(\vec{x}_0 + \theta_1 k\vec{e}_j)\},$$

for some $\theta_1 \in (0, 1)$.

Let $H(\vec{x}) = f_{x_j}(\vec{x} + \theta_1 k\vec{e}_j)$. Then, $\triangle(h, k) = \frac{1}{h}[H(\vec{x}_0 + h\vec{e}_i) - H(\vec{x}_0)]$ which by the Mean Value Theorem we may write as

$$\triangle(h, k) = H_{x_i}(\vec{x}_0 + \theta_2 h\vec{e}_i) = f_{x_j x_i}(\vec{x}_0 + \theta_2 h\vec{e}_i + \theta_1 k\vec{e}_j),$$

for some $\theta_2 \in (0, 1)$.

Define $A(\vec{x}) = f(\vec{x} + k\vec{e}_j) - f(\vec{x})$, then by the Mean Value Theorem we have
$$\triangle(h,k) = \frac{1}{k}\frac{A(\vec{x}_0 + h\vec{e}_i) - A(\vec{x}_0)}{h} = \frac{1}{k}A_{x_i}(\vec{x}_0 + \theta_3 h\vec{e}_i)$$
$$= \frac{1}{k}\{f_{x_i}(\vec{x}_0 + \theta_3 h\vec{e}_i + k\vec{e}_j) - f_{x_i}(\vec{x}_0 + \theta_3 h\vec{e}_i)\},$$
for some $\theta_3 \in (0,1)$.

Define $B(\vec{x}) = f_{x_i}(\vec{x} + \theta_3 h\vec{e}_i)$, then by the Mean Value Theorem we have
$$\triangle(h,k) = \frac{B(\vec{x}_0 + k\vec{e}_j) - B(\vec{x}_0)}{k} = B_{x_j}(\vec{x}_0 + \theta_4 k\vec{e}_j) = f_{x_i x_j}(\vec{x}_0 + \theta_3 h\vec{e}_i + \theta_4 k\vec{e}_j)$$
for some $\theta_4 \in (0,1)$.

Thus we have shown that
$$f_{x_j x_i}(\vec{x}_0 + \theta_2 h\vec{e}_i + \theta_1 k\vec{e}_j) = f_{x_i x_j}(\vec{x}_0 + \theta_3 h\vec{e}_i + \theta_4 k\vec{e}_j),$$
for all h, k small. Letting $(h, k) \to (0, 0)$, by continuity of the second partials $f_{x_i x_j}$ and $f_{x_j x_i}$ in $B_r(\vec{x}_0)$ we get
$$f_{x_j x_i}(\vec{x}_0) = f_{x_i x_j}(\vec{x}_0).$$
□

Of course our discussion would not be complete if we did not have an example where $f_{x_i x_j} \neq f_{x_j x_i}$.

Consider the function
$$f(x,y) = \begin{cases} xy\frac{x^2-y^2}{x^2+y^2} & \text{for } (x,y) \neq (0,0) \\ 0 & \text{at } (0,0) \end{cases}.$$

Notice that $f_x(0, y) = -y$, and thus $f_{xy}(0, 0) = -1$. Also, $f_y(x, 0) = x$, and thus $f_{yx}(0, 0) = 1$.

Differentiability and Continuity

Having just partial derivatives exist at a point is not sufficient for continuity. To see this note the example:

$$f(x,y) = \begin{cases} \frac{xy}{x^2+y^2} & \text{for } (x,y) \neq (0,0) \\ 0 & \text{at } (0,0) \end{cases}$$

Here we have $f_x(0,0) = 0 = f_y(0,0)$. However $f(x,x) = \frac{1}{2}$ and thus

$$\frac{1}{2} = \lim_{x \to 0} f(x,x) \neq f(0,0) = 0.$$

Thus f is not continuous at the origin.

By theorem (8.3.2) we know that differentiability implies continuity. So simply having all partial derivatives exist at a point is not sufficient to imply differentiability at that point. Using the definition of differentiability to show that a given function is differentiable is not convenient to check in practice, and in light of this we note the following sufficient condition for differentiability.

Theorem 8.3.5 *Suppose that* $f : D \to \mathbb{R}$, *where* $D \subseteq \mathbb{R}^n$, *and* f_{x_i}, $i = 1, 2, \cdots, n$ *exist on* $B_r(\vec{x}_0)$ *and are continuous at* $\vec{x}_0$, *then* f *is differentiable at* $\vec{x}_0$.

Proof: Let $\varepsilon > 0$ be given. Then there are $\delta_j > 0$ for $j = 1, 2, \cdots, n$ so that for $\vec{x} \in B_{\delta_j}(\vec{x}_0)$ we have $f_{x_j}(\vec{x}) \in B_{\frac{\varepsilon}{\sqrt{n}}}(f_{x_j}(\vec{x}_0))$. We define $\delta = \min\{r, \delta_1, \delta_2, \cdots, \delta_n\}$. Fix $\vec{x} \in B_\delta(\vec{x}_0) \setminus \{\vec{x}_0\}$. To prove this theorem we need to show

$$|f(\vec{x}) - f(\vec{x}_0) - <\nabla f(\vec{x}_0), \vec{x} - \vec{x}_0>| < \varepsilon \|\vec{x} - \vec{x}_0\|,$$

where here $\|\ \|$ is the standard metric $\mathbb{R}^n$ for which $B_r(\vec{x}_0) = \{\vec{x} : \|\vec{x} - \vec{x}_0\| < r\}$.

For simplicity of notation we let $\vec{\xi} = \frac{1}{\|\vec{x} - \vec{x}_0\|}(\vec{x} - \vec{x}_0)$. Let $(\xi^1, \xi^2, \cdots, \xi^n) = \vec{\xi} = \sum_{i=1}^n \xi^i \vec{e}_i$ and $h := \|\vec{x} - \vec{x}_0\|$.

For $j \in \{1, 2, 3, \cdots, n\}$ we define a collection of points $\vec{x}_j$ by $\vec{x}_j = \vec{x}_{j-1} + h\xi^j \vec{e}_j$, and observe that $\vec{x}_n = \vec{x}$.

Note that $\vec{x}_j \in B_\delta(\vec{x}_0)$, since

$$\|\vec{x}_j - \vec{x}_0\| = \|\sum_{i=1}^{j} h\xi^i \vec{e}_i\| = h\sqrt{\sum_{i=1}^{j}(\xi^i)^2} \leq h < \delta.$$

By the mean value theorem (5.3.1) we have

$$f(\vec{x}) - f(\vec{x}_0) = \sum_{j=1}^{n}(f(\vec{x}_j) - f(\vec{x}_{j-1}))$$

$$= \sum_{j=1}^{n}(f(\vec{x}_{j-1} + h\xi^j \vec{e}_j) - f(\vec{x}_{j-1}))$$

$$= \sum_{j=1}^{n} f_{x_j}(\vec{x}_{j-1} + t_j \vec{e}_j)h\xi_j,$$

where $\vec{x}_{j-1} + t_j \vec{e}_j$ lies in the line segment joining $\vec{x}_{j-1}$ and $\vec{x}_j$, and thus lies in $B_\delta(\vec{x}_0)$.

By the Cauchy-Schwarz inequality we see

$$|f(\vec{x}) - f(\vec{x}_0) - <\nabla f(\vec{x}_0), \vec{x} - \vec{x}_0>| = |\sum_{j=1}^{n}(f_{x_j}(\vec{x}_{j-1} + t_j \vec{e}_j) - f_{x_j}(\vec{x}_0))h\xi^j|$$

$$\leq \sqrt{\sum_{j=1}^{n}(f_{x_j}(\vec{x}_{j-1} + t_j \vec{x}_j) - f_{x_j}(\vec{x}_0))^2} \cdot h < h\varepsilon,$$

since for each j we have $|f_{x_j}(\vec{x}_{j-1} + t_j \vec{e}_j) - f_{x_j}(\vec{x}_0)| < \frac{\varepsilon}{\sqrt{n}}$.
□

8.3.3 Taylor's Theorem, the Second Derivative and Extreme Values

Taylor's Theorem in $\mathbb{R}^n$

In the setting of $f : \mathbb{R}^n \to \mathbb{R}$ we wish to have a version of Taylor's theorem (5.6.1). Here we should expect that our Taylor polynomial is a finite linear

8.3. DIFFERENTIATION

combinations of terms of the form $x_1^{r_1} x_2^{r_2} \cdots x_n^{r_n}$, $r_j \in \{0\} \cup \mathbb{N}$, which is the natural analogue to a polynomial of one variable.

For simplicity of notation, for $f : \mathbb{R}^n \to \mathbb{R}$ with $\alpha = (\alpha_1, \alpha_2, \cdots, \alpha_n) \in (\mathbb{N} \cup \{0\})^n$ and $|\alpha| = \alpha_1 + \alpha_2 + \cdots + \alpha_n$, we let

$$D^\alpha f = \frac{\partial^{|\alpha|}}{\partial^{\alpha_1} x_1 \partial^{\alpha_2} x_2 \cdots \partial^{\alpha_n} x_n} f.$$

We also let $\alpha! = \alpha_1! \alpha_2! \cdots \alpha_n!$ and $\vec{x}^\alpha = x_1^{\alpha_1} x_2^{\alpha_2} \cdots x_n^{\alpha_n}$. Thus we have the following analogue of the one variable Taylor's theorem.

Taylor's Theorem in $\mathbb{R}^n$ 8.3.6 *Let $f : D \to \mathbb{R}$, where $D \subseteq \mathbb{R}^n$ an open set, and let f be in $C^m(D)$. That is, f has all partial derivatives up to order m on D, and they are all continuous on D. Let $\vec{x}_0 \in D$ be fixed. Then for the multivariable polynomial*

$$T_m(\vec{x}; \vec{x}_0) := \sum_{|\alpha| < m} \frac{1}{\alpha!} D^\alpha f(\vec{x}_0) \cdot (\vec{x} - \vec{x}_0)^\alpha,$$

we have

$$f(\vec{x}) = T_m(\vec{x}; \vec{x}_0) + \sum_{|\alpha| = m} \frac{1}{\alpha!} D^\alpha f(\vec{x}_t) \cdot (\vec{x} - \vec{x}_0)^\alpha,$$

for some $t \in (0, 1)$, where $\vec{x}_t := t\vec{x} + (1-t)\vec{x}_0$.

Proof: The method used in this proof is to reduce the multivariable function to the one variable function, and then use Taylor's Theorem (5.6.1). For $f : B_r(\vec{x}_0) \subseteq D$, with $B_r(\vec{x}_0)$ an open ball in $\mathbb{R}^n$, we fix $\vec{x} \in B_r(\vec{x}_0)$ and define $g(t) := f(\vec{x}_t)$, where $\vec{x}_t := t\vec{x} + (1-t)\vec{x}_0$. Since $f \in C^m(B_r(\vec{x}_0))$, by the chain rule (8.3.3) we see that $g(t)$ is m times differentiable. Thus by Taylor's Theorem (5.6.1), for $t \in [0, 1]$

$$g(t) = \sum_{k=0}^{m-1} \frac{g^{(k)}(0)}{k!} t^k + \frac{1}{m!} g^{(m)}(\tau) t^m,$$

where $g^{(0)}(t) = g(t)$ and $g^{(k)}(t) = \frac{d^k}{dt^k} g(t)$.

Evaluating at $t = 1$ gives us a representation for $f(\vec{x})$ as

$$f(\vec{x}) = g(1) = \sum_{k=0}^{m-1} \frac{g^{(k)}(0)}{k!} + \frac{1}{m!} g^{(m)}(\tau),$$

for some $\tau \in (0,1)$. Now, by the chain rule

$$g^{(1)}(0) = \frac{d}{dt}|_{t=0} f(\vec{x}_t) = \frac{d}{dt}|_{t=0} f(t\vec{x} + (1-t)\vec{x}_0)$$

$$= <\nabla f(\vec{x}_0), \vec{x} - \vec{x}_0> = \sum_{j=1}^{n} f_{x_j}(\vec{x}_0)(x_j - x_j^0) = \sum_{|\alpha|=1} \frac{1}{\alpha!} D^\alpha f(\vec{x}_0)(\vec{x} - \vec{x}_0)^\alpha,$$

where here $\vec{x} = (x_1, x_2, \cdots, x_n)$ and $\vec{x}_0 = (x_1^0, x_2^0, \cdots, x_n^0)$. In general we have

$$g^{(j)}(t) =$$

$$\sum_{k_j=1}^{n} \sum_{k_{j-1}=1}^{n} \cdots \sum_{k_1=1}^{n} \frac{\partial^j}{\partial x_{k_j} \partial x_{k_{j-1}} \cdots \partial x_{k_1}} f(\vec{x}_t)(x_{k_j} - x_{k_j}^0)(x_{k_{j-1}} - x_{k_{j-1}}^0) \cdots (x_{k_1} - x_{k_1}^0)$$

$$= \sum_{|\alpha|=j} \frac{j!}{\alpha!} D^\alpha f(\vec{x}_t)(\vec{x} - \vec{x}_0)^\alpha.$$

So evaluating at $t = 0$ we get the desired result.
□

Critical Values, Extreme Values and the Second Derivative Hessian Matrix

Let f be a real valued C^3 function whose domain is an open subset of $\mathbb{R}^n$. Suppose f attains a relative maximum or relative minimum at $\vec{x}_0$, we wish to determine what we can say about $Df(\vec{x}_0)$ and $D^2 f(\vec{x}_0)$. However, before we proceed any further, there is another definition that we will make, and that is for the matrix $(f_{x_i x_j}(\vec{x}_0))_{1 \le i,j \le n}$, which is often called the **Hessian** or **Hessian matrix** of f at $\vec{x}_0$.

Since f has a relative extremum at $\vec{x}_0$, observe that $Df(\vec{x}_0)$ satisfies

$$0 = \lim_{\vec{x} \to \vec{x}_0} \frac{|f(\vec{x}) - f(\vec{x}_0) - Df(\vec{x}_0)(\vec{x} - \vec{x}_0)|}{\|\vec{x} - \vec{x}_0\|},$$

or

$$0 = \lim_{\vec{x} \to \vec{x}_0} \frac{|f(\vec{x}) - f(\vec{x}_0) - <\nabla f(\vec{x}_0), \vec{x} - \vec{x}_0>|}{\|\vec{x} - \vec{x}_0\|}.$$

By restricting ourselves to $\vec{x} \in \{\vec{x}_0 + t\vec{e}_j : t \in \mathbb{R}\}$ we easily see that $\nabla f = (f_{x_1}, f_{x_2}, \cdots, f_{x_n})$. If f has a relative extrema at $\vec{x}_0$ then all of the functions

8.3. DIFFERENTIATION

$g_j(t) = f(\vec{x}_0 + t\vec{e}_j)$ have a relative extrema at 0. Hence $\frac{d}{dt}g_j(0) = 0$ for all $j \in \{1, 2, \cdots, n\}$. However, $\frac{d}{dt}g_j(0) = f_{x_j}(\vec{x}_0)$, and so we see that the gradient of f is the zero vector, i.e $\nabla f(\vec{x}_0) = \vec{0}$. Thus the linear operator $Df(\vec{x}_0) = 0$ as well.

Applying Taylor's Theorem (8.3.6), we see that for $\vec{x}$ near $\vec{x}_0$ we have

$$f(\vec{x}) =$$

$$f(\vec{x}_0) + \frac{1}{2}\begin{bmatrix} x_1 - x_1^0 \\ x_2 - x_2^0 \\ \vdots \\ x_n - x_n^0 \end{bmatrix}^T \begin{bmatrix} f_{x_1 x_1} & f_{x_1 x_2} & \cdots & f_{x_1 x_n} \\ f_{x_2 x_1} & f_{x_2 x_2} & \cdots & f_{x_2 x_n} \\ \vdots & \vdots & \ddots & \vdots \\ f_{x_n x_1} & f_{x_n x_2} & \cdots & f_{x_n x_n} \end{bmatrix} \begin{bmatrix} x_1 - x_1^0 \\ x_2 - x_2^0 \\ \vdots \\ x_n - x_n^0 \end{bmatrix} + \epsilon(\vec{x})$$

where $\frac{\epsilon(\vec{x})}{\|\vec{x}-\vec{x}_0\|^2} \to 0$ as $\vec{x} \to \vec{x}_0$, and here a superscript of T on a matrix means the transpose of the matrix.

Recall that an $n \times n$ matrix P is **positive definite** if for any column vector $\vec{y}$ we have

$$\vec{y}^T P \vec{y} > 0$$

Likewise, an $n \times n$ matrix N is **negative definite** if $-N$ is positive definite. It is an exercise to show that positive definite matrices have all positive eigenvalues. Moreover, by looking at the mapping $\vec{y} \mapsto \vec{y}^T P \vec{y}$ on the $\{\vec{y} : \|\vec{y}\| = 1\}$, by compactness and continuity of this mapping, we may see that $\vec{y}^T P \vec{y} > \lambda \|\vec{y}\|^2$ for some $\lambda > 0$.

Suppose that the Hessian matrix of f at $\vec{x}_0$ is positive definite. Then there is a $\lambda > 0$ so that[10] $(\vec{x} - \vec{x}_0)^T (f_{x_i x_j}(\vec{x}_0))_{1 \leq i,j \leq n}(\vec{x} - \vec{x}_0) > \lambda \|\vec{x} - \vec{x}_0\|^2$.

Choose r so that $\vec{x} \in B_r(\vec{x}_0)$ implies $|\epsilon(\vec{x})| < \frac{1}{4}\lambda \|\vec{x} - \vec{x}_0\|^2$, and so that our Taylor polynomial approximation applies. Then for such $\vec{x}$ we have

$$f(\vec{x}) > f(\vec{x}_0) + \frac{\lambda}{2}\|\vec{x} - \vec{x}_0\|^2 - \frac{\lambda}{4}\|\vec{x} - \vec{x}_0\|^2$$

$$= f(\vec{x}_0) + \frac{\lambda}{4}\|\vec{x} - \vec{x}_0\|^2.$$

Thus we see that f attains a relative minimum at $\vec{x}_0$. Similarly, if the Hessian is negative definite at $\vec{x}_0$, we see that f attains a relative maximum at $\vec{x}_0$.

---
[10] by viewing vectors as column vectors

The Path of Steepest Descent or Ascent, and Level Sets

We will now examine the problem of determining the direction of a unit vector $\vec{u}$ for which $D_{\vec{u}}f$ is a maximum or a minimum. The direction for which $D_{\vec{u}}f$ is a minimum will be call the **direction of steepest descent**, and the direction for which $D_{\vec{u}}f$ is a maximum is the **direction of steepest ascent**. To determine these directions we examine $D_{\vec{u}}f(\vec{x}_0) = <\nabla f(\vec{x}_0), \vec{u}>$. By the Cauchy-Schwarz inequality we have $|D_{\vec{u}}f(\vec{x}_0)| \leq \|\nabla f(\vec{x}_0)\| \cdot \|\vec{u}\| = \|\nabla f(\vec{x}_0)\|$, and equality is attained if and only if[11] $\vec{u}$ is a multiple of $\nabla f(\vec{x}_0)$. Thus if $\vec{u} = -\frac{\nabla f(\vec{x}_0)}{\|\nabla f(\vec{x}_0)\|}$, then $D_{\vec{u}}f(\vec{x}_0) = -\|\nabla f(\vec{x}_0)\|$, and this is the direction of steepest descent. Likewise, if $\vec{u} = \frac{\nabla f(\vec{x}_0)}{\|\nabla f(\vec{x}_0)\|}$, then $D_{\vec{u}}f(\vec{x}_0) = \|\nabla f(\vec{x}_0)\|$, and this is the direction of steepest ascent.

Consider a function $f : D \to \mathbb{R}$, with $D \subseteq \mathbb{R}^n$. Then a set of the form $\{\vec{x} : f(\vec{x}) = c\}$ is called a **level set** of f. Note that if we think of the graph of f as a surface in $\mathbb{R}^{n+1}$ we see that for any parametrized curve $\vec{\gamma}(t)$ that lies on this graph with $\vec{\gamma}(0) = (\vec{x}_0, f(\vec{x}_0))$ has $\frac{d}{dt}\vec{\gamma}(0)$ parallel to the tangent plane to the graph at $\vec{\gamma}(0)$.

Suppose that $\vec{\gamma}(t)$ is an curve on the level set given by $f(\vec{x}) = c$. Then $f(\vec{\gamma}(t)) = c$ for all t. By the chain rule, we see that $0 = \frac{d}{dt}f(\vec{\gamma}(t)) = <\nabla f(\vec{\gamma}(t)), \frac{d}{dt}\vec{\gamma}(t)>$. So for any $\vec{x}$ so that $f(\vec{x}) = c$ we have $\nabla f(\vec{x})$ is perpendicular[12] to level set given by $f = c$.

The Method of Lagrange Multipliers

Consider two functions $f, g : D \to \mathbb{R}$ with $D \subseteq \mathbb{R}^n$. We consider the problem of maximizing or minimizing f subject to the constraint $g = C$, where C is a constant in the range of g. Let $\vec{\gamma}(t)$ be a curve on the surface $g = c$ with $\vec{\gamma}(0) = \vec{x}_0$. At a place $\vec{x}_0$ where f has a relative extremum we must have

$$0 = \frac{d}{dt}|_{t=0}f(\vec{\gamma}(t)) = <\nabla f(\vec{x}_0), \frac{d}{dt}\vec{\gamma}(0)>.$$

Also, we must have

$$0 = \frac{d}{dt}|_{t=0}g(\vec{\gamma}(t)) = <\nabla g(\vec{0}), \frac{d}{dt}\vec{\gamma}(0)>.$$

---

[11] It is an exercise to show the only if portion of this biconditional statement.
[12] Meaning the inner product of ∇f with any tangent vector to $f = c$ is zero when at a particular point.

8.3. DIFFERENTIATION

If f and g are functions of two variables, then from this condition we may conclude that $\nabla f(\vec{x}_0)$ and $\nabla g(\vec{x}_0)$ are parallel and so there is a λ so that $\nabla f(\vec{x}_0) = \lambda \nabla g(\vec{x}_0)$. Examining this condition along with $g = C$ to solve for $\vec{x}_0$ is often called the **method of Lagrange multipliers**.

8.3.4 Homework Exercises

1. For $f : \mathbb{R}^2 \to \mathbb{R}$ given by $f(x, y) = |xy|$. Show $f_x(0, 0) = 0 = f_y(0, 0)$. Is f differentiable at $(0, 0)$?

2. Let
$$f(x, y) = \begin{cases} x^2 \arctan \frac{y}{x} - y^2 \arctan \frac{x}{y}, & x, y \neq 0, \\ 0, & x = 0 \text{ or } y = 0. \end{cases}$$

Show $f_{xy}(0, 0) = 1$, but $f_{yx}(0, 0) = -1$.

3. Consider the function
$$f(x, y) = \begin{cases} \frac{xy^2}{x^2+y^2}, & (x, y) \neq \vec{0}, \\ 0, & (x, y) = \vec{0}. \end{cases}$$

Show that $\nabla_{\vec{u}} f(0, 0)$ exists for any unit vector $\vec{u}$, however, the map $\vec{u} \mapsto \nabla_{\vec{u}} f(0, 0)$ is not linear. Conclude that f is not differentiable at $(0, 0)$.

4. Consider the function
$$f(x, y) = \begin{cases} \frac{xy}{x^4+y^4}, & (x, y) \neq \vec{0}, \\ 0, & (x, y) = \vec{0}. \end{cases}$$

Show that f has partial derivatives exist at all points in $\mathbb{R}^2$. However, f is not only discontinuous at $(0, 0)$, but $\lim_{x \to 0} f(x, x) = \infty$.

5. Consider the function
$$f(x, y) = \begin{cases} \frac{xy}{\sqrt{x^2+y^2}} \sin \frac{1}{\sqrt{x^2+y^2}}, & (x, y) \neq (0, 0), \\ 0, & (x, y) = (0, 0). \end{cases}$$

Find all points (x, y) for which f_x and f_y both exist. Also, show that f is not differentiable at $(0, 0)$.

6. We say that a function $f : D \to \mathbb{R}$, with $D \subseteq \mathbb{R}^2$ is **harmonic** if
$$\Delta f := f_{xx} + f_{yy} = 0$$
on D. Find all harmonic polynomials of degree 3 or less. Note that a function is a polynomial of degree 3 or less if it is of the form
$$a_{0,0}+a_{1,0}x+a_{0,1}y+a_{2,0}x^2+a_{1,1}xy+a_{0,2}y^2+a_{3,0}x^3+a_{2,1}x^2y+a_{1,2}xy^2+a_{0,3}y^3,$$
where $a_{i,j}$ is a constant for each i, j.

7. Suppose $f : D \to \mathbb{R}^m$, where $D \subseteq \mathbb{R}^n$ is convex. Suppose f is differentiable in D and there is a number M so that $\|Df\|_{\mathcal{L}(\mathbb{R}^n,\mathbb{R}^m)} \leq M$ (when f is restricted to D). Show that on D we have
$$\|f(\vec{x}) - f(\vec{y})\| \leq M \|\vec{x} - \vec{y}\|.$$

8. Suppose that f is differentiable on an open convex set D in $\mathbb{R}^n$, and here its derivative is zero everywhere. Show that f is constant on D.

9. Suppose f, g are functions from $\mathbb{R}^n$ to $\mathbb{R}$. Show that
$$D(f \cdot g) = f \cdot Dg + g \cdot Df.$$
This is the so-called **product rule**, and if $g \neq 0$ prove the **quotient rule**
$$D(\frac{f}{g}) = \frac{Df \cdot g - f \cdot Dg}{g^2}.$$

10. Find the local extrema of the following functions, and classify them as relative maximums, minimums, or neither.

    (a) $f(x, y) = (4x^2 + y^2)e^{-x^2-4y^2}$
    (b) $f(x, y) = x^2 - y^2 + 10$
    (c) $f(x, y) = x^3 + y^3 - x - y$
    (d) $f(x, y) = \frac{x^4}{32} + x^2y^2 - x - y^2$
    (e) $f(x, y) = x^4 + x^2y^2 - y$
    (f) $f(x, y) = \frac{x}{1+x^2+y^2}$
    (g) $f(x, y) = x^4 + y^4 - x^3$

11. Maximize $f(x,y) = xy$ subject to $x + y = 20$ using the method of Lagrange multipliers.

12. Minimize $x^2 + y^2$ subject to $xy = 1$ using the method of Lagrange multipliers.

8.4 The Inverse Function and Implicit Function Theorems

8.4.1 The Contraction Mapping Principle

Let (M, d) be a metric space, a function $f : M \to M$ is called a **contraction mapping** if there is a $c \in (0, 1)$ so that

$$d(f(\vec{x}), f(\vec{y})) \leq cd(\vec{x}, \vec{y}) \quad \text{for any} \quad \vec{x}, \vec{y} \in M. \tag{8.11}$$

A point $\vec{x}$ is called a **fixed point** of a function f if $f(x) = x$. The following theorem tells us that contraction mappings have unique fixed points.

Theorem 8.4.1 *Let $f : M \to M$ be a contraction mapping from a complete metric space M to itself. Then $f(\vec{x}) = \vec{x}$ has a unique solution.*

Proof: Let $\vec{x}_0 \in M$ be any element of M. We define a sequence by $\vec{x}_n = f(\vec{x}_{n-1})$ for $n \in \mathbb{N}$. Then for any $n > 1$ we have

$$d(\vec{x}_n, \vec{x}_{n-1}) = d(f(\vec{x}_{n-1}), f(\vec{x}_{n-2})) \leq cd(\vec{x}_{n-1}, \vec{x}_{n-2}).$$

Inductively, this gives

$$d(\vec{x}_n, \vec{x}_{n-1}) \leq c^{n-1} d(\vec{x}_1, \vec{x}_0)$$

for any $n > 1$. Thus for any $n < m$ we have

$$d(\vec{x}_n, \vec{x}_m) \leq \sum_{i=n+1}^{m} d(\vec{x}_i, \vec{x}_{i-1}) \leq (\sum_{i=n}^{m-1} c^i) d(\vec{x}_1, \vec{x}_0) \leq \frac{c^n}{1-c} d(\vec{x}_1, \vec{x}_0).$$

We see that $\{\vec{x}_n\}_{n=0}^{\infty}$ is a Cauchy sequence, and so it has a limit value $\vec{x}_\infty \in M$. Moreover, since contraction mappings are clearly continuous functions, we have

$$f(\vec{x}_\infty) = \lim_{n \to \infty} f(\vec{x}_n) = \lim_{n \to \infty} \vec{x}_{n+1} = \vec{x}_\infty.$$

Thus we have established that f has a fixed point. We will now show that this is unique. For suppose $\vec{x}, \vec{y}$ are two fixed points of f. Then

$$d(\vec{x}, \vec{y}) = d(f(\vec{x}), f(\vec{y})) \leq cd(\vec{x}, \vec{y}).$$

If $d(\vec{x}, \vec{y}) > 0$ then this implies $d(\vec{x}, \vec{y}) < d(\vec{x}, \vec{y})$, which is impossible.
□

8.4.2 The Inverse Function Theorem

The **inverse function theorem** tells us that a C^1 function[13] is locally invertible around any point for which its derivative is an invertible linear transformation.

Inverse Function Theorem 8.4.2 *Let* $f : D \to \mathbb{R}^n$, *with* $D \subseteq \mathbb{R}^n$ *open,* $f \in C^1(D)$ *and* $\vec{x}_0 \in D$. *If* $Df(\vec{x}_0)$ *is invertible, and* $\vec{y}_0 = f(\vec{x}_0)$, *then there are open sets* $U, V \subseteq \mathbb{R}^n$ *so that* $\vec{x}_0 \in U$, $\vec{y}_0 \in V$, *and* f *is injective on* U *with* $f(U) = V$. *Moreover,* f^{-1} *is in* $C^1(V)$.

Proof: For simplicity of notation, we let $L := Df(\vec{x}_0)$ and $r = \frac{1}{2\|L^{-1}\|}$. By continuity of Df at $\vec{x}_0$, there is an open neighbourhood of $\vec{x}_0$ U so that for $\vec{x} \in U$

$$\|Df(\vec{x}) - L\| < r.$$

Given $\vec{y} \in \mathbb{R}^n$, we define a function $z : D \to \mathbb{R}^n$ by

$$z(\vec{x}) = \vec{x} + L^{-1}(\vec{y} - f(\vec{x})).$$

Observe that the function z has the property that $z(\vec{x}) = \vec{x}$ if and only if $f(\vec{x}) = \vec{y}$.

Let I be the identity mapping from $\mathbb{R}^n$ to itself. Observe that

$$Dz(\vec{x}) = I - L^{-1}Df(\vec{x}) = L^{-1}[L - Df(\vec{x})].$$

By our assumption on U, for $\vec{x} \in U$ we have

$$\|Dz(\vec{x})\| < \frac{1}{2}.$$

---
[13] Meaning the function f is differentiable and has a continuous derivative Df.

8.4. THE INVERSE FUNCTION AND IMPLICIT FUNCTION THEOREMS

Before we proceed any further, we shall prove a result which guarantees Lipschitz continuity whenever our function is C^1. We note that in the one dimensional case this easily follows from the mean value theorem (5.3.1). Suppose $g : E \to \mathbb{R}^m$, where $E \subseteq \mathbb{R}^n$ is convex and g satisfies

$$\|Dg(\vec{x})\| \leq M, \quad \vec{x} \in E.$$

Then for any $\vec{x}, \vec{y} \in E$ we have

$$\|g(\vec{x}) - g(\vec{y})\| \leq M \|\vec{x} - \vec{y}\|.$$

To see this result we shall use the one dimensional mean value theorem. For any $\vec{x}, \vec{y} \in E$ we define $\vec{\gamma}(t) = (1-t)\vec{x} + t\vec{y}$ for $t \in [0, 1]$. By convexity of E we know $\vec{\gamma}(t) \in E$ for all $t \in [0, 1]$. We define

$$k(t) := g(\vec{\gamma}) \quad t \in [0, 1].$$

By the chain rule (8.3.3) we know

$$Dk(t) = Dg(\vec{\gamma}(t)) D\vec{\gamma}(t) = Dg(\vec{\gamma}(t))(\vec{y} - \vec{x}).$$

Thus we see that for any $t \in [0, 1]$

$$\|Dk(t)\| \leq \|Dg(\vec{\gamma})\| \|\vec{y} - \vec{x}\| \leq M \|\vec{y} - \vec{x}\|.$$

Let $\vec{\xi} := g(\vec{y}) - g(\vec{x})$ and consider the function $\phi(t) = \vec{\xi} \cdot k(t)$. Then the mean value theorem (5.3.1) guarantees that there is a $\tau \in (0, 1)$ so that

$$\phi(1) - \phi(0) = D\phi(\tau) = \vec{\xi} \cdot Dk(\tau).$$

However, we also have

$$\phi(1) - \phi(0) = \|\vec{\xi}\|^2.$$

Moreover, the Cauchy-Schwarz inequality gives us

$$\|\vec{\xi}\|^2 = \|\vec{\xi} \cdot Dk(\tau)\| \leq \|\vec{\xi}\| \cdot \|Dk(\tau)\|$$

and so

$$\|\vec{\xi}\| \leq \|Dk(\tau)\|.$$

Putting all this together, we get

$$\|g(\vec{y}) - g(\vec{x})\| \leq \|Dk(\tau)\| \leq M \|\vec{y} - \vec{x}\|.$$

Applying this result, we may conclude that

$$\|z(\vec{x}) - z(\vec{y})\| \le \frac{1}{2}\|\vec{x} - \vec{y}\|$$

for $\vec{x}, \vec{y} \in U$. By the contraction mapping theorem, we know that z has a unique fixed point. Thus we see that f is injective in U.

Let $V := \{f(\vec{x}) : \vec{x} \in U\}$, and let $\vec{y}_0 \in V$. Then there is a unique $\vec{x}_0 \in U$ so that $f(\vec{x}_0) = \vec{y}_0$. Let $\rho > 0$ be such that the closure of $B_\rho(\vec{x}_0)$ is in U. We will now show V is open, and we will do this by showing $\vec{y} \in V$ if $\|\vec{y} - \vec{y}_0\| < r\rho$.

Let $\vec{y}$ be so that $\|\vec{y} - \vec{y}_0\| < r\rho$. Then

$$\|z(\vec{x}_0) - \vec{x}_0\| = \|L^{-1}(\vec{y} - \vec{y}_0)\| \le \|L^{-1}\| r\rho = \frac{\rho}{2}.$$

If $\vec{x} \in \overline{B_\rho(\vec{x}_0)}$, then

$$\|z(\vec{x}) - \vec{x}_0\| \le \|z(\vec{x}) - z(\vec{x}_0)\| + \|z(\vec{x}_0) - \vec{x}_0\|$$

$$< \frac{1}{2}\|\vec{x} - \vec{x}_0\| + \frac{\rho}{2} \le \rho.$$

Hence $z(\vec{x}) \in B_\rho(\vec{x}_0)$.

Thus z is a contraction mapping from $\overline{B_\rho(\vec{x}_0)}$ to itself, and so z has a fixed point $\vec{x} \in \overline{B_\rho(\vec{x}_0)}$. Moreover, for this $\vec{x}$, $f(\vec{x}) = \vec{y}$ and $\vec{y} \in V$.

Before we proceed, we will prove another general result. Suppose that we have two linear operators A, B from $\mathbb{R}^n$ to itself, and A^{-1} exists. We will now show that if

$$\|B - A\| \cdot \|A^{-1}\| < 1,$$

then B^{-1} exists.

Now,

$$\frac{1}{\|A^{-1}\|}\|\vec{x}\| = \frac{1}{\|A^{-1}\|} \cdot \|A^{-1} A\vec{x}\| \le \frac{1}{\|A^{-1}\|} \cdot \|A^{-1}\|\|A\vec{x}\|$$

$$= \|A\vec{x}\| \le \|(A - B)\vec{x}\| + \|B\vec{x}\| \le \|B - A\|\|\vec{x}\| + \|B\vec{x}\|.$$

8.4. THE INVERSE FUNCTION AND IMPLICIT FUNCTION THEOREMS

So
$$(\frac{1}{\|A^{-1}\|} - \|B - A\|)\|\vec{x}\| \leq \|B\vec{x}\|.$$

Since $\frac{1}{\|A^{-1}\|} - \|B - A\| > 0$ we see that $B\vec{x} \neq \vec{0}$ if $\vec{x} \neq \vec{0}$. So we see that B is injective, and hence invertible.

Moreover, we claim that the mapping $A \mapsto A^{-1}$ is a continuous mapping on the set of invertible linear operators on $\mathbb{R}^n$. By replacing $\vec{x}$ in
$$(\frac{1}{\|A^{-1}\|} - \|B - A\|)\|\vec{x}\| \leq \|B\vec{x}\|$$
by $B^{-1}\vec{y}$ we see that
$$(\frac{1}{\|A^{-1}\|} - \|B - A\|)\|B^{-1}\vec{y}\| \leq \|\vec{y}\|$$
and so $\|B^{-1}\| \leq (\frac{1}{\|A^{-1}\|} - \|B - A\|)^{-1}$. Observing that $B^{-1} - A^{-1} = B^{-1}(A - B)A^{-1}$, we see that
$$\|B^{-1} - A^{-1}\| \leq \|B^{-1}\| \cdot \|A - B\| \cdot \|A^{-1}\| \leq \frac{\|B - A\| \cdot \|A^{-1}\|}{\frac{1}{\|A^{-1}\|} - \|B - A\|}.$$

Thus the continuity is established since $\|B - A\| \to 0$ as $B \to A$.

From our assumption on $\vec{x}$, namely
$$\|Df(\vec{x}) - L\| < \frac{1}{2\|L^{-1}\|},$$
we see that the linear operator $Df(\vec{x})$ is invertible, and we let Λ denote its inverse.

We will now show that the inverse function of f is a C^1 function on V. Let $\vec{y}, \vec{y} + \vec{k} \in V$, then there are $\vec{x}, \vec{x} + \vec{h} \in U$ so that $f(\vec{x}) = \vec{y}$ and $f(\vec{x} + \vec{h}) = \vec{y} + \vec{k}$. Then
$$z(\vec{x} + \vec{h}) - z(\vec{x}) = \vec{h} + L^{-1}[f(\vec{x}) - f(\vec{x} + \vec{h})] = \vec{h} - L^{-1}\vec{k},$$
and so we see
$$\|\vec{h} - L^{-1}\vec{k}\| \leq \frac{1}{2}\|\vec{h}\|.$$

Hence
$$\|L^{-1}\vec{k}\| \geq \frac{1}{2}\|\vec{h}\|,$$
and so
$$\|\vec{h}\| \leq 2\|L^{-1}\| \cdot \|\vec{k}\| = \frac{\|\vec{k}\|}{r}.$$
Since
$$f^{-1}(\vec{y}+\vec{k}) - f^{-1}(\vec{y}) - \Lambda\vec{k} = \vec{h} - \Lambda\vec{k} = -\Lambda[f(\vec{x}+\vec{h}) - f(\vec{x}) - Df(\vec{x})\vec{h}],$$
we see that
$$\frac{\|f^{-1}(\vec{y}+\vec{k}) - f^{-1}(\vec{y}) - \Lambda\vec{k}\|}{\|\vec{k}\|} \leq \frac{\|\Lambda\|}{r} \cdot \frac{\|f(\vec{x}+\vec{h}) - f(\vec{x}) - Df(\vec{x})\vec{h}\|}{\|\vec{h}\|}.$$

As $\vec{k} \to \vec{0}$ we see that $\vec{h} \to \vec{0}$, and so we see that $D(f^{-1})(\vec{y}) = \Lambda$. Moreover, $D(f^{-1})(\vec{y}) = (Df(f^{-1}(\vec{y})))^{-1}$ for $\vec{y} \in V$. The remainder of the proof follows from general results we established above.
□

Example

The following example shows that we can not omit the hypothesis that our function is C^1 in the hypothesis of the inverse function theorem.

Let $\alpha \in (0, 1)$, and consider the function $f(x) = \alpha x + x^2 \sin \frac{1}{x}$ for $x \neq 0$ and $f(0) = 0$.

It is clear that $f'(x)$ exists for $x \neq 0$, and in such a case $f'(x) = \alpha + 2x \sin \frac{1}{x} - \cos \frac{1}{x}$. Moreover $f'(0) = \alpha + \lim_{h \to 0} h \sin \frac{1}{h} = \alpha \neq 0$.

Since $\lim_{x \to 0} \cos \frac{1}{x}$ does not exist, we clearly have f' is not continuous at $x = 0$, and so f is not C^1. We will show that f does not have an inverse in any neighbourhood of 0. To see this note that at a point where $f' = 0$ but $f'' \neq 0$ there cannot be a local inverse in a neighbourhood of this point. Claim that there are infinitely any such points where $f' = 0$ but $f'' \neq 0$ in any neighbourhood of 0.

8.4. THE INVERSE FUNCTION AND IMPLICIT FUNCTION THEOREMS

Suppose $f' = 0$ and $f'' = 0$ at some point $x \neq 0$. Since $f''(x) = (2 - \frac{1}{x^2})\sin\frac{1}{x} - \frac{2}{x}\cos\frac{1}{x}$, at such a point where $f' = f'' = 0$ we have

$$2x\sin\frac{1}{x} - \cos\frac{1}{x} = -\alpha, \quad (2 - \frac{1}{x^2})\sin\frac{1}{x} - \frac{2}{x}\cos\frac{1}{x} = 0.$$

Thus we have

$$\sin\frac{1}{x} = \alpha\frac{-2x}{1+2x^2}, \quad \cos\frac{1}{x} = \alpha\frac{1-2x^2}{1+2x^2},$$

and so

$$\sin^2\frac{1}{x} + \cos^2\frac{1}{x} = \alpha^2\frac{1+4x^4}{(1+2x^2)^2} < 1.$$

Thus we see that $f' = 0$ and $f'' = 0$ can not hold at the same time. However, there are infinitely many points x near 0 so that $f' = 0$.

8.4.3 The Implicit Function Theorem

Suppose $f(\vec{x}, \vec{y}) = 0$, where f is a C^1 function, then we may wish to know when we may solve the equation $f(\vec{x}, \vec{y}) = 0$ for $\vec{y}$ in terms of $\vec{x}$ in any neighbourhood of a point which is a solution to this equation. The so-called **implicit function theorem** addresses this question. For instance, for the equation $f(x, y) = x^2 + y^2 - 1 = 0$ we may solve for y in terms of x at all points for which $(\frac{\partial f}{\partial x})^{-1}$ exists. In this example, this can be done in a neighbourhood of all points except $(\pm 1, 0)$. Moreover, in this case we may explicitly write y as either $\sqrt{1-x^2}$ or $-\sqrt{1-x^2}$ depending on the sign of y_0 for which $f(x_0, y_0) = 0$. Note however that the expression

$$y^5 + 16y - 32x^3 + 32x = 0 \tag{8.12}$$

gives rise for a subset of $\mathbb{R}^2$ which passes the so-called vertical line test, and thus may be seen as the graph of a function. However, no algebraic formula $y = f(x)$ exists. Thus we see the need for a test to see if an expression of the form $f(x, y) = 0$ implicitly defines a function of x, even if we cannot solve for $y = f(x)$ explicitly. The so-called **implicit function theorem** gives us such a condition.[14]

In the following discussion we shall use the notation

$$f(\vec{x}, \vec{y}) = f(x_1, x_2, \cdots, x_n, y_1, y_2, \cdots, y_m),$$

---
[14] For a very nice reference on the implicit function theorem, see [5].

with $(\vec{x}, \vec{y}) \in \mathbb{R}^{n+m}$.

Implicit Function Theorem 8.4.3 *Let $f : D \to \mathbb{R}^n$, where $D \subseteq \mathbb{R}^{n+m}$ is an open set, and $f \in C^1(D)$. Suppose $f(\vec{x}_0, \vec{y}_0) = \vec{0}$. Let $L := Df(\vec{x}_0, \vec{y}_0)$ and $L_x = L(\vec{x}, 0)$ and $L_y = L(0, \vec{y})$. Suppose L_x is invertible, then there exists open sets $U \subseteq \mathbb{R}^{n+m}$ and $W \subseteq \mathbb{R}^m$, with $(\vec{x}_0, \vec{y}_0) \in U$ and $\vec{y}_0 \in W$, and having the property: To each $\vec{y} \in W$, there is a unique $\vec{x}$ so that $(\vec{x}, \vec{y}) \in U$ and $f(\vec{x}, \vec{y}) = \vec{0}$. Letting $\vec{x} = g(\vec{y})$, then $g \in C^1(W)$, $g(\vec{y}_0) = \vec{x}_0$, and $f(g(\vec{y}), \vec{y}) = \vec{0}$ for any $\vec{y} \in W$, and $Dg(\vec{y}_0) = -(L_x)^{-1} L_y$.*

Proof: Define $F(\vec{x}, \vec{y}) := (f(\vec{x}, \vec{y}), \vec{y}) \in \mathbb{R}^{n+m}$, for $(\vec{x}, \vec{y}) \in D \subseteq \mathbb{R}^{n+m}$. Then F is a continuously differentiable function from D into $\mathbb{R}^{n+m}$. If DF represents the derivative of F, then we shall now show that $DF(\vec{x}_0, \vec{y}_0)$ is an invertible linear transformation from $\mathbb{R}^{n+m}$ to $\mathbb{R}^{n+m}$.

For simplicity, let $L = Df(\vec{x}_0, \vec{y}_0)$. Since $f(\vec{x}_0, \vec{y}_0) = \vec{0}$ we have that

$$f(\vec{x}_0 + \vec{h}, \vec{y}_0 + \vec{k}) = L(\vec{h}, \vec{k}) + \epsilon(\vec{h}, \vec{k}),$$

where $\frac{\|\epsilon(\vec{h}, \vec{k})\|}{\|(\vec{h}, \vec{k})\|} \to 0$ as $(\vec{h}, \vec{k}) \to \vec{0}$.

Then,

$$F(\vec{x}_0 + \vec{h}, \vec{y}_0 + \vec{k}) - F(\vec{x}_0, \vec{y}_0) = (f(\vec{x}_0 + \vec{h}, \vec{y}_0 + \vec{k}), \vec{k})$$

$$= (L(\vec{h}, \vec{k}), \vec{k}) + (\epsilon(\vec{h}, \vec{k}), \vec{0}).$$

Thus we see that $DF(\vec{x}_0, \vec{y}_0)$ is a linear transformation on $\mathbb{R}^{n+m}$ which maps $(\vec{h}, \vec{k})$ to $(L(\vec{h}, \vec{k}), \vec{k})$.

Note that if $(L(\vec{h}, \vec{k}), \vec{k}) = \vec{0}$, then $\vec{k} = \vec{0}$ and $L(\vec{h}, \vec{0}) = \vec{0}$. We may split $L(\cdot, \cdot)$ as follows, namely: $L(\vec{h}, \vec{k}) = L_x \vec{h} + L_y \vec{k}$. We see that if L_x has an inverse, then $L(\vec{h}, \vec{k}) = \vec{0}$ if and only if $L_x \vec{h} = -L_y \vec{k}$ and so $\vec{h} = -(L_x)^{-1} L_y \vec{k}$. If $\vec{k} = \vec{0}$, we see that $\vec{h} = \vec{0}$ must also hold.

It follows that $DF(\vec{x}_0, \vec{y}_0)$ is injective, and hence invertible. Thus we may apply the inverse function theorem (8.4.2) to F, and so there are open sets $U, V \subseteq \mathbb{R}^{n+m}$ with $(\vec{x}_0, \vec{y}_0) \in U$ and $(\vec{0}, \vec{y}_0) \in V$ so that $F : U \to V$ is a

8.4. THE INVERSE FUNCTION AND IMPLICIT FUNCTION THEOREMS

bijection.

Define $W := \{\vec{y} \in \mathbb{R}^m : (\vec{0}, \vec{y}) \in V\}$, and observe that $\vec{y}_0 \in W$. Since V is open, we see that W is open. Moreover, if $\vec{y} \in W$, then $(\vec{0}, \vec{y}) = F(\vec{x}, \vec{y})$ for some $(\vec{x}, \vec{y}) \in U$. By our definition of F we see that $f(\vec{x}, \vec{y}) = \vec{0}$.

Suppose that for the same $\vec{y}$ there is a $\vec{z}$ so that $(\vec{z}, \vec{y}) \in U$ with $f(\vec{z}, \vec{y}) = \vec{0}$. Then $F(\vec{z}, \vec{y}) = (f(\vec{z}, \vec{y}), \vec{y}) = (f(\vec{x}, \vec{y}), \vec{y}) = F(\vec{x}, \vec{y})$, and since F is injective, it follows that $\vec{x} = \vec{z}$.

Define $g(\vec{y})$, for $\vec{y} \in W$, so that $(g(\vec{y}), \vec{y}) \in U$ and $f(g(\vec{y}), \vec{y}) = \vec{0}$. If $G : V \to U$ inverts F, then G is C^1 by the inverse function theorem (8.4.2), and thus $(g(\vec{y}), \vec{y}) = G(\vec{0}, \vec{y})$ for $y \in W$. Since G is C^1 we have that g is also C^1.

Let $H(\vec{y}) := (g(\vec{y}), \vec{y})$. Then $DH(\vec{y})\vec{k} = (Dg(\vec{y})\vec{k}, \vec{k})$ for $\vec{y} \in W$ and $\vec{k} \in \mathbb{R}^m$. Now $f(H(\vec{y})) = \vec{0}$ in W, and so by the chain rule we have $Df(H(\vec{y}))DH(\vec{y}) = 0$.

Observe that $H(\vec{y}_0) = (\vec{x}_0, \vec{y}_0)$, and thus $Df(H(\vec{y}_0)) = L$, as defined above. Thus $L(DH(\vec{y}_0)) = 0$. So we see that for all $\vec{k} \in \mathbb{R}^m$

$$L_x Dg(\vec{y}_0)\vec{k} + L_y \vec{k} = L(Dg(\vec{y}_0)\vec{k}, \vec{k}) = L(DH(\vec{y}_0)\vec{k}) = \vec{0}.$$

Therefore $L_x Dg(\vec{y}_0) + L_y$ is the zero linear transformation, which gives us the desired result since L_x is invertible.
□

A Remark on the Conditions in the Implicit Function Theorem

Recall that $f(\vec{x}, \vec{y}) = f(x_1, \cdots, x_n, y_1, \cdots, y_m)$, $f : D \to \mathbb{R}^n$. Given $f(\vec{x}_0, \vec{y}_0) = \vec{0}$, we wish to find $\vec{x} = g(\vec{y})$ so that $f(g(\vec{y}), \vec{y}) = \vec{0}$, for g defined on some open set containing $\vec{y}_0$. Seeing that this is a vector (possibly multidimensional) condition, we will write the condition $f(\vec{x}, \vec{y}) = \vec{0}$ without vector notation as

$$f_1(x_1, \cdots, x_n, y_1, \cdots, y_m) = 0$$

$$\vdots$$

$$f_j(x_1, \cdots, x_n, y_1, \cdots, y_m) = 0$$

CHAPTER 8. FUNCTIONS OF SEVERAL REAL VARIABLES

$$\vdots$$
$$f_n(x_1, \cdots, x_n, y_1, \cdots, y_m) = 0.$$

We also assume that

$$f_1(x_1^0, \cdots, x_n^0, y_1^0, \cdots, y_m^0) = 0$$
$$\vdots$$
$$f_j(x_1^0, \cdots, x_n^0, y_1^0, \cdots, y_m^0) = 0$$
$$\vdots$$
$$f_n(x_1^0, \cdots, x_n^0, y_1^0, \cdots, y_m^0) = 0.$$

We hope to find a collection of functions $g_1, \cdots, g_n$ so that for each $k = 1, \cdots, n$ we have $g_k(y_1, \cdots, y_m) = x_k$ and $g_k(y_1^0, \cdots, y_m^0) = x_k^0$. Taking the linearization of the left-hand-side of each equation in the system

$$f_1(x_1, \cdots, x_n, y_1, \cdots, y_m) = 0$$
$$\vdots$$
$$f_j(x_1, \cdots, x_n, y_1, \cdots, y_m) = 0$$
$$\vdots$$
$$f_n(x_1, \cdots, x_n, y_1, \cdots, y_m) = 0$$

we get for each $k = 1, \cdots, n$

$$\frac{\partial f_k}{\partial x_1}(\vec{x}_0, \vec{y}_0) \cdot (x_1 - x_1^0) + \cdots + \frac{\partial f_k}{\partial x_n}(\vec{x}_0, \vec{y}_0) \cdot (x_n - x_n^0)$$

$$+ \frac{\partial f_k}{\partial y_1}(\vec{x}_0, \vec{y}_0) \cdot (y_1 - y_1^0) + \cdots + \frac{\partial f_k}{\partial y_m}(\vec{x}_0, \vec{y}_0) \cdot (y_m - y_m^0) + f_k(\vec{x}_0, \vec{y}_0) = 0.$$

Using that $f_k(\vec{x}_0, \vec{y}_0) = 0$ for each $k = 1, \cdots, n$, we may rewrite this system as

$$\begin{bmatrix} \frac{\partial f_1}{\partial x_1}(\vec{x}_0, \vec{y}_0) & \cdots & \frac{\partial f_1}{\partial x_n}(\vec{x}_0, \vec{y}_0) \\ \vdots & \ddots & \vdots \\ \frac{\partial f_n}{\partial x_1}(\vec{x}_0, \vec{y}_0) & \cdots & \frac{\partial f_n}{\partial x_n}(\vec{x}_0, \vec{y}_0) \end{bmatrix} \cdot \begin{bmatrix} x_1 \\ \vdots \\ x_n \end{bmatrix} = \begin{bmatrix} \frac{\partial f_1}{\partial x_1}(\vec{x}_0, \vec{y}_0) & \cdots & \frac{\partial f_1}{\partial x_n}(\vec{x}_0, \vec{y}_0) \\ \vdots & \ddots & \vdots \\ \frac{\partial f_n}{\partial x_1}(\vec{x}_0, \vec{y}_0) & \cdots & \frac{\partial f_n}{\partial x_n}(\vec{x}_0, \vec{y}_0) \end{bmatrix} \cdot \begin{bmatrix} x_1^0 \\ \vdots \\ x_n^0 \end{bmatrix}$$

8.4. THE INVERSE FUNCTION AND IMPLICIT FUNCTION THEOREMS

Assuming that
$$-\begin{bmatrix} \frac{\partial f_1}{\partial y_1}(\vec{x}_0, \vec{y}_0) & \cdots & \frac{\partial f_1}{\partial y_m}(\vec{x}_0, \vec{y}_0) \\ \vdots & \ddots & \vdots \\ \frac{\partial f_n}{\partial y_1}(\vec{x}_0, \vec{y}_0) & \cdots & \frac{\partial f_n}{\partial y_m}(\vec{x}_0, \vec{y}_0) \end{bmatrix} \cdot \begin{bmatrix} y_1 - y_1^0 \\ \vdots \\ y_m - y_m^0 \end{bmatrix}.$$

$$\begin{bmatrix} \frac{\partial f_1}{\partial x_1}(\vec{x}_0, \vec{y}_0) & \cdots & \frac{\partial f_1}{\partial x_n}(\vec{x}_0, \vec{y}_0) \\ \vdots & \ddots & \vdots \\ \frac{\partial f_n}{\partial x_1}(\vec{x}_0, \vec{y}_0) & \cdots & \frac{\partial f_n}{\partial x_n}(\vec{x}_0, \vec{y}_0) \end{bmatrix}$$
is invertible, we get

$$\begin{bmatrix} x_1 \\ \vdots \\ x_n \end{bmatrix} = \begin{bmatrix} x_1^0 \\ \vdots \\ x_n^0 \end{bmatrix}$$

$$-\begin{bmatrix} \frac{\partial f_1}{\partial x_1}(\vec{x}_0, \vec{y}_0) & \cdots & \frac{\partial f_1}{\partial x_n}(\vec{x}_0, \vec{y}_0) \\ \vdots & \ddots & \vdots \\ \frac{\partial f_n}{\partial x_1}(\vec{x}_0, \vec{y}_0) & \cdots & \frac{\partial f_n}{\partial x_n}(\vec{x}_0, \vec{y}_0) \end{bmatrix}^{-1} \begin{bmatrix} \frac{\partial f_1}{\partial y_1}(\vec{x}_0, \vec{y}_0) & \cdots & \frac{\partial f_1}{\partial y_m}(\vec{x}_0, \vec{y}_0) \\ \vdots & \ddots & \vdots \\ \frac{\partial f_n}{\partial y_1}(\vec{x}_0, \vec{y}_0) & \cdots & \frac{\partial f_n}{\partial y_m}(\vec{x}_0, \vec{y}_0) \end{bmatrix} \cdot \begin{bmatrix} y_1 - y_1^0 \\ \vdots \\ y_m - y_m^0 \end{bmatrix}$$

Thus the condition
$$\begin{bmatrix} \frac{\partial f_1}{\partial x_1}(\vec{x}_0, \vec{y}_0) & \cdots & \frac{\partial f_1}{\partial x_n}(\vec{x}_0, \vec{y}_0) \\ \vdots & \ddots & \vdots \\ \frac{\partial f_n}{\partial x_1}(\vec{x}_0, \vec{y}_0) & \cdots & \frac{\partial f_n}{\partial x_n}(\vec{x}_0, \vec{y}_0) \end{bmatrix}$$
is invertible is precisely the condition L_x is invertible in the statement of the inverse function theorem.

Example

Consider the equation
$$y - xe^y = 1.$$
We will show that near $(-1, 0)$ we may write $y = g(x)$, and we will compute $g'(-1)$.

Let $f(x, y) = y - xe^y$. Then $\frac{\partial f}{\partial x} = -e^y$ and $\frac{\partial f}{\partial y} = 1 - xe^y$, and so $\frac{\partial f}{\partial x}(-1, 0) = -1$ and $\frac{\partial f}{\partial y}(-1, 0) = 2$. The implicit function theorem guarantees the existence of a C^1 function $y = g(x)$, which however can not be defined explicitly. Now g satisfies
$$g(x) - xe^{g(x)} = 1$$
and so $g'(x)$ satisfies
$$g'(x) - e^{g(x)} - xg'(x)e^{g(x)} = 0$$

which gives us
$$g'(x) = \frac{e^{g(x)}}{1 - xe^{g(x)}}$$
for x near -1. Moreover,
$$g'(-1) = \frac{1}{2}.$$

8.4.4 Homework Exercises

1. Show that the continuity of Df is needed at $\vec{x}_0$ in the hypothesis of the inverse function theorem (8.4.2), even in $n = 1$ dimensions. Let $f(x) = x + 2x^2 \sin \frac{1}{x}$ if $x \neq 0$ and $f(0) = 0$. Then show that $Df(0) = f'(0) = 1$, Df is bounded in $(-1, 1)$, but f is not injective in any neighbourhood of 0.

2. Consider the system of equations
$$3x + y - z + u^2 = 0,$$
$$x - y + 2z + u = 0,$$
$$2x + 2y - 3z + 2u = 0.$$
Show that this system can be solved for x, y, u in terms of z, for x, z, u in terms of y and for y, z, u in terms of x, but not for x, y, z in terms of u.

3. Let $f : \mathbb{R}^3 \to \mathbb{R}$ be given by $f(x, y, z) = x^2 y + e^x + z$. Show that $f(0, 1, -1) = 0$, $D_x f(0, 1, -1) \neq 0$, and that there exists a differentiable g defined in a neighbourhood of $(1, -1)$ in $\mathbb{R}^2$ so that $g(1, -1) = 0$ and $f(g(y, z), y, z) = 0$. Moreover, find $D_y g(1, -1)$ and $D_z g(1, -1)$.

Chapter 9

The Lebesgue Integral

In this chapter we will examine the so-called Lebesgue integral, which generalizes the concept of the Riemann integral. We begin with a discussion of the Lebesgue outer measure for subsets of $\mathbb{R}^n$, and a discussion on the Lebesgue measure for sets in terms of the outer measure. Following this we'll discuss Lebesgue measurable functions, and define the Lebesgue integral for such functions. In this chapter we will also define the Riemann integral in $\mathbb{R}^n$, and we will show that the Lebesgue integral generalizes this. We conclude this chapter with a discussion on iterated integration and Fubini's theorem, which in some sense is comparable to the fundamental theorem of calculus in one dimension to calculate the Riemann integral.

9.1 Lebesgue Outer Measure and Measurable Sets

9.1.1 The Lebesgue Outer Measure

Let $I = [a_1, b_1] \times [a_2, b_2] \times \cdots \times [a_n, b_n]$ be an closed interval in $\mathbb{R}^n$. We define the volume [1] of I by[2]

$$V(I) := \prod_{j=1}^{n} [b_j - a_j]. \tag{9.1}$$

---

[1] In 2 dimensions we often call it the area, and in 1 dimension, the length.
[2] Note that for I, an interval, any of the $[a_i, b_i]$'s may be replaced by either (a_i, b_i), $(a_i, b_i]$, or $[a_i, b_i)$, and the definition of $V(I)$ is the same.

Let $E \subseteq \mathbb{R}^n$ and let $\mathcal{C}$ be a countable collection of closed intervals so that $E \subseteq \{x \in C : C \in \mathcal{C}\} = \bigcup_{C \in \mathcal{C}} C$. For such a collection $\mathcal{C}$, we say that $\mathcal{C}$ **covers** E, and we define

$$\sigma(\mathcal{C}) = \sum_{C \in \mathcal{C}} V(C).$$

The **Lebesgue outer measure** of E is defined to be

$$\mu_e(E) = \inf\{\sigma(\mathcal{C}) : \mathcal{C} \text{ is a collection of intervals that covers } E\}. \quad (9.2)$$

Thus we see $\mu_e(E) \geq 0$, with $\mu_e(E) = \infty$ as a possibility.

Our first observation is that for an interval I, not necessarily closed, we have $\mu_e(I) = V(I)$. To see this observe that the closure of I, $\overline{I}$, covers I. Also, observe that $V(\overline{I}) = V(I)$, and thus $\mu_e(I) \leq V(I)$. Now suppose $\mathcal{C} := \{I_k\}_{k=1}^{\infty}$ is a collection of intervals that covers $\overline{I}$, and of course I. Let $\varepsilon > 0$, and let I_k^ε be an interval that has $I_k \subseteq (I_k^\varepsilon)^\circ$, also satisfying $V(I_k^\varepsilon) \leq \max\{(1+\varepsilon)V(I_k), \frac{\varepsilon}{2^k}\}$. We see that $\overline{I} \subseteq \bigcup_{k=1}^{\infty}(I_k^\varepsilon)^\circ$, and since $\overline{I}$ is compact, there is a finite subcover so that $\overline{I} \subseteq \bigcup_{k=1}^{N}(I_k^\varepsilon)^\circ$. Thus

$$V(I) \leq \sum_{k=1}^{N} V(I_k^\varepsilon) \leq (1+\varepsilon)\sum_{k=1}^{N} V(I_k) + \sum_{k=1}^{N} \frac{\varepsilon}{2^k} \leq (1+\varepsilon)\sigma(\mathcal{C}) + \varepsilon.$$

Since $\varepsilon > 0$ is arbitrary, we have $V(I) \leq \sigma(\mathcal{C})$; and since $\mathcal{C}$ is arbitrary, we have $V(I) \leq \mu_e(I)$.

Another observation to note is: if $E \subseteq F$, then $\mu_e(E) \leq \mu_e(F)$. This follows since any covering of F is also a covering of E.

Theorem 9.1.1 *Let $\mathcal{E} := \{E_a : a \in \mathcal{A}\}$ be a collection of sets where $\mathcal{A}$ is a countable index set. Then, for $E = \bigcup_{a \in \mathcal{A}} E_a$, we have*

$$\mu_e(E) \leq \sum_{a \in \mathcal{A}} \mu_e(E_a).$$

Proof: We may assume $\mu_e(E_a) < \infty$ for all $a \in \mathcal{A}$, since otherwise $\mu_e(E) = \infty = \sum_{a \in \mathcal{A}} \mu_e(E_a)$, and clearly we have the desired inequality. Let $\{a_1, a_2, \cdots\}$ be an enumeration of $\mathcal{A}$.

9.1. LEBESGUE OUTER MEASURE AND MEASURABLE SETS

Fix $\varepsilon > 0$. For each $a_k \in \mathcal{A}$, let $\{I_j^k\}_{j \in \mathbb{N}}$ be a collection of intervals that covers E_{a_k}, which satisfies $\sum_{j \in \mathbb{N}} V(I_j^k) < \mu_e(E_{a_k}) + \frac{\varepsilon}{2^k}$. Since $E \subseteq \bigcup_{j,k \in \mathbb{N}} I_j^k$ we have

$$\mu_e(E) \leq \sum_k \sum_j V(I_j^k) \leq \sum_k (\mu_e(E_{a_k}) + \frac{\varepsilon}{2^k}) = \sum_k \mu_e(E_{a_k}) + \varepsilon.$$

Since $\varepsilon > 0$ was arbitrary, by letting $\varepsilon \to 0$ we get the desired result.
□

From the above results, we see that the boundary of any interval has outer measure 0, any subset of a set with outer measure zero has outer measure zero, and any set which is the countable union of sets of outer measure zero has outer measure zero.

We claim that any countable set in $\mathbb{R}^n$ has outer measure zero. Let $\{x_1, x_2, \cdots\}$ be a countable set. If $x_k = (x_k^1, x_k^2, \cdots, x_k^n)$, then we let $I_k = [x_k^1 - \frac{1}{2}\sqrt[n]{\frac{\varepsilon}{2^k}}, x_k^1 + \frac{1}{2}\sqrt[n]{\frac{\varepsilon}{2^k}}] \times \cdots \times [x_k^n - \frac{1}{2}\sqrt[n]{\frac{\varepsilon}{2^k}}, x_k^n + \frac{1}{2}\sqrt[n]{\frac{\varepsilon}{2^k}}]$. Thus $V(I_k) = \frac{\varepsilon}{2^k}$, and so

$$\mu_e(\{x_1, x_2, \cdots\}) \leq \sum_{k=1}^{\infty} \frac{\varepsilon}{2^k} = \varepsilon.$$

Since $\varepsilon > 0$ is arbitrary, we see that $\mu_e(\{x_1, x_2, \cdots\}) = 0$.

Hence, we see that $\mu_e(\mathbb{Q}) = 0$, and thus there are dense subsets of Euclidean space which have outer measure zero. More generally, in n dimensions, the set of points with rational coordinates is dense in $\mathbb{R}^n$, is countable, and thus has outer measure zero.

The next theorem provides a comparison result between an arbitrary set and an open set containing it who has close outer measure. In particular, it allows us to realize the outer measure of a set E by the formula

$$\mu_e(E) = \inf\{\mu_e(\mathcal{O}) : \mathcal{O} \text{ is open}, E \subseteq \mathcal{O}\}. \tag{9.3}$$

Theorem 9.1.2 *Let $E \subseteq \mathbb{R}^n$. For each $\varepsilon > 0$, there is an open set $\mathcal{O}_\varepsilon$, with $E \subseteq \mathcal{O}_\varepsilon$, and*

$$\mu_e(\mathcal{O}_\varepsilon) \leq \mu_e(E) + \varepsilon.$$

Proof: Let $\varepsilon > 0$ be fixed. Choose a covering of intervals $\{I_k\}$ so that $E \subseteq \bigcup_k I_k$ and $\sum_k V(I_k) \leq \mu_e(E) + \frac{\varepsilon}{2}$. For each k, let J_k be an interval so that $I_k \subseteq J_k^\circ$ and $V(J_k) \leq V(I_k) + \frac{\varepsilon}{2^{k+1}}$. We define $\mathcal{O}_\varepsilon := \bigcup_k J_k^\circ$. Then by construction, $E \subseteq \mathcal{O}_\varepsilon$ and

$$\mu_e(\mathcal{O}_\varepsilon) \leq \sum_k V(J_k) \leq \sum_k V(I_k) + \frac{\varepsilon}{2}\sum_k \frac{1}{2^k} \leq \mu_e(E) + \varepsilon.$$

□

A set G is called a **set of type G_δ** if it is the intersection of a countable number of open sets.[3] The following is a corollary of the above theorem.

Corollary 9.1.3 *Let $E \subseteq \mathbb{R}^n$, then there is a set G of type G_δ with $E \subseteq G$ and $\mu_e(E) = \mu_e(G)$.*

Proof: By the above theorem (9.1.2), for each natural number n there is an open $\mathcal{O}_n$ containing E so that $\mu_e(\mathcal{O}_n) \leq \mu_e(E) + \frac{1}{n}$. Define $G := \bigcap_{n=1}^\infty \mathcal{O}_n$, and observe that $E \subseteq G$. For each natural number n we have

$$\mu_e(E) \leq \mu_e(G) \leq \mu_e(\mathcal{O}_n) \leq \mu_e(E) + \frac{1}{n}.$$

Letting $n \to \infty$ we see that $\mu_e(E) = \mu_e(G)$.
□

9.1.2 The Lebesgue Measure

A set $E \subseteq \mathbb{R}^n$ is **Lebesgue measurable**, or simply **measurable**, if for each $\varepsilon > 0$ there is an open set $\mathcal{O}_\varepsilon$ so that $E \subseteq \mathcal{O}_\varepsilon$ and $\mu_e(\mathcal{O}_\varepsilon \setminus E) < \varepsilon$. In such a case we define the **Lebesgue measure** of E, denoted by $\mu(E)$, to be the Lebesgue outer measure of E. That is, $\mu(E) = \mu_e(E)$. Clearly we have that all open sets are Lebesgue measurable. Also, sets of outer measure zero are Lebesgue measurable. To see this, use the result of theorem (9.1.2), and observe that

$$\mu_e(\mathcal{O}_\varepsilon \setminus E) \leq \mu_e(\mathcal{O}_\varepsilon) \leq \mu_e(E) + \varepsilon = \varepsilon.$$

The question we now must address is: which sets are Lebesgue measurable?

Recall theorem (9.1.2), which says that for each $\varepsilon > 0$ there is an open set $\mathcal{O}_\varepsilon$ so that $E \subseteq \mathcal{O}_\varepsilon$ and $\mu_e(\mathcal{O}_\varepsilon) \leq \mu_e(E) + \varepsilon$. Note that if E is measurable,

---
[3] Recall that the intersection of countably infinite many open sets may not be an open set. However, any open set is a set of type G_δ.

9.1. LEBESGUE OUTER MEASURE AND MEASURABLE SETS

then for $\varepsilon > 0$ there is an open $\mathcal{O}_\varepsilon$ so that $E \subseteq \mathcal{O}_\varepsilon$, and $\mu_e(\mathcal{O}_\varepsilon \setminus E) < \varepsilon$. Writing $\mathcal{O}_\varepsilon = E \cup (\mathcal{O}_\varepsilon \setminus E)$ we see that

$$\mu_e(\mathcal{O}_\varepsilon) \leq \mu_e(E) + \mu_e(\mathcal{O}_\varepsilon \setminus E) < \mu_e(E) + \varepsilon.$$

Hence we see that the $\mathcal{O}_\varepsilon$ given by the measurability of E also guarantees the result given in theorem (9.1.2). However, if $E \subseteq \mathcal{O}_\varepsilon$ satisfies $\mu_e(\mathcal{O}_\varepsilon) \leq \mu_e(E) + \varepsilon$, then it may not be the case that $\mu_e(\mathcal{O}_\varepsilon \setminus E) < \varepsilon$.

Further Examples of Lebesgue Measurable Sets

We now examine the question: which sets are Lebesgue measurable? We will show that in addition to open sets and sets of outer measure zero, closed sets also are measurable, intervals are also measurable, the complement of a measurable set is also measurable, and the union or intersection of countably many measurable sets is also measurable.

Theorem 9.1.4 *Let $\{S_k\}$ be a countable collection of measurable sets, then $S = \bigcup_k S_k$ is also measurable, and*

$$\mu(S) \leq \sum_k \mu(S_k).$$

Proof: Let $\varepsilon > 0$ be given, then for each S_k find an open set $\mathcal{O}_k$ so that $S_k \subseteq \mathcal{O}_k$ and $\mu_e(\mathcal{O}_k \setminus S_k) < \frac{\varepsilon}{2^k}$. Then $\mathcal{O} = \bigcup_k \mathcal{O}_k$ is open, and $S \subseteq \mathcal{O}$. Also, $\mathcal{O} \setminus S \subseteq \bigcup_k (\mathcal{O}_k \setminus S_k)$, and so

$$\mu_e(\mathcal{O} \setminus S) \leq \sum_k \mu_e(\mathcal{O}_k \setminus S_k) < \varepsilon.$$

Hence S is measurable. Moreover,

$$\mu(S) = \mu_e(S) \leq \sum_k \mu_e(S_k) = \sum_k \mu(S_k).$$

□

It follows from this theorem that intervals are measurable, since an interval is a union of its interior, an open set, and possibly portions of its boundary, a set of measure 0.

Lemma 9.1.5 *Suppose E_1 and E_2 are two disjoint sets with $d(E_1, E_2) = \inf\{d(\vec{x}, \vec{y}) : \vec{x} \in E_1, \vec{y} \in E_2\} > 0$, then*

$$\mu_e(E_1 \cup E_2) = \mu_e(E_1) + \mu_e(E_2).$$

Proof: We have $\mu_e(E_1 \cup E_2) \leq \mu_e(E_1) + \mu_e(E_2)$. Next we will show the opposite inequality. Let $\varepsilon > 0$. Choose intervals $\{I_k\}$ so that $E_1 \cup E_2 \subseteq \bigcup_k I_k$ and $\sum_k \mu(I_k) \leq \mu_e(E_1 \cup E_2) + \varepsilon$.

Observe that for any closed interval $I = \{\vec{x} : a_i \leq x_i \leq b_i, i = 1, 2, \cdots, n\}$, and any $c \in (a_j, b_j)$ for some j, we may bisect I into two intervals I', I'', so that $I = I' \cup I''$, with $I' = \{\vec{x} : a_i \leq x_i \leq b_i \text{ for } i \neq j, \text{ and } a_j \leq x_j \leq c\}$, and $I'' = \{\vec{x} : a_i \leq x_i \leq b_i \text{ for } i \neq j, \text{ and } c \leq x_j \leq b_j\}$. Moreover, $\mu(I) = \mu(I') + \mu(I'')$. To see this, note that

$$\mu(I) = v(I) = (b_1 - a_1) \cdots (b_{j-1} - a_{j-1}) \cdot (b_j - a_j) \cdot (b_{j+1} - a_{j+1}) \cdots (b_n - a_n)$$

$$= (b_1 - a_1) \cdots (b_{j-1} - a_{j-1}) \cdot (b_j - c + c - a_j) \cdot (b_{j+1} - a_{j+1}) \cdots (b_n - a_n)$$

$$= (b_1 - a_1) \cdots (b_{j-1} - a_{j-1}) \cdot [(b_j - c) + (c - a_j)] \cdot (b_{j+1} - a_{j+1}) \cdots (b_n - a_n)$$

$$= (b_1 - a_1) \cdots (b_{j-1} - a_{j-1}) \cdot (b_j - c) \cdot (b_{j+1} - a_{j+1}) \cdots (b_n - a_n)$$
$$+ (b_1 - a_1) \cdots (b_{j-1} - a_{j-1}) \cdot (c - a_j) \cdot (b_{j+1} - a_{j+1}) \cdots (b_n - a_n)$$

$$= v(I') + v(I'') = \mu(I') + \mu(I'').$$

Because of this, by otherwise repeatedly bisecting our initial intervals in this manner finitely many times, we may assume that the diameter of each I_k is less than $d(E_1, E_2)$. Hence $\{I_k\}$ splits into two disjoint collections $\{I_{1,j}\}$ and $\{I_{2,j}\}$, where $\{I_{1,j}\}$ covers E_1 and $\{I_{2,j}\}$ covers E_2.

Thus,

$$\mu_e(E_1) + \mu_e(E_2) \leq \sum_j \mu(I_{1,j}) + \sum_j \mu(I_{2,j}) = \sum_k \mu(I_k) \leq \mu_e(E_1 \cup E_2) + \varepsilon.$$

Since $\varepsilon > 0$, we have $\mu_e(E_1) + \mu_2(E_2) \leq \mu_e(E_1 \cup E_2)$.
□

9.1. LEBESGUE OUTER MEASURE AND MEASURABLE SETS

Corollary 9.1.6 *Let $\{I_k\}_{k=1}^N$ be a finite collection of pairwise disjoint, closed and bounded intervals. Then $I = \bigcup_{k=1}^N I_k$ is measurable and*

$$\mu(I) = \sum_{k=1}^N \mu(I_k).$$

Proof: Since intervals are measurable, and the countable union of measurable sets is measurable, we clearly have I is measurable. The remainder of the proof follows by by repeated use of 9.1.5

$$\mu(I) = \mu(I_1 \cup \cdots \cup I_{N-1} \cup I_N) = \mu(I_1 \cup \cdots \cup I_{N-1}) + \mu(I_N)$$

$$\leq \cdots \leq \mu(I_1) + \cdots + \mu(I_{N-1}) + \mu(I_N).$$

□

In the following lemma, a cube is simply an interval of the form $[a_1, a_1 + b] \times [a_2, a_2 + b] \times \cdots \times [a_n, a_n + b]$, where $b > 0$.

Lemma 9.1.7 *Every open set in $\mathbb{R}^n$ can be written as a countable union of nonoverlapping[4] closed cubes.*

Proof: Let C_0 be the collection of closed cubes in $\mathbb{R}^n$ of the form $[j_1, j_1 + 1] \times \cdots \times [j_n, j_n + 1]$, where $j_1, \cdots j_n \in \mathbb{Z}$. That is, C_0 is the collection of unit cubes whose vertices are all points with coordinates that are integer valued. For each cube in C_0, bisect the edges of this cube to obtain 2^n subcubes each with edge length $\frac{1}{2}$, and let C_1 be the collection of all such subcubes obtained from the cubes by C_0 in this manner. We may then take every cube in C_1 and bisect it into 2^n subcubes, each with edge length $\frac{1}{4}$, and let C_2 be the cubes obtained in this manner from the cubes in C_1. We continue in this manner to obtain C_3, C_4, etc.

Let $\mathcal{O}$ be any open set, and let S_0 be the collection of all cubes in C_0 that lie entirely within $\mathcal{O}$. Let S_1 be the cubes in C_1 which lie inside of $\mathcal{O}$ but are not subcubes of any of the cubes in S_0. More generally, for $j \geq 1$, let S_j be the cubes in C_j which lie entirely in $\mathcal{O}$ but are not subcubes of any of the cubes in $S_0, S_1, \cdots, S_{j-1}$. If $S = \bigcup_k S_k$, then S is a countable collection of cubes since each S_j contains countably many cubes. Since $\mathcal{O}$ is open and

---
[4]Meaning that their boundaries may intersect, but their interiors are pairwise disjoint.

the elements of C_j have their diameter go to 0 as $j \to \infty$, we have that each point of $\mathcal{O}$ will eventually be in some cube of some C_k. Hence $\mathcal{O} = \bigcup_{Q \in S} Q$.
□

The following is a more general version of corollary 9.1.6.

Lemma 9.1.8 *If $\{I_k\}_{k=1}^N$ is a collection of non-overlapping intervals each with finite diameter, then the set $\bigcup_{k=1}^N I_k$ is measurable, and*

$$\mu(\bigcup_{k=1}^N I_k) = \sum_{k=1}^N \mu(I_k).$$

Proof: As in the above corollary, we have $\bigcup_k I_k$ is measurable. It remains to show $\mu(\bigcup_{k=1}^N I_k) = \sum_{k=1}^N \mu(I_k)$. Clearly the collection $\{\overline{I_k}\}$ covers $\bigcup_k I_k$, and so $\mu(\bigcup_k I_k) \le \sum_k \mu(I_k)$. Thus we only need to show the reverse inequality. For simplicity, we let $I := \bigcup_{k=1}^N I_k$.

Let $\varepsilon > 0$ be given. For each I_k we may find a closed interval J_k so that $J_k \subseteq I_k^\circ$ with $(1-\varepsilon)\mu(\bigcup_k I_k) \le \mu(\bigcup_k J_k) + \varepsilon$. Since each point $\vec{x}$ of J_k is an element of I_k°, there is an $r(\vec{x}) > 0$ so that $B_{r(\vec{x})}(\vec{x}) \subseteq I_k^\circ$. The collection $\{B_{r(\vec{x})}(\vec{x}) : \vec{x} \in J_k\}$ is an open cover of J_k, and so there is a finite subcover $\{B_{r(\vec{x}_j)}(\vec{x}_j) : j = 1, 2, \cdots, M\}$. Letting $r_k = \min\{r(\vec{x}_j) : j = 1, 2, \cdots, M\}$ we see that the distance between the boundary of J_k and the boundary of I_k is positive. Hence, the collection of J_k's are pairwise disjoint compact intervals. So by corollary 9.1.6 we have $\mu(\bigcup_k J_k) = \sum_k \mu(J_k)$. Thus

$$(1-\varepsilon)\mu(\bigcup_k I_k) \le \mu(\bigcup_k J_k) + \varepsilon = \sum_k \mu(J_k) + \varepsilon \le \sum_k \mu(I_k) + \varepsilon.$$

and so

$$(1-\varepsilon)\mu(\bigcup_k I_k) \le \sum_k \mu(I_k) + \varepsilon.$$

Since $\varepsilon > 0$ is arbitrary, we get the desired inequality by letting $\varepsilon \to 0$.
□

Theorem 9.1.9 *Every closed set is measurable.*

9.1. LEBESGUE OUTER MEASURE AND MEASURABLE SETS

Proof: Let E be a compact set and let $\varepsilon > 0$ be fixed. Then by theorem 9.1.2, there is an open set $\mathcal{O}$ so that $\mu(\mathcal{O}) < \mu_e(E) + \varepsilon$. Since $\mathcal{O} \setminus E$ is open, by lemma 9.1.7 we know there exists a collection of non-overlapping cubes $\{I_k\}_{k=1}^{\infty}$ so that $\mathcal{O} \setminus E = \bigcup_{k=1}^{\infty} I_k$. Hence $\mu(\mathcal{O} \setminus E) \leq \sum_{k=1}^{\infty} \mu(I_k)$. So it suffices to show that $\sum_{k=1}^{\infty} \mu(I_k) \leq \varepsilon$.

For each N we have

$$E \cup \left(\bigcup_{k=1}^{N} I_k\right) \subseteq E \cup \left(\bigcup_{k=1}^{\infty} I_k\right) = \mathcal{O}.$$

Thus, since E and $\bigcup_{k=1}^{N} I_k$ are disjoint and compact (and thus the distance between them is positive), we have

$$\mu_e(E) + \mu\left(\bigcup_{k=1}^{N} I_k\right) = \mu_e\left(E \cup \left(\bigcup_{k=1}^{N} I_k\right)\right) \leq \mu(\mathcal{O}).$$

However, since $\mu(\bigcup_{k=1}^{N} I_k) = \sum_{k=1}^{N} \mu(I_k)$, we see that for all N

$$\sum_{k=1}^{N} \mu(I_k) \leq \mu(\mathcal{O}) - \mu_e(E) < \varepsilon.$$

Since this is true for all N, we get

$$\sum_{k=1}^{\infty} \mu(I_k) \leq \varepsilon.$$

This proves the theorem when E is compact. In the general case, write

$$E = \bigcup_{k=1}^{\infty} [E \cap \{\vec{x} : \|\vec{x}\| \leq k\}]$$

and observe that E is the countable union of compact [5] sets.
□

Theorem 9.1.10 *The complement of a measurable set is measurable.*

---

[5]Compact sets are measurable, and the countable union of measurable sets is measurable.

Proof: Let E be a measurable set. For each natural number m, choose an open set $\mathcal{O}_m$ so that $E \subseteq \mathcal{O}_m$ and $\mu(\mathcal{O}_m \setminus E) < \frac{1}{m}$. By theorem 9.1.9, $\mathcal{O}_m^c$ is measurable, and so is $\bigcup_{m=1}^{\infty} \mathcal{O}_m^c$. Moreover $(\bigcap_{m=1}^{\infty} \mathcal{O}_m)^c = \bigcup_{m=1}^{\infty} \mathcal{O}_m^c \subseteq E^c$. Write

$$E^c = (\bigcup_{m=1}^{\infty} \mathcal{O}_m^c) \cup (E^c \setminus (\bigcup_{m=1}^{\infty} \mathcal{O}_m^c)).$$

Let $S = E^c \setminus (\bigcup_{m=1}^{\infty} \mathcal{O}_m^c)$. Then $S \subseteq E^c \setminus \mathcal{O}_m^c = \mathcal{O}_m \setminus E$, and so $\mu_e(S) < \frac{1}{m}$ for every m. Thus S has measure zero. Since $E^c = (\bigcup_{m=1}^{\infty} \mathcal{O}_m^c) \cup S$ is the countable union of measurable sets, it too is measurable. $\square$

Using theorem 9.1.10 we see that the countable intersection of measurable sets is measurable, for if $\{E_k\}$ are measurable, then

$$\bigcap_k E_k = ((\bigcap_k E_k)^c)^c = (\bigcup_k E_k^c)^c.$$

Since the E_k^c's are measurable, so is their union, and so is the complement of their union.

Likewise, for two measurable sets E, F, we have $E \setminus F = E \cap F^c$ is measurable. Hence we see that the collection of Lebesgue measurable sets is closed under countable unions, countable intersections and complements. Such a collection of sets is often called a σ-**algebra**.

Further Properties of Lebesgue Measure

Lemma 9.1.11 *A set $E \subseteq \mathbb{R}^n$ is Lebesgue measurable if and only if for all $\varepsilon > 0$, there is a closed set $\mathcal{C}$ so that $\mathcal{C} \subseteq E$ and $\mu_e(E \setminus \mathcal{C}) < \varepsilon$.*

Proof: Let $\varepsilon > 0$. E is measurable if and only if E^c is measurable by theorem 9.1.10. Assuming E^c is measurable, we know that there is an open set $\mathcal{O}$ so that $E^c \subseteq \mathcal{O}$ and $\mu_e(\mathcal{O} \setminus E^c) < \varepsilon$. However, letting $\mathcal{C} = \mathcal{O}^c$, and using $\mathcal{O} \setminus E^c = E \setminus \mathcal{C}$ we get one direction of the biconditional in the statement of this theorem.

Conversely, if for any $\varepsilon > 0$ there is a closed $\mathcal{C}$ so that $\mathcal{C} \subseteq E$ and $\mu_e(E \setminus \mathcal{C}) < \varepsilon$. Using the observation above, with $\mathcal{O} = \mathcal{C}^c$, we see that $\mathcal{O}$ is an open set containing E^c and $\mu_e(\mathcal{O} \setminus E^c) < \varepsilon$. Hence, we see that E^c is

9.1. LEBESGUE OUTER MEASURE AND MEASURABLE SETS

measurable, and thus E is as well.
□

The following theorem generalizes lemma 9.1.8 and corollary 9.1.6.

Theorem 9.1.12 *If $\{E_k\}$ is a countable collection of pairwise disjoint measurable sets, then $\mu(\bigcup_k E_k) = \sum_k \mu(E_k)$.*

Proof: First, consider the case where each E_k is bounded. Let $\varepsilon > 0$ be given, then by lemma 9.1.11, for each E_k, we may find a closed set C_k, a subset of E_k, satisfying $\mu(E_k \setminus C_k) < \frac{\varepsilon}{2^k}$. Thus $\mu(E_k) \leq \mu(C_k) + \frac{\varepsilon}{2^k}$. The collection $\{C_k\}$ is a collection of compact disjoint sets. Hence, by lemma 9.1.5 we have $\mu(\bigcup_{k=1}^m C_k) = \sum_{k=1}^m \mu(C_k)$ for any m. Now, since $\bigcup_{k=1}^m C_k \subseteq \bigcup_k E_k$ we know $\sum_{k=1}^m \mu(C_k) \leq \mu(\bigcup_k E_k)$. Hence,

$$\sum_k \mu(E_k) - \varepsilon \leq \sum_k (\mu(E_k) - \frac{\varepsilon}{2^k}) \leq \sum_k \mu(C_k) \leq \mu(\bigcup_k E_k).$$

Since $\varepsilon > 0$ is arbitrary, we get $\mu(\bigcup_k E_k) \geq \sum_k \mu(E_k)$. Since the opposite inequality always holds, we have $\mu(\bigcup_k E_k) = \sum_k \mu(E_k)$ whenever $\{E_k\}$ is a collection of bounded, measurable and pairwise disjoint sets.

In the general case we let $\{I_k\}$ be a sequence of nested intervals so that

$$I_1 \subseteq I_2 \subseteq I_3 \subseteq \cdots \subseteq I_k \subseteq \cdots$$

and $\bigcup_{k=1}^\infty I_k = \mathbb{R}^n$. Let $J_1 = I_1$, and for $k > 1$ let $J_k = I_k \setminus I_{k-1}$. Then the collection $\{E_k \cap J_i\}_{k,i}$ are bounded, disjoint and measurable, $E_k = \bigcup_i (E_k \cap J_i)$, and by the argument above, $\mu(E_k) = \mu(\bigcup_i (E_k \cap J_i)) = \sum_i \mu(E_k \cap J_i)$. Thus,

$$\mu(\bigcup_k E_k) = \mu(\bigcup_{k,i}(E_k \cap J_i)) = \sum_{k,i} \mu(E_k \cap J_i) = \sum_k (\sum_i \mu(E_k \cap J_i)) = \sum_k \mu(E_k).$$

□

Applying this theorem to any collection of non-overlapping intervals $\{I_k\}$ we get $\mu(\bigcup_k I_k) = \sum_k \mu(I_k)$.

Corollary 9.1.13 *Suppose $A \subseteq B$ and A, B are measurable sets with $\mu(A)$ finite in value, then*

$$\mu(B \setminus A) = \mu(B) - \mu(A).$$

Proof: Write $B = A \cup (B \setminus A)$. Then A and $B \setminus A$ are disjoint measurable sets, and so $\mu(B) = \mu(A) + \mu(B \setminus A)$.
□

Convergence in Measure for Sets

For a collection of nested measurable sets $\{E_k\}$ satisfying

$$E_1 \subseteq E_2 \subseteq E_3 \subseteq \cdots \subseteq E_k \subseteq \cdots,$$

we call E the limit $\bigcup_{k=1}^{\infty} E_k$, and denote the convergence of E_k to E by $E_k \nearrow E$. Likewise, for

$$\cdots \subseteq E_k \subseteq \cdots \subseteq E_3 \subseteq E_2 \subseteq E_1,$$

we call E the limit $\bigcap_{k=1}^{\infty} E_k$, and denote the convergence of E_k to E by $E_k \searrow E$.

In either of these cases, we have $\lim_{k \to \infty} \mu(E_k) = \mu(E)$, provided some $\mu(E_k) < \infty$ in the second case.

Consider the case where $E_k \nearrow E$ and write

$$E = E_1 \cup (E_2 \setminus E_1) \cup \cdots \cup (E_k \setminus E_{k-1}) \cup \cdots.$$

Assume that $\mu(E_k) < \infty$ for all k, otherwise $\mu(E_k) \to \infty$ and $\mu(E) = \infty$, and we are done. Use theorem 9.1.12 and corollary 9.1.13 to get

$$\mu(E) = \mu(E_1) + \mu(E_2 \setminus E_1) + \cdots + \mu(E_{k+1} \setminus E_k) + \cdots$$
$$= \mu(E_1) + (\mu(E_2) - \mu(E_1)) + \cdots + (\mu(E_{k+1}) - \mu(E_k)) + \cdots$$
$$= \lim_{k \to \infty} \mu(E_k).$$

For the second case, assume $\mu(E_1) < \infty$. Here we write

$$E_1 = E \cup (E_1 \setminus E_2) \cup \cdots \cup (E_k \setminus E_{k+1}) \cup \cdots.$$

Then,

$$\mu(E_1) = \mu(E) + (\mu(E_1) - \mu(E_2)) + \cdots + (\mu(E_k) - \mu(E_{k+1})) + \cdots$$
$$= \mu(E) + \mu(E_1) - \lim_{k \to \infty} \mu(E_k).$$

In the second case, to see that the $\mu(E_k)$ must be finite for at least some k, consider the case $E_k = B_k(\vec{0})^c$. Here $\mu(E_k) = \infty$ and $E_k \searrow \emptyset$. Hence our assumption $\mu(E_1) < \infty$, for simplicity of our argument.

The Existence of Non-measurable Sets

We will now see that not all subsets of $\mathbb{R}$ are measurable, and to do this we will only need to consider subsets of $[0, 1)$.

We define a relation on $[0, 1)$ by $x \sim_R y$ if $x - y \in \mathbb{Q}$. Clearly $\sim_R$ is reflexive since $0 \in \mathbb{Q}$ and it is symmetric since $q \in \mathbb{Q} \iff -q \in \mathbb{Q}$. Suppose $x \sim_R y$ and $y \sim_R z$, then $x - y = r, y - z = q$ with $r, q \in \mathbb{Q}$. Since $\mathbb{Q}$ is closed under $+$ and $x - z = (x - y) + (y - z) = q + r$ we have $\sim_R$ is transitive. Hence this relation $\sim_R$ is an equivalence relation on $[0, 1)$ and thus it defines a partition of $[0, 1)$, with one equivalence class being $\mathbb{Q} \cap [0, 1)$, and all others being irrational translates of $\mathbb{Q} \cap [0, 1)$. Clearly there is an uncountable number of equivalence classes. Let $\{A_\alpha\}$ represent the equivalence classes in $[0, 1)$.

By the axiom of choice, we may define a set E by selecting one element from each equivalence class A_α. We claim that E is not measurable.

For $q \in \mathbb{Q}$ we let $F_q := E \cap [0, 1 - q)$ and $F_q^* := E \cap [1 - q, 1)$. For any set $A \subseteq \mathbb{R}$ and any $x \in \mathbb{R}$, we let $A + x := \{a + x : a \in A\}$. It is an exercise to show that this sort of translation of a set preserves measurability, and the outer measure of $A + x$ is the same as that of A. Using this notation, we let $G_q := (F_q + q) \cup (F_q^* + (q - 1))$. The collection $\{G_q\}$ for $q \in \mathbb{Q} \cap [0, 1)$ are pairwise disjoint since $x \in G_q$ implies $x = z + q$ where $z \in E$. If $x \in G_q \cap G_p$, then $x = z' + p$ where $z' \in E$. Then $z + q = z' + p$ and so $z - z' = p - q \in \mathbb{Q}$. Thus $z \sim_R z'$. Now this forces $z = z'$ and so $p = q$.

We also claim that $\bigcup_{q \in \mathbb{Q} \cap [0,1)} G_q = [0, 1)$. To see this, let $x \in [0, 1)$, then there is a $z \in E$ so that $x \sim_R z$. Thus $x = z + q$ for some rational q. Hence $x \in G_q$. If E were measurable we have $\mu(E) = \mu(G_q)$ for each rational q, and by theorem 9.1.12 we would have

$$1 = \mu([0, 1)) = \sum_{q \in \mathbb{Q} \cap [0,1)} \mu(G_q) = \sum_{q \in \mathbb{Q} \cap [0,1)} \mu(E),$$

which is infinite if $\mu(E) > 0$ and 0 if $\mu(E) = 0$. Either way, we would have a contradiction to theorem 9.1.12 if E were measurable.

9.1.3 Homework Exercises

1. Show that the Cantor set has outer measure zero.

2. Suppose E is a set so that $E^\circ \neq \emptyset$. Show $\mu_e(E) > 0$.

3. Suppose $E \subseteq [a,b]$ which satisfies $\mu_e(E) = 0$. Show that E^c is dense in $[a,b]$.

4. The following is a construction of a two dimensional Cantor set: Consider the unit square $\{(x,y) : 0 \leq x, y \leq 1\}$ and subdivide this set into 9 equal parts and keep only the 4 closed corner squares, removing the remaining regions (in the shape of a cross). Repeat this process with the remaining 4 closed squares, resulting in a collection of 16 closed squares. Continue along these lines, ad infinitum. Show that the remaining region is a perfect set, has $\mathbb{R}^2$ Lebesgue measure zero, and is the cross product of the $\mathbb{R}^1$ Cantor set with itself.

5. If $\{E_k\}_{k=1}^\infty$ is a collection of measurable sets so that $\sum_{k=1}^\infty \mu(E_k) < \infty$, then
$$\limsup_{k \to \infty} E_k := \bigcap_{j=1}^\infty (\bigcup_{k=j}^\infty E_k), \quad \liminf_{k \to \infty} E_k := \bigcup_{j=1}^\infty (\bigcap_{k=j}^\infty E_k)$$
are both measurable, and both have measure zero.

6. Show that the outer measure of a set is unchanged if in the definition of outer measure we use coverings of the set by parallelepipeds with a fixed orientation (with edges parallel to a fixed set of n linearly independent, pairwise orthogonal vectors), rather than by intervals. Note that if $P = \{\vec{x} = \sum_{k=1}^n c_k \vec{e}_k, 0 \leq c_k \leq 1\}$ is a parallelepiped ($\{\vec{e}_1, \vec{e}_2, \cdots, \vec{e}_n\}$ linearly independent vectors emanating from a fixed point in $\mathbb{R}^n$), then the volume of P is given by the determinant of the matrix having the $\vec{e}_k$'s as rows. Note that this exercise (in addition to the fact that Lebesgue outer measure is translation invariant - see other exercise in this HW set) shows that the definition of Lebesgue outer measure is independent of the orthonormal basis you choose for $\mathbb{R}^n$.

7. Let E be a set and define $E + \vec{x} := \{\vec{e} + \vec{x} : \vec{e} \in E\}$. Show that $E + \vec{x}$ satisfies: $\mu_e(E + \vec{x}) = \mu_e(E)$. Moreover, show that E measurable implies $E + \vec{x}$ is as well.

8. Let E be a non-measurable subset of $[0,1)$ as developed in the previous section. Let $\{r_k\}_{k=1}^{\infty}$ be an enumeration of the rational numbers in $(0,1)$ and define $E_k = E + r_k$. Show that $\{E_k\}$ are pairwise disjoint, and by the above exercise have $\mu_e(E_k) = \mu_e(E)$. Show that $\mu_e(\bigcup_k E_k) < \sum_k \mu_e(E_k)$.

9.2 Lebesgue Measurable Functions

Let $f : D \to \mathbb{R} \cup \{\pm\infty\}$, with domain D a measurable set in $\mathbb{R}^n$. We say that f is **measurable** if for each $\alpha \in \mathbb{R}$ the set

$$\{\vec{x} \in D : f(\vec{x}) > \alpha\} \tag{9.4}$$

is a measurable set in $\mathbb{R}^n$. Since

$$D = \{\vec{x} \in D : f(\vec{x}) = \infty\} \cup [\bigcup_{k \in \mathbb{Z}} \{\vec{x} \in D : f(\vec{x}) > k\}],$$

we see that if f is measurable, then $\{\vec{x} \in D : f(\vec{x}) = \infty\}$ must also be measurable. Also, $\{\vec{x} : f(\vec{x}) = -\infty\} = \bigcap_{k \in \mathbb{Z}} \{\vec{x} : f(\vec{x}) > k\}$ is also measurable if f is measurable. From the definition of measurability for functions, we see that for each $\alpha \in \mathbb{R}$ the set $\{\vec{x} \in D : f(\vec{x}) \le \alpha\} = \{\vec{x} \in D : f(\vec{x}) > \alpha\}^c$ is measurable, $\{\vec{x} \in D : f(\vec{x}) \ge \alpha\} = \bigcap_{k \in \mathbb{N}} \{\vec{x} \in D : f(\vec{x}) > \alpha - \frac{1}{k}\}$ is measurable, and so $\{\vec{x} \in D : f(\vec{x}) = \alpha\} = \{\vec{x} \in D : f(\vec{x}) \le \alpha\} \cap \{\vec{x} \in D : f(\vec{x}) \ge \alpha\}$ is also measurable. Thus it follows that for all $\alpha < \beta$ we have $\{\vec{x} \in D : \alpha \le f(\vec{x}) \le \beta\} = \{\vec{x} \in D : f(\vec{x}) \ge \alpha\} \cap \{\vec{x} \in D : f(\vec{x}) \le \beta\}$ is measurable. Moreover, since the difference between two measurable sets is measurable, and $\{\vec{x} \in D : f(\vec{x}) = \alpha\}$ is measurable for any $\alpha \in \mathbb{R}$, we see that we can replace any non-strict inequality in $\{\vec{x} \in D : \alpha \le f(\vec{x}) \le \beta\}$ by a strict inequality and obtain a measurable set.

Suppose that f and g are measurable functions and are defined on a common set D, then the set $\{\vec{x} \in D : f(\vec{x}) > g(\vec{x})\}$ is measurable. To see this, suppose $\{q_k\}_{k=1}^{\infty}$ is an enumeration of $\mathbb{Q}$, then

$$\{\vec{x} \in D : f(\vec{x}) > g(\vec{x})\} = \bigcup_{k=1}^{\infty} [\{\vec{x} \in D : f(\vec{x}) > q_k\} \cap \{\vec{x} \in D : g(\vec{x}) < q_k\}].$$

It is an exercise to show that $h(\vec{x}) = f(\vec{x})+c$ and $k(\vec{x}) = cf(\vec{x})$ are measurable for any $c \in \mathbb{R}$. From this, and the observation we just made, we see that $f + g$ is measurable whenever f and g are measurable. This follows since

$$\{\vec{x} : f(\vec{x}) + g(\vec{x}) > \alpha\} = \{\vec{x} : f(\vec{x}) > \alpha - g(\vec{x})\}.$$

More generally, we have that f is measurable if and only if for $\mathcal{O} \subset \mathbb{R}$, any open set, we have $f^{-1}(\mathcal{O}) := \{\vec{x} \in D : f(\vec{x}) \in \mathcal{O}\}$ is measurable. To see this we note that if $f^{-1}(\mathcal{O})$ is measurable for every open $\mathcal{O}$, then by taking $\mathcal{O} = (\alpha, \infty)$ we see that f is measurable. To see the converse, observe that for any open set $\mathcal{O} \subseteq \mathbb{R}$ we may write $\mathcal{O}$ as the countable union of disjoint open intervals $\bigcup_k (a_k, b_k)$. To see that this is possible we let $I(x)$ be the intersection of all open intervals contained in $\mathcal{O}$ that contain x. If $x, y \in \mathcal{O}$, then either $I(x) = I(y)$ or $I(x) \cap I(y) = \emptyset$. Moreover, for any $x \in \mathcal{O}$, there is a rational y so that $I(x) = I(y)$, and so we see that $\mathcal{O} = \bigcup_{x \in \mathbb{Q} \cap \mathcal{O}} I(x)$. Hence we see that every open set is the countable union of disjoint open intervals. Since, $f^{-1}((a_k, b_k)) = \{\vec{x} \in D : a_k < f(\vec{x}) < b_k\}$ is measurable, we see $f^{-1}(\mathcal{O}) = \bigcup_k f^{-1}((a_k, b_k))$ is measurable.

Another useful observation to make is that f is measurable if and only if for any countable dense set $\mathcal{D} \subseteq \mathbb{R}$ and any $\alpha \in \mathcal{D}$ we have that $\{\vec{x} \in D : f(\vec{x}) > \alpha\}$ is measurable. Observe that for any $\beta \in \mathbb{R}$ we may find a sequence $\{\beta_k\}_{k=1}^{\infty}$ with $\beta_k \geq \beta$ and $\{\beta_k\}_{k=1}^{\infty}$ decreasing to β. Then $\{\vec{x} \in D : f(\vec{x}) > \beta\} = \bigcup_k \{\vec{x} \in D : f(\vec{x}) > \beta_k\}$. Hence on any of our initial statements about measurablility, we can restrict our assumptions on the parameters α, β, etc. to be elements of a countable dense set such as $\mathbb{Q}$.

If f is a measurable function on D, and Z is a subset of the domain of f of Lebesgue measure zero, then any property that holds on $D \setminus Z$ is said to hold **almost everywhere**, which is often abbreviated a.e. For instance, if $f = g$ a.e., then for any α we have

$$\mu(\{\vec{x} \in D : f(\vec{x}) > \alpha\}) = \mu(\{\vec{x} \in D : g(\vec{x}) > \alpha\}).$$

In fact, the set of functions $[f]$ which agree with a given function f a.e. form an equivalence class in the set of all measurable functions. Moreover, most properties we discuss for measurable functions f are statements that hold for each function in the equivalence class $[f]$.

9.2. LEBESGUE MEASURABLE FUNCTIONS

The following theorem allows us to create new measurable functions via composition of a given measurable function with a continuous function.

Theorem 9.2.1 *Let f be a real valued measurable function defined on a measurable set $D \subseteq \mathbb{R}^n$ which is finite a.e., and let g be a continuous function defined so that $g(f)$ is defined a.e. Then $g(f)$ is measurable on E.*

Proof: By otherwise replacing f with another element in $[f]$, we may assume f is finite on E. Since g is continuous on its domain, we know that by theorem 4.2.10 that for any open $\mathcal{O} \subseteq \mathbb{R}$ we have $g^{-1}(\mathcal{O})$ is open relative to the domain of g. That is, $g^{-1}(\mathcal{O}) = U \cap D$, where D is the domain of g and U is an open set in $\mathbb{R}$. However, the range of f is a subset of the domain of g and so $f^{-1}(U \cap D) = f^{-1}(U)$, so the result follows. So by measurability of f, we know that $f^{-1}(g^{-1}(\mathcal{O}))$ is measurable in $\mathbb{R}^n$.
□

From the above theorem we see that for a measurable f, for which the following are defined, we know that $\cos f$, $\sin f$, e^f, $|f|$, f^2, $\sqrt{f}$, $\frac{1}{f}$, etc. are all measurable. In fact, the functions $f^+(\vec{x}) := \max\{f(\vec{x}), 0\}$ and $f^-(\vec{x}) := -\min\{f(\vec{x}), 0\}$ are also measurable. We leave it as an exercise to show that this follows from theorem 9.2.1. To do this it is enough to show $x^+ := \max\{x, 0\}$ and $x^- := -\min\{x, 0\}$ are continuous. Since $f \pm g$ and cf, $c \in \mathbb{R}$, are measurable whenever f, g are measurable, and f^2 is measurable whenever f is measurable, we have that fg is also measurable. To see this, observe that $fg = \frac{1}{4}[(f+g)^2 - (f-g)^2]$.

9.2.1 Sequences of Measurable Functions

Point-wise Limit of a Sequence of Measurable Functions

We will now discuss how measurability is preserved under convergence for sequences of measurable functions.

Suppose $\{f_k\}_{k=1}^{\infty}$ is a sequence of measurable functions all defined on a measurable set $D \subseteq \mathbb{R}^n$. We will now show that $F(\vec{x}) := \sup_{k \in \mathbb{N}} f_k(\vec{x})$ and $f(\vec{x}) := \inf_{k \in \mathbb{N}} f_k(\vec{x})$ are measurable functions. To see this, we simply observe that

$$\{\vec{x} \in D : F(\vec{x}) > \alpha\} = \bigcup_{k=1}^{\infty} \{\vec{x} \in D : f_k(\vec{x}) > \alpha\}$$

and
$$\{\vec{x} \in D : f(\vec{x}) > \alpha\} = \bigcap_{k=1}^{\infty}\{\vec{x} \in D : f_k(\vec{x}) > \alpha\}$$
are measurable sets.

Theorem 9.2.2 *If $\{f_k\}_{k=1}^{\infty}$ is a sequence of measurable functions defined on a measurable set D, then $\limsup_{k\to\infty} f_k$ and $\liminf_{k\to\infty} f_k$ are measurable, and thus if $\lim_{k\to\infty} f_k$ exists, then it too is measurable.*

Proof: Observe that
$$\liminf_{k\to\infty} f_k = \sup_{j\in\mathbb{N}}\{\inf_{k\geq j} f_k\},$$
and
$$\limsup_{k\to\infty} f_k = \inf_{j\in\mathbb{N}}\{\sup_{k\geq j} f_k\}.$$
By our work above, we get the desired result.
□

It is a homework exercise to show that characteristic functions of measurable sets are measurable functions, and thus a simple function
$$f(\vec{x}) = \sum_{k=1}^{N} a_k \chi_{E_k}(\vec{x})$$
is measurable if each E_k is measurable. The following theorem allows us to realize any measurable function as the point-wise limit of a sequence of measurable simple functions. Moreover, by our proof, we have that for each $\vec{x}$, the sequence $\{f_k(\vec{x})\}_{k=1}^{\infty}$ is increasing if $f \geq 0$.

Theorem 9.2.3 *Suppose f is a measurable function defined on a measurable set D, then there exists a sequence of measurable simple functions $\{f_k\}_{k=1}^{\infty}$ so that $f_k \to f$ point-wise on D.*

Proof: We may assume that $f \geq 0$ in our proof, since otherwise we may prove the result for f^+ and f^- and use $f = f^+ - f^-$.

9.2. LEBESGUE MEASURABLE FUNCTIONS

Let
$$f_k = \sum_{i=1}^{k2^k} \frac{i-1}{2^k} \chi_{\{\vec{y}: \frac{i-1}{2^k} \leq f(\vec{y}) < \frac{i}{2^k}\}} + k\chi_{\{\vec{x}: f(\vec{x}) \geq k\}}.$$

Then $f_k \to f$ point-wise as $k \to \infty$.
□

Egorov's and Lusin's Theorems

We will now discuss a result which is often called Egorov's theorem, and it says that a sequence of measurable functions which converges point-wise a.e. also converges almost uniformly, as long as the set of convergence is finite in measure and the limit function is finite a.e.

Egorov's Theorem 9.2.4 *Suppose $\{f_k\}_{k=1}^{\infty}$ is a sequence of measurable functions defined on a set D, a set of finite Lebesgue measure, so that the point-wise limit $f := \lim_{k \to \infty} f_k$ exists and is finite a.e. on D. Then for each $\varepsilon > 0$ there is a closed subset $C_\varepsilon \subseteq D$, with $\mu(D \setminus C_\varepsilon) < \varepsilon$, so that $f_k \to f$ uniformly on C_ε.*

Proof: Let $F_{k,n} := \{\vec{x} : |f_n(\vec{x}) - f(\vec{x})| > \frac{1}{k}\}$, and $E_{k,m} := \bigcup_{n \geq m} F_{k,n}$. So we see that $\bigcap_{k,m} E_{k,m}$ is the set of all points for which $\{f_n(\vec{x})\}_{n=1}^{\infty}$ does not converge to $f(\vec{x})$, and thus $\mu(\bigcap_{k,m} E_{k,m}) = 0$. Moreover, $\lim_{n \to \infty} \mu(E_{k,n}) = 0$.

Let $\varepsilon > 0$ be given. For each $k \in \mathbb{N}$ there is a $m_k \in \mathbb{N}$ so that $\mu(E_{k,m_k}) < \frac{\varepsilon}{2^{k+1}}$. Define $F := D \setminus (\bigcup_k E_{k,m_k})$, and so $\mu(D \setminus F) \leq \sum_k \mu(E_{k,m_k}) < \frac{\varepsilon}{2}$.

For $\vec{x} \in F$ we have $\vec{x} \notin E_{k,m_k}$ for any $k \in \mathbb{N}$. So for any k, $x \notin F_{k,n}$ if $n > m_k$, i.e. if $n > m_k$, then $|f_n(\vec{x}) - f(\vec{x})| \leq \frac{1}{k}$. To complete the proof, we choose C_ε a closed subset of F, which satisfies $\mu(F \setminus C_\varepsilon) < \frac{\varepsilon}{2}$.
□

At this point we comment that almost uniform convergence which is guaranteed by Egorov's theorem is not equivalent to uniform convergence a.e. To see this we take $D = [0, 1]$ and define $f_n(x)$ to be 0 on $[\frac{1}{n}, 1]$ and to be $1 - nx$ on $[0, \frac{1}{n})$. Here $f_n \to 0$ a.e., $f_n \to 0$ almost uniformly, but $f_n \not\to 0$ uniformly a.e. However, note that almost uniform convergence does imply convergence a.e. because there is a set F_j with $\mu(D \setminus F_j) < \frac{1}{j}$ so that $f_k \to f$ uniformly on F_j. So $f_k \to f$ on $\bigcup_j F_j$ and $\mu(D \setminus (\bigcup_j F_j)) = 0$ since $D \setminus (\bigcup_j F_j) \subseteq D \setminus F_j$

for all j.

Our next result is often called Lusin's theorem, and it gives a characterization of measurability of functions as being functions which are continuous when restricted to a set of almost full measure.

Lusin's Theorem 9.2.5 *Let f be defined and finite on a measurable set D. Then f is measurable if and only if for all $\varepsilon > 0$ there is a closed $F_\varepsilon \subseteq D$ so that $\mu(D \setminus F_\varepsilon) < \varepsilon$ and $f|_{F_\varepsilon}$ is continuous.*

Proof: Suppose that f is measurable on D, and let's first suppose $\mu(D) < \infty$. Let f_k be defined by $f_k = \sum_{j=1}^{N_k} a_{j,k} E_{j,k}$ with $k = 1, 2, \cdots$ and $D = \bigcup_{j=1}^{N_k} E_{j,k}$, so that $\{f_k\}_{k=1}^\infty$ is a sequence of simple functions which converge to f point-wise.

Given $\varepsilon > 0$, by Egorov's theorem we may find a closed set $F_1 \subseteq D$ with $\mu(D \setminus F_1) < \frac{\varepsilon}{2}$ so that $f_k \to f$ uniformly on F_1. Also, let $F_{j,k} \subseteq E_{j,k}$ be closed sets so that $\mu(E_{j,k} \setminus F_{j,k}) < \frac{\varepsilon}{2^{j+k+1}}$. Then f_k is constant on $F_{j,k}$, and so f_k is continuous on $F_{j,k}$. Hence we must have that f_k is continuous on $\bigcup_j F_{j,k}$.

On $F_\varepsilon := F_1 \cap [\bigcap_k (\bigcup_{j=1}^{N_k} F_{j,k})]$ each f_k is continuous and $f_k \to f$ uniformly, and so $f|_{F_\varepsilon}$ is continuous. Moreover,

$$\mu(D \setminus F_\varepsilon) \leq \mu(D \setminus F_1) + \sum_k \mu(D \setminus (\bigcup_{j=1}^{N_k} F_{j,k}))$$

$$\leq \frac{\varepsilon}{2} + \sum_{j,k} \frac{\varepsilon}{2^{1+j+k}} = \varepsilon.$$

If $\mu(D) = \infty$, we write $D = \bigcup_k D_k$, where $D_k := D \cap \{\vec{x} : k-1 \leq \|\vec{x}\| \leq k\}$. Find $F_k \subseteq D_k$ closed and so that $\mu(D_k \setminus F_k) < \frac{\varepsilon}{2^k}$, with $f|_{F_k}$ continuous. So $F_\varepsilon := \bigcup_k F_k$ satisfies $f|_{F_\varepsilon}$ is continuous and $\mu(D \setminus F_\varepsilon) < \varepsilon$.
indent
Conversely, for each $k \in \mathbb{N}$ find a closed set F_k with $\mu(D \setminus F_k) < \frac{1}{k}$ and so that $f|_{F_k}$ is continuous. Let $E_k := \bigcup_{j \leq k} F_j$. Then $f|_{E_k}$ is continuous and $\mu(D \setminus (\bigcup_k E_k)) = 0$. We define $f_k(\vec{x}) = f(\vec{x})$ if $\vec{x} \in E_k$ and $f(\vec{x}) = 0$ otherwise. Therefore f_k is measurable and $f_k \to f$ a.e. Therefore, f is measurable.
□

9.2. LEBESGUE MEASURABLE FUNCTIONS

Convergence in Measure

Let $\{f_k\}_{k=1}^\infty$ be a sequence of measurable functions each defined on a measurable set D and each is finite a.e. We say that $\{f_k\}_{k=1}^\infty$ **converges in measure** to a function f, often denoted by

$$f_k \to_m f \quad k \to \infty,$$

if for each $\varepsilon > 0$,

$$\lim_{k \to \infty} \mu(\{\vec{x} \in D : |f_k(\vec{x}) - f(\vec{x})| > \varepsilon\}) = 0.$$

Theorem 9.2.6 *Suppose $\{f_k\}_{k=1}^\infty$ are measurable and finite a.e. in a measurable set D of finite measure. If $f_k \to f$ a.e., then $f_k \to_m f$.*

Proof: Given $\varepsilon > 0$ we let $E_k := \{\vec{x} : |f_k(\vec{x}) - f(\vec{x})| \leq \varepsilon\}$. For $F_j := \bigcap_{k \geq j} E_k$ observe that $\vec{x} \in F_j$ if and only if $|f_k(\vec{x}) - f(\vec{x})| \leq \varepsilon$ for $k \geq j$. If $f_k(\vec{x}) \to f(\vec{x})$, then $\vec{x} \in F_j$ for some j. So $\bigcup_j F_j = \{\vec{x} : f_k(\vec{x}) \to f(\vec{x})\} = D \setminus Z$, where $\mu(Z) = 0$. So $F_j \to \bigcup_k F_k$ and hence $\mu(F_j) \to \mu(\bigcup_j F_j) = \mu(D)$. Since $\mu(D) < \infty$, we have $\mu(D \setminus F_j) \to 0$ as $j \to \infty$. However, $\{\vec{x} : |f_k(\vec{x}) - f(\vec{x})| > \varepsilon\} = D \setminus E_k \subseteq D \setminus F_k$, and so $\mu(\{\vec{x} : |f_k(\vec{x}) - f(\vec{x})| > \varepsilon\}) \to 0$ as $k \to \infty$.
□

Note that the converse of theorem 9.2.6 does not hold. To see this we consider $D = [0,1]$ and $f_1 = \chi_D$, $f_2 = \chi_{[0,\frac{1}{2}]}$, $f_3 = \chi_{[\frac{1}{2},1]}$, $f_4 = \chi_{[0,\frac{1}{4}]}$, $f_5 = \chi_{[\frac{1}{4},\frac{1}{2}]}$, $f_6 = \chi_{[\frac{1}{2},\frac{3}{4}]}$, $f_7 = \chi_{[\frac{3}{4},1]}$, $f_8 = \chi_{[0,\frac{1}{8}]}$, $f_9 = \chi_{[\frac{1}{8},\frac{1}{4}]}$, $f_{10} = \chi_{[\frac{1}{4},\frac{3}{8}]}$, etc. We have $f_j \to_m 0$ but for no $x \in D$ does $f_j(x) \to 0$. That is, for $n \in \{0\} \cup \mathbb{N}$ and $k \in \{0, 1, 2, \cdots 2^n - 1\}$ let $E_{k+2^n} = [\frac{k}{2^n}, \frac{k+1}{2^n}]$. Let $f_j = \chi_{E_j}$. Now $f_j \to_m 0$ since $\mu(\{x : |f_{k+2^n}(x) - 0| \geq \varepsilon\}) = \frac{1}{2^n}$ for $0 < \varepsilon < 1$. But $\{f_j\}$ does not converge pointwise to 0, or even pointwise a.e. to 0. In fact, each $x \in [0,1]$ is a limit of a sequence of dyadic rational numbers[6], and so we see that $\limsup_{j \to \infty} f_j(x) = 1$.

Also, if $\mu(D) = \infty$ in theorem 9.2.6, then the conclusion of theorem 9.2.6 may not hold. To see this we take $D = \mathbb{R}^n$ and $f_k = \chi_{\{\vec{x}: \|\vec{x}\| < k\}}$, $f = \chi_{\mathbb{R}^n}$. Another example along these lines is the sequence $f_n = \chi_{[n,n+1]}$ which

---
[6]Rational Numbers of the form $\frac{k}{2^n}$

converges to 0 pointwise on $[0, \infty)$, where as $\mu(\{x : |f_n(x) - 0| \geq \varepsilon\}) = 1$ for any $0 < \varepsilon < 1$. Still another example is $g_n(x) = \frac{x}{n}$, which converges to 0 pointwise on $[0, \infty)$, but $\mu(\{x : |g_n(x) - 0| \geq \varepsilon\}) = \infty$ for any $\varepsilon > 0$.

The following theorem tells us a partial converse of theorem 9.2.6.

Theorem 9.2.7 *Suppose $\{f_n\}_{n=1}^{\infty}$ is a sequence of measurable functions defined on a measurable set D, so that each f_k is finite a.e. Suppose $f_n \to_m f$ on D, then there is a subsequence $\{f_{n_k}\}_{k=1}^{\infty}$ so that $f_{n_k} \to f$ a.e. as $k \to \infty$.*

Proof: Let $F_{k,n} := \{\vec{x} : |f_n(\vec{x}) - f(\vec{x})| > \frac{1}{k}\}$. Convergence in measure guarantees that $\lim_{n \to \infty} \mu(F_{n,k}) = 0$. So for each $k \in \mathbb{N}$ we choose n_k so that $\mu(F_{k,n_k}) < \frac{1}{2^k}$. Let $G_j := \bigcup_{k \geq j} F_{k,n_k}$, and let $Z = \bigcup_{j=1}^{\infty} G_j$, which is actually $\limsup_{k \to \infty} F_{k,n_k}$. Then we have $\mu(G_j) \leq \sum_{k=j}^{\infty} \mu(F_{k,n_k}) = \frac{1}{2^{j+1}}$. So $G_j \searrow Z$ and $\mu(G_j) \searrow 0$ as $j \to \infty$, and thus $\mu(Z) = 0 = \lim_{j \to \infty} \mu(G_j)$.

Take $\vec{x} \notin Z$, then $\vec{x} \notin G_j$ for some j, and so $\vec{x} \notin F_{k,n_k}$ for $k \geq j$, i.e. for $k \geq j$ we have $|f_{n_k}(\vec{x}) - f(\vec{x})| < \frac{1}{k}$.

So given $\varepsilon > 0$ we choose j so that $\frac{1}{j} \leq \varepsilon$. Then for $k \geq j$ we have $|f_{n_k}(\vec{x}) - f(\vec{x})| \leq \frac{1}{k} \leq \frac{1}{j} \leq \varepsilon$, and so we see that $\{f_{n_k}\}_{k=1}^{\infty}$ is a subsequence so that $f_{n_k} \to f$ on $D \setminus Z$.
□

9.2.2 Homework Exercises

1. Suppose that f is a real valued measurable function and $c \in \mathbb{R}$, then cf and $f + c$ are measurable functions.

2. Suppose that $f = \sum_{k=1}^{n} c_k \chi_{E_k}$ is a simple function. Show that f is measurable if and only if each E_k is measurable.

3. Suppose that f and g are measurable functions defined on a measurable set D. Show that $M(\vec{x}) := \max\{f(\vec{x}), g(\vec{x})\}$ and $m(\vec{x}) = \min\{f(\vec{x}), g(\vec{x})\}$ are measurable functions.

4. Suppose that f is upper or lower semicontinuous on a measurable set D, then show that f is measurable. To do this first, show that f is usc (lsc) if and only if $\{\vec{x} \in D : f(\vec{x}) \geq (\leq)\alpha\}$ is relatively closed in D for each $\alpha \in \mathbb{R}$.

9.2. LEBESGUE MEASURABLE FUNCTIONS

5. Suppose $\{f_k\}_{k=1}^\infty$ is a sequence of measurable functions. Show that this sequence converges in measure to a function f on a set D if and only if for all $\varepsilon > 0$, there is an $N \in \mathbb{N}$ so that if $n > N$ we have
$$\mu(\{\vec{x} \in D : |f_n(\vec{x}) - f(\vec{x})| > \varepsilon\}) < \varepsilon.$$

6. In the above exercise, give an analogous Cauchy Convergence Criterion for convergence in measure of the sequence $\{f_k\}_{k=1}^\infty$. Show that if $f_k \to_m f$, then $\{f_k\}_{k=1}^\infty$ is Cauchy in measure. Also, if $\{f_n\}_{n=1}^\infty$ is Cauchy in measure, then $f_n \to_m f$.

7. Suppose that $\{f_k\}_{k=1}^\infty$ and $\{g_k\}_{k=1}^\infty$ are sequences of measurable functions that are defined on a measurable set D. Suppose there exists measurable functions f, g defined on D so that $f_k \to f$ in measure and $g_k \to g$ in measure on D, as $k \to \infty$. Show that $f_k + g_k \to f + g$ in measure on D.

8. Suppose f is a measurable function on a measurable set D, with $\mu(D) < \infty$. Suppose f is finite a.e. Show that for any $\varepsilon > 0$ there is an a so that $\mu(\{\vec{x} : |f(\vec{x})| > a\}) < \varepsilon$.

9. Suppose that $\{f_k\}_{k=1}^\infty$ and $\{g_k\}_{k=1}^\infty$ are sequences of measurable functions defined on a measurable set D satisfying $\mu(D) < \infty$. Suppose there exists measurable functions f, g defined on D so that $f_k \to_m f$ and $g_k \to_m g$ on D, as $k \to \infty$. Suppose as well that f, g are finite a.e. Show that $f_k \cdot g_k \to_m f \cdot g$ on D.

10. Suppose $\{f_k\}_{k=1}^\infty$ is a sequence of measurable functions, prove that the set of points $\vec{x}$ for which $\{f_k(\vec{x})\}_{k=1}^\infty$ converges is measurable.

11. Prove that $f_n \to_m f$ if and only if $(f_m - f) \to_m 0$.

12. Give an example of $f_n \to_m f$ on D and $g_n \to_m g$ on D, with $\mu(D) = \infty$ for which $f_n g_n \to_m fg$ on D fails to hold.

13. If $f_n \to_m f$, then does $|f_n| \to_m |f|$ necessarily hold? If so, prove it.

14. If $f_n \to_m f$, show that every subsequence has $f_{n_k} \to_m f$ as well.

9.3 The Riemann and Lebesgue Integrals

9.3.1 The Riemann Integral in $\mathbb{R}^n$

In this subsection we will discuss the Riemann integral for functions whose domains are intervals in $\mathbb{R}^n$.

Let $I = [a_1, b_1] \times [a_2.b_2] \times \cdots \times [a_n, b_n]$. A **partition** of I is a collection of non-overlapping intervals $\{[x_{1,i}, y_{1,i}] \times [x_{2,i}, y_{2,i}] \times \cdots \times [x_{n,i}, y_{n,i}] : i = 1, 2, \cdots, N\} = \{I_k : k = 1, 2, \cdots, N\}$ so that $\bigcup_{i=1}^{N}[x_{1,i}, y_{1,i}] \times [x_{2,i}, y_{2,i}] \times \cdots \times [x_{n,i}, y_{n,i}] = \bigcup_{k=1}^{N} I_k = I$. A partition of I is commonly denoted by $\mathcal{P} = \{I_k\}_{k=1}^{N}$.

Recall that for a given set S, the diameter of S, often denoted by diam S, is given by diam $S = \sup\{\|\vec{x} - \vec{y}\| : \vec{x}, \vec{y} \in S\}$. For a given partition $\mathcal{P} = \{I_k\}_{k=1}^{N}$ of I, we define the **norm** or **mesh** of $\mathcal{P}$ by

$$\|\mathcal{P}\| := \max\{\text{diam } I_k : k = 1, 2, \cdots, N\}. \tag{9.5}$$

Also, for a partition $\mathcal{P}$ we often use the notation

$$\Delta I_k := V(I_k), \quad i = 1, 2, \cdots, N,$$

where $V(I_k)$ is the volume of I_k as defined in the beginning of this chapter.

For a given partition $\mathcal{P} = \{I_k\}_{k=1}^{N}$ of I and a collection of points $\{\vec{t}_k\}_{k=1}^{N}$, with $\vec{t}_k \in I_k$, $k = 1, 2, \cdots, N$, the quantity

$$\sum_{k=1}^{N} f(\vec{t}_k)\Delta I_k, \tag{9.6}$$

is often called a **Riemann sum** for f on I. The partition $\mathcal{P}$ along with a collection of points $\{\vec{t}_k\}_{k=1}^{N}$ is often called a **marked partition** and shall be denoted by $\mathcal{P}^\dagger$. For convenience we shall use $S(f, I, \mathcal{P}^\dagger)$ to represent $\sum_{k=1}^{N} f(\vec{t}_k)\Delta I_k$.

If there exists a quantity $\int_I f(\vec{x})dV$ so that for any $\varepsilon > 0$, there is a $\delta > 0$ so that if $\mathcal{P}^\dagger$ is a marked partition of I with $\|\mathcal{P}\| < \delta$, then

$$\left|S(f, I, \mathcal{P}^\dagger) - \int_I f(\vec{x})dV\right| < \varepsilon, \tag{9.7}$$

9.3. THE RIEMANN AND LEBESGUE INTEGRALS

we say that f is **Riemann integrable** on I. The quantity $\int_I f(\vec{x})dV$ is called the **Riemann integral** of f on I, or simply the integral of f on I.[7]

We often think of $\int_I f(\vec{x})dV$ as a limit of Riemann sums, and we will often use the notation

$$\int_I f(\vec{x})dV = \lim_{\|\mathcal{P}\| \to 0} S(f, I, \mathcal{P}^\dagger). \tag{9.8}$$

If $f : I \to [0, \infty)$, then we may see $S(f, I, \mathcal{P}^\dagger)$ as an approximation for the volume of the region in $\mathbb{R}^{n+1}$ given by $\{(\vec{x}, x_{n+1}) : 0 \leq x_{n+1} \leq f(\vec{x})\}$, if such a volume exists. If $\int_I f(\vec{x})dV$ exists, then the volume of this region is $\int_I f(\vec{x})dV$.

Theorem 9.3.1 *If there is a a quantity that serves as $\int_I f(\vec{x})dV$, then it is unique.*

Proof: Suppose there are two quantities S and L that serve as limits of the Riemann sums. Then given any $\varepsilon > 0$ there is a $\delta_1 > 0$ and a $\delta_2 > 0$ so that for any marked partition $\mathcal{P}^\dagger$ that satisfies $\|\mathcal{P}\| < \delta_1$ we have

$$|L - S(f, I, \mathcal{P}^\dagger)| < \frac{\varepsilon}{2},$$

and for $\|\mathcal{P}\| < \delta_2$ we have

$$|S - S(f, I, \mathcal{P}^\dagger)| < \frac{\varepsilon}{2}.$$

Moreover, for $\|\mathcal{P}\|$ less than the minimum of δ_1 and δ_2, we have

$$|S - L| = |S - S(f, I, \mathcal{P}^\dagger) + S(f, I, \mathcal{P}^\dagger) - L|$$

$$\leq |S - S(f, I, \mathcal{P}^\dagger)| + |S(f, I, \mathcal{P}^\dagger) - L| < \frac{\varepsilon}{2} + \frac{\varepsilon}{2} = \varepsilon.$$

Since $\varepsilon > 0$ was arbitrary, we must have $S = L$.
□

---
[7]Sometimes we may use $\int_I f dV$ or even $\int_I f$ to represent $\int_I f(\vec{x})dV$.

Linearity and Comparative Properties of the Integral

We now introduce a notational convention for Riemann integrability. If f is Riemann integrable on I, we shall use the notation $f \in \mathcal{R}(I)$.

We will now discuss the linearity properties of the Riemann integral, and they are given in the theorem below, along with a comparison result.

Theorem 9.3.2 *If $f, g \in \mathcal{R}(I)$, and $k \in \mathbb{R}$, then $f + g, kf \in \mathcal{R}(I)$. Moreover,*

$$\int_I f(\vec{x}) + g(\vec{x}) dV = \int_I f(\vec{x}) dV + \int_I g(\vec{x}) dV,$$

$$\int_I kf(\vec{x}) dV = k \int_I f(\vec{x}) dV.$$

Also, if $f(\vec{x}) \leq g(\vec{x})$ on I, then

$$\int_I f(\vec{x}) dV \leq \int_I g(\vec{x}) dV.$$

Proof: Let $\varepsilon > 0$ be given. Then there exists a $\delta > 0$ so that for $\|\mathcal{P}\| < \delta$ and any $\{\vec{t}_k\}_{k=1}^N$, $|\sum_{k=1}^N f(\vec{t}_k) \Delta I_k - \int_I f(\vec{x}) dV| < \frac{\varepsilon}{2}$ and $|\sum_{k=1}^N g(\vec{t}_k) \Delta I_k - \int_I g(\vec{x}) dV| < \frac{\varepsilon}{2}$. Since $S(f+g, I, \mathcal{P}^\dagger) = \sum_{k=1}^N (f(\vec{t}_k) + g(\vec{t}_k)) \Delta I_k = \sum_{k=1}^N f(\vec{t}_k) \Delta I_k + \sum_{k=1}^N g(\vec{t}_k) \Delta I_k$, we have

$$|\sum_{k=1}^N (f(\vec{t}_k) + g(\vec{t}_k)) \Delta I_k - \int_I f(\vec{x}) dA - \int_I g(\vec{x}) dA|$$

$$\leq |\sum_{k=1}^N f(\vec{t}_k) \Delta I_k - \int_I f(\vec{x}) dA| + |\sum_{k=1}^N g(\vec{t}_k) \Delta I_k - \int_I g(\vec{x}) dA|$$

$$< \frac{\varepsilon}{2} + \frac{\varepsilon}{2} = \varepsilon.$$

If $k = 0$, then $0 \cdot f(\vec{x}) \equiv 0$, and $0 \cdot \int_I f(\vec{x}) dA = 0$. We leave it an an exercise to show that for any constant c, and any bounded interval I we have $\int_I c\, dV = c \cdot V(I)$. So if $k = 0$, we are done. Thus we consider $k \neq 0$. Choosing

9.3. THE RIEMANN AND LEBESGUE INTEGRALS

$\delta > 0$ so that for $\|\mathcal{P}\| < \delta$ and taking any collection $\{\vec{t}_j\}_{j=1}^N$, $\vec{t}_j \in I_j$, we have $|\sum_{j=1}^N f(\vec{t}_j)\Delta I_j - \int_I f(\vec{x})dV| < \frac{\varepsilon}{|k|}$. Thus we have

$$|\sum_{j=1}^N kf(\vec{t}_j)\Delta I_j - k\int_I f(\vec{x})dV| = |k||\sum_{j=1}^N f(\vec{t}_j)\Delta I_j - \int_I f(\vec{x})dV| < |k| \cdot \frac{\varepsilon}{|k|} = \varepsilon.$$

Assume $f \leq g$ on I. For any partition $\mathcal{P}$ with $\|\mathcal{P}\| < \delta$, and for any $\{\vec{t}_k\}_{k=1}^N$ which gives rise to the marked partition $\mathcal{P}^\dagger$, we have

$$|\sum_{k=1}^N f(\vec{t}_k)\Delta I_k - \int_I f(\vec{x})dV| < \frac{\varepsilon}{2}, \quad |\sum_{k=1}^N g(\vec{t}_k)\Delta I_k - \int_I g(\vec{x})dV| < \frac{\varepsilon}{2}.$$

Clearly $f(\vec{t}_k) \leq g(\vec{t}_k)$, and so $\sum_{k=1}^N f(\vec{t}_k)\Delta I_k \leq \sum_{k=1}^N g(\vec{t}_k)\Delta I_k$. Moreover,

$$\int_I f(\vec{x})dV - \frac{\varepsilon}{2} < \sum_{k=1}^N f(\vec{t}_k)\Delta I_k \leq \sum_{k=1}^N g(\vec{t}_k)\Delta I_k < \int_I g(\vec{x})dV + \frac{\varepsilon}{2}.$$

Thus
$$\int_I f(\vec{x})dV < \int_I g(\vec{x})dV + \varepsilon.$$

Since $\varepsilon > 0$ is arbitrary, we have

$$\int_I f(\vec{x})dV \leq \int_I g(\vec{x})dV.$$

□

Corollary 9.3.3 *If $f \geq 0$ on I, and $f \in \mathcal{R}(I)$, then*

$$\int_I f(\vec{x})dV \geq 0.$$

Corollary 9.3.4 *If f is bounded on I,[8] and $f \in \mathcal{R}(I)$, then*

$$\inf_{x \in I} f(\vec{x}) \cdot V(I) \leq \int_I f(\vec{x})dV \leq \sup_{x \in I} f(\vec{x}) \cdot V(I).$$

Corollary 9.3.5 *Suppose $f, |f| \in \mathcal{R}(I)$, then*

$$|\int_I f(\vec{x})dV| \leq \int_I |f(\vec{x})|dV.$$

---
[8]We shall later show that $f \in \mathcal{R}(I)$ implies f is bounded.

A Cauchy Convergence Criterion for Integrability

As in the case of limits of sequences, we have a **Cauchy criterion** for convergence of Riemann sums to the Riemann integral.

Theorem (Cauchy Criterion for Integrability in $\mathbb{R}^n$) 9.3.6 $f \in \mathcal{R}(I)$ if and only if for any $\varepsilon > 0$, there is a $\delta > 0$ so that if $\mathcal{P} = \{I_j\}_{j=1}^{M}$ and $\mathcal{Q} = \{J_k\}_{k=1}^{M}$ are two partitions with $\|\mathcal{P}\|, \|\mathcal{Q}\| < \delta$ and $\{\vec{t}_k\}_{k=1}^{N}$ and $\{\vec{\tau}_j\}_{j=1}^{M}$ are any points with $\vec{t}_k \in I_k$ for $k = 1, 2, \cdots, N$ and $\vec{\tau}_k \in J_k$, for $k = 1, 2, \cdots, M$, which give rise to the marked partitions $\mathcal{P}^\dagger$ and $\mathcal{Q}^\dagger$ respectively, then we have

$$|S(f, I, \mathcal{P}^\dagger) - S(f, I, \mathcal{Q}^\dagger)| < \varepsilon. \tag{9.9}$$

Proof: Suppose $f \in \mathcal{R}(I)$. Then for any $\varepsilon > 0$ there is a $\delta > 0$ so that if $\mathcal{P}$ is any partition satisfying $\|\mathcal{P}\| < \delta$, with $\{\vec{t}_k\}_{k=1}^{N}$ satisfying $\vec{t}_k \in I_k$, which gives rise to the marked partition $\mathcal{P}^\dagger$, we have

$$|S(f, I, \mathcal{P}^\dagger) - \int_I f(\vec{x}) dV| < \frac{\varepsilon}{2}.$$

Thus for any two partitions $\mathcal{P} = \{I_k\}_{k=1}^{N}$, $\mathcal{Q} = \{J_k\}_{k=1}^{M}$ whose mesh is less than δ, and any corresponding collections of points $\{\vec{t}_k\}_{k=1}^{N}$ and $\{\vec{\tau}_k\}_{k=1}^{M}$ which give rise to the marked partitions $\mathcal{P}^\dagger$ and $\mathcal{Q}^\dagger$ respectively, then we have

$$|\sum_{k=1}^{N} f(\vec{t}_k) \Delta I_k - \sum_{i=1}^{M} f(\vec{\tau}_i) \Delta J_i|$$

$$\leq |\sum_{k=1}^{N} f(\vec{t}_k) \Delta I_k - \int_I f(\vec{x}) dV| + |\int_I f(\vec{x}) dV - \sum_{i=1}^{M} f(\vec{\tau}_i) \Delta J_i|$$

$$\leq \frac{\varepsilon}{2} + \frac{\varepsilon}{2} = \varepsilon.$$

Conversely, suppose f satisfies: for any $\varepsilon > 0$, there is a $\delta > 0$ so that if $\mathcal{P} = \{I_k\}_{k=i}^{n}$ and $\mathcal{Q} = \{J_i\}_{i=1}^{M}$ are two partitions with $\|\mathcal{P}\|, \|\mathcal{Q}\| < \delta$, and letting $\{\vec{t}_k\}_{k=1}^{N}$ and $\{\vec{\tau}_j\}_{j=1}^{M}$ be any collections of points with $\vec{t}_k \in I_k$ for $k = 1, 2, \cdots, N$ and $\vec{\tau}_i \in J_i$, for $i = 1, 2, \cdots, M$, which give rise to the marked partitions $\mathcal{P}^\dagger$ and $\mathcal{Q}^\dagger$, then we have

$$|S(f, I, \mathcal{P}^\dagger) - S(f, I, \mathcal{Q}^\dagger)| < \varepsilon.$$

9.3. THE RIEMANN AND LEBESGUE INTEGRALS

For each natural number K, there is a $\delta_K > 0$ so that for any two marked partitions $\mathcal{P}^\dagger$ and $\mathcal{Q}^\dagger$ whose mesh is less that δ_K, we have

$$|S(f, I, \mathcal{P}^\dagger) - S(f, I, \mathcal{Q}^\dagger)| < \frac{1}{K}.$$

Moreover, we may assume that $\{\delta_K\}_{K=1}^\infty$ is a decreasing sequence, since otherwise we may use $\delta_K' = \min\{\delta_1, \delta_2, \cdots, \delta_K\}$ instead. For each K we fix a marked partition $\mathcal{P}_K^\dagger$, and for $L > K$ we have $\|\mathcal{P}_K\|, \|\mathcal{P}_L\| < \delta_K$. So

$$|S(f, I, \mathcal{P}_K^\dagger) - S(f, I, \mathcal{P}_L^\dagger)| < \frac{1}{K}.$$

Thus we see that

$$\{S(f, I, \mathcal{P}_k^\dagger)\}_{k=1}^\infty$$

is a Cauchy sequence of real numbers, and hence it has a limit value $\mathcal{I}$. Moreover, for any natural number k, and $r \geq k$ we have

$$\||S(f, I, \mathcal{P}_k^\dagger) - \mathcal{I}| - |\mathcal{I} - S(f, I, \mathcal{P}_r^\dagger)|\| \leq |S(f, I, \mathcal{P}_k^\dagger) - S(f, I, \mathcal{P}_r^\dagger)| \leq \frac{1}{k}.$$

By then letting $r \to \infty$ we get

$$|S(f, I, \mathcal{P}_k^\dagger) - \mathcal{I}| \leq \frac{1}{k}.$$

Let $s \in \mathbb{N}$ satisfy $s > \frac{2}{\varepsilon}$, where $\varepsilon > 0$ was any preassigned number. For any marked partition $\mathcal{Q}^\dagger$ satisfying $\|\mathcal{Q}\| < \delta_s$, we have

$$|S(f, I, \mathcal{Q}^\dagger) - \mathcal{I}|$$

$$\leq |S(f, I, \mathcal{Q}^\dagger) - S(f, I, \mathcal{P}_s^\dagger)|$$

$$+ |S(f, I, \mathcal{P}_s^\dagger) - \mathcal{I}|$$

$$\leq \frac{1}{s} + \frac{1}{s} < \varepsilon.$$

Thus we see that $\mathcal{I} = \int_I f(\vec{x}) dV$.
□

Integrability Implies Boundedness

Theorem 9.3.7 *If $f \in \mathcal{R}(I)$, then f is bounded on I.*

Proof: Suppose there is an $f \in \mathcal{R}(I)$ which is not bounded on I and let $\mathcal{I} = \int_I f(\vec{x})dV$. Then there is a $\delta > 0$ so that for any marked partition $\mathcal{P}^\dagger$ satisfying $\|\mathcal{P}\| < \delta$ we have $|S(f, I, \mathcal{P}^\dagger) - \mathcal{I}| < 1$, and thus

$$|S(f, I, \mathcal{P}^\dagger)| < |\mathcal{I}| + 1.$$

Let $\mathcal{P}$ be any partition of norm less than δ. Since f is unbounded on I there is a subinterval of I defined by the partition $\mathcal{P}$ for which f is unbounded on this subinterval. Let I_j be such a subinterval. Thus for any $M > 0$ there exists $\vec{x} \in I_j$ so that $|f(\vec{x})| > M$.

We now pick $\vec{t}_k$'s to create a marked partition $\mathcal{P}^\dagger$. Let $\vec{t}_k$, for $k \neq j$, be any points in the corresponding subintervals I_k defined by the partition $\mathcal{P}$. Let $\vec{t}_j \in I_j$ be chosen so that

$$|f(\vec{t}_j)| > \frac{1 + |\mathcal{I}| + |\sum_{k \neq j} f(\vec{t}_k) \Delta I_k|}{V(I_j)}.$$

Thus we see that

$$|S(f, I, \mathcal{P}^\dagger)| = |\sum_k f(\vec{t}_k)\Delta I_k| \geq |f(\vec{t}_j)V(I_j)| - |\sum_{k \neq j} f(\vec{t}_k)\Delta I_k| > |\mathcal{I}| + 1,$$

which is a contradiction.
□

Upper and Lower Sums and the Darboux Integral

Another approach to defining the Riemann integral is through what are commonly called **upper** and **lower sums**. For f defined on I and a partition $\mathcal{P} = \{I_k\}_{k=1}^N$ of I, the upper sum is defined by

$$U(f, I, \mathcal{P}) = \sum_{k=1}^N \sup_{\vec{x} \in I_k} f(\vec{x}) \cdot \Delta I_k, \qquad (9.10)$$

9.3. THE RIEMANN AND LEBESGUE INTEGRALS

and the lower sum is defined by

$$L(f, I, \mathcal{P}) = \sum_{k=1}^{N} \inf_{\vec{x} \in I_k} f(\vec{x}) \cdot \Delta I_k. \tag{9.11}$$

If $\inf_{\mathcal{P}} U(f, I, \mathcal{P}) = \sup_{\mathcal{P}} L(f, I, \mathcal{P})$, with the inf and sup taken over all partitions $\mathcal{P}$ of I, we define $(D) \int_I f(\vec{x}) dV$ to be this quantity, and this is often called the **Darboux integral**.

For a given partition $\mathcal{P} = \{I_j\}_{j=1}^{N}$ of I we say that a partition $\mathcal{Q} = \{J_k\}_{k=1}^{M}$ of I is a **refinement** of $\mathcal{P}$ if each J_k is a subset of some I_j. If we have two different partitions $\mathcal{P}_1$ and $\mathcal{P}_2$, we may create a **common refinement** of both partitions. We leave it an an exercise to construct a common refinement of two given partitions.

Theorem 9.3.8 *Let $\mathcal{P}$ be a partition of I, $f : I \to \mathbb{R}$, and $\mathcal{Q}$ a refinement of $\mathcal{P}$. Then*

$$L(f, I, \mathcal{P}) \leq L(f, I, \mathcal{Q}),$$

and

$$U(f, I, \mathcal{Q}) \leq U(f, I, \mathcal{P}).$$

Proof: Let $\mathcal{P} = \{I_j\}_{j=1}^{N}$. We will first prove this in the case where

$$\mathcal{Q} = \{I_1, I_2, \cdots, I_{k-1}, I_{k+1}, \cdots, I_N\} \cup \{I_{k,1}, I_{k,2}, \cdots, I_{k,M}\}$$

where $\{I_{k,1}, I_{k,2}, \cdots, I_{k,M}\}$ is a partition of I_k.

Let $m_j = \inf_{\vec{x} \in I_j} f(\vec{x})$, $M_j = \sup_{\vec{x} \in I_j} f(\vec{x})$, for $j = 1, 2, \cdots, N$. Let $m_{k,j} = \inf_{\vec{x} \in I_{k,j}} f(\vec{x})$, and $M_{k,j} = \sup_{\vec{x} \in I_{k,j}} f(\vec{x})$, for $j = 1, 2, 3, \cdots, M$. Then clearly $m_k \leq \min\{m_{k,j} : j = 1, 2, \cdots, M\}$ and $M_k \geq \max\{M_{k,i} : i = 1, 2, \cdots, M\}$.

Hence we see

$$L(f, I, \mathcal{Q}) - L(f, I, \mathcal{P}) = \sum_{i=1}^{M} (m_{k,i} - m_k) V(I_{k,i}) \geq 0$$

and

$$U(f, [a, b], \mathcal{P}) - U(f, [a, b], \mathcal{Q}) = \sum_{i=1}^{M} (M_k - M_{k,i}) V(I_{k,i}) \geq 0.$$

In the general case, for $\mathcal{Q}$ any refinement of $\mathcal{P}$, we simply repeat this type of reasoning over and over to get the desired result.
□

Theorem 9.3.9 *Let $f : I \to \mathbb{R}$, then*

$$\inf_{\mathcal{P}} U(f, I, \mathcal{P}) \geq \sup_{\mathcal{P}} L(f, I, \mathcal{P})$$

where the inf *and* sup *are taken over all partitions of I.*

Proof: For any partitions $\mathcal{P}_1$ and $\mathcal{P}_2$ of I, for any common refinement $\mathcal{P}$ of $\mathcal{P}_1$ and $\mathcal{P}_2$ we have

$$L(f, I, \mathcal{P}_1) \leq L(f, I, \mathcal{P}) \leq U(f, I, \mathcal{P}) \leq U(f, I, \mathcal{P}_2),$$

and so

$$L(f, I, \mathcal{P}_1) \leq U(f, I, \mathcal{P}_2)$$

for any two partitions of I. Keeping $\mathcal{P}_2$ fixed, upon taking the least upper bound of the quantities $L(f, I, \mathcal{P}_1)$ over all possible partitions $\mathcal{P}_1$ of I we get

$$\sup_{\mathcal{Q}} L(f, I, \mathcal{Q}) \leq U(f, I, \mathcal{P}_2).$$

Next, we take the greatest lower bound of the quantities $U(f, I, \mathcal{P}_2)$ over all possible partitions $\mathcal{P}_2$ of I to get

$$\sup_{\mathcal{P}} L(f, I, \mathcal{P}) \leq \inf_{\mathcal{P}} U(f, I, \mathcal{P}).$$

□

Theorem 9.3.10 $(D) \int_I f dV$ *exists if and only if for any $\varepsilon > 0$ there is a partition $\mathcal{P}$ so that*

$$U(f, I, \mathcal{P}) - L(f, I, \mathcal{P}) < \varepsilon.$$

Proof: Suppose $(D) \int_I f(\vec{x}) dV$ exists and let $\varepsilon > 0$ be given. Then there partitions $\mathcal{P}_1$ and $\mathcal{P}_2$ of I so that

$$0 \leq (D) \int_I f(\vec{x}) dV - L(f, I, \mathcal{P}_1) < \frac{\varepsilon}{2},$$

9.3. THE RIEMANN AND LEBESGUE INTEGRALS

$$0 \leq U(f, I, \mathcal{P}_2) - (D)\int_I f(\vec{x})dV < \frac{\varepsilon}{2}.$$

Let $\mathcal{P}$ be a common refinement of $\mathcal{P}_1$ and $\mathcal{P}_2$. By theorem (9.3.8) we have

$$U(f, I, \mathcal{P}) \leq U(f, I, \mathcal{P}_2) \leq (D)\int_I f(\vec{x})dV + \frac{\varepsilon}{2}$$

$$< L(f, I, \mathcal{P}_1) + \varepsilon \leq L(f, I, \mathcal{P}) + \varepsilon.$$

Clearly, this implies

$$0 \leq U(f, I, \mathcal{P}) - L(f, I, \mathcal{P}) < \varepsilon.$$

Conversely, if for any $\varepsilon > 0$ there is a partition $\mathcal{P}$ so that

$$U(f, I, \mathcal{P}) - L(f, I, \mathcal{P}) < \varepsilon$$

then

$$0 \leq \inf_{\mathcal{Q}} U(f, I, \mathcal{Q}) - \sup_{\mathcal{Q}} L(f, I, \mathcal{Q}) < \varepsilon.$$

since $\varepsilon > 0$ is arbitrary, we have $(D)\int_I f(\vec{x})dV$ exists.
□

Theorem 9.3.11 *If $f : I \to \mathbb{R}$ and $\int_I f(\vec{x})dV$ and $(D)\int_I f(\vec{x})dV$ both exist, then*

$$(D)\int_I f(\vec{x})dV = \int_I f(\vec{x})dV.$$

Proof: Suppose that both $(D)\int_I f(\vec{x})dV$ and $\int_I f(\vec{x})dV$ exist, given $\varepsilon > 0$ we first choose a partition $\mathcal{Q}$ of I so that

$$U(f, I, \mathcal{Q}) - L(f, I, \mathcal{Q}) < \frac{\varepsilon}{3}.$$

We observe that for any refinement $\mathcal{P}$ of $\mathcal{Q}$ we also have

$$U(f, I, \mathcal{P}) - L(f, I, \mathcal{P}) < \frac{\varepsilon}{3}.$$

Let $\delta > 0$ be so that if $\mathcal{P}$ has norm less than δ we have

$$\left|\int_I f(\vec{x})dV - S(f, I, \mathcal{P}^\dagger)\right| < \frac{\varepsilon}{3}.$$

Let $\mathcal{P} = \{I_k\}_{k=1}^{N}$ be a refinement of $\mathcal{Q}$ which has norm less than δ. For each interval I_k, $k = 1, 2, \cdots, N$, pick $\vec{t}_k \in I_k$ so that $f(\vec{t}_k) > \sup_{\vec{x} \in I_k} f(\vec{x}) - \frac{\varepsilon}{3V(I)}$. Let $\mathcal{P}^\dagger$ be the corresponding marked partition using the $\vec{t}_k$'s as our marked points. Then

$$|(D)\int_I f(\vec{x})dV - \int_I f(\vec{x})dV|$$

$$\leq |(D)\int_I f(\vec{x})dV - U(f, I, \mathcal{P})| + |U(f, I, \mathcal{P}) - S(f, I, \mathcal{P}^\dagger)|$$

$$+ |S(f, I, \mathcal{P}^\dagger) - \int_I f(\vec{x})dV|$$

$$< \frac{\varepsilon}{3} + \sum_{k=1}^{N}(\sup_{\vec{x} \in I_k} f(\vec{x}) - f(\vec{t}_k))\Delta I_k + \frac{\varepsilon}{3}$$

$$< \frac{\varepsilon}{3} + \frac{\varepsilon}{3V(I)}\sum_{k=1}^{N}\Delta I_k + \frac{\varepsilon}{3}$$

$$= \frac{\varepsilon}{3} + \frac{\varepsilon}{3} + \frac{\varepsilon}{3} = \varepsilon.$$

Since $\varepsilon > 0$ is arbitrary, we see that

$$(D)\int_I f(\vec{x})dV = \int_I f(\vec{x})dV.$$

□

We leave it as an exercise to show that $(D)\int_I f dV$ exists if and only if $\int_I f dV$ exists. We conclude this subsection with a sufficient condition for Riemann integrability.

Theorem 9.3.12 *Suppose $f : I \to \mathbb{R}$ is bounded and is continuous on I, or is continuous at all but finitely many points in I, then $\int_I f(\vec{x})dV$ exists.*

Proof: We assume that f is continuous except at $\vec{c}_1, \vec{c}_2, \cdots, \vec{c}_N \in I$.

Let $M := \sup_{\vec{x} \in I}|f(\vec{x})|$, and let $\varepsilon > 0$ be given. Let I_j be so that it contain $\vec{c}_j$ as an interior point and let I_j be a cube so that $V(I_j)$ has volume $\frac{\varepsilon}{4NM}$. Define $D := I \setminus \cup_{k=1}^{n} I_k$. Then D is compact, and thus f is uniformly continuous on D. Therefore there is a $\delta > 0$ so that for $\vec{x}, \vec{y} \in D$ so that

9.3. THE RIEMANN AND LEBESGUE INTEGRALS

$\|\vec{x} - \vec{y}\| < \delta$ we have $|f(\vec{x}) - f(\vec{y})| < \frac{\varepsilon}{2V(I)}$.

Let $\mathcal{P}_1 = \{J_r\}_{r=1}^K$ be a partition of D of norm less than δ. Extend $\mathcal{P}_1$ to a partition of I in the obvious way by adding the cubes $\{I_s\}_{s=1}^N$. Let $\mathcal{P}$ be such a partition, and denote it by $\{\mathcal{I}_t\}_{t=1}^L$.

Let $\mathcal{J} \subseteq \{1, 2, \cdots, L\}$ be the collection of j's for which $\mathcal{I}_j = I_k$ for some $k \in \{1, 2, \cdots, N\}$. Then $i \in \{1, 2, \cdots, L\} \setminus \mathcal{J}$ if and only if $\mathcal{I}_i \subseteq D$.

Let $M_i := \sup_{\vec{x} \in \mathcal{I}_i} f(\vec{x})$ and $m_i := \inf_{\vec{x} \in \mathcal{I}_i} f(\vec{x})$ for $i = 1, 2, \cdots, L$. Then

$$U(f, I, \mathcal{P}) - L(f, I, \mathcal{P}) = \sum_{i=1}^L (M_i - m_i) \Delta \mathcal{I}_i$$

$$= \sum_{i \in \mathcal{J}} (M_i - m_i) \Delta \mathcal{I}_i + \sum_{i \notin \mathcal{J}} (M_i - m_i) \Delta \mathcal{I}_i$$

$$< N \cdot 2M \cdot \frac{\varepsilon}{4NM} + \frac{\varepsilon}{2V(I)} \sum_{j \notin \mathcal{J}} \Delta \mathcal{I}_j$$

$$\leq \frac{\varepsilon}{2} + \frac{\varepsilon}{2V(I)} \cdot V(I) = \varepsilon.$$

Thus we see $(D) \int_I f \, dV$ exists. Since it was an exercise to show $\int_I f \, dV$ exists if and only if $(D) \int_I f \, dV$ exists, and in such a case they both agree, we see that $f \in \mathcal{R}(I)$.
□

9.3.2 The Lebesgue Integral of Non-Negative Functions in $\mathbb{R}^n$

Let $D \subseteq \mathbb{R}^n$ be a measurable set and let $f : D \to \mathbb{R}$ be a non-negative function. Let $\mathcal{P}$ be a **partition** of D, that is $\mathcal{P} = \{D_k\}_{k=1}^N$, with $\{D_k\}_{k=1}^N$ pairwise disjoint measurable sets satisfying $\bigcup_{k=1}^N D_k = D$. We define the **Lebesgue sum** of f on D to be

$$\mathcal{L}(f, D, \mathcal{P}) := \sum_{k=1}^N \inf_{\vec{x} \in D_k} f(\vec{x}) \mu(D_k) \tag{9.12}$$

and
$$\int_D f d\mu = \sup_{\mathcal{P}} \mathcal{L}(f, D, \mathcal{P}) \qquad (9.13)$$
with the supremum taken over all possible partitions of D.

Using the convention $0 \cdot \infty = \infty \cdot 0 = 0$ we see that if $\mu(D) = 0$ or $f \equiv 0$, then we have $\int_D f d\mu = 0$. We define a **simple function** to be a function of the form
$$g(\vec{x}) = \sum_{k=1}^{m} c_k \chi_{E_k}(\vec{x}), \qquad (9.14)$$
where the collection of sets $\{E_k\}$ we shall assume are measurable sets and $c_k \in \mathbb{R}$.

Theorem 9.3.13 *Suppose $\{E_k\}_{k=1}^{m}$ is a collection of pairwise disjoint measurable sets with $E_k \subseteq D$ for each k, and $\{a_k\}_{k=1}^{m}$ is a collection of nonnegative real numbers, then for $g := \sum_{k=1}^{m} a_k \chi_{E_k}$ we have*
$$\int_D g d\mu = \sum_{k=1}^{m} a_k \mu(E_k).$$

Proof: Let $L := \sum_{k=1}^{m} a_k \mu(E_k)$, and define a partition $\mathcal{Q}$ of D by $\mathcal{Q} := \{E_k\}_{k=1}^{m} \cup \{D \setminus \bigcup_{k=1}^{m} E_k\}$. Then
$$\mathcal{L}(g, D, \mathcal{Q}) = \sum_{k=1}^{m} a_k \mu(E_k) + 0 \cdot \mu(D \setminus \bigcup_{k=1}^{m} E_k) = L.$$
Hence we have
$$\int_D g d\mu \geq L.$$
Suppose $\mathcal{P} = \{D_k\}_{k=1}^{N}$ is any partition of D, then
$$L = \sum_{k=1}^{m} a_k \mu(E_k)$$
$$= \sum_{k=1}^{m} a_k \sum_{j=1}^{N} \mu(E_k \cap D_j)$$

9.3. THE RIEMANN AND LEBESGUE INTEGRALS

$$\geq \sum_{k,j} (\inf_{D_j} g)\mu(E_k \cap D_j)$$

since $a_k \geq \inf_{D_j} g$ if $E_k \cap D_j \neq \emptyset$. Moreover,

$$\sum_{k,j} (\inf_{D_j} g)\mu(E_k \cap D_j) = \sum_{j=1}^{N} (\inf_{D_j} g)\mu(D_j)$$

$$= \mathcal{L}(g, D, \mathcal{P}).$$

So,

$$L \geq \sup_{\mathcal{P}} \mathcal{L}(g, D, \mathcal{P}) = \int_D g\, d\mu.$$

□

Theorem 9.3.14 *Suppose $D = E \cup F$ where E, F are measurable sets and $E \cap F = \emptyset$. Suppose $f \geq 0$ is a measurable function, then*

$$\int_D f\, d\mu = \int_E f\, d\mu + \int_F f\, d\mu.$$

Proof: If $\mathcal{P}_1$ is a partition of E and $\mathcal{P}_2$ is a partition of F, then $\mathcal{P} := \mathcal{P}_1 \cup \mathcal{P}_2$ is a partition of D and

$$\mathcal{L}(f, D, \mathcal{P}) = \mathcal{L}(f, E, \mathcal{P}_1) + \mathcal{L}(f, F, \mathcal{P}_2).$$

So we see that

$$\mathcal{L}(f, E, \mathcal{P}_1) + \mathcal{L}(f, F, \mathcal{P}_2) \leq \int_D f\, d\mu,$$

and thus

$$\int_E f\, d\mu + \int_F f\, d\mu \leq \int_D f\, d\mu,$$

since we may choose $\mathcal{P}_1$ and $\mathcal{P}_2$ independently.

Conversely, if $\mathcal{P} = \{D_k\}_{k=1}^m$ is a partition of D, then $\mathcal{P}_1 := \{D_k \cap E\}_{k=1}^m$ and $\mathcal{P}_2 := \{D_k \cap E\}_{k=1}^m$ are partitions of E and F respectively, and

$$\mathcal{L}(f, D, \mathcal{P}) = \sum_{j=1}^{m} \inf_{D_j} f\mu(D_j)$$

$$\leq \sum_{j=1}^{m} [(\inf_{D_j \cap E} f)\mu(D_j \cap E) + (\inf_{D_j \cap F} f)\mu(D_j \cap F)]$$

$$\leq \int_E f d\mu + \int_F f d\mu.$$

So we see that

$$\int_D f d\mu \leq \int_E f d\mu + \int_F f d\mu.$$

□

Corollary 9.3.15 *Suppose f, g are non-negative measurable functions defined on a measurable set D. Suppose $f \leq g$ a.e. on D, then $\int_D f d\mu \leq \int_D g d\mu$.*

Proof: Suppose $f \leq g$ on a set F, where $D = F \cup Z$ and $\mu(Z) = 0$. Then for any partition $\mathcal{P}$ of F, we have $\mathcal{L}(f, F, \mathcal{P}) \leq \mathcal{L}(g, F, \mathcal{P})$. Thus we see that $\int_F f d\mu \leq \int_F g d\mu$. However, by theorem 9.3.14 we have

$$\int_D f d\mu = \int_F f d\mu + \int_Z f d\mu = \int_F f d\mu$$

$$\leq \int_F g d\mu = \int_F g d\mu + \int_Z g d\mu = \int_D g d\mu.$$

□

Corollary 9.3.16 *For a non-negative measurable function f defined on a measurable set D of positive measure we have $\int_D f d\mu = 0$ if and only if $f = 0$ a.e.*

Proof: Clearly if $f \equiv 0$ on a measurable set, we must have its integral be zero on this set. This follows easily from the definition of the Lebesgue integral. Suppose $f = 0$ a.e. on D, then write $D = F \cup Z$ where $\mu(Z) = 0$ and $f|_F = 0$. Then, by theorem 9.3.14 we know

$$\int_D f d\mu = \int_F f d\mu + \int_Z f d\mu = 0 + 0 = 0.$$

Conversely, suppose $f \geq 0$ on D and $\int_D f d\mu = 0$. Let $D_n := \{\vec{x} \in D : f(\vec{x}) \geq \frac{1}{n}\}$. Then, by corollary 9.3.15

$$0 = \int_D f d\mu = \int_{D \setminus D_n} f d\mu + \int_{D_n} f d\mu$$

9.3. THE RIEMANN AND LEBESGUE INTEGRALS

$$\geq \int_{D_n} f d\mu \geq \int_{D_n} \frac{1}{n} d\mu = \frac{1}{n} \mu(D_n) \geq 0.$$

Hence we see that $\mu(D_n) = 0$ for each n. Since $\{\vec{x} \in D : f(\vec{x}) > 0\} = \bigcup_{n=1}^{\infty} D_n$ we see that $\mu(\{\vec{x} \in D : f(\vec{x}) > 0\}) \leq \sum_{n=1}^{\infty} \mu(D_n) = 0$.

□

From the above corollary we see that $\int_D \chi_\mathbb{Q} d\mu = 0$ for any measurable $D \subseteq \mathbb{R}$. Note however that the Riemann integral of $\chi_\mathbb{Q}$ did not exist on any non-trivial interval $[a,b]$.

Tchebyshev's Inequality 9.3.17 *Let $f \geq 0$ be measurable on D, then for any $\alpha > 0$ we have*

$$\mu(\{\vec{x} \in D : f(\vec{x}) > \alpha\}) \leq \frac{1}{\alpha} \int_D f d\mu.$$

Proof: Let $E := \{\vec{x} \in D : f(\vec{x}) > \alpha\}$, and write $D = E \cup (D \setminus E)$. Then

$$\int_D f d\mu = \int_E f d\mu + \int_{D \setminus E} f d\mu \geq \int_E f d\mu \geq \int_E \alpha d\mu = \alpha \mu(E).$$

□

Approximation of a Measurable Function by Simple Functions with Close Integral Values

The following theorem tells us that we may always approximate a measure function f from below by a simple function whose Lebesgue integral is close in value to the Lebesgue integral of f.

Theorem 9.3.18 *Suppose $f : D \to \mathbb{R} \cup \{\infty\}$ is non-negative and measurable. If $\int_D f d\mu < \infty$, then for any $\varepsilon > 0$ there exists a simple function σ defined of D so that $\sigma \leq f$ and $\int_D \sigma d\mu > \int_D f d\mu - \varepsilon$. If $\int_D f d\mu = \infty$, then for any $N > 0$ there is a simple function σ defined on D so that $\sigma \leq f$ and $\int_D \sigma d\mu > N$.*

Proof: In the case where $\int_D f d\mu < \infty$, given $\varepsilon > 0$, there is a partition $\mathcal{P}$ of D so that

$$\mathcal{L}(f, D, \mathcal{P}) \leq \int_D f d\mu \leq \mathcal{L}(f, D, \mathcal{P}) + \varepsilon.$$

If $\mathcal{P} = \{D_k\}_{k=1}^m$, then in such a case we define

$$\sigma := \sum_{k=1}^m (\inf_{D_k} f)\chi_{D_k}.$$

Here $\int_D \sigma d\mu = \mathcal{L}(f, D, \mathcal{P})$.

In the case where $\int_D f d\mu = \infty$, let $N > 0$. Here we take a partition $\mathcal{P}$ so that $\mathcal{L}(f, D, \mathcal{P}) > N$ and define $\sigma := \sum_{k=1}^m (\inf_{D_k} f)\chi_{D_k}$, unless $\inf_{D_k} f = \infty$ for some D_k in our partition $\mathcal{P}$. If $\inf_{D_k} f = \infty$ and $\mu(D_k) = \infty$ then we define $\sigma = \chi_{D_k}$. Otherwise we take $\sigma := \frac{N+1}{\mu(D_k)}\chi_{D_k}$.
□

The Monotone Convergence Theorem

Monotone Convergence Theorem 9.3.19 *If $\{f_k\}_{k=1}^\infty$ is a sequence of nonnegative measurable functions with each f_k defined on a set D, and for each $\vec{x} \in D$ the sequence $\{f_k(\vec{x})\}_{k=1}^\infty$ is an increasing sequence which converges to $f(\vec{x})$, then*

$$\lim_{j\to\infty} \int_D f_j d\mu = \int_D f d\mu.$$

Proof: Since $f_k \to f$ point-wise, we know f is measurable, and thus $\int_D f d\mu$ exists. Moreover, $\int_D f_k d\mu \leq \int_D f d\mu$ for all k, and thus

$$\lim_{k\to\infty} \int_D f_k d\mu \leq \int_D f d\mu.$$

First suppose $f = a\chi_D$ for some $a \geq 0$. Given $\varepsilon > 0$, we let $F_j := \{\vec{x} : f_j(\vec{x}) > f(\vec{x}) - \varepsilon\}$. Then $F_j \nearrow D$, and so $\mu(F_j) \to \mu(D)$ as $j \to \infty$. But $f_j \geq (f - \varepsilon)\chi_{F_j} = (a - \varepsilon)\chi_{F_j}$, and so

$$\int_D f_j d\mu \geq \int_D (a - \varepsilon)\chi_{F_j} d\mu = (a - \varepsilon)\mu(F_j).$$

Thus

$$\lim_{j\to\infty} \int_D f_j d\mu \geq \lim_{j\to\infty} (a - \varepsilon)\mu(F_j) = (a - \varepsilon)\mu(D).$$

Since $\varepsilon > 0$ is arbitrary, we have

$$\lim_{j\to\infty} \int_D f_j d\mu \geq a\mu(D) = \int_D f d\mu.$$

9.3. THE RIEMANN AND LEBESGUE INTEGRALS

Next we suppose that f is a simple function. That is, $f = \sum_{k=1}^{m} a_k \chi_{D_k}$, where $\{D_k\}_{k=1}^{m}$ is a partition of D. Then using the result above we see that

$$\int_D f_j d\mu = \sum_k \int_{D_k} f_j d\mu \to \sum_k \int_{D_k} f d\mu = \int_D f d\mu.$$

In the case of a general non-negative measurable f, first we suppose $\int_D f d\mu < \infty$. By theorem 9.3.18 we know that for any $\varepsilon > 0$ there is a simple function σ so that $\sigma \leq f$ and $\int_D \sigma d\mu \geq \int_D f d\mu - \varepsilon$. Define $g_j := \min\{f_j, \sigma\}$. Then g_j is measurable, and $g_j \to \sigma$. So by the above case, we have $\int_D g_j d\mu \to \int_D \sigma d\mu \geq \int_D f d\mu - \varepsilon$. Also, $g_j \leq f_j$, so $\int_D f_j d\mu \geq \int_D g_j d\mu$, and thus $\lim_{j \to \infty} \int_D f_j d\mu \geq \int_D f d\mu - \varepsilon$. But $\varepsilon > 0$ was arbitrary, and so $\lim_{j \to \infty} \int_D f_j d\mu \geq \int_D f d\mu$.

Suppose $\int_D f d\mu = \infty$. By theorem 9.3.18 we know that for any $N > 0$ there is a simple function σ so that $\sigma \leq f$ and $\int_D \sigma d\mu \geq N$. Define $g_j := \min\{f_j, \sigma\}$. Then g_j is measurable, and $g_j \to \sigma$. So, $\int_D g_j d\mu \to \int_D \sigma d\mu \geq N$. Also, $g_j \leq f_j$, so $\int_D f_j d\mu \geq \int_D g_j d\mu$, and thus $\lim_{j \to \infty} \int_D f_j d\mu \geq N$. But $N > 0$ was arbitrary, and therefore $\lim_{j \to \infty} \int_D f_j d\mu = \infty$.
□

Note that the assumption $f_j \to f$ can be replaced by $f_j \to f$ a.e. in the above theorem. Moreover, the following corollary gives us linearity for the Lebesgue integral.

Theorem 9.3.20 *If $f, g : D \to \mathbb{R} \cup \{\infty\}$ are non-negative and measurable and $\alpha, \beta \geq 0$, then*

$$\int_D (\alpha f + \beta g) d\mu = \alpha \int_D f d\mu + \beta \int_D g d\mu.$$

Proof: First suppose f, g are simple functions with $f = \sum_{j=1}^{m} a_j \chi_{A_j}$, $g = \sum_{j=1}^{M} b_j \chi_{B_j}$, where $\{A_j\}_{j=1}^{m}$ and $\{B_j\}_{j=1}^{M}$ are partitions of D. Then

$$\alpha f + \beta g = \sum_{j,k} (\alpha a_j + \beta b_k) \chi_{A_j \cap B_k}.$$

So,

$$\int_D (\alpha f + \beta g) d\mu = \sum_{j,k} (\alpha a_j + \beta b_k) \mu(A_j \cap B_k)$$

$$= \alpha \sum_{j,k} a_j \mu(A_j \cap B_k) + \beta \sum_{j,k} b_k \mu(A_j \cap B_k) = \alpha \int_D f d\mu + \beta \int_D g d\mu.$$

In the general case there is a sequence f_k of simple functions with $f_k \to f$ monotonically from below and there is a sequence of simple functions $g_k \to g$ monotonically from below. Then, $(\alpha f_k + \beta g_k) \to (\alpha f + \beta g)$, and so by the monotone convergence theorem 9.3.19 we have

$$\int_D (\alpha f + \beta g) d\mu = \lim_{k \to \infty} \int_D (\alpha f_k + \beta g_k) d\mu$$

$$= \lim_{k \to \infty} (\alpha \int_D f_k d\mu + \beta \int_D g_k d\mu) = \alpha \int_D f d\mu + \beta \int_D g d\mu.$$

□

Fatou's Lemma and Lebesgue's Dominated Convergence Theorem

Fatou's Lemma 9.3.21 *If $\{f_k\}_{k=1}^\infty$ is a sequence of non-negative measurable functions on D, then*

$$\int_D (\liminf_{k \to \infty} f_k) d\mu \le \liminf_{k \to \infty} \int_D f_k d\mu.$$

Proof: Since $\liminf_{k \to \infty} f_k$ is measurable and non-negative we know that $\int_D (\liminf_{k \to \infty} f_k) d\mu$ exists. Define $g_k := \inf\{f_k, f_{k+1}, f_{k+2}, \cdots\}$ for $k \in \mathbb{N}$. Then we have $\{g_k\}_{k=1}^\infty$ is monotone increasing and $g_k \to \liminf_{j \to \infty} f_j$. Moreover, $0 \le g_k \le f_k$ for all $k \in \mathbb{N}$. Thus by the monotone convergence theorem 9.3.19 we have $\int_D g_k d\mu \to \int_E (\liminf_{j \to \infty} f_j) d\mu$. Moreover, by corollary 9.3.15 we have $\int_D g_k d\mu \le \int_D f_k d\mu$ for all $k \in \mathbb{N}$. Putting this together, we have

$$\int_D (\liminf_{j \to \infty} f_j) d\mu = \lim_{k \to \infty} \int_D g_k d\mu \le \liminf_{j \to \infty} \int_D f_j d\mu.$$

□

Clearly if $\{f_k\}_{k=1}^\infty$ is a sequence of measurable functions whose integrals $\int_D |f_k| d\mu \le M$ for all k[9], so if $f_k \to f$ a.e., then $\int_D |f| d\mu \le M$ as well. To see this, we use Fatou's lemma, observing that $\int_D \liminf_{k \to \infty} |f_k| d\mu \le M$ and

---

[9]The notation that is often used is $\|f\|_{L^1(D)} = \int_D |f| d\mu$, and thus we may say that $\{f_k\}_{k=1}^\infty$ is a sequence of functions whose L^1 norms are bounded by M.

9.3. THE RIEMANN AND LEBESGUE INTEGRALS

$\liminf_{j\to\infty} f_j = \lim_{j\to\infty} f_j = f$ a.e. in D.

An example of a function for which there is a strict inequality in the conclusion of Fatou's lemma is given by taking $D = \mathbb{R}$ and $f_j = \chi_{[j,\infty)}$. Here we have $\liminf_{j\to\infty} f_j = \lim_{j\to\infty} f_j \equiv 0$. Thus $\int_D \liminf_{j\to\infty} f_j d\mu = 0$, but $\liminf_{j\to\infty} \int_D f_j d\mu = \infty$.

The following theorem gives us a result about interchanging the limit operations of integration and limits of sequences of functions that converge a.e.

Lebesgue's Dominated Convergence Theorem 9.3.22 *If $\{f_k\}_{k=1}^{\infty}$ is a sequence of non-negative measurable functions on D that converge a.e. to a function f in D. If there is a measurable function g on D so that $f_n \leq g$ on D for all $n \in \mathbb{N}$, with $\int_D g d\mu < \infty$, then $\int_D f_n d\mu \to \int_D f d\mu$ as $n \to \infty$.*

Proof: Using that $f_k \to f$ a.e. on D, and thus $\liminf_{k\to\infty} f_k = f$, by Fatou's lemma 9.3.21 we know that

$$\int_D f d\mu = \int_D \liminf_{k\to\infty} f_k d\mu \leq \liminf_{k\to\infty} \int_D f_k d\mu.$$

Thus, it is only necessary to show $\int_D f d\mu \geq \limsup_{k\to\infty} \int_D f_k d\mu$. Consider the non-negative measurable functions $g_k := g - f_k$. Then by Fatou's lemma we know that

$$\int_D \liminf_{k\to\infty}(g-f_k) d\mu = \int_D \liminf_{k\to\infty} g_k d\mu \leq \liminf_{k\to\infty} \int_D g_k d\mu = \liminf_{k\to\infty} \int_D (g-f_k) d\mu.$$

Since $f_k \to f$ a.e. on D, we know $g_k = g - f_k \to g - f$ a.e. on D as well. Thus we have

$$\int_D (g-f) d\mu \leq \liminf_{k\to\infty} \int_D (g-f_k) d\mu = \int_D g d\mu - \limsup_{k\to\infty} \int_D f_k d\mu.$$

Thus we obtain the desired inequality

$$\int_D f d\mu \geq \limsup_{k\to\infty} \int_D f_k d\mu.$$

□

9.3.3 Lebesgue Integral for a Measurable Function of Any Sign

For an arbitrary real-valued measurable function f defined on a measurable set D, we write
$$f = f^+ - f^-$$
where
$$f^+(\vec{x}) = \max\{f(\vec{x}), 0\}, \quad f^-(\vec{x}) = \max\{-f(\vec{x}), 0\}. \tag{9.15}$$
In such a case we observe that if $f \geq 0$ a.e. then $f = f^+$ a.e. and if $f \leq 0$ a.e., then $f = -f^-$ a.e. Also, $|f| = f^+ + f^-$. Moreover, f is measurable if and only if f^+ and f^- are both measurable. So, $\int_D f^+ d\mu$ and $\int_D f^- d\mu$ both exist. If at least one of these is finite, we may define $\int_D f d\mu$ by

$$\int_D f d\mu = \int_D f^+ d\mu - \int_D f^- d\mu. \tag{9.16}$$

If $\int_D f d\mu$ exists and is finite, then we say that f is **Lebesgue integrable** on D.

If $\int_D f d\mu$ exists, observe that

$$\left| \int_D f d\mu \right| \leq \int_D f^+ d\mu + \int_D f^- d\mu = \int_D (f^+ + f^-) d\mu = \int_D |f| d\mu.$$

Also, f is Lebesgue integrable if and only if both $\int_D f^+ d\mu$ and $\int_D f^- d\mu$ are finite, and thus if and only if $\int_D |f| d\mu$ is finite. Thus the collection of Lebesgue integrable functions on D, which shall be denoted by $L^1(D)$, is the set of measurable functions f so that $\int_D |f| d\mu < \infty$. Moreover, it can be shown that

$$\|f\|_{L^1(D)} := \int_D |f| d\mu \tag{9.17}$$

defines a norm on the set of functions $L^1(D)$. Moreover, $f \in L^1(D)$ with $\mu(D) > 0$ if and only if f is finite a.e. Also, for $p > 1$ we may also define the set

$$L^p(D) = \{f : \int_D |f|^p d\mu < \infty\} \tag{9.18}$$

and on this set we may define a norm

$$\|f\|_{L^p(D)} = (\int_D |f|^p d\mu)^{\frac{1}{p}}. \tag{9.19}$$

9.3. THE RIEMANN AND LEBESGUE INTEGRALS

It is an exercise to show these are norms and $L^q(D) \subseteq L^p(D)$ if $q > p$ when $\mu(D) < \infty$.

So many of the results for non-negative functions carry over for arbitrary functions. For instance $f \leq g$ a.e. on D implies $\int_D f d\mu \leq \int_D g d\mu$ provided both exist. To see this observe that $f \leq g$ if and only of $0 \leq f^+ \leq g^+$ and $0 \leq g^- \leq f^-$. Also, if D is written as a countable union of disjoint measurable sets, $D = \bigcup_n D_n$, then $\int_D f d\mu$ exists implies $\int_{D_n} f d\mu$ exist for all n, and $\int_D f d\mu = \sum_n \int_{D_n} f d\mu$. Moreover, if $f = 0$ a.e or $\mu(D) = 0$, then $\int_D f d\mu = 0$. The linearity properties of integration also hold. For $f, g \in L^1(D)$ and $c, k \in \mathbb{R}$, then

$$\int_D (cf + kg) d\mu = c \int_D f d\mu + k \int_D g d\mu. \tag{9.20}$$

For this general linearity result, first we will show $cf \in L^1(D)$ and $\int_D cf d\mu = c \int_D f d\mu$. To see this, first suppose $c > 0$. Then $(cf)^+ = cf^+$ and $(cf)^- = cf^-$, and so $\int_D (cf)^+ d\mu = \int_D cf^+ d\mu = c \int_D f^+ d\mu$. Likewise, $\int_D (cf)^- d\mu = c \int_D f^- d\mu$. Next, consider $c = -1$ and observe that $-f = -f^+ + f^-$, and so $(-f)^+ = f^-$ and $(-f)^- = f^+$. Thus $\int_D (-f) d\mu = \int_D f^- d\mu - \int_D f^+ d\mu = -\int_D f d\mu$. Using this and $c = -|c|$ when $c \leq 0$, we get $\int_D cf d\mu = c \int_D f d\mu$ for any $c \in \mathbb{R}$. To get the general statement (9.20), we will now show that $f, g \in L^1(D)$ implies $f + g \in L^1(D)$ and $\int_D (f+g) d\mu = \int_D f d\mu + \int_D g d\mu$.

To see that $f, g \in L^1(D)$ implies $f + g \in L^1(D)$ we observe that

$$\|f + g\|_{L^1(D)} = \int_D |f + g| d\mu \leq \int_D (|f| + |g|) d\mu$$

$$= \int_D |f| d\mu + \int_D |g| d\mu = \|f\|_{L^1(D)} + \|g\|_{L^1(D)}.$$

To establish $\int_D f + g d\mu = \int_D f d\mu + \int_D g d\mu$, we partition D into six disjoint measurable subsets, namely $D_1 = \{\vec{x} : f(\vec{x}), g(\vec{x}) \geq 0\}$, $D_2 = \{\vec{x} : f(\vec{x}), f(\vec{x}) + g(\vec{x}) \geq 0, g(\vec{x}) < 0\}$, $D_3 = \{\vec{x} : f(\vec{x}) \geq 0, f(\vec{x}) + g(\vec{x}), g(\vec{x}) < 0\}$, $D_4 = \{\vec{x} : g(\vec{x}), f(\vec{x}) + g(\vec{x}) \geq 0, f(\vec{x}) < 0\}$, $D_5 = \{\vec{x} : g(\vec{x}) \geq 0, f(\vec{x}) + g(\vec{x}), f(\vec{x}) < 0\}$, and $D_6 = \{\vec{x} : f(\vec{x}), g(\vec{x}) < 0\}$.

On D_1, we have $\int_{D_1} (f+g) d\mu = \int_{D_1} f + \int_{D_1} g$ from the integrable properties for non-negative functions established above, namely theorem 9.3.20. On D_6

we use $\int_{D_6}(-f-g)d\mu = \int_{D_6}(-f)d\mu + \int_{D_6}(-g)d\mu$; also $-\int_{D_6}(f+g)d\mu = \int_{D_6}(-f-g)d\mu$, $-\int_{D_6}fd\mu = \int_{D_6}(-f)d\mu$ and $-\int_{D_6}gd\mu = \int_{D_6}(-g)d\mu$, and so $\int_{D_6}(f+g)d\mu = \int_{D_6}fd\mu + \int_{D_6}gd\mu$.

On D_2 we use that $f, f+g \geq 0$ and $-g > 0$ and so

$$\int_{D_2} fd\mu = \int_{D_2}[(f+g) + (-g)]d\mu$$

$$= \int_{D_2}(f+g)d\mu + \int_{D_2}(-g)d\mu = \int_{D_2}(f+g)d\mu - \int_{D_2}gd\mu.$$

On D_3 we use $f \geq 0$ and $-(f+g), -g > 0$ and so

$$-\int_{D_3}gd\mu = \int_{D_3}(-g)d\mu = \int_{D_3}[f-(f+g)]d\mu$$

$$= \int_{D_3}fd\mu + \int_{D_3}[-(f+g)]d\mu = \int_{D_3}fd\mu - \int_{D_3}(f+g)d\mu.$$

On D_4 we use $-f > 0$ and $g, f+g \geq 0$, and so

$$\int_{D_4}gd\mu = \int_{D_4}[(f+g) + (-f)]d\mu$$

$$= \int_{D_4}(f+g)d\mu + \int_{D_4}(-f)d\mu = \int_{D_4}(f+g)d\mu - \int_{D_4}fd\mu.$$

On D_5 we use that $-f, -(f+g) > 0$ and $g \geq 0$, and so

$$-\int_{D_5}fd\mu = \int_{D_5}(-f)d\mu = \int_{D_5}[g-(f+g)]d\mu$$

$$= \int_{D_5}gd\mu + \int_{D_5}[-(f+g)]d\mu = \int_{D_5}gd\mu - \int_{D_5}(f+g)d\mu.$$

Thus for each i we have shown that $\int_{D_i}(f+g)d\mu = \int_{D_i}fd\mu + \int_{D_i}gd\mu$. Hence

$$\int_D (f+g)d\mu = \int_{\cup_{i=1}^6 D_i}(f+g)d\mu = \sum_{i=1}^6 \int_{D_i}(f+g)d\mu = \sum_{i=1}^6 [\int_{D_i}fd\mu + \int_{D_i}gd\mu]$$

9.3. THE RIEMANN AND LEBESGUE INTEGRALS

$$= \sum_{i=1}^{6} \int_{D_i} f d\mu + \sum_{i=1}^{6} \int_{D_i} g d\mu = \int_{\cup_{i=1}^{6} D_i} f d\mu + \int_{\cup_{i=1}^{6} D_i} g d\mu = \int_D f d\mu + \int_D g d\mu.$$

Putting all things together, we have established the general statement (9.20).

One final result for this subsection is a result that tells us that $f, g \in L^1(D)$ implies $fg \in L^1(D)$ in the case where one of the functions f or g is bounded. To see this, let us suppose that $|g| < M$ a.e. on D. In fact, if we define the **essential supremum** of $|f|$ and the L^∞ norm by $\|f\|_{L^\infty(D)} = \text{ess sup}_D |f| = \inf\{\alpha : \mu(\{\vec{x} \in D : |f(\vec{x})| > \alpha\}) = 0\}$, then we may assume $\|g\|_{L^\infty(D)} < M$. Since $|fg| \leq \|g\|_{L^\infty(D)}|f|$ a.e. we have by corollary 9.3.15 and theorem 9.3.20

$$\int_D |fg| d\mu \leq \int_D \|g\|_{L^\infty(D)} |f| d\mu = \|g\|_{L^\infty(D)} \int_D |f| d\mu < M \int_D |f| d\mu$$

and so $fg \in L^1(D)$. Moreover, we actually have shown

$$\|f \cdot g\|_{L^1(D)} \leq \|f\|_{L^1(D)} \cdot \|g\|_{L^\infty(D)}. \quad (9.21)$$

Convergence Theorems

Once again we return to the question of whether or not $\lim_{n \to \infty} \int_D f_n d\mu = \int_D \lim_{n \to \infty} f_n d\mu$, but now we require no special conditions on the sign of the f_n's and the limit function $\lim_{n \to \infty} f_n$ when it exists.

Monotone Convergence Theorem (General Case) 9.3.23 *Let $\{f_n\}_{n=1}^\infty$ be a sequence of measurable functions of D, and for each $\vec{x} \in D$ the sequence $\{f_k(\vec{x})\}_{k=1}^\infty$ is an increasing sequence which converges to $f(\vec{x})$ and there exists a $g \in L^1(D)$ so that $f_k \geq g$ a.e.[10], or for each $\vec{x} \in D$ the sequence $\{f_k(\vec{x})\}_{k=1}^\infty$ is an decreasing sequence which converges to $f(\vec{x})$ and there exists a $g \in L^1(D)$ so that $f_k \leq g$ a.e.[11], then, in either case,*

$$\lim_{j \to \infty} \int_D f_j d\mu = \int_D f d\mu.$$

---

[10] Or $\int_D f_k d\mu > -\infty$ for some k.
[11] Or $\int_D f_k d\mu < \infty$ for some k.

Proof: By otherwise considering the sequence $\{-f_k\}_{k=1}^\infty$ we may assume that the sequence $\{f_k(\vec{x})\}_{k=1}^\infty$ is an increasing sequence which converges to $f(\vec{x})$ and there exists a $g \in L^1(D)$ so that $f_k \geq g$ a.e. Consider the sequence $f_k - g$, then by the monotone convergence theorem in the nonnegative case 9.3.19 we know $\int_D (f_k - g) d\mu \to \int_D (f - g) d\mu$. However, $\int_D (f_k - g) d\mu = \int_D f_k d\mu - \int_D g d\mu$ and $\int_D (f - g) d\mu = \int_D f d\mu - \int_D g d\mu$. □

Uniform Convergence Theorem 9.3.24 *Suppose $\{f_k\}_{k=1}^\infty$ is a sequence of functions in $L^1(D)$ so that $f_k \to f$ uniformly in D, where $\mu(D) < \infty$. Then $f \in L^1(D)$ and $\int_D f_k d\mu \to \int_D f d\mu$ as $k \to \infty$.*

Proof: We know that

$$|f| = |f - f_k + f_k| \leq |f - f_k| + |f_k| < 1 + |f_k|$$

for k sufficiently large. Thus, we see that $f \in L^1(D)$. Moreover,

$$\left| \int_D f_k d\mu - \int_D f d\mu \right| = \left| \int_D (f_k - f) d\mu \right|$$

$$\leq \int_D |f - f_k| d\mu \leq \sup_{\vec{x} \in D} |f_k(\vec{x}) - f(\vec{x})| \cdot \mu(D) \to 0$$

as $k \to \infty$. □

Fatou's Lemma (General Case) 9.3.25 *If $\{f_k\}_{k=1}^\infty$ is a sequence of measurable functions on D so that $f_k \geq g$ for all $k \in \mathbb{N}$, where $g \in L^1(D)$, then*

$$\int_D (\liminf_{k \to \infty} f_k) d\mu \leq \liminf_{k \to \infty} \int_D f_k d\mu.$$

Proof: We may apply Fatou's lemma in the non-negative case 9.3.21 to the sequence of functions given by $f_k - g$, for $k \in \mathbb{N}$. Then

$$\int_D \liminf_{k \to \infty} (f_k - g) d\mu \leq \liminf_{k \to \infty} \int_D (f_k - g) d\mu.$$

However, $\liminf_{k \to \infty} (f_k - g) = \liminf_{k \to \infty} f_k - g$ and so

$$\int_D \liminf_{k \to \infty} (f_k - g) d\mu = \int_D (\liminf_{k \to \infty} f_k - g) d\mu = \int_D \liminf_{k \to \infty} f_k d\mu - \int_D g d\mu.$$

9.3. THE RIEMANN AND LEBESGUE INTEGRALS

Also,

$$\liminf_{k\to\infty} \int_D (f_k - g)d\mu = \liminf_{k\to\infty} (\int_D f_k d\mu - \int_D g d\mu) = \liminf_{k\to\infty} \int_D f_k d\mu - \int_D g d\mu.$$

Thus,

$$\int_D \liminf_{k\to\infty} f_k d\mu - \int_D g d\mu \leq \liminf_{k\to\infty} \int_D f_k d\mu - \int_D g d\mu,$$

and hence

$$\int_D \liminf_{k\to\infty} f_k d\mu \leq \liminf_{k\to\infty} \int_D f_k d\mu.$$

□

Lebesgue's Dominated Convergence Theorem (General Case) 9.3.26
If $\{f_k\}_{k=1}^\infty$ is a sequence of measurable functions on D that converge a.e. to a function f in D. If there is a measurable function $g \in L^1(D)$ so that $|f_n| \leq g$ a.e. in D for all $n \in \mathbb{N}$, then $\int_D f_n d\mu \to \int_D f d\mu$ as $n \to \infty$.

Proof: Since $-g \leq f_k \leq g$ a.e. we have $0 \leq f_k + g \leq 2g \leq 2|g|$ a.e. in D. Also, $2g \in L^1(D)$ and so by Lebesgue's Dominated Convergence Theorem 9.3.22, we know that $\int_D (f_k + g) d\mu \to \int_D (f + g) d\mu$. Since f, g and all the f_k's are in $L^1(D)$, we get the desired result since $\int_D (f_k + g) d\mu = \int_D f_k d\mu + \int_D g d\mu$ and $\int_D (f + g) d\mu = \int_D f d\mu + \int_D g d\mu$.
□

We also have a corollary of theorem 9.3.26, often called the bounded convergence theorem.

Corollary– Bounded Convergence Theorem 9.3.27 *If $\{f_k\}_{k=1}^\infty$ is a sequence of measurable functions on D that converge a.e. to a function f in D. If $\mu(D) < \infty$ and there is a $M \in \mathbb{R}$ so that $|f_n| \leq M$ a.e. in D for all $n \in \mathbb{N}$, then $\int_D f_n d\mu \to \int_D f d\mu$ as $n \to \infty$.*

Relationship between Riemann and Lebesgue Integrals

We will now discuss the relationship between $\int_a^b f dx$ and $\int_{[a,b]} f d\mu$ when both exist. We claim that if $\int_a^b f dx$ exists, then so does $\int_{[a,b]} f d\mu$ exist, and in such a case,

$$\int_a^b f dx = \int_{[a,b]} f d\mu.$$

Let $\mathcal{P} = \{x_i\}_{i=0}^N$ be any partition of $[a,b]$. Then we define the simple functions – which are also step functions – $L_\mathcal{P}$ and $U_\mathcal{P}$ by

$$L_\mathcal{P} = \sum_{i=1}^N \inf_{x \in [x_{i-1},x_i]} f(x) \cdot \chi_{[x_i,x_{i-1})}$$

and

$$U_\mathcal{P} = \sum_{i=1}^N \sup_{x \in [x_{i-1},x_i]} f(x) \cdot \chi_{[x_i,x_{i-1})}.$$

Then,

$$\int_{[a,b]} L_\mathcal{P} d\mu = \sum_{i=1}^N \inf_{[x_{i-1},x_i]} f \cdot \mu([x_{i-1},x_i]) = L(f,[a,b],\mathcal{P}),$$

and

$$\int_{[a,b]} U_\mathcal{P} d\mu = \sum_{i=1}^N \sup_{[x_{i-1},x_i]} f \cdot \mu([x_{i-1},x_i]) = U(f,[a,b],\mathcal{P}).$$

Let $\{\mathcal{P}_k\}_{k=1}^\infty$ be a sequence of partitions, so that $\mathcal{P}_{k+1}$ is a refinement of $\mathcal{P}_k$, and whose norms are going to zero as $k \to \infty$. Then $\{L_{\mathcal{P}_n}\}_{n=1}^\infty$ and $\{U_{\mathcal{P}_n}\}_{n=1}^\infty$ are both bounded increasing and decreasing sequences of functions respectively, and so there exists limiting functions L and U respectively so that $L_{\mathcal{P}_n} \nearrow L$ and $U_{\mathcal{P}_n} \searrow U$ as $n \to \infty$. Then by theorem 9.3.27 we know that

$$L(f,[a,b],\mathcal{P}_n) = \int_{[a,b]} L_{\mathcal{P}_n} d\mu \to \int_{[a,b]} L d\mu$$

and

$$U(f,[a,b],\mathcal{P}_n) = \int_{[a,b]} U_{\mathcal{P}_n} d\mu \to \int_{[a,b]} U d\mu.$$

Since f is Riemann integrable we know that both $L(f,[a,b],\mathcal{P}_n)$ and $U(f,[a,b],\mathcal{P}_n)$ converge to $\int_a^b f dx$ as $n \to \infty$. Thus, $\int_{[a,b]} L d\mu = \int_{[a,b]} U d\mu = \int_a^b f dx$. Moreover, $L_{\mathcal{P}_n} \leq f \leq U_{\mathcal{P}_n}$, and so $L \leq f \leq U$ on $[a,b]$. But $\int_{[a,b]} (U - L) d\mu = \int_{[a,b]} U d\mu - \int_{[a,b]} L d\mu = 0$, and therefore $U - L = 0$ a.e., and so $U = f = L$ a.e. So f is measurable, and $\int_{[a,b]} f d\mu = \int_a^b f dx$.

9.3. THE RIEMANN AND LEBESGUE INTEGRALS

A Necessary and Sufficient Condition for Riemann Integrability

Recall that earlier we proved theorem 6.5.5. However, here we will provide another proof of this result which says that $f \in \mathcal{R}([a,b]) \iff f$ is continuous a.e. in $[a,b]$. First we suppose that f is bounded and in $\mathcal{R}([a,b])$. As in the above discussion let $\mathcal{P}_k$, $L_{\mathcal{P}_k}$, $U_{\mathcal{P}_k}$ be a sequence of partitions and associated simple functions. Let Z be a set of measure zero outside of which we have the limiting functions agree with f, that is $U = L = f$ a.e. If x is not a partition point for any of the $\mathcal{P}_k$'s or a point in Z, we claim that f is continuous at x. For if f were not continuous at such an x, there is $\varepsilon > 0$, which depends on x, but not on k so that $U_{\mathcal{P}_k}(x) - L_{\mathcal{P}_k}(x) \geq \varepsilon$. Thus $U(x) - L(x) \geq \varepsilon$ as well, but this is impossible, since $x \in [a,b] \setminus Z$. Hence f must be continuous at x.

Suppose f is bounded and continuous a.e. in $[a,b]$. Let $\{\mathcal{Q}_k\}_{k=1}^{\infty}$ be a sequence of partitions of $[a,b]$ whose norms are tending to zero, and define $L_{\mathcal{Q}_k}$ and $U_{\mathcal{Q}_k}$ for the $\mathcal{Q}_k$'s as we did above for the $\mathcal{P}_k$'s. Since $\mathcal{Q}_{k+1}$ may not be a refinement of $\mathcal{Q}_k$, we may not have that $\{L_{\mathcal{Q}_k}\}_{k=1}^{\infty}$ and $\{U_{\mathcal{Q}_k}\}_{k=1}^{\infty}$ are monotone. However, since f is continuous a.e., both sequences $\{L_{\mathcal{Q}_k}\}_{k=1}^{\infty}$ and $\{U_{\mathcal{Q}_k}\}_{k=1}^{\infty}$ converge to f a.e. Hence by the bounded convergence theorem 9.3.27 we know that $\int_{[a,b]} L_{\mathcal{Q}_k} d\mu$ and $\int_{[a,b]} U_{\mathcal{Q}_k} d\mu$ converge to $\int_{[a,b]} f d\mu$. Moreover, we know that $L(f, [a,b], \mathcal{Q}_k) = \int_{[a,b]} L_{\mathcal{Q}_k} d\mu$ and $U(f, [a,b], \mathcal{Q}_k) = \int_{[a,b]} U_{\mathcal{Q}_k} d\mu$ by construction, and thus they too converge to $\int_{[a,b]} f d\mu$, and thus $f \in \mathcal{R}([a,b])$.

9.3.4 Homework Exercises

1. Let
$$g(x) = \begin{cases} 0 & 0 \leq x \leq \frac{1}{2}, \\ 1 & \frac{1}{2} < x \leq 1. \end{cases}$$
and define a sequence of function by $f_{2k}(x) = g(x)$ and $f_{2k+1}(x) = g(1-x)$. Show that $\liminf_{k \to \infty} f_k = 0$ but $\int_{[0,1]} f_k d\mu = \frac{1}{2}$.

2. Suppose that f is Riemann integrable on $[a,b]$ and define $F(x) = \int_{[a,x]} f d\mu$. Prove that $F'(x) = f(x)$ a.e. on $[a,b]$.

3. Let
$$f_n(x) = \begin{cases} \frac{1}{2n} & -n \leq x \leq n, \\ 0 & |x| > n. \end{cases}$$

Show that $f_n \to 0$ uniformly on $\mathbb{R}$, but $\int_{\mathbb{R}} f_n d\mu = 1$.

4. Show that
$$f(x) = \frac{1}{1+|x|}$$
is not in $L^1(\mathbb{R})$ but is in $L^2(\mathbb{R})$.

5. On $L^1(D)$ we define distance between f, g to be $\|f - g\|_{L^1(D)}$. Show that $L^1(D)$ is a complete metric space under this metric.

6. On $L^2([-\pi, \pi])$ consider the family of functions given by $f_n(x) = \sin nx$. Prove that this set is closed and bounded, but is not compact.

7. Suppose for positive p we have $f_k \in L^p(D)$ for each k and $\int_D |f - f_k|^p d\mu \to 0$ as $k \to \infty$, then $f_k \to_m f$ on D. In particular, there is a subsequence which converges point-wise a.e. D.

8. Suppose that for positive p we have $f_k \in L^p(D)$ for each k, $\int_D |f - f_k|^p d\mu \to 0$ and $\int_D |f_k|^p d\mu \le M$ for all $k \in \mathbb{N}$, then $\int_D |f|^p d\mu \le M$ as well.

9. Suppose that f is measurable and $\int_A f d\mu = 0$ for all measurable $A \subseteq D$, D measurable, then $f = 0$ a.e. on D.

10. Prove the so-called **Cauchy's inequality**
$$ab \le \frac{a^2}{2} + \frac{b^2}{2},$$
and then use this to show that for any $\varepsilon > 0$ there is $C(\varepsilon) > 0$ so that for any positive a, b we have
$$ab \le \varepsilon a^2 + C(\varepsilon) b^2.$$

11. Prove the so-called **Young's inequality**: For $1 < p, q < \infty$ so that $\frac{1}{p} + \frac{1}{q} = 1$ and any $a, b > 0$ we have
$$ab \le \frac{a^p}{p} + \frac{b^q}{q}.$$
To show this, use that the function $f(x) = e^x$ is convex and use $x = e^{\log x}$ for $x > 0$.

9.3. THE RIEMANN AND LEBESGUE INTEGRALS

12. Use Young's inequality to show that for $1 < p, q < \infty$ as above and each $\varepsilon > 0$, there is a $C(\varepsilon)$ so that for any $a, b > 0$ we have
$$ab \leq \varepsilon a^p + C(\varepsilon) b^q.$$

13. Prove the so-called **Hölder's inequality** using Young's inequality: For $1 < p, q < \infty$, so that $\frac{1}{p} + \frac{1}{q} = 1$, and $f \in L^p(D)$ and $g \in L^q(D)$, we have
$$\|fg\|_{L^1(D)} \leq \|f\|_{L^p(D)} \|g\|_{L^q(D)}.$$
    Hint: Use homogeneity to assume the norms of f and g in the respective spaces are 1 and use Young's inequality on $|fg|$.

14. For $1 < p < \infty$ prove the so-called **Minkowski's inequality**: For $f, g \in L^p(D)$ we have
$$\|f + g\|_{L^p(D)} \leq \|f\|_{L^p(D)} + \|g\|_{L^p(D)}.$$

15. For $p \in [1, \infty)$ and D so that $\mu(D) \in (0, \infty)$ define $G(f; p) = \left(\frac{1}{\mu(D)} \int_D |f|^p d\mu\right)^{\frac{1}{p}}$. Use Hölder's inequality to show that for f fixed, G is an increasing function in p. Also, show that $G(f + g; p) \leq G(f; p) + G(g; p)$.

16. Suppose f, g, h are measurable and $f \leq g \leq h$ a.e. If $f, h \in L^1(D)$, does it follow that $g \in L^1(D)$? Please explain.

17. If $f \in L^1(D)$ and the elements of the sequence of functions $\{f_n\}_{n=1}^{\infty}$ are in $L^1(D)$, and suppose $\{f_n(\vec{x})\}_{n=1}^{\infty}$ increases up to its limit value $f(\vec{x})$ for each $\vec{x} \in D$. Does it follow that $\int_D f_n d\mu \to \int_D f d\mu$? Please explain.

18. Construct a sequence of Lebesgue integrable functions $\{f_n\}_{n=1}^{\infty}$ so that $f_n \to 0$ a.e. but $\int_D |f_n| d\mu \not\to 0$.

19. Construct a sequence of Lebesgue integrable functions $\{f_n\}_{n=1}^{\infty}$ so that $f_n \not\to 0$ a.e. but $\int_D |f_n| d\mu \to 0$.

20. Suppose $f, \{f_n\}_{n=1}^{\infty}$ are all Lebesgue integrable, and suppose $\int_D |f_n - f| d\mu \to 0$. Show that $\int_D f_n d\mu \to \int_D f d\mu$ and $\int_D |f_n| d\mu \to \int_D |f| d\mu$.

21. Use Egorov's theorem to prove the following: Suppose $f, \{f_n\}_{n=1}^{\infty}$ are all measurable and uniformly bounded on $[a, b]$. If $f_n \to f$ on $[a, b]$, prove that $\int_{[a,b]} |f_n - f| d\mu \to 0$.

22. Suppose that $f \in L^1(D)$. Show that for any $\varepsilon > 0$, there is a $\delta > 0$ so that for measurable $E \subseteq D$ with $\mu(D) < \delta$, then $\int_E |f| d\mu < \varepsilon$.

23. Suppose $f \in L^1(\mathbb{R})$, and define $F(x) = \int_{(-\infty,x]} f d\mu$, show that F is continuous on $\mathbb{R}$.

24. Find a sequence of Lebesgue integrable functions so that $f_n \to 0$ uniformly but $\int_D |f_n| d\mu = 1$ for all n.

25. Prove **Liapounov's inequality** : If $1 \leq p, q < \infty$ and $0 \leq \alpha \leq 1$, and $r = \alpha p + (1-\alpha)q$ we have
$$\|f\|_{L^r(D)}^r \leq \|f\|_{L^p(D)}^{\alpha p} \|f\|_{L^q(D)}^{(1-\alpha)q}.$$

26. Given $1 \leq p < \infty$, construct functions $f, g \in L^p(\mathbb{R})$ so that $fg \notin L^p(\mathbb{R})$.

27. Compute the following integral:
$$\lim_{n \to \infty} \int_{[0,\infty)} \frac{\sin \frac{x}{n}}{(1 + \frac{x}{n})^n} d\mu.$$

9.4 Iterated Integration and Fubini's Theorem

In this section we would like to establish a familiar result from multivariable calculus that allows us to compute integrals of the form $\int_I f(\vec{x}) dV$, where $I = [a_1, b_1] \times [a_2, b_2] \times \cdots \times [a_n, b_n]$,[12] by

$$\int_I f(\vec{x}) dV = \int_{a_1}^{b_1} [\int_{a_2}^{b_2} \cdots [\int_{a_n}^{b_n} f(x_1, x_2, \cdots, x_n) dx_n] \cdots dx_2] dx_1, \quad (9.22)$$

which is an **iterated integral** of single variable integrals.

Here we shall prove what is commonly referred to as **Fubini's theorem**, which states that for measurable functions so that $f \in L^1(I)$ that (9.22) holds. Note that even in the case of continuous f, the result may not hold in

---
[12]Where $a_j = -\infty$ and $b_j = \infty$ is allowed and any of the [or]'s may be replaced by (or).

9.4. ITERATED INTEGRATION AND FUBINI'S THEOREM

general. To see this consider the example of $f(x,y) := e^{-xy} - 2e^{-2xy}$. Here we have

$$\int_0^1 [\int_1^\infty f(x,y)dx]dy > 0$$

but

$$\int_1^\infty [\int_0^1 f(x,y)dy]dx < 0.$$

Hence we see that in this case we can't have (9.22) hold.

To establish equation (9.22) we shall first prove a lemma and a result commonly called Tonelli's theorem. Moreover, we may commonly use the notation $d\vec{x} = dV$ for the differential volume element in $\mathbb{R}^n$.

Lemma 9.4.1 *For any measurable set $E \subseteq \mathbb{R}^{n+m}$, the set $E_{\vec{x}} = \{\vec{y} : (\vec{x},\vec{y}) \in E\} \subseteq \mathbb{R}^n$ is measurable for a.e. $\vec{x}$, $\mu(E_{\vec{x}})$ is a measurable function on $\mathbb{R}^n$, and*

$$\mu(E) = \int_{\mathbb{R}^n} \mu(E_{\vec{x}})d\vec{x} = \int_{\mathbb{R}^{n+m}} \chi_E d\mu = \int_{\mathbb{R}^n} [\int_{\mathbb{R}^m} \chi_E d\vec{y}]d\vec{x}.$$

Proof: We proceed by cases. **Case 1:** Suppose that $E = I_1 \times I_2$, where I_1, I_2 are intervals. Then in such a case we have that $E_{\vec{x}} = I_2$ if $\vec{x} \in I_1$ and is otherwise the empty set. So, $\mu(E_{\vec{x}}) = \mu(I_2)$ if $\vec{x} \in I_1$ and is otherwise 0. Thus $\mu(E_{\vec{x}})$ is measurable. Moreover,

$$\int_{\mathbb{R}^n} \mu(E_{\vec{x}})d\vec{x} = \int_{I_1} \mu(I_2)d\vec{x} = \mu(I_2) \cdot \mu(I_1) = \mu(I_1 \times I_2).$$

Case 2: Suppose that E is an open subset of $\mathbb{R}^{n+m}$. In such a case, we may use lemma 9.1.7 to write $E = \bigcup_{k=1}^\infty I_k$, where $\{I_k\}_{k=1}^\infty$ is a family of non-overlapping closed cubes. Therefore, $E_{\vec{x}} = \bigcup_k (I_k)_{\vec{x}}$, and so $E_{\vec{x}}$ is measurable. Thus

$$\mu(E) = \sum_k \mu(I_k) = \sum_k \int_{\mathbb{R}^n} \mu((I_k)_{\vec{x}})d\vec{x}$$

$$= \int_{\mathbb{R}^n} \sum_k \mu((I_k)_{\vec{x}})d\vec{x} = \sum_k \int_{\mathbb{R}^n} \mu(E_{\vec{x}})d\vec{x}$$

by case 1, the monotone convergence theorem, and $\mu(E_{\vec{x}}) = \sum_k \mu((I_k)_{\vec{x}})$ for a.e. $\vec{x}$. To see this last statement, write $I_k = I_1^k \times I_2^k$ for each k. To see that $\mu(E_{\vec{x}}) = \sum_k \mu((I_k)_{\vec{x}})$ for a.e. $\vec{x}$, we note that for $\vec{x} \notin \bigcup_k \partial I_1^k$ this holds.

This is enough, since we know $\mu(\partial I_1^k) = 0$. Thus we take $\vec{x} \notin \bigcup_k \partial I_1^k$ and we will show that $\{(I_k)_{\vec{x}} : (I_k)_{\vec{x}} \neq \emptyset\}$ is a non-overlapping collection. Take $(I_k)_{\vec{x}} = I_2^k$ and $(I_j)_{\vec{x}} = I_2^j$ with $\vec{y} \in (I_2^k)^\circ \cap (I_2^j)^\circ$, then $\vec{x} \in (I_1^k)^\circ \cap (I_1^j)^\circ$. So $(\vec{x}, \vec{y}) \in (I_k)^\circ \cap (I_j)^\circ$, which is impossible unless $j = k$.

Case 3: Now we assume E is a bounded set of type G_δ. Then $E = \bigcap_i G_i$ where G_i are all open and $G_i \searrow E$. Then $(G_i)_{\vec{x}}$ is open and $(G_i)_{\vec{x}} \searrow E_{\vec{x}}$. So $E_{\vec{x}}$ is measurable and $\mu(E_{\vec{x}}) = \lim_{i \to \infty} \mu(G_i)_{\vec{x}}$, and by either the bounded convergence theorem or the monotone convergence theorem we see that

$$\mu(E) = \lim_{i \to \infty} \mu(G_i) = \lim_{i \to \infty} \int_{\mathbb{R}^n} \mu((G_i)_{\vec{x}}) d\vec{x}$$

$$= \int_{\mathbb{R}^n} \lim_{i \to \infty} \mu((G_i)_{\vec{x}}) d\vec{x} = \int_{\mathbb{R}^n} \mu(E_{\vec{x}}) d\vec{x}.$$

Case 4: Suppose $\mu(E) = 0$. Here we take $E \subseteq G$ a set of type G_δ satisfying $\mu(G) = 0$. Then

$$0 = \mu(G) = \int_{\mathbb{R}^n} \mu(G_{\vec{x}}) d\vec{x}$$

which implies that $\mu(G_{\vec{x}}) = 0$ for a.e. $\vec{x}$. Since $E_{\vec{x}} \subseteq G_{\vec{x}}$ we have that $\mu(E_{\vec{x}}) = 0$ for a.e. $\vec{x}$. So $E_{\vec{x}}$ is measurable for a.e. $\vec{x}$. Thus $\mu(E_{\vec{x}})$ is a measurable function and

$$\int_{\mathbb{R}^n} \mu(E_{\vec{x}}) d\vec{x} = 0 = \mu(E).$$

Case 5: Suppose E is a bounded measurable set. In this case we may write $E = G \setminus Z$ where G is a bounded set of type G_δ and $\mu(Z) = 0$, with Z bounded as well. So $E_{\vec{x}} = G_{\vec{x}} \setminus Z_{\vec{x}}$ for a.e. $\vec{x}$ and $\mu(E_{\vec{x}}) = \mu(G_{\vec{x}})$ for a.e. $\vec{x}$. Therefore,

$$\mu(E) = \mu(G) = \int_{\mathbb{R}^n} \mu(G_{\vec{x}}) d\vec{x} = \int_{\mathbb{R}^n} \mu(E_{\vec{x}}) d\vec{x}.$$

Case 6: Finally E is an arbitrary measurable set. Here we let $E_k = E \cap \{\vec{x} : \|\vec{x}\| < k\}$. Then $E_k \nearrow E$, and so $(E_k)_{\vec{x}} \nearrow E_{\vec{x}}$ for all $\vec{x}$. So, by the monotone convergence theorem we have

$$\mu(E) = \lim_{k \to \infty} \mu(E_k) = \lim_{k \to \infty} \int_{\mathbb{R}^n} \mu((E_k)_{\vec{x}}) d\vec{x}$$

9.4. ITERATED INTEGRATION AND FUBINI'S THEOREM

$$= \int_{\mathbb{R}^n} \lim_{k\to\infty} \mu((E_k)_{\vec{x}}) d\vec{x} = \int_{\mathbb{R}^n} \mu(E_{\vec{x}}) d\vec{x}.$$

□

Next we shall discuss what is commonly called Tonelli's theorem, which is an important ingredient in the proof of Fubini's theorem.

Tonelli's Theorem 9.4.2 *If $f(\vec{x},\vec{y}) \geq 0$ and is measurable on $\mathbb{R}^{n+m}$, then $f(\vec{x},\vec{y})$ is measurable in $\vec{y}$ for a.e. $\vec{x} \in \mathbb{R}^n$, $\int_{\mathbb{R}^m} f(\vec{x},\vec{y}) d\vec{y}$ is non-negative and measurable in $\vec{x}$ for a.e. $\vec{x}$ and*

$$\int_{\mathbb{R}^{n+m}} f d\mu = \int_{\mathbb{R}^n} [\int_{\mathbb{R}^m} f(\vec{x},\vec{y}) d\vec{y}] d\vec{x}.$$

Proof: For the special case of $f = \chi_E$, with E measurable, this is lemma 9.4.1, and so by linearity of integration the result also holds for simple functions which are non-negative in value.

For a general non-negative measurable f we take a sequence of simple functions $\{f_k\}_{k=1}^{\infty}$ where $f_k \nearrow f$. That is, point-wise f_k monotonically increases up to f. Write $f_k = \sum_j a_{k,j} \chi_{A_{j,k}}$, where for each k the collection $\{A_{k,j}\}_j$ are pairwise disjoint and measurable. Thus

$$f_k(\vec{x},\vec{y}) = \sum_j a_{k,j} \chi_{(A_{k,j})_{\vec{x}}}(\vec{y})$$

for a.e. $\vec{x}$. For each $\vec{x}$ fixed, by lemma 9.4.1 we have

$$\int_{\mathbb{R}^{n+m}} f_k d\mu = \int_{\mathbb{R}^{n+m}} \sum_j a_{k,j} \chi_{A_{k,j}} d\mu$$

$$= \sum_j a_{k,j} \mu(A_{k,j}) = \sum_j a_{k,j} \int_{\mathbb{R}^n} \mu((A_{k,j})_{\vec{x}}) d\vec{x} = \sum_j \int_{\mathbb{R}^n} [\int_{\mathbb{R}^m} \chi_{(A_{k,j})_{\vec{x}}}(\vec{y}) d\vec{y}] d\vec{x}$$

$$= \int_{\mathbb{R}^n} [\int_{\mathbb{R}^m} \sum_j a_{k,j} \chi_{(A_{k,j})_{\vec{x}}}(\vec{y}) d\vec{y}] d\vec{x} = \int_{\mathbb{R}^n} [\int_{\mathbb{R}^m} f_k(\vec{x},\vec{y}) d\vec{y}] d\vec{x}.$$

For each k there is a set Z_k of $\mathbb{R}^n$ measure zero for which $\int_{\mathbb{R}^m} f_k(\vec{x},\vec{y}) d\vec{y}$ is undefined. $f_k(\vec{x},\vec{y})$ is measurable in $\vec{y}$ for a.e. $\vec{x}$, and by the monotone convergence theorem we have $\int_{\mathbb{R}^m} f_k(\vec{x}, \vec{y}) d\vec{y} \to \int_{\mathbb{R}^m} f(\vec{x}, \vec{y}) d\vec{y}$ for a.e. $\vec{x}$. So by the monotone convergence theorem, we have

$$\int_{\mathbb{R}^{n+m}} f \, d\mu = \lim_{k \to \infty} \int_{\mathbb{R}^{n+m}} f_k \, d\mu = \lim_{k \to \infty} \int_{\mathbb{R}^n} [\int_{\mathbb{R}^m} f_k \, d\vec{y}] d\vec{x}$$

$$= \int_{\mathbb{R}^n} [\lim_{k \to \infty} \int_{\mathbb{R}^m} f_k \, d\vec{y}] d\vec{x} = \int_{\mathbb{R}^n} [\int_{\mathbb{R}^m} f \, d\vec{y}] d\vec{x}.$$

□

A corollary of Tonelli's theorem is that for E_1 and E_2 measurable subsets of $\mathbb{R}^n$ and $\mathbb{R}^m$ respectively, endowed with the usual Lebesgue measures, then $E_1 \times E_2$ is Lebesgue measurable in $\mathbb{R}^{n+m}$ and in particular

$$\mu(E_1 \times E_2) = \mu(E_1) \cdot \mu(E_2)$$

with the measures μ be evaluated in the $\mathbb{R}^{n+m}$, $\mathbb{R}^n$ and $\mathbb{R}^m$ respectively as we go from left to right in the above equation.

At last, we arrive at what is usually referred to as Fubini's theorem.

Fubini's Theorem 9.4.3 *If $f \in L^1(\mathbb{R}^{n+m})$, then $f(\vec{x}, \vec{y}) \in L^1(\mathbb{R}^m)$ for a.e. fixed $\vec{x}$, $\int_{\mathbb{R}^m} f(\vec{x}, \vec{y}) d\vec{y} \in L^1(\mathbb{R}^n)$, and*

$$\int_{\mathbb{R}^{n+m}} f \, d\mu = \int_{\mathbb{R}^n} [\int_{\mathbb{R}^m} f(\vec{x}, \vec{y}) d\vec{y}] d\vec{x}.$$

Proof: We apply Tonelli's theorem 9.4.2 to f^+ and f^- to get

$$\int_{\mathbb{R}^{n+m}} f^+ d\mu = \int_{\mathbb{R}^n} [\int_{\mathbb{R}^m} f^+ d\vec{y}] d\vec{x},$$

$$\int_{\mathbb{R}^{n+m}} f^- d\mu = \int_{\mathbb{R}^n} [\int_{\mathbb{R}^m} f^- d\vec{y}] d\vec{x}.$$

Since $f \in L^1(\mathbb{R}^{n+m})$ we know $\int_{\mathbb{R}^{n+m}} f^+ d\mu < \infty$ and so $\int_{\mathbb{R}^m} f^+(\vec{x}, \vec{y}) d\vec{y} < \infty$ for a.e. $\vec{x}$. Similarly, $\int_{\mathbb{R}^m} f^-(\vec{x}, \vec{y}) d\vec{y} < \infty$ for a.e. $\vec{x}$. So when viewed as a function of $\vec{y}$, $f = f^+ - f^-$ is in $L^1(\mathbb{R}^m)$ for a.e. $\vec{x}$, and so

$$\int_{\mathbb{R}^m} f(\vec{x}, \vec{y}) d\vec{y} = \int_{\mathbb{R}^m} f^+(\vec{x}, \vec{y}) d\vec{y} - \int_{\mathbb{R}^m} f^-(\vec{x}, \vec{y}) d\vec{y}.$$

9.4. ITERATED INTEGRATION AND FUBINI'S THEOREM

Since
$$\int_{\mathbb{R}^m} [\int_{\mathbb{R}^n} f^+(\vec{x},\vec{y})d\vec{y}]d\vec{x} < \infty$$

and
$$\int_{\mathbb{R}^m} [\int_{\mathbb{R}^n} f^-(\vec{x},\vec{y})d\vec{y}]d\vec{x} < \infty$$

we have
$$\int_{\mathbb{R}^{n+m}} f d\mu = \int_{\mathbb{R}^n} [\int_{\mathbb{R}^m} f d\vec{y}]d\vec{x}.$$

□

Note that although our results are done in terms of $\mathbb{R}^n$, $\mathbb{R}^m$ and $\mathbb{R}^{n+m}$, it is easy to establish similar results for general measurable sets instead.

9.4.1 Convolutions

For $f, g \in L^1(\mathbb{R}^n)$ we define the **convolution** of f and g to be

$$(f * g)(\vec{x}) = \int_{\mathbb{R}^n} f(\vec{x} - \vec{t}) g(\vec{t}) d\vec{t}, \tag{9.23}$$

provided the integral exists. We first claim that $(f * g)(\vec{x}) = (g * f)(\vec{x})$ for each $\vec{x} \in \mathbb{R}^n$ for which the convolutions exist. This follows from a simple change of variables $\vec{t} \mapsto \vec{x} - \vec{t}$. Moreover, by Fubini's theorem 9.4.3

$$\int_{\mathbb{R}^n} |(f * g)(\vec{x})| d\vec{x} = \int_{\mathbb{R}^n} |\int_{\mathbb{R}^n} f(\vec{t}) g(\vec{x} - \vec{t}) d\vec{t}| d\vec{x}$$

$$\leq \int_{\mathbb{R}^n} [\int_{\mathbb{R}^n} |f(\vec{t}) g(\vec{x} - \vec{t})| d\vec{t}] d\vec{x} = \int_{\mathbb{R}^n} |f(\vec{t})| \int_{\mathbb{R}^n} |g(\vec{x} - \vec{t})| d\vec{x} d\vec{t}$$

$$= \int_{\mathbb{R}^n} |f(\vec{t})| d\vec{t} \cdot \int_{\mathbb{R}^n} |g(\vec{y})| d\vec{y},$$

where $\vec{y} = \vec{x} - \vec{t}$ for any $\vec{t}$ fixed. Thus we see that $f * g$ is in $L^1(\mathbb{R}^n)$ and so for $f, g \in L^1(\mathbb{R}^n)$, we have $(f * g)(\vec{x})$ exists for a.e. $\vec{x}$.

9.4.2 Homework Exercises

1. Use Fubini's theorem to deduce that $\int_{-\infty}^{\infty} e^{-x^2} dx = \sqrt{\pi}$. Do this by using polar coordinates and using that

$$\int_{-\infty}^{\infty} e^{-x^2} dx \cdot \int_{-\infty}^{\infty} e^{-y^2} dy = \int_{-\infty}^{\infty} [\int_{-\infty}^{\infty} e^{-x^2-y^2} dx] dy = \int_{0}^{\infty} [\int_{0}^{2\pi} e^{-r^2} r \, dr] d\theta.$$

2. Let V_n be the volume (Lebesgue measure) of the unit ball in $\mathbb{R}^n$. Use Fubini's theorem to write

$$V_n = 2V_{n-1} \int_0^1 (1-t^2)^{\frac{n-1}{2}} dt.$$

3. For $a > 0$ show that

$$\int_{\mathbb{R}^n} e^{-a\|\vec{x}\|^2} dV = \left(\frac{\pi}{a}\right)^{\frac{n}{2}}.$$

4. Letting $\sigma(\mathbb{S}^{n-1})$ be the hypersurface area of $\mathbb{S}^{n-1} = \{\vec{x} : \|\vec{x}\| = 1\}$ in $\mathbb{R}^n$, establish that

$$\sigma(\mathbb{S}^{n-1}) = \frac{2\pi^{\frac{n}{2}}}{\Gamma(\frac{n}{2})},$$

where $\Gamma(x) = \int_0^\infty t^{x-1} e^{-t} dt$ is the so-called **Gamma function** defined for $x \in \mathbb{C}$ with positive real parts. Here you may assume the so-called polar coordinates or co-area formula that allows one to compute $\int_{\mathbb{R}^n} f dV = \int_0^\infty [\int_{\partial B_r(\vec{x}_0)} f d\sigma] dr$. Then of course you have $\sigma(\mathbb{S}^{n-1}) = \int_{\partial B_1(\vec{0})} 1 d\sigma$ as a surface integral in the calculus sense. Start by using the above exercise's result with $a = 1$, and use polar coordinates to rewrite this integral.

5. Using that $\Gamma(\frac{1}{2}n + 1) = \frac{n}{2}\Gamma(\frac{1}{2}n)$, show that

$$\mu(\{\vec{x} : \|\vec{x}\| < 1\}) = \frac{\pi^{\frac{n}{2}}}{\Gamma(\frac{1}{2}n + 1)}.$$

Show this by establishing that $\mu(\{\vec{x} : \|\vec{x}\| < 1\}) = n^{-1}\sigma(\mathbb{S}^{n-1})$.

6. Suppose f is measurable on $\mathbb{R}^{n+m}$ and $\int_{\mathbb{R}^n}[\int_{\mathbb{R}^m}|f(\vec{x},\vec{y})|d\vec{y}]d\vec{x} < \infty$. Then apply Tonelli's theorem to $|f|$, and use Fubini's theorem to conclude that $f \in L^1(\mathbb{R}^{n+m})$ and

$$\int_{\mathbb{R}^{n+m}} f d\mu = \int_{\mathbb{R}^n}[\int_{\mathbb{R}^m} f d\vec{y}]d\vec{x} = \int_{\mathbb{R}^m}[\int_{\mathbb{R}^n} f d\vec{x}]d\vec{y}.$$

7. Let

$$f(x,y) = \begin{cases} y^{-2} & 0 < x < y < 1, \\ -x^{-2} & 0 < y < x < 1, \\ 0 & \text{otherwise.} \end{cases}$$

Show that

$$\int_0^1 \int_0^1 f(x,y)dxdy = 1,$$

and

$$\int_0^1 \int_0^1 f(x,y)dydx = -1.$$

Appendix A

Preliminary Materials

A.1 Sets and Related Notation

- $\mathbb{N}$ is the set of natural number; $\mathbb{Z}$ is the set of integers; $\mathbb{Q}$ is the set of rational numbers; $\mathbb{R}$ is the set of real numbers; $\mathbb{C}$ is the set of complex numbers

- $\in$ Reads "is in" or "is an element of." For example: $3 \in \mathbb{N}$ reads "3 is in the set of natural numbers."

- $\subseteq$ reads "subset of." For example: $\{1\} \subseteq \{1,2\}$.

- $\subset$ reads is a "proper subset of." For example: $\mathbb{N} \subset \mathbb{Z}$.

- $/$ reads "not" and simply negates the symbol after it. For example: $\neq$ reads "is not equal to"; $\notin$ reads "is not in" or "is not an element of"; $\nsubseteq$ reads "is not a subset of."

- $\cup$ reads "union" and $\cap$ reads "intersect" and is used in relation to sets. For example $A \cap B$ reads "A intersect B" and $A \cup B$ reads "A union B." $A \cup B$ is the set you get when you combine all the elements of A with those of B (i.e. the elements of the universal set that are in A or in B). $A \cap B$ is the set of elements that are in both A and B.

- ^{c} reads "complement." For example A^c reads "A complement." This is the set of elements of the universal set that are not in A.

A.2 Notation in Symbolic Logic

- $\Rightarrow$ reads "implies." This is used with propositions. Here $A \Rightarrow B$ reads "A implies B" or "If A, then B."

- $\iff$ reads "if and only if" or "is equivalent to." This is also used in the context of propositions. For example, $A \iff B$ reads "A if and only if B" or "A is equivalent to B."

- $\sim$ reads "not" and negates a proposition. For example $\sim A$ reads "not A." If A is true then $\sim A$ is false, and if A is false $\sim A$ is true.

- $\forall$ reads "for all" and is a quantifier. For example $\forall x$ reads "for all x."

- $\exists$ reads "there exists" and is a quantifier. For example $\exists x$ reads "there exists an x."

- $\vee$ reads "or." This is used in the context of propositions. For example $A \vee B$ reads "A or B."

- $\wedge$ reads "and." This is used in the context of propositions. For example $A \wedge B$ reads "A and B."

A.3 The Basics in Symbolic Logic

- DeMorgan's Laws for propositions: $\sim (A \vee B)$ is logically equivalent to $(\sim A) \wedge (\sim B)$. That is the negation of "A or B" is "not A and not B." We also note that the negation of "A or B" is often read as "neither A nor B." $\sim (A \wedge B)$ is logically equivalent to $(\sim A) \vee (\sim B)$. That is the negation of "A and B" is "not A or not B."

- DeMorgan's laws for sets: $(A \cap B)^c = A^c \cup B^c$ and $(A \cup B)^c = A^c \cap B^c$.

- Some items to note about certain compound propositions:

    - $A \Rightarrow B$ is false only when A is true and B is false.

A.3. THE BASICS IN SYMBOLIC LOGIC

A	B	$A \Rightarrow B$
T	T	T
T	F	F
F	T	T
F	F	T

- $A \wedge B$ is true only when both A and B are true.

A	B	$A \wedge B$
T	T	T
T	F	F
F	T	F
F	F	F

- $A \vee B$ is false only when both A and B are false.

A	B	$A \vee B$
T	T	T
T	F	T
F	T	T
F	F	F

- $A \iff B$ is true if A and B have the same truth value. That is, both are true or both are false.

A	B	$A \iff B$
T	T	T
T	F	F
F	T	F
F	F	T

- A and $\sim A$ have opposite truth values (one is true and the other is false).

A	$\sim A$
T	F
F	T

- Use
$$P \Rightarrow Q$$
to translate:

If P, then Q.
P implies Q.
P is sufficient for Q.
P only if Q.
Q, if P.
Q whenever P.
Q is necessary for P.
Q, when P.

- Use
$$P \iff Q$$
to translate:

P if and only if Q.
P if, but only if, Q.
P is equivalent to Q.
P is necessary and sufficient for Q.

- **The Order of Operations We Follow in Symbolic Logic:**

  - You examine the propositions in parentheses first, working from the inside out. In the innermost parentheses, or in the absence of parentheses, follow the remaining steps.
  - $\sim$ is performed first, going from left to right.
  - $\wedge$ is performed second, going from left to right.
  - $\vee$ is performed third, going from left to right.
  - $\Rightarrow$ is performed fourth, going from left to right.
  - $\iff$ is performed last, going from left to right. [1]

- We say that two propositions P and Q are equivalent if $P \iff Q$ is a tautology, meaning it is always true.

- Given a conditional
$$P \Rightarrow Q,$$

---

[1] When in doubt, use parentheses!

the **converse** is
$$Q \Rightarrow P,$$
the **contrapositive** is
$$\sim Q \Rightarrow \sim P,$$
and the **inverse** is
$$\sim P \Rightarrow \sim Q.$$

The conditional and its contrapositive are logically equivalent.

The inverse and the converse are logically equivalent.

The conditional and its converse are not logically equivalent.

- Another logical equivalence to observe that relates a conditional proposition to a disjunction is
$$P \Rightarrow Q \equiv (\sim P) \vee Q.$$

To see this, note the truth table below:

P	Q	$\sim P$	$\sim P \vee Q$	$P \Rightarrow Q$
T	T	F	T	T
T	F	F	F	F
F	T	T	T	T
F	F	T	T	T

A.4 Quantified Sentences in Symbolic Logic

Quantifiers are applied to elements in the universe of discourse U for a given open sentence $P(x)$. Upon fixing a value of $x \in U$, $P(x)$ is either a true or false proposition.

- **All, every, none, no** refer to every element in your reference set or universe of discourse, and thus they are commonly referred to as **universal quantifiers**.

- **Some, there exists, not all** refer to one or more elements of your reference set or universe of discourse, and thus they are commonly referred to as **existential quantifiers.**

- **All, every, each** have the same mathematical meaning; **some, there exists at least one** also have the same mathematical meaning.

An open sentence becomes a proposition when a quantifier is applied to it, and this type of proposition is called a **quantified sentence.**

The quantifier $\forall$ reads **for all** and the quantifier $\exists$ reads **there exists.** Thus
$$(\forall x)P(x)$$
would read **for all** x, $P(x)$.

Also,
$$(\exists x)P(x)$$
reads **there exists an** x **such that** $P(x)$ or **for some** x, $P(x)$.

The proposition
$$(\forall x)P(x)$$
is true if and only if the truth set of $P(x)$ is the universe of discourse.

The proposition
$$(\exists x)P(x)$$
is true if and only if the truth set is non-empty.

Given two open sentences $P(x)$ and $Q(x)$ with a common universe of discourse, we read
$$(\forall x)[P(x) \Rightarrow Q(x)]$$
as **All** $P(x)$ **are** $Q(x)$. and
$$(\exists x)[P(x) \wedge Q(x)]$$
as **Some** $P(x)$ **are** $Q(x)$. A special case of this is when $P(x)$ is $x \in A$, that is x is in A. **If** x **is in** A, **then** $P(x)$. or **For all** x **in** A, $P(x)$. Note that this is written
$$(\forall x)[x \in A \Rightarrow P(x)].$$

A.4. QUANTIFIED SENTENCES IN SYMBOLIC LOGIC

Often this is abbreviated as

$$(\forall x \in A)P(x).$$

Likewise, **There is an x in A such that $P(x)$.** is written

$$(\exists x)[x \in A \wedge P(x)].$$

Often this is abbreviated as

$$(\exists x \in A)P(x).$$

Two quantified sentences are **equivalent in a given universe** of discourse if and only if they have the same truth sets within that universe. Moreover, two quantified sentences are **equivalent** if and only if they are equivalent in a universe for all universes of discourse.

A.4.1 Negations of Quantified Sentences

If $P(x)$ is an open sentence with a variable x in a universe of discourse U, then

$\sim [(\forall x)P(x)]$ is equivalent to $(\exists x)[\sim P(x)]$,

and

$\sim [(\exists x)P(x)]$ is equivalent to $(\forall x)[\sim P(x)]$.

To see a proof of these suppose $\sim [(\forall x)P(x)]$ is true. Then $(\forall x)P(x)$ is false. So, the truth set of $P(x)$ is not all of U. Thus the truth set of $\sim P(x)$ is non-empty. Hence $(\exists x)[\sim P(x)]$ is true.

Next, suppose $\sim [(\exists x)P(x)]$ is true. Then $(\exists x)P(x)$ is false. Thus the truth set of $P(x)$ is empty. Hence, the truth set of $\sim P(x)$ is all of U. Thus $(\forall x)[\sim P(x)]$ is true.

A.5 Equivalence Relations

Let R be a relation on a set S (where formally we see R as a subset of $S \times S$, where x is R-related to y if and only if $(x, y) \in R$). R is said to be an **equivalence relation** if R is *reflexive, symmetric, and transitive*. R is reflexive on S means that every $x \in S$ satisfies x is R-related to x. R is symmetric on S means that if x is R-related to y (for $x, y \in S$), then y is also R-related to x. R is transitive on S means that for $x, y, z \in S$ if we have x is R-related to y and y is R-related to z, then x is R-related to z.

An important result to note is equivalence relations give rise to a **partition** of S into so-called **equivalence classes**. We use $\sim_R$ to mean R-related. For an equivalence relation R on S we define

$$[x]_R = \{y \in S : x \sim_R y\}.$$

$[x]_R$ is often called the **equivalence class** of x, and sometimes $\frac{x}{R}$ is used to denote this instead.

The collection of equivalence classes is often called S **modulo** R, and denoted by $\frac{S}{R}$.

Theorem: Equivalence relations on a set S give rise to partitions of S and vice versa.

Proof: Since R is transitive $x \in [x]_R$, so the equivalence class of an element is non-empty. Moreover, every element $x \in S$ is in an equivalence class, namely $[x]_R$.

Now consider two equivalence classes $[x]_R$ and $[y]_R$, and suppose $z \in [x]_R \cap [y]_R$. Thus $x \sim_R z, y \sim_R z$ and so $z \sim_R y$ as well, by symmetry.

By transitivity we get $x \sim_R y$, and of course by symmetry $y \sim_R x$.

Any element $w \in [y]_R$ satisfies $y \sim_R w$, and thus by transitivity, $x \sim_R w$. Thus $[y]_R \subseteq [x]_R$. Likewise, any $v \in [x]_R$ satisfies $x \sim_R v$, and by symmetry $v \sim_R x$. Using $x \sim_R y$ and transitivity, we get $v \sim_R y$, and by symmetry $y \sim_R v$. Thus $[x]_R \subseteq [y]_R$.

A.6. THE NATURAL NUMBERS AND INDUCTION

Hence
$$[x]_R \cap [y]_R \neq \emptyset \quad \Rightarrow \quad [x]_R = [y]_R.$$

Hence $\{[x]_R : x \in S\}$ forms a partition of S. That is $\{[x]_R : x \in S\}$ is a family of disjoint subsets whose union is A.

Now suppose $\{A_\alpha : \alpha \in \Delta\}$ is a partition of S.

We define a relation R on S by $x \sim_R y$ if and only if $x, y \in A_\beta$ for some $\beta \in \Delta$.

Clearly for any $x \in S$, $x \sim_R x$ since x is in one of the elements of this family of sets. Hence R is reflexive.

Moreover, $x \sim_R y$ if and only if $x, y \in A_\beta$ for some $\beta \in \Delta$. So clearly $y \sim_R x$ also holds. Hence R is symmetric.

Now suppose $x \sim_R y$ and $y \sim_R z$. Then $x, y \in A_\beta$ for some $\beta \in \Delta$. Moreover, $y, z \in A_\gamma$ for some $\gamma \in \Delta$. Thus $y \in A_\beta \cap A_\gamma$ and so $\beta = \gamma$ Hence, $x, z \in A_\beta$, which yields $x \sim_R z$. Thus R is transitive. Hence R is an equivalence relation on S.
□

A.6 The Natural Numbers and Induction

There is an important set of numbers called the **natural numbers**,[2] denoted by $\mathbb{N}$, and which satisfies the property that $1 \in \mathbb{N}$, and if $n \in \mathbb{N}$, then $n + 1 \in \mathbb{N}$. Moreover, there does not exists an $m \in \mathbb{N}$ so that $m + 1 = 1$. Here of course $\mathbb{N}$ has two binary operations $+, \cdot$.

Moreover, the set of natural numbers is closed under $+$ and $\cdot$ and satisfies the following list of properties for any $k, m, n \in \mathbb{N}$:

---

[2]The natural numbers $\mathbb{N}$ can actually be characterized by the following five axioms: A1: $1 \in \mathbb{N}$. A2: $\forall x \in \mathbb{N}$ there exists a unique $(x+1) \in \mathbb{N}$, called the successor of x. A3: $\forall x \in \mathbb{N}$, $x + 1 \neq 1$. That is 1 is not the successor of any natural number. A4: If $x + 1 = y + 1$, then $x = y$. A5 (the axiom of induction): Suppose $A \subseteq \mathbb{N}$ satisfies: $1 \in A$; $x \in A$ implies $(x+1) \in A$. Then $A = \mathbb{N}$.

1. **Associativity:**
$$k + (m + n) = (k + m) + n, \quad k \cdot (m \cdot n) = (k \cdot m) \cdot n$$

2. **Commutativity:**
$$k + m = m + k, \quad k \cdot m = m \cdot k$$

3. **Existence of a Unity:** There is an element $1 \in \mathbb{N}$, called **one**, so that
$$1 \cdot k = k$$
for any $k \in \mathbb{N}$.

4. **Cancellation Property:**
$$k + m = k + n \Rightarrow m = n$$
$$k \cdot m = k \cdot n \Rightarrow m = n$$

5. **The Distributive Property:**
$$k \cdot (m + n) = (k \cdot m) + (k \cdot n)$$

6. For a given pair of natural numbers m, n one and only one of the following holds:
   - $m = n$,
   - $m = n + k$ for some $k \in \mathbb{N}$,
   - $n = m + k$ for some $k \in \mathbb{N}$.

   This in turn allows us to define an ordering on $\mathbb{N}$. The following are **order properties:**

   (a) $m < n$ if and only if there exists $j \in \mathbb{N}$ so that $n = m + j$.
   (b) $m \leq n$ if and only if $m < n$ or $m = n$.
   (c) $k < m$ and $m < n$ implies $k < n$.
   (d) $m \leq n$ and $n \leq m$ implies $m = n$
   (e) $m < n$ implies $m + k < n + k$ and $k \cdot m < k \cdot n$.

A.6. THE NATURAL NUMBERS AND INDUCTION

Moreover, we also have the following axiom for the set $\mathbb{N}$ of natural numbers:

The Axiom of Induction:[3] Let $S \subseteq \mathbb{N}$. Suppose S satisfies

1. $1 \in S$.

2. $n \in S$ implies $n + 1 \in S$.

Then $S = \mathbb{N}$.

From this characterization of $\mathbb{N}$ in terms of the above properties, we see that $\mathbb{N} = \{1, 2, 3, 4, \cdots\}$.

A.6.1 The Well-Ordering Principle

The **well-ordering property (WOP)** of $\mathbb{N}$ says that

Every non-empty subset of $\mathbb{N}$ has a smallest element.

It turns out that this is equivalent to the principle of mathematical induction.

(WOP) implies (PMI):

Proof: (WOP) $\Rightarrow$ (PMI):

Suppose $S \subseteq \mathbb{N}$ satisfies

1. $1 \in S$.

2. $n \in S$ implies $n + 1 \in S$.

If $S \neq \mathbb{N}$, then $\mathbb{N} \setminus S \neq \emptyset$.

By the Well-Ordering Principle, $\mathbb{N} \setminus S$ has a least element, say m. Since $1 \in S$, $m > 1$. Also, $m - 1 \notin \mathbb{N} \setminus S$, thus $m - 1 \in S$. Hence $m \in S$. Contradiction! Since $m \in \mathbb{N} \setminus S$. Thus $S = \mathbb{N}$. □

---

[3]Often called the **principle of mathematical induction**, or **PMI** for short.

(PMI) implies (WOP):

Proof: (PMI) $\Rightarrow$ (WOP):

Suppose that for each $S \subseteq \mathbb{N}$, if S satisfies

1. $1 \in S$.

2. $n \in S$ implies $n + 1 \in S$.

Then $S = \mathbb{N}$.

We will use this to show the well-ordering principle. That is, every non-empty subset of $\mathbb{N}$ has a smallest element.

Let $W \subseteq \mathbb{N}$ be a non-empty subset. Let $S = \mathbb{N} \setminus W$. Then since $W \neq \emptyset$ we know $S \neq \mathbb{N}$. We proceed with a proof by contradiction.

Since $1 \in \mathbb{N}$ is the smallest element of $\mathbb{N}$, and W does not have a smallest element, then $1 \notin W$.

Suppose $n \in S$. Then none of $\{1, 2, 3, \cdots, n-1\}$ could be in W, since otherwise W would have a smallest element, namely the smallest element of $\{1, 2, \cdots, n-1\} \cap W$, since this set is finite, and finite sets all have smallest elements.

Moreover, every $m \geq n$ is larger than the elements of $\{1, 2, 3, \cdots, n-1\}$.

Also $n \notin W$ since $n \in S = \mathbb{N} \setminus W$. If $n + 1 \in W$ than W would have this as its smallest element. Thus $n + 1 \in S$.

Thus by PMI we see that $S = \mathbb{N}$, and this is a contradiction.
$\square$

A.6.2 Inductive Sets

A set S is called an **inductive set** if S is non-empty and $n \in S \Rightarrow n+1 \in S$.

A.6. THE NATURAL NUMBERS AND INDUCTION

If S is an inductive set, and $1 \in S$, we see that $S = \mathbb{N}$ by (PMI).

If $n_0 \in S$ is the smallest element of S and S is inductive, then we see that

$$S = \{n_0, n_0 + 1, n_0 + 2, \cdots\}.$$

A.6.3 How the Principle of Mathematical Induction is Used in Proofs

Let $n_0 \in \mathbb{N}$ and let

$$P(n)$$

be a statement for each natural number $n \geq n_0$. Suppose that

- $P(n_0)$ is true.

- $\forall k \geq n_0$, the truth of $P(k)$ implies the truth of $P(k+1)$.

Then $P(n)$ is true $\forall n \geq n_0$.

A.6.4 The Principle of Complete Induction

Another version[4] of the Principle of Mathematical Induction is called the **Principle of Complete Induction (PCI)**:

Let $S \subseteq \mathbb{N}$ be such that: *If $\{1, 2, \cdots, k-1\} \subseteq S$, then $k \in S$.*

Then $S = \mathbb{N}$.

(PMI) implies (PCI):

Proof: (PMI) $\Rightarrow$ (PCI):
Assume (PMI) holds for $\mathbb{N}$. Suppose $S \subseteq \mathbb{N}$ satisfies the property: *For any $m \in \mathbb{N}$, $\{1, 2, \cdots, m-1\} \subseteq S$ implies $m \in S$.*
First we will show that for all $n \in \mathbb{N}$, we have $\{1, 2, \cdots, n\} \subseteq S$, and we will use (PMI) to show this.

---

[4]That is this version is equivalent to the original version of the Principle of Mathematical Induction.

1. Case $n = 1$: Here $\{1, 2, \cdots, n-1\} = \emptyset$ and thus $\{1, 2, \cdots, n-1\} \subseteq S$. Hence $1 \in S$.

2. Assume $\{1, 2 \cdots, n-1\} \subseteq S$. Then by assumption $n \in S$ and thus $\{1, 2, \cdots, n\} \subseteq S$.

3. By steps 1, and 2, and (PMI), we see
$$\{1, 2, 3 \cdots, m\} \subseteq S, \quad \forall m \in \mathbb{N}.$$

4. Let $k \in \mathbb{N}$. By step 3 we see that $k \in S$. Thus $\mathbb{N} \subseteq S$. Since $S \subseteq \mathbb{N}$ was assumed, we have $S = \mathbb{N}$.

□

(PCI) implies (PMI):

Proof: (PCI) $\Rightarrow$ (PMI):
Here we are assume (PCI) holds for $\mathbb{N}$.
Suppose S is a subset of $\mathbb{N}$ which satisfies

1. $1 \in S$

2. $n \in S$ implies $n + 1 \in S$.

We need to show $S = \mathbb{N}$. We will do this by show S satisfies the hypotheses of (PCI). Namely, for $m \in \mathbb{N}$, if $\{1, 2, \cdots, m-1\} \subseteq S$, then $m \in S$.

If $m = 1$. Clearly $\emptyset \subseteq S$ and thus $1 \in S$ by assumption. Thus the hypothesis of (PCI) holds for $m = 1$.

Next, suppose $m > 1$ and $\{1, 2, \cdots, m-1\} \subseteq S$. Then $m-1, 1 \in S$ and by our assumptions on S, we have $m \in S$.

Hence, by (PCI) we have $S = \mathbb{N}$

□

Appendix B

Homework Solutions

B.1 Homework Solutions 1.2.3

1. Suppose that $n < m < n+1$ are all natural numbers. Then $m = n+k$ for some $k \in \mathbb{N}$, since $n < m$. Also, $n+1 = m+j$ for some $j \in \mathbb{N}$, since $m < n+1$. Thus $n+1 = m+j = n+k+j$, and so $1 = k+j$. If $j = 1$, then this is a contradiction. Otherwise $j > 1$ and so $j = (j-1) + 1$, with $j-1 \in \mathbb{N}$. In such a case, $1 = k + (j-1) + 1$, a contradiction.

2. Suppose $ri = \frac{a}{b} \in \mathbb{Q}$, and $r = \frac{m}{n} \in \mathbb{Q}$. Since $r \neq 0$ we have $m \neq 0$, and so $i = \frac{a/b}{m/n} = \frac{an}{bm} \in \mathbb{Q}$, a contradiction. Likewise, let $r + i = \frac{c}{d} \in \mathbb{Q}$. Then $i = \frac{c}{d} - \frac{m}{n} = \frac{cn-md}{dn} \in \mathbb{Q}$, a contradiction.

3. Suppose $\sqrt{3} = \frac{a}{b} \in \mathbb{Q}$. Without loss of generality, $\gcd(a,b) = 1$, and $b > 0$. Then $3 = \frac{a^2}{b^2}$ and so $3b^2 = a^2$. Thus 3 divides a^2, and so 3 divides a. Hence $a = 3k$. Hence $3b^2 = (3k)^2 = 9k^2$ and so $b^2 = 3k^2$. Thus 3 divides b^2 and so 3 divides b, a contradiction to $\gcd(a,b) = 1$.

4. Observe that $\frac{1}{a} + \frac{1}{-a} = \frac{1 \cdot (-a) + 1 \cdot a}{a \cdot (-a)} = \frac{-a+a}{a \cdot (-a)} = \frac{0}{a \cdot (-a)} = 0 \cdot \frac{1}{a \cdot (-a)} = 0$. By uniqueness of additive inverses, $\frac{1}{-a} = -\frac{1}{a}$.

5. Observe that $\frac{-a}{b} + \frac{a}{b} = \frac{-a+a}{b} = \frac{0}{b} = 0 \cdot \frac{1}{b} = 0$. Also, $\frac{a}{-b} + \frac{a}{b} = \frac{ab+a(-b)}{b(-b)} = \frac{a(-b+b)}{b(-b)} = \frac{a \cdot 0}{b(-b)} = \frac{0}{b(-b)} = 0$. By uniqueness of additive inverses, we get $-\frac{a}{b} = \frac{-a}{b} = \frac{a}{-b}$.

6. Observe that $(ab) \cdot (\frac{1}{a} \cdot \frac{1}{b}) = a \cdot \frac{1}{a} \cdot b \cdot \frac{1}{b} = 1 \cdot 1 = 1$, and so by uniqueness of multiplicative inverses, we get $\frac{1}{ab} = \frac{1}{a} \cdot \frac{1}{b}$.

7. $a^2 = a \iff a^2 - a = 0 \iff a(a-1) = 0 \iff a = 0$ or $a - 1 = 0$ $\iff a = 0$ or $a = 1$.

8. Proof of $x^m x^n = x^{m+n}$ by induction on n: $x^m x^1 = x^{m+1}$ by definition of x^{m+1}. Suppose $x^m x^n = x^{m+n}$. Then $x^m x^{n+1} = x^m x^n x^1 = x^{m+n} x^1 = x^{m+n+1}$. Proof of $(x^n)^m = x^{nm}$ by induction on m: $(x^n)^1 = x^n = x^{n \cdot 1}$. Suppose $(x^n)^m = x^{nm}$. Then $(x^n)^{m+1} = (x^n)^m (x^n)^1 = x^{nm} x^n = x^{nm+n} = x^{n(m+1)}$.

9. $|x| = |(x - y) + y| \leq |x - y| + |y|$ and so $|x| - |y| \leq |x - y|$. Likewise, $|y| = |(y - x) + x| \leq |y - x| + |x|$ and so $|y| - |x| \leq |y - x|$. By the exercise below, we may assume $|y - x| = |x - y|$, and since $||x| - |y||$ equals either $|x| - |y|$ or $|y| - |x|$, we get the desired result.

10. Let $x = \frac{m}{n}$, with m, n integers. Since $\frac{-a}{b} = \frac{a}{-b}$ and $-(-a) = a$, we may assume that $n > 0$. Thus $x > 0$ implies $m > 0$ and $x < 0$ implies $m < 0$. Clearly $x = 0$ implies $m = 0$ and in such a case $-0 = 0$ and so $0 = |0| = |-0|$. Thus we assume $x \neq 0$. If $x > 0$, then $|x| = x$. Then $|-x| = |-\frac{m}{n}| = |\frac{-m}{n}| = -\frac{-m}{n} = \frac{-(-m)}{n} = \frac{m}{n} = x$. If $x < 0$, then apply the above analysis to $y = -x$, and use $-y = -(-x) = x$.

11. $d(x, y) = |x - y| = 0 \iff x - y = 0 \iff x = y$. Using $|-z| = |z|$ we have $d(x, y) = |x - y| = |-(x - y)| = |y - x| = d(y, x)$. By the triangle inequality, $d(x, z) = |x - z| = |(x - y) + (y - z)| \leq |x - y| + |y - z| = d(x, y) + d(y, z)$.

12. Use $z = \frac{x+y}{2}$. And use an exercise below that is proven independently.

13. If $x = 0$ or $y = 0$, then $|xy| = 0$ and $|x| \cdot |y| = 0$, so we may assume $x, y \neq 0$. By symmetry we shall only consider three cases. Assume $x, y < 0$. Then $xy > 0$ and $|xy| = xy$. Also, $|x||y| = (-x)(-y) = (-1)^2 xy = xy$. If $x, y > 0$, then $xy > 0$ and so $|xy| = xy = |x||y|$. If $x < 0$ and $y > 0$, then $xy < 0$, and so $|xy| = -xy = (-x)y = |x||y|$.

14. Suppose $|x| < y$, with $y > 0$. Then if $x \geq 0$ we have $0 \leq x < y$. Hence $-y < x < -y$ since $-y < 0$. If $x < 0$, then we know $0 < -x < y$ and so $-y < -x < 0$. Since $y > 0$ we have $-y < x < y$. If $|x| = y$ then

B.1. HOMEWORK SOLUTIONS 1.2.3

either $x = y$ or $x = -y$. Thus if we assume $|x| \le y$, then $-y \le x \le y$ follows.

15. For $x \in \mathbb{Q} \setminus \{0\}$, we let $x = \frac{a}{b}$, and a, b are integers. Then $x^{-1} = \frac{b}{a}$. If $x > 0$, then so is x^{-1}, and in such a case $|x^{-1}| = x^{-1} = |x|^{-1}$. If $x < 0$, then so is $x^{-1} < 0$. Thus $|x| = -x = -\frac{a}{b} = \frac{-a}{b} = \frac{a}{-b}$. Thus $|x|^{-1} = \frac{b}{-a} = \frac{-b}{a} = -\frac{b}{a} = -x^{-1} = |x^{-1}|$.

16. This follows from $|xy| = |x| \cdot |y|$ and $|z^{-1}| = |z|^{-1}$ for $z \ne 0$: $|\frac{a}{b}| = |a \cdot b^{-1}| = |a| \cdot |b^{-1}| = |a| \cdot |b|^{-1} = \frac{|a|}{|b|}$.

17. For $n = 2$ this is the triangle inequality, and for $n = 1$, it is trivial. Suppose this holds for a specific n, for any n-tuple $(a_1, \cdots, a_n)$. Then consider a $(n+1)$-tuple of rational numbers $(a_1, \cdots, a_n, a_{n+1})$. Thus, $|a_1 + \cdots + a_n + a_{n+1}| = |(a_1 + \cdots + a_n) + a_{n+1}| \le |a_1 + \cdots + a_n| + |a_{n+1}| \le |a_1| + \cdots + |a_n| + |a_{n+1}|$, where the first inequality is true by the $n = 2$ case, and the second inequality is true by the induction hypothesis along with $x < y$ implies $x + z < y + z$, which we will prove independently in another exercise.

18. We are supposing $x < y$ which is equivalent to $y - x$ is positive. Since $\frac{1}{2}$ is positive, and the positive numbers are closed under multiplication, we get $\frac{1}{2} \cdot (y - x) = \frac{y}{2} - \frac{x}{2}$ is positive. Hence $\frac{x}{2} < \frac{y}{2}$. However, $x < \frac{x+y}{2}$ is true $\iff \frac{x+y}{2} - x$ is positive, but $\frac{x+y}{2} - x = \frac{y}{2} - \frac{x}{2}$. Likewise, $\frac{x+y}{2} < y$ is true $\iff y - \frac{x+y}{2}$ is positive, but $y - \frac{x+y}{2} = \frac{y}{2} - \frac{x}{2}$.

19. Observe $x < y \iff y - x$ is positive. Also, $y - x = (y + z) - (x + z)$ and so $x + z < y + z$.

20. Observe $x < y \iff y - x$ is positive. If z is positive, then so is $z(y - x) = zy - zx$, and so $xz < yz$.

21. If $x < y$ and $z < 0$, then $-z > 0$. Thus $(-z)x < (-z)y$. Add $zx + zy$ to both sides to get the result from the two above exercises results.

22. Suppose $a > 1$, then $a > 0$ since $1 > 0$. Also, we must have $a^{-1} > 0$ since $a \cdot a^{-1} = 1$. Observe that $a > 1 \iff a - 1$ is positive. Thus $\frac{1}{a} \cdot (a - 1) = 1 - \frac{1}{a}$ is positive, which gives us the result.

23. Since positive times positive is positive, and negative times negative is positive, we have squares are always non-negative. Thus $(a-b)^2 \geq 0$. This implies $a^2+b^2-2ab \geq 0$. Multiplying both sides by $\frac{1}{2}$, then adding ab to both sides gives the result.

24. Suppose $xy \geq 0$. Then if $xy = 0$, one of the two numbers x, y must be zero, and so $|x+y| = |x|+|y|$. If $xy > 0$, then if $x, y > 0$ we have $x+y > 0$ and so $|x+y| = x+y = |x|+|y|$. If $x, y < 0$, then $x+y < 0$ and so $|x+y| = -(x+y) = -x-y = |x|+|y|$. Suppose $xy < 0$. Without loss of generality, suppose $x < 0$ and $y > 0$. Then if $x+y < 0$ we have $|x+y| = -x-y = |x|-|y| < |x|+|y|$. If $x+y \geq 0$, then $|x+y| = x+y = -|x|+|y| < |x|+|y|$.

B.2 Homework Solutions 1.3.4

1. Let $-s \in -S$ be arbitrary, then $s \in S$, and so $s \leq \sup S$. Thus we have $-s \geq -\sup S$. Thus we see that $-\sup S$ is a lower bound for $-S$. Hence $\inf(-S) \geq -\sup S$. Let $s \in S$ be arbitrary, then $-s \in -S$, and so $-s \geq \inf(-S)$. So $s \leq -\inf(-S)$. Since $s \in S$ was arbitrary, we see that $-\inf(-S)$ is an upper bound for S. So $\sup S \leq -\inf(-S)$, which gives us $-\sup S \geq \inf(-S)$. To get the other identity, we use this argument and the fact that $-(-S) = S$.

2. If $s \in S$ is an upper bound for S then since it's an upper bound we have $s \geq \sup S$, using here that $\sup S$ is the smallest upper bound. Since it's an element of S, we have $s \leq \sup S$.

3. If a set S is no empty, then there exists an $s \in S$, and so $\inf S \leq s$ and $s \leq \sup S$, and so $\inf S \leq \sup S$ for a non-empty set. Supposing $A \subseteq B \subset \mathbb{R}$ are non-empty, we have $\inf A \leq \sup A$. Since $a \in A$ implies $a \in B$, we have $\inf B$ is a lower bound for A and $supB$ is an upper bound for A. Thus $\inf B \leq \inf A$ and $\sup A \leq \sup B$.

4. We will now show that every non-empty finite subset A of $\mathbb{R}$ contains its maximum. The other result follows from $\inf(-S) = -\sup(S)$. To prove this we proceed by induction on the cardinality of the set. If $A = \{x\}$, then clearly $x = \sup A$, since x is an upper bound for A. Suppose that for any set with n elements the result is true, and let

$A = \{x_1, \cdots, x_n, x_{n+1}\}$. Then $A \setminus \{x_{n+1}\}$ contains its maximum, say x_k. Then either $x_{n+1} > x_k$, in which case x_{n+1} is the maximum element of A, or $x_{n+1} < x_k$, in which case x_k is the maximum of A.

5. Here we are assuming of course that $\sup \mathbb{N} = \infty$. since every element $n \in \mathbb{N}$ satisfies $n \geq 1$, we know $0 < \frac{1}{n} \leq 1$, and so $0 \leq 1 - \frac{1}{n} < 1$. Hence 1 is an upper bound for $S = \{1 - \frac{1}{n} : n \in \mathbb{N}\}$. We will now show that for any positive ε, $1 - \varepsilon$ is not an upper bound for S. To do this, pick $k \in \mathbb{N}$ so that $k > \frac{1}{\varepsilon}$. Then $1 - \frac{1}{k} > 1 - \varepsilon$.

6. To show that $x^3 = 3$ has a real solution, we use the theorem that insures the existence of nth roots of positive numbers (i.e. using that the theorem guarantees $\sqrt[3]{3}$ exists as a real number).

7. To show that x, which satisfies $x^3 = 3$, is not rational we suppose it is. If so, suppose $x = \frac{m}{n}$ where $m, n \in \mathbb{N}$ and the gcd of m and n is 1. Then $\frac{m^3}{n^3} = 3$ and so $m^3 = 3n^3$. So $3|m^3$. Since 3 is prime, we know $3|m$. Hence $m = 3k$. Then $(3k)^3 = 3n^3$ implies $9k^3 = n^3$, and so $3|n^3$. Hence $3|n$, a contradiction.

8. Let $a > 0$, S a bounded set, and $aS = \{as : s \in S\}$. Let $as \in aS$ be arbitrary, then $s \in S$, and so $\inf S \leq s \leq \sup S$. Hence $a \inf S \leq as \leq a \sup S$. Hence $a \sup S$ is an upper bound for aS and $a \inf S$ is a lower bound for aS. So $\sup(aS) \leq a \sup S$ and $a \inf S \leq \inf(aS)$. Let $s \in S$ be arbitrary, and so $as \in aS$. so $\inf(aS) \leq as \leq \sup(aS)$. Thus $\frac{\inf(aS)}{a} \leq s \leq \frac{\sup(aS)}{a}$, and so we see $\frac{\inf(aS)}{a}$ is a lower bound for S and $\frac{\sup(aS)}{a}$ is an upper bound for S. Thus $\frac{\inf(aS)}{a} \leq \inf S$ and $\sup S \leq \frac{\sup(aS)}{a}$.

9. For $a < 0$, S a bounded set, and $aS = \{as : s \in S\}$, we consider $|a|S$ and by the above exercise we have $|a| \inf S = \inf(|a|S)$ and $|a| \sup S = \sup(|a|S)$. Observe that $aS = -|a|S$, and from an earlier exercise, we know $\sup(-|a|S) = -\inf(|a|S)$ and $\inf(-|a|S) = -\sup(|a|S)$.

10. Suppose $\sqrt{3} = \frac{a}{b}$ with a, b natural numbers whose gcd is 1. Then $3 = \frac{a^2}{b^2}$ and so $a^2 = 3b^2$. Since $3|a^2$ and 3 is prime we know $3|a$. Thus $a = 3c$, $9c^2 = 3b^2$, and so $b^2 = 3c^2$, and thus $3|b^2$. Since 3 is prime, we know $3|b$. Thus 3 divides the gcd of a, b which is 1. This is a contradiction.

11. Suppose $\sqrt[3]{5} = \frac{a}{b}$ with a, b natural numbers whose gcd is 1. Then $5 = \frac{a^3}{b^3}$ and so $a^3 = 5b^3$. Since $5|a^3$ and 5 is prime we know $5|a$. Thus $a = 5c$, $125c^3 = 5b^3$, and so $b^3 = 25c^3$, and thus $5|b^3$. Since 5 is prime, we know $5|b$. Thus 5 divides the gcd of a, b which is 1. This is a contradiction.

12. Observe that for any real number x and any natural numbers n, m that $x^{n \cdot m} = (x^n)^m = (x^m)^n$. This follows from $x^k = \underbrace{x \cdot x \cdots x}_{k \text{ times}}$ for $k \in \mathbb{N}$.
If $b > 0$, then $x := \sqrt[n]{b} > 0$ is defined. Likewise, $b^m > 0$, and so $y := \sqrt[n]{b^m} > 0$ is defined. Now $x^m = b$ and so $x^{n \cdot m} = b^m$. Likewise, $y^n = b^m$ and so $(x^m)^n = y^n$, which implies $x^m = y$ by taking n^{th} roots. For the second part of this exercise, we assume $\frac{i}{j} = \frac{m}{n}$. Thus $in = jm$. Let $z = b^{\frac{m}{n}}$. Then $z^n = b^m$, and so $z^{in} = b^{im}$. However, $z^{jm} = z^{in}$, and so $z^{jm} = b^{im}$. By taking m^{th} roots, we get $z^j = b^i$, and so $z = b^{\frac{i}{j}}$.

13. For $r, s \in \mathbb{N}$ we have $b^r \cdot b^s = \underbrace{b \cdot b \cdots b}_{r \text{ times}} \cdot \underbrace{b \cdot b \cdots b}_{s \text{ times}} = \underbrace{b \cdot b \cdots b}_{r+s \text{ times}}$. Also, as observed above, $(b^r)^s = \underbrace{b^r \cdot b^r \cdots b^r}_{s \text{ times}} = \underbrace{b \cdot b \cdots b}_{r \text{ times}} \cdot \underbrace{b \cdot b \cdots b}_{r \text{ times}} \cdots \underbrace{b \cdot b \cdots b}_{r \text{ times}}$
$$\underbrace{}_{s \text{ times}}$$
$= \underbrace{b \cdot b \cdots b}_{r \cdot s \text{ times}}$. For the general case, let's assume $r = \frac{i}{j}$ and $s = \frac{m}{n}$. Then, letting $b^{\frac{i}{j}} \cdot b^{\frac{m}{n}} = x$, we see that from the above exercise and the case just considered, that $b^{in} \cdot b^{mj} = x^{jn}$ (possibly also using that $(a \cdot b)^k = a^k \cdot b^k$ for any a, b and any $k \in \mathbb{N}$). Hence $b^{in+mj} = x^{jn}$, and so by taking $(jn)^{th}$ roots we get $b^{\frac{in+mj}{jn}} = x$.

14. (a) $\sqrt[7]{\frac{2}{5}}$ is a solution of $5x^7 - 2 = 0$.

   (b) For a general $\frac{a}{b}$, with a, b integers, $b \neq 0$, consider $bx - a = 0$, for which $\frac{a}{b}$ is a solution.

   (c) Consider a polynomial $p(x) = a_n x^n + \cdots + a_2 x^2 + a_1 x + a_0$ with integer coefficients $a_0, a_1, \cdots a_n$. There are at most n solutions to $p(x) = 0$ for a given $p(x)$. Define $h(p) = n + |a_n| + \cdots + |a_2| + |a_1| + |a_0|$ for such a $p(x)$. We see that for a given $m \in \mathbb{N}$ there is at most $(2m+1)^{m+2}$ possible polynomials p so that $h(p) \leq m$, and each one has at most m roots. So if we define A_m to be the set of algebraic numbers x whose irreducible polynomial p has $p(x) = 0$, then A_m has at most $(2m+1)^{m+3}$ elements in it. Using that the set of algebraic numbers is $\bigcup_{m \in \mathbb{N}} A_m$, which is a countable union of finite sets, we see that the collection of algebraic numbers is countable. (To see this first list the elements of A_1, then list the elements of $A_2 \setminus (A_1)$, then list the elements of $A_3 \setminus (A_1 \cup A_2)$, then list the elements in $A_4 \setminus (A_1 \cup A_2 \cup A_3)$, etc).

B.3 Homework Solutions 1.5.1

1. Here we examine when $(a, 0) \cdot (c, d) = (1, 0)$ is true. For this, we need $ac = 1$ and $ad = 0$. The first equation requires $a \neq 0$ and $c = \frac{1}{a}$. The second equation in conjunction with this yields $d = 0$. Thus $(a, 0)^{-1} = (\frac{1}{a}, 0)$.

2. $(0, 1) \cdot (0, 1) = (0 \cdot 0 - 1 \cdot 1, 0 \cdot 1 + 0 \cdot 1) = (-1, 0)$.

3. $(a, 0) + (b, o) \cdot (0, 1) = (a, 0) + (b \cdot 0 - 0 \cdot 1, b \cdot 1 + 0 \cdot 0) = (a, 0) + (0, b) = (a, b)$.

4. (a) $\overline{(a, b) + (c, d)} = \overline{(a+c, b+d)} = (a+c, -(b+d)) = (a+c, -b-d) = (a, -b) + (c, -d) = \overline{(a, b)} + \overline{(c, d)}$.

   (b) $\overline{(a, b) \cdot (c, d)} = \overline{(ac - bd, ad + bc)} = (ac - bd, -(ad + bc)) = (ac - bd, -ad - bc) = (ac - (-b)(-d), a(-d) + (-b)c) = (a, -b) \cdot (c, -d) = \overline{(a, b)} \cdot \overline{(c, d)}$.

   (c) $(a, b) + \overline{(a, b)} = (a, b) + (a, -b) = (a + a, b - b) = (2a, 0)$, and $(a, b) - \overline{(a, b)} = (a, b) - (a, -b) = (a - a, b - (-b)) = (0, 2b)$.

5. (a) $|\overline{(a, b)}| = |(a, -b)| = \sqrt{a^2 + (-b)^2} = \sqrt{a^2 + b^2} = |(a, b)|$.

   (b) $|(a, b) \cdot (c, d)| = |(ac - bd, ad + bc)| = \sqrt{(ac - bd)^2 + (ad + bc)^2} = \sqrt{a^2c^2 + b^2d^2 - 2abcd + a^2d^2 + b^2c^2 + 2abcd} = \sqrt{a^2c^2 + a^2d^2 + b^2c^2 + b^2d^2} = \sqrt{(a^2 + b^2)(c^2 + d^2)} = \sqrt{a^2 + b^2} \cdot \sqrt{c^2 + d^2} = |(a, b)| \cdot |(c, d)|$.

   (c) Observe that $|(a, b) + (c, d)| = |(a+c, b+d)| = \sqrt{(a+c)^2 + (b+d)^2} = \sqrt{a^2 + c^2 + 2ac + b^2 + d^2 + 2bd}$. Also, $|(a, b)| + |(c, d)| = \sqrt{a^2 + b^2} + \sqrt{c^2 + d^2}$. So we need to show $\sqrt{a^2 + c^2 + 2ac + b^2 + d^2 + 2bd} \leq \sqrt{a^2 + b^2} + \sqrt{c^2 + d^2}$. However, for non-negative numbers x, we know $x \leq y \iff x^2 \leq y^2$, and so it suffices to show $a^2 + c^2 + 2ac + b^2 + d^2 + 2bd \leq (\sqrt{a^2 + b^2} + \sqrt{c^2 + d^2})^2$, which is $a^2 + c^2 + 2ac + b^2 + d^2 + 2bd \leq a^2 + b^2 + c^2 + d^2 + 2\sqrt{a^2 + b^2} \cdot \sqrt{c^2 + d^2}$.

We see that this is equivalent to $ac + bd \le \sqrt{(a^2+b^2)(c^2+d^2)}$. Again, this is equivalent to $(ac+bd)^2 \le (a^2+b^2)(c^2+d^2)$, or simply $a^2c^2 + b^2d^2 + 2abcd \le a^2c^2 + b^2c^2 + a^2d^2 + b^2d^2$. This is equivalent to $2acbd \le b^2c^2 + a^2d^2$. Letting $e = ad$ and $f = bc$ this is $2ef \le e^2 + f^2$, which we know is true since $e^2 + f^2 - 2ef = (e - f)^2 \ge 0$.

6. $(c,d) \cdot (c,d) = (\sqrt{\frac{\sqrt{a^2+b^2}+a}{2}}, \sqrt{\frac{\sqrt{a^2+b^2}-a}{2}}) \cdot (\sqrt{\frac{\sqrt{a^2+b^2}+a}{2}}, \sqrt{\frac{\sqrt{a^2+b^2}-a}{2}}) = (\frac{\sqrt{a^2+b^2}+a}{2} - \frac{\sqrt{a^2+b^2}-a}{2}, 2\sqrt{\frac{\sqrt{a^2+b^2}+a}{2}} \cdot \sqrt{\frac{\sqrt{a^2+b^2}-a}{2}}) = (a, \sqrt{(\sqrt{a^2+b^2}+a)(\sqrt{a^2+b^2}-a)})$
$= (a, \sqrt{b^2}) = (a, |b|)$. If $b \ge 0$, then this is equal to (a,b). Also, $\overline{(c,d) \cdot (c,d)} = \overline{(c,d) \cdot (c,d)} = \overline{(a, |b|)} = (a, -|b|)$ which equals (a,b) if $b < 0$.

B.4 Homework Solutions 2.3.4

1. $\{\frac{1}{n} : n \in \mathbb{N}\}$

2. $\{\frac{1}{n} : n \in \mathbb{N}\} \cup \{0\}$

3. $\{\frac{1}{n} : n \in \mathbb{N}\} \cup \{2 + \frac{1}{n} : n \in \mathbb{N}\}$

4. $\{\frac{1}{n} : n \in \mathbb{N}\} \cup \{2 + \frac{1}{n} : n \in \mathbb{N}\} \cup \{0, 2\}$

5. $\{\frac{1}{n} : n \in \mathbb{N}\} \cup \{2 + \frac{1}{n} : n \in \mathbb{N}\} \cup \{4 + \frac{1}{n} : n \in \mathbb{N}\}$

6. $\{\frac{1}{n} : n \in \mathbb{N}\} \cup \{2 + \frac{1}{n} : n \in \mathbb{N}\} \cup \{4 + \frac{1}{n} : n \in \mathbb{N}\} \cup \{0, 2, 4\}$

7. For 1, 3, 5 the sets are neither open nor closed, and the complements are neither open nor closed as well. In 2, 4, 6, the sets are closed, and the complements are open. The complements are as follows: For 1 it is $(-\infty, 0] \cup (1, \infty) \cup (\bigcup_{n \in \mathbb{N}}(\frac{1}{n+1}, \frac{1}{n}))$; for 2 it is $(-\infty, 0) \cup (1, \infty) \cup (\bigcup_{n \in \mathbb{N}}(\frac{1}{n+1}, \frac{1}{n}))$; for 3 it is $(-\infty, 0] \cup (1, 2] \cup (\bigcup_{n \in \mathbb{N}}(\frac{1}{n+1}, \frac{1}{n})) \cup (3, \infty) \cup (\bigcup_{n \in \mathbb{N}}(2 + \frac{1}{n+1}, 2 + \frac{1}{n}))$; for 4 it is $(-\infty, 0) \cup (1, 2) \cup (\bigcup_{n \in \mathbb{N}}(\frac{1}{n+1}, \frac{1}{n})) \cup (3, \infty) \cup (\bigcup_{n \in \mathbb{N}}(2 + \frac{1}{n+1}, 2 + \frac{1}{n}))$; for 5 it is $(-\infty, 0] \cup (1, 2] \cup (\bigcup_{n \in \mathbb{N}}(\frac{1}{n+1}, \frac{1}{n})) \cup (3, 4] \cup (\bigcup_{n \in \mathbb{N}}(2 + \frac{1}{n+1}, 2 + \frac{1}{n})) \cup (5, \infty) \cup (\bigcup_{n \in \mathbb{N}}(4 + \frac{1}{n+1}, 4 + \frac{1}{n}))$; for 6 it is $(-\infty, 0) \cup (1, 2) \cup (\bigcup_{n \in \mathbb{N}}(\frac{1}{n+1}, \frac{1}{n})) \cup (3, 4) \cup (\bigcup_{n \in \mathbb{N}}(2 + \frac{1}{n+1}, 2 + \frac{1}{n})) \cup (5, \infty) \cup (\bigcup_{n \in \mathbb{N}}(4 + \frac{1}{n+1}, 4 + \frac{1}{n}))$.

8. (a) $(-1, 1)$

(b) $(-1, 1)$

(c) $(-\frac{1}{N}, \frac{1}{N})$

(d) $\{0\}$

9. (a) $[-1, 1]$

   (b) $[-1, 1]$

   (c) $[-\frac{1}{N}, \frac{1}{N}]$

   (d) $\{0\}$

10. (a) $[-N, N+1]$

    (b) $(-\infty, \infty)$

    (c) $\emptyset$

    (d) $\emptyset$

11. (a) $d(x, y) = 1$ if $x \neq y$ and $d(x, y) = 0$ if $x = y$ clearly satisfies 1: $d(x, y) \geq 0$ and 2: $d(x, y) = 0$ if and only if $x = y$. Also, 3: $d(x, y) = d(y, x)$ since $x = y \iff y = x$, and $x \neq y \iff y \neq x$. For the triangle inequality 4: $d(x, z) \leq d(x, y) + d(y, z)$ we observe that if $x = y = z$ then both sides are zero, i.e $d(x, z) = 0$ and $d(x, y) + d(y, z) = 0$. If $x = y$ and $y \neq z$, then both sides are 1, i.e $d(x, z) = 1$ and $d(x, y) + d(y, z) = 1$. If $y = z$ and $x \neq y$ then both sides are one as well, i.e $d(x, z) = 1$ and $d(x, y) + d(y, z) = 1$. In the last case $x \neq y \neq z$ we have $d(x, z) = 1$ and $d(x, y) + d(y, z) = 2$. In all cases we have $d(x, z) \leq d(x, y) + d(y, z)$.

    (b) All non-empty subsets of S are bounded, since for any set $A \subseteq S$ and any $x, y \in A$, we have $d(x, y) \leq 1$.

    (c) There are no non-empty unbounded sets.

    (d) Let $A \subset S$ be non-empty. Let $x \in A$ and $0 < r < 1$. Then $B_r(x) = \{x\} \subseteq A$, so A is open. Trivially, $\emptyset, S$ are open as well as well (since this is true for all metric spaces). So all sets are open.

    (e) If $B \subseteq S$, and $B^c := A$ we have A is open, so B is closed. Thus all sets are both open and closed in this metric space.

12. Suppose S does not contain $\inf S$ or S does not contain $\sup S$, where S is a bounded non-empty set. Then if $\inf S \notin S$, we will show $\inf S$ is

a limit point for S. Fix $\varepsilon > 0$ and consider $\inf S + \varepsilon$. Then this is not a lower bound for S, so there is an $s \in S$ so that $s < \inf S + \varepsilon$. Then $s \in (\inf S, \inf S + \varepsilon)$. Thus $\inf S$ is a limit point for S. If $\sup S \notin S$, we will show $\sup S$ is a limit point for S. Fix $\varepsilon > 0$ and consider $\sup S - \varepsilon$. Then this is not an upper bound for S, so there is an $s \in S$ so that $s > \sup S - \varepsilon$. Then $s \in (\sup S - \varepsilon, \sup S)$. Thus $\sup S$ is a limit point for S.

13. Let $\vec{x} = (x_1, x_2, \cdots, x_n)$ and $\vec{y} = (y_1, y_2, \cdots, y_n)$ and $d(\vec{x}, \vec{y}) = \max\{|x_1 - y_1|, |x_2 - y_2|, \cdots, |x_n - y_n|\}$. Clear $d(\vec{x}, \vec{y})$ is the maximum of a collection of non-negative numbers in any instance, and this is always greater than or equal to zero. Moreover, if $d(\vec{x}, \vec{y}) = 0$, then the maximum of $|x_1 - y_1|, |x_2 - y_2|, \cdots, |x_n - y_n|$ is zero. Thus each is zero, since each is already non-negative. This implies $\vec{x} = \vec{y}$. Also, since $|x_k - y_k| = |y_k - x_k|$ for any k, we have $d(\vec{x}, \vec{y}) = d(\vec{y}, \vec{x})$. Lastly, suppose $\vec{z} = (z_1, z_2, \cdots, z_n)$. Consider $d(\vec{x}, \vec{z}) = \max\{|x_1 - z_1|, \cdots, |x_n - z_n|\}$. For any k, we have $|x_k - z_k| = |x_k - y_k + y_k - z_k| \leq |x_k - y_k| + |y_k - z_k|$ by the triangle inequality for the absolute value on the set of real numbers. This holds for each k, so $d(\vec{x}, \vec{z}) = \max\{|x_1 - z_1|, \cdots, |x_n - z_n|\}$ $\leq \max\{|x_1 - y_1| + |y_1 - z_1|, \cdots, |x_n - y_n| + |y_n - z_n|\}$. Now $\max\{|x_1 - y_1| + |y_1 - z_1|, \cdots, |x_n - y_n| + |y_n - z_n|\} \leq \max\{|x_1 - y_1|, \cdots, |x_n - y_n|\} + \max\{|y_1 - z_1|, \cdots, |y_n - z_n|\} = d(\vec{x}, \vec{y}) + d(\vec{y}, \vec{z})$. To see this last one, one must observe that if $\max\{|x_1 - y_1| + |y_1 - z_1|, \cdots, |x_n - y_n| + |y_n - z_n|\} = |x_i - y_i| + |y_i - z_i|$, then $|x_i - y_i| \leq \max\{|x_1 - y_1|, \cdots, |x_n - y_n|\}$ and $|y_i - z_i| \leq \max\{|y_1 - z_1|, \cdots, |y_n - z_n|\}$.

14. Let $\vec{x} = (x_1, x_2, \cdots, x_n)$ and $\vec{y} = (y_1, y_2, \cdots, y_n)$, $d(\vec{x}, \vec{y}) = \sqrt{|x_1 - y_1|^2 + \cdots + |x_n - y_n|^2}$ and $d_1(\vec{x}, \vec{y}) = |x_1 - y_1| + \cdots + |x_n - y_n|$. Then $(d(\vec{x}, \vec{y}))^2 = |x_1 - y_1|^2 + \cdots + |x_n - y_n|^2 \leq |x_1 - y_1|^2 + \cdots + |x_n - y_n|^2 + \sum_{i<j; i,j \in \{1,2,\cdots,n\}} 2|x_i - y_i| \cdot |x_j - y_j| = (|x_1 - y_1| + |x_2 - y_2| + \cdots + |x_n - y_n|)^2 = (d_1(\vec{x}, \vec{y}))^2$. Thus we see $d(\vec{x}, \vec{y}) \leq d_1(\vec{x}, \vec{y})$. Now, $(d_1(\vec{x}, \vec{y}))^2 = |x_1 - y_1|^2 + \cdots + |x_n - y_n|^2 + \sum_{i<j; i,j \in \{1,2,\cdots,n\}} 2|x_i - y_i| \cdot |x_j - y_j| \leq |x_1 - y_1|^2 + \cdots + |x_n - y_n|^2 + \sum_{i<j; i,j \in \{1,2,\cdots,n\}} [|x_i - y_i|^2 + |x_j - y_j|^2] \leq |x_1 - y_1|^2 + \cdots + |x_n - y_n|^2 + \sum_{i<j; i,j \in \{1,2,\cdots,n\}} [|x_1 - y_1|^2 + \cdots + |x_n - y_n|^2]$ $= (1 + \sum_{i<j; i,j \in \{1,2,\cdots,n\}} 1)(d(\vec{x}, \vec{y}))^2 \leq (1 + \binom{n}{2})(d(\vec{x}, \vec{y}))^2$. So $d_1(\vec{x}, \vec{y}) \leq \sqrt{1 + \binom{n}{2}} d(\vec{x}, \vec{y})$.

15. Let (M, d) be a metric space. We claim that $\delta : M \times M \to [0, \infty)$ given

by $\delta(x,y) = \frac{d(x,y)}{1+d(x,y)}$ is also a metric on M. In such a case, we may create infinitely many metrics on M, namely $d_1 = d$ and $d_{i+1} = \frac{d_i}{1+d_i}$, for $i = 1, 2, \cdots$. Observe that $\delta(x,y) = 0$ if and only if $d(x,y) = 0$, and this happens if and only if $x = y$. Also, $\delta(x,y) = \frac{d(x,y)}{1+d(x,y)} = \frac{d(y,x)}{1+d(y,x)} = \delta(y,x)$. $\delta(x,y) \geq 0$. Suppose that there is $x, y, z \in M$ so that $\delta(x,z) > \delta(x,y) + \delta(y,z)$, that is $\frac{d(x,z)}{1+d(x,z)} > \frac{d(x,y)}{1+d(x,y)} + \frac{d(y,z)}{1+d(y,z)}$. Then $d(x,z)(1+d(x,y))(1+d(y,z)) > d(x,y)(1+d(x,z))(1+d(y,z)) + d(y,z)(1+d(x,z))(1+d(x,y))$. It follows that $d(x,z) > d(x,y) + d(x,y)d(y,z) + d(y,z) + d(y,z)d(x,y) + d(y,z)d(x,z)d(x,y) \geq d(x,y) + d(y,z)$, a violation of the triangle inequality for d. Hence δ satisfies the triangle inequality.

B.5 Homework Solutions 2.4.1

1. Take $\{(\frac{1}{n}, 1+\frac{1}{n}) : n \in \mathbb{N}\}$. This is an open cover of $(0,1]$ that does not have a finite subcover for $(0,1]$.

2. Take $\{(\frac{1}{n}, 1) : n \in \mathbb{N}\}$. This is an open cover of $(0,1)$ that does not have a finite subcover for $(0,1)$.

3. Take $\{(0, n) : n \in \mathbb{N}\}$. This is an open cover for $(0, \infty)$ that does not have a finite subcover for $(0, \infty)$.

4. $\bigcup_{n \in \mathbb{N}}[\{\frac{1}{n+1} + \frac{1}{2n(n+1)m} : m \in \mathbb{N}\} \cup \{\frac{1}{n+1}\}] \cup \{0\}$ is bounded since it's a subset of $[0,1]$. It also contains its limit points, namely $\{\frac{1}{n+1} : n \in \mathbb{N}\} \cup \{0\}$, and so it's closed.

5. Here we take $K_i = [i-1, i]$. Then $\bigcup_{i=1}^{\infty} K_i = [0, \infty)$, which is not compact.

6. Yes, since $\bigcup_{i=1}^{N} K_i = [0, N]$, and in general, the union of finitely many closed sets is closed, and the union of finitely many bounded sets is still bounded.

7. Here we may take $K_i = [0, \frac{1}{i}]$. Then $\bigcup_{i=1}^{\infty} = K_1 = [0,1]$ is still compact.

8. By an exercise from the previous set of homework exercises, we have that a non-empty bounded sets S has $\inf S$ and $\sup S$ as either elements

of S or limit points of S. Since compact implies bounded (and closed), this holds for a compact set S. If either $\inf S$ or $\sup S$ is a limit point, then it must be an element of S as well, since a closed set contains its limit points.

B.6 Homework Solutions 2.6.3

1. In our construction of the Cantor set as $C = \bigcap_{k=0}^{\infty} C_k$, we have $C_0 = [0,1]$, $C_1 = [0, \frac{1}{3}] \cup [\frac{2}{3}, 1]$, $C_2 = [0, \frac{1}{9}] \cup [\frac{2}{9}, \frac{3}{9}] \cup [\frac{6}{9}, \frac{7}{9}] \cup [\frac{8}{9}, 1]$, etc. We see that $x \in C$ requires $x \in C_k$ for all k. If we take $x \in C_0$ and use a ternary expansion $= \sum_{k=1}^{\infty} \frac{x_k}{3^k}$, where $x_k \in \{0, 1, 2\}$, we may write $x = 0.x_1 x_2 x_3 \cdots$ like we do with decimal expansions. Also, we assume that no ternary expansion of a non-zero number terminates. For instead of writing $0.1 = 0.10000\cdots$, we write $0.02222\cdots$. In general, if $x = 0.x_1 x_2 \cdots x_{n-1} x_n 000 \cdots$ with $x_n > 0$, we may instead write $x = 0.x_1 x_2 \cdots x_{n-1}(x_n - 1)222 \cdots$. The only exception to this is $0 = 0.000\cdots$ With this convention, for x to be in C_1 we have $x_1 \in \{0, 2\}$, where $x_1 = 0$ if $x \in [0, \frac{1}{3}]$, and $x_1 = 2$ if $x \in [\frac{2}{3}, 1]$. If $x \in C_2$, we see that $x_2 \in \{0, 2\}$, since $x \in [0, \frac{1}{9}]$ or $x \in [\frac{6}{9}, \frac{7}{9}]$ implies $x_2 = 0$, and $x \in [\frac{2}{9}, \frac{3}{9}]$ or $x \in [\frac{8}{9}, 1]$ implies $x_2 = 2$. We continue this line of reasoning. For each interval in the disjoint union of closed intervals that gives us C_m, we remove the middle open one-third to get two disjoint closed intervals in the disjoint union of closed intervals that is C_{m+1}. The one interval will have $x_{m+1} = 0$ and the other will have $x_{m+1} = 2$.

2. We see that any real number x in $[0, 1]$ may be expanded in a binary expansion $x = \sum_{k=1}^{\infty} \frac{b_k}{2^k}$ where $b_k \in \{0, 1\}$. If we have $0.b_1 b_2 \cdots b_n 1000 \cdots$ we will instead write $0.b_1 b_2 \cdots b_n 0111 \cdots$. The only exception to this is $0 = 0.000\cdots$. Then there is a bijection between $[0, 1]$ and C, the Cantor set, given as follows: If $0.b_1 b_2 \cdots$ is the binary expansion of a number in $[0, 1]$, we associate to it element of the Cantor set whose ternary expansion is $0.t_1 t_2 \cdots$, where $b_k = 0 \mapsto t_k = 0$ and $b_k = 1 \mapsto t_k = 2$.

B.7 Homework Solutions 3.1.1

1. Let $\varepsilon > 0$ be given. Then for any natural number $N > \frac{1}{\varepsilon}$ (which is guaranteed by the Archimedian property) we have $\frac{1}{n} = |\frac{1}{n} - 0| =$

B.8. HOMEWORK SOLUTIONS 3.1.4

$|a_n - 0| < \varepsilon$ if $n > N$.

2. Let $\varepsilon > 0$ be given. Then for any natural number $N > \frac{1}{\sqrt{\varepsilon}}$ (which is guaranteed by the Archimedian property) we have $\frac{1}{n^2} = |\frac{1}{n^2} - 0| = |a_n - 0| < \varepsilon$ if $n > N$.

3. Let $\varepsilon > 0$ be given. Then for any natural number $N > \frac{1}{\varepsilon^2}$ (which is guaranteed by the Archimedian property) we have $\frac{1}{\sqrt{n}} = |\frac{1}{\sqrt{n}} - 0| = |a_n - 0| < \varepsilon$ if $n > N$.

4. Let $\varepsilon > 0$ be given. Then for any natural number N we have $0 = |c - c| = |a_n - c| < \varepsilon$ if $n > N$.

B.8 Homework Solutions 3.1.4

1. We will show for non-negative x_n that $x_n \to x$ as $n \to \infty \iff \sqrt{x_n} \to \sqrt{x}$ as $n \to \infty$. Observe that $x \geq 0$, since otherwise, we may take $\varepsilon = \frac{|x|}{2}$ and observe that there are no elements of the sequence $\{x_n\}$ in $(x - \frac{|x|}{2}, x + \frac{|x|}{2}) = (\frac{3x}{2}, \frac{x}{2})$. First we consider the case $x = 0$. If we assume $x_n \to x = 0$. Let $\varepsilon > 0$ be given. We must show $\sqrt{x_n} \to \sqrt{x} = 0$ as $n \to \infty$. Since $x_n \to x = 0$ we have that there is an $N \in \mathbb{N}$ so that $n \geq N$ we have $x_n = |x_n - 0| < \varepsilon^2$ and so for such n we also have $\sqrt{x_n} = |\sqrt{x_n} - 0| < \varepsilon$. Next we assume $\sqrt{x_n} \to \sqrt{x} = 0$. Let $\varepsilon > 0$ be given. We must show that $x_n \to x = 0$ as $n \to \infty$. Since $\sqrt{x_n} \to 0$, we know that there is an $N \in \mathbb{N}$ so that $n \geq N$ implies $\sqrt{x_n} = |\sqrt{x_n} - 0| < \sqrt{\varepsilon}$. Thus, for such n we have $x_n = |x_n - 0| < \varepsilon$. Lastly, we assume $x > 0$. Assume first that $x_n \to x$ as $n \to \infty$, and let $\varepsilon > 0$ be given. Next, we consider $|x_n - x| = |\sqrt{x_n} - \sqrt{x}| \cdot |\sqrt{x_n} + \sqrt{x}|$, and so $|\sqrt{x_n} - \sqrt{x}| = \frac{|x_n - x|}{|\sqrt{x_n} + \sqrt{x}|}$. Let $N \in \mathbb{N}$ be so that $n \geq N$ implies $|x_n - x| < \sqrt{x} \cdot \varepsilon$. Then, for such an n we have $|\sqrt{x_n} - \sqrt{x}| = \frac{|x_n - x|}{|\sqrt{x_n} + \sqrt{x}|} < \frac{\sqrt{x}\varepsilon}{\sqrt{x_n} + \sqrt{x}} \leq \varepsilon$. Assume now that $\sqrt{x_n} \to \sqrt{x}$ as $n \to \infty$, and let $\varepsilon > 0$ be given. Since $\{\sqrt{x_n}\}_{n=1}^{\infty}$ converges, there is a C so that $\sqrt{x_n} \leq C$ for all n. Moreover, there is a $N \in \mathbb{N}$ so that for all $n \geq N$ we have $|\sqrt{x_n} - \sqrt{x}| \leq \frac{\varepsilon}{C + \sqrt{x}}$. Then for such a n we have $|x_n - x| = |\sqrt{x_n} - \sqrt{x}| \cdot |\sqrt{x_n} + \sqrt{x}| < \frac{\varepsilon}{C + \sqrt{x}} \cdot |\sqrt{x_n} + \sqrt{x}| < \varepsilon$.

2. Let $f(x) = a_0 + a_1 x + a_2 x^2 + \cdots + a_n x^n$ and suppose $x_k \to x$ as $k \to \infty$.

Then by limit laws we see $\lim_{k\to\infty} f(x_k) = \lim_{k\to\infty}(a_0 + a_1 x_k + a_2 x_k^2 + \cdots + a_n x_k^n) = \lim_{k\to\infty} a_0 + \lim_{k\to\infty} a_1 x_k + \lim_{k\to\infty} a_2 x_k^2 + \cdots + \lim_{k\to\infty} a_n x_k^n = \lim_{k\to\infty} a_0 + a_1 \lim_{k\to\infty} x_k + a_2 \lim_{k\to\infty} x_k^2 + \cdots + a_n \lim_{k\to\infty} x_k^n = \lim_{k\to\infty} a_0 + a_1 \lim_{k\to\infty} x_k + a_2 (\lim_{k\to\infty} x_k)^2 + \cdots + a_n (\lim_{k\to\infty} x_k)^n = a_0 + a_1 x + a_2 x^2 + \cdots + a_n x^n = f(x)$. Here we used that $\lim_{k\to\infty} a_0 = a_0$, which was established in an earlier homework exercise.

3. $a_n = \sqrt{n+1} - \sqrt{n} = (\sqrt{n+1} - \sqrt{n}) \cdot \frac{\sqrt{n+1}+\sqrt{n}}{\sqrt{n+1}+\sqrt{n}} = \frac{1}{\sqrt{n+1}+\sqrt{n}}$. Clearly we see that $a_n > 0$ and $a_n < \frac{1}{\sqrt{n}}$. So given $\varepsilon > 0$ we may pick $N > \frac{1}{\varepsilon^2}$, so that if $n \geq N$ we have $|a_n - 0| < \varepsilon$.

4. Suppose L satisfies $L = \sqrt{2 + \sqrt{L}}$, so $L \approx 1.831177207$. Let $s_1 = \sqrt{2}$ and $s_{n+1} = \sqrt{2 + \sqrt{s_n}}$. Claim that $s_n \leq L$ for all n. Since $2 \leq 2 + \sqrt{L}$ we know $s_1 = \sqrt{2} \leq \sqrt{2 + \sqrt{L}} = L$. Assume $s_n < L$, then $\sqrt{s_n} \leq \sqrt{L}$ and so $2 + \sqrt{s_n} \leq 2 + \sqrt{L}$, and so $s_{n+1} = \sqrt{2 + \sqrt{s_n}} \leq \sqrt{2 + \sqrt{L}} = L$. Next, we claim that $\{s_n\}$ is increasing. $s_2 = \sqrt{2 + \sqrt{2}} \geq \sqrt{2 + 0} = s_1$. Suppose $s_n \geq s_{n-1}$. Then $\sqrt{s_n} \geq \sqrt{s_{n-1}}$ and so $2 + \sqrt{s_n} \geq 2 + \sqrt{s_{n-1}}$. Thus $s_{n+1} = \sqrt{2 + \sqrt{s_n}} \geq \sqrt{2 + \sqrt{s_{n-1}}} = s_n$. Thus $\{s_n\}$ is monotone and bounded. Hence it has a limit. The limit value s is obtained by replacing s_n and s_{n-1} by s is the formula that relates the two. In this case, s is L.

5. Consider $a_n = (1 + \frac{1}{n})^n$. Then $a_n = 1 + \binom{n}{1} \cdot \frac{1}{n} + \binom{n}{2} \cdot \frac{1}{n^2} + \cdots + \binom{n}{k} \cdot \frac{1}{n^k} + \cdots + \binom{n}{n-1} \cdot \frac{1}{n^{n-1}} + \frac{1}{n^n} = 1 + 1 + \frac{n(n-1)}{2!n^2} + \frac{n(n-1)(n-2)}{3!n^3} + \cdots + \frac{n(n-1)\cdots(n-k+1)}{k!n^k} + \cdots + \frac{n!}{n!n^n}$. Observing that $n^k > n(n-1)\cdots(n-k+1)$, we see that $a_n < 1 + 1 + \frac{1}{2!} + \cdots + \frac{1}{n!}$. However, $n! = n(n-1) \cdots 2 \cdot 1 > \underbrace{2 \cdot 2 \cdots 2}_{n-1 \text{ times}} = 2^{n-1}$, and so we see that $a_n < 1 + 1 + \frac{1}{2^1} + \frac{1}{2^2} + \cdots + \frac{1}{2^{n-1}} < 3$. This last inequality is established by an exercise below.

6. $a_n = \sqrt{n^2 + n} - n = (\sqrt{n^2 + n} - n) \cdot \frac{\sqrt{n^2+n}+n}{\sqrt{n^2+n}+n} = \frac{n}{\sqrt{n^2+n}+n} = \frac{1}{\sqrt{1+\frac{1}{n}}+1}$. Since $b_n = 1 + \frac{1}{n}$ converges, so does $\sqrt{1 + \frac{1}{n}}$ by the first exercise above. The limit laws and this observation show that $a_n \to \frac{1}{2}$.

7. Let $a_n = 1 + \frac{1}{2^2} + \frac{1}{3^2} + \cdots + \frac{1}{n^2}$. Then for a given n we observe that $2^{k-1} \leq n < 2^k$ for some $k \in \mathbb{N}$. Since $a_n = a_{n-1} + \frac{1}{n^2} > a_{n-1}$ we know $\{a_n\}$ is strictly increasing. By this, and the earlier observation,

B.9. HOMEWORK SOLUTIONS 3.2.1

it suffices to examine $\{a_{2^k}\}$. Observe that $a_{2^k} = 1 + \frac{1}{2^2} + \frac{1}{3^2} + \frac{1}{4^2} + \cdots + \frac{1}{(2^k)^2}$. We observe that $\frac{1}{(2^{j-1})^2} + \frac{1}{(2^{j-1}+1)^2} + \cdots + \frac{1}{(2^{j-1}+2^{j-1}-1)^2} <$
$\underbrace{\frac{1}{(2^{j-1})^2} + \cdots + \frac{1}{(2^{j-1})^2}}_{2^{j-1} \text{ times}} = \frac{1}{2^{j-1}}$, and so $a_{2^k} < 1 + \frac{1}{2} + \frac{1}{2^2} + \cdots + \frac{1}{2^k}$.

Then the convergence of $\{a_{2^k}\}$ is reduced to the convergence of $b_k = 1 + \frac{1}{2} + \frac{1}{2^2} + \cdots + \frac{1}{2^k}$. This is true because we will know a_n is monotone and bounded. The convergence of $\{b_n\}$ is the next exercise.

8. Let $a_n = 1 + \frac{1}{2} + \frac{1}{2^2} + \cdots + \frac{1}{2^n}$. Observe that $a_n = a_{n-1} + \frac{1}{2^n} > a_{n-1}$. So a_n is strictly increasing. Moreover, $a_n = \frac{1 - \frac{1}{2^{n+1}}}{1 - \frac{1}{2}} < 2$. So $\{a_n\}$ is bounded and monotone, and hence converges.

9. Let $c > 1$ and define $a_n = \sqrt[n]{c}$. Since $c > 1$ we know $a_n = \sqrt[n]{c} > 1$ (because, if $\sqrt[n]{c} \leq 1$, then $c = (\sqrt[n]{c})^n \leq 1^n = 1$). Thus $\sqrt[n+1]{a_n} > 1$ as well, and so $a_{n+1} = \frac{a_n}{\sqrt[n+1]{a_n}} < a_n$. Thus $\{a_n\}_{n=1}^\infty$ is decreasing, and bounded below by 1. Hence $x = \lim_{n \to \infty} a_n$ exists and satisfies $1 \leq x$. Suppose $x > 1$. Let $\varepsilon \leq \frac{x-1}{2}$. Then $\exists N \in \mathbb{N}$ so that $n \geq N$ implies $|a_n - x| < \varepsilon$. So $1 < x - \varepsilon < \sqrt[n]{c}$ and so $1 < (x - \varepsilon)^n < c$. However, $x - \varepsilon = 1 + z$, for $z > 0$, and $1 + nz < (1 + z)^n = (x - \varepsilon)^n < c$ implies $n < \frac{c-1}{z}$. This is a contradiction, since $n \geq N$ can be as large as one wishes. So $x = 1$.

10. $\frac{1}{c} > 1$, and so by the above exercise we have $\lim_{n \to \infty} \sqrt[n]{\frac{1}{c}} = 1$. We use this along with $\lim_{n \to \infty} \sqrt[n]{\frac{1}{c}} = \lim_{n \to \infty} \frac{1}{\sqrt[n]{c}} = \frac{1}{\lim_{n \to \infty} \sqrt[n]{c}}$.

B.9 Homework Solutions 3.2.1

1. Let $a_n = (-1)^{n+1}$. Then $a_{2k} = -1$ for any $k \in \mathbb{N}$, and so $a_{2k} \to -1$ as $k \to \infty$. Also, $a_{2k-1} = 1$ for any $k \in \mathbb{N}$, and so $a_{2k-1} \to 1$ as $k \to \infty$.

2. Consider a_n given by $a_{3k} = 0$ for $k \in \mathbb{N}$, $a_{3k+1} = 1$ for $k = 0, 1, 2, \cdots$ and $a_{3k+2} = 2$ for $k = 0, 1, 2, \cdots$.

3. Consider a_n given by $a_{4k} = 0$ for $k \in \mathbb{N}$, $a_{4k+1} = 1$ for $k = 0, 1, 2, \cdots$, $a_{4k+2} = 2$ for $k = 0, 1, 2, \cdots$ and $a_{4k+3} = 3$ for $k = 0, 1, 2, \cdots$.

4. Consider a_n given by $a_{5k} = 0$ for $k \in \mathbb{N}$, $a_{5k+1} = 1$ for $k = 0, 1, 2, \cdots$, $a_{5k+2} = 2$ for $k = 0, 1, 2, \cdots$, $a_{5k+3} = 3$ for $k = 0, 1, 2, \cdots$ and $a_{5k+4} = 4$ for $k = 0, 1, 2, \cdots$.

5. Consider a_m given by $a_{nk} = 0$ for $k \in \mathbb{N}$, $a_{nk+1} = 1$ for $k = 0, 1, 2, \cdots$, $a_{nk+2} = 2$ for $k = 0, 1, 2, \cdots$, etc. and in general, for $j = 1, 2, \cdots, n-1$ $a_{nk+j} = j$ for $k = 0, 1, 2, \cdots$.

6. Consider the sequence $1, 1, 2, 1, 2, 3, 1, 2, 3, 4, \cdots, 1, 2, 3, \cdots, n, 1, 2, 3, \cdots, n, n+1, \cdots$. Every natural number n is a subsequential limit value.

7. If $\{r_k\}_{k=1}^{\infty}$ is an enumeration of the rational numbers, consider the sequence $\{a_n\}$ given by $r_1, r_1, r_2, r_1, r_2, r_3, r_1, r_2, r_3, r_4, \cdots, r_1, r_2, r_3, \cdots, r_n,$ $r_1, r_2, r_3, \cdots, r_n, r_{n+1}, \cdots$. Since $\mathbb{Q} \subseteq \mathbb{R}$ is dense in $\mathbb{R}$, the set $\mathbb{R}$ is the set of subsequential limit values. To see this, suppose $x \in \mathbb{R}$. Let $\{\rho_n\}$ be a sequence of rational numbers converging to x. The be define a subsequence of $\{a_n\}$ by: let n_1 be so that $a_{n_1} = \rho_1$, let $n_2 > n_1$ be so that $a_{n_2} = \rho_2$, and in general, let $n_k > n_{k-1}$ be so that $a_{n_k} = \rho_k$.

8. Consider the sequence $1, 1, 2, 1, 2, 3, 1, 2, 3, 4, \cdots, 1, 2, 3, \cdots, n, 1, 2, 3, \cdots, n, n+1, \cdots$. Every natural number n is a subsequential limit value.

B.10 Homework Solutions 3.3.1

1. First we will show that given any real number x, there is an increasing sequence of rational numbers and a decreasing sequence of rational numbers, each whose limit value is x. Let w_1 be any rational number in $(x-1, x)$ and y_1 be any rational number in $(x, x+1)$. Let w_2 be a rational number in $(w_1, x) \cap (x - \frac{1}{2}, x)$ and let y_2 be a rational number in $(x, y_1) \cap (x, x + \frac{1}{2})$. Assuming $w_1, w_2, \cdots, w_k$ and $y_1, y_2, \cdots, y_k$ have been chosen, let w_{k+1} be a rational number in $(w_k, x) \cap (x - \frac{1}{2^k}, x)$ and let y_{k+1} be a rational number in $(x, y_k) \cap (x, x + \frac{1}{2^k})$. Clearly, by construction, $|x - y_j|, |x - w_j| < \frac{1}{2^j} \to 0$ as $j \to \infty$, $\{w_j\}$ is strictly increasing, and $\{y_j\}$ is strictly decreasing. An observation to note is that for any x and any positive ε, both $(x - \varepsilon, x)$ and $(x, x + \varepsilon)$ contain infinitely many distinct rational numbers. Let $\{r_k\}$ be an enumeration

of $\mathbb{Q}$. Let $k_1 \geq 1$ be so that $r_{k_1} \in (x-1, x)$. Let $k_2 > k_1$ be chosen so that $r_{k_2} \in (x - \frac{1}{2}, x) \cap (r_{k_1}, x) = I_2$. This is possible since $I_2 \cap (\mathbb{Q} \setminus \{r_1, r_2, \cdots, r_{k_1}\}) \neq \emptyset$. Assuming $k_1 < k_2 < \cdots < k_j$ have been chosen, we choose $k_{j+1} > k_j$ so that $r_{k_{j+1}} \in (x - \frac{1}{2^j}, x) \cap (r_{k_j}, x) = I_{j+1}$. He we use that $I_{j+1} \cap (\mathbb{Q} \setminus \{r_1, r_2, \cdots, r_{k_j}\}) \neq \emptyset$. By construction $|x - r_{k_i}| < \frac{1}{2^{i-1}}$, and so $r_{k_i} \to x$.

2. For this we use the above exercise, which says that given any real number, there is a subsequence of $\{r_k\}$ that converges to x. Thus $\liminf_{k \to \infty} r_k \leq x$ and $\limsup_{k \to \infty} r_k \geq x$. Since x is arbitrary, we get the desired result.

3. Assume first that $b_n \to b$. Then given any $\varepsilon > 0$, we know that there is an N so that $n \geq N$ implies $|b_n - b| < \varepsilon$. Now suppose $\{b_{k_n}\}_{n=1}^\infty$ is any subsequence of $\{b_n\}$ Then for $n > N$ we know $k_n \geq n > N$ and so $|b_{k_n} - b| < \varepsilon$. This says that every subsequence converges to b, and so $\limsup_{n \to \infty} b_n = b = \liminf_{n \to \infty} b_n$. Now assume $\limsup_{n \to \infty} b_n = b = \liminf_{n \to \infty} b_n$. By the two exercises just below this one, for any $\varepsilon > 0$ we know that there is N so that $n \geq N$ implies $b_n > b - \varepsilon$. Likewise, there is an M so that for $n \geq M$ we have $b_n < b + \varepsilon$. Hence, for $n \geq \max\{N, M\}$ we have $|b_n - b| < \varepsilon$. So $b_n \to b$.

4. Suppose $x < \liminf_{n \to \infty} a_n$, and $\forall k \in \mathbb{N}, \exists n_k > k$ so that $a_{n_k} \leq x$. Thus we have a subsequence $\{a_{n_k}\}_{k=1}^\infty$ so that $\{a_{n_k} : k \in \mathbb{N}\} \subseteq (-\infty, x]$. Consider first where $\inf\{a_{n_k} : k \in \mathbb{N}\} = -\infty$. Then there exists $n_{k_1} < n_{k_2} < \cdots < n_{k_j} < \cdots$ so that $|a_{n_{k_j}}| > j$. Thus $a_{n_{k_j}} \to -\infty$ as $j \to \infty$, and this contradicts $\liminf_{n \to \infty} > -\infty$. If $\inf\{a_{n_k} : k \in \mathbb{N}\} = i > -\infty$, then the sequence $\{a_{n_k}\}_{k=1}^\infty$ has its values lie in a compact set $[i, x]$ and so it has a convergent subsequence. In this case, it converges to some subsequential limit value, whose value is at most $x > \liminf_{n \to \infty} a_n$. This also provides a contradiction, since there is a subsequence of the original sequence with a subsequential limit value less than the infimum of the set of all subsequential limit values.

5. For this we use $b_n = -a_n$. If L is the set of subsequntial limits of $\{a_n\}$, then $-L$ is the subsequential limits of $\{b_n\}$. An earlier exercise showed $\sup(-L) = -\inf L$ and $\inf(-L) = -\sup L$. So we see that $\liminf_{n \to \infty}(-a_n) = -\limsup_{n \to \infty} a_n$ and $\limsup_{n \to \infty}(-a_n) = -\liminf_{n \to \infty} a_n$. If $y > \limsup_{n \to \infty} a_n$, then $x = -y$ satisfies $x = -y <$

$-\limsup_{n\to\infty} a_n = \liminf_{n\to\infty}(-a_n) = \liminf_{n\to\infty} b_n$. From the above exercise, we know that there is an N so that $n > N$ implies $b_n > x$. So, for $n > N$ we have $a_n < y$.

6. Let $\{a_n\}$ be any sequence of real numbers. We define $\alpha_n := \inf\{a_k : k \geq n\}$ and $A_n := \sup\{a_k : k \geq n\}$. From the definition, we see $\{\alpha_n\}$ is increasing, and $\{A_n\}$ is decreasing. Suppose that $\{a_{n_k}\}_{k=1}^\infty$ is any convergent subsequence of $\{a_n\}_{n=1}^\infty$ (allowing $\pm\infty$ as limit values). Then $\alpha_{n_k} \leq a_{n_k} \leq A_{n_k}$ for any k. Letting $k \to \infty$ we see that $\lim_{k\to\infty} \alpha_{n_k} \leq \lim_{k\to\infty} a_{n_k} \leq \lim_{k\to\infty} A_{n_k}$. Since our convergent subsequence was arbitrary, and $\{\alpha_n\}$ and $\{A_n\}$ are convergent (allowing of course $\pm\infty$ as limit values to extend the notion of convergence), we see that $\lim_{n\to\infty} \alpha_n \leq \liminf_{n\to\infty} a_n \leq \limsup_{n\to\infty} a_n \leq \lim_{n\to\infty} A_n$. If $\liminf_{n\to\infty} a_n = -\infty$, then $\lim_{n\to\infty} \alpha_n = -\infty$ as well. If $\limsup_{n\to\infty} a_n = \infty$, then $\lim_{n\to\infty} A_n = \infty$ as well. If $\liminf_{n\to\infty} a_n = \infty$, then it must be that α_n is increasing without bound, and so $\lim_{n\to\infty} \alpha_n = \infty$. If $\limsup_{n\to\infty} a_n = -\infty$, then in must be A_n is decreasing without bound, and so $\lim_{n\to\infty} A_n = -\infty$. In the cases were they both are not infinite, we proceed as is described in the remainder of this solution. Let ε be any positive number. For each n, there is an $k_n \geq n$ so that $\alpha_n \leq a_{k_n} \leq \alpha_n + \varepsilon$. Likewise there is a $j_n \geq n$ so that $A_n - \varepsilon \leq a_{j_n} \leq A_n$. We may restrict to subsequences n_i and n_m so that k_{n_i} is strictly increasing, and j_{n_m} is strictly increasing. Moreover, $\alpha_1 \leq a_{k_{n_i}} \leq \lim_{n\to\infty} \alpha_n + \varepsilon$ and $\lim_{n\to\infty} A_n - \varepsilon \leq a_{j_{n_m}} \leq A_1$. Since $\{a_{k_{n_i}}\}_{i=1}^\infty$ and $\{a_{j_{n_m}}\}_{m=1}^\infty$ are bounded, there are convergent subsequences $\{a_{k_{n_{i_r}}}\}_{r=1}^\infty$ and $\{a_{j_{n_{m_s}}}\}_{s=1}^\infty$. In such a case $\lim_{n\to\infty} \alpha_n \leq \lim_{r\to\infty} a_{k_{n_{i_r}}} \leq \lim_{n\to\infty} \alpha_n + \varepsilon$, and $\lim_{n\to\infty} A_n - \varepsilon \leq \lim_{s\to\infty} a_{j_{n_{m_s}}} \leq \lim_{n\to\infty} A_n$. Thus we have established $\liminf_{n\to\infty} a_m \leq \lim_{n\to\infty} \alpha_n + \varepsilon$, and $\lim_{n\to\infty} A_n - \varepsilon \leq \limsup_{n\to\infty} a_n$. Since $\varepsilon > 0$ is arbitrary, we get $\liminf_{n\to\infty} a_n \leq \lim_{n\to\infty} \alpha_n$, and $\lim_{n\to\infty} A_n \leq \limsup_{n\to\infty} a_n$.

7. Here we will make use of the above exercise, namely $\liminf_{k\to\infty} a_k = \lim_{j\to\infty}(\inf\{a_i : i \geq j\})$ and $\limsup_{k\to\infty} a_k = \lim_{j\to\infty}(\sup\{a_i : i \geq j\})$. Note also that the sequences $\alpha_j = \inf\{a_i : i \geq j\}$ and $A_j = \sup\{a_i : i \geq j\}$ are increasing and decreasing, respectively. We also define $\beta_j = \inf\{b_i : i \geq j\}$ and $B_j = \sup\{b_i : i \geq j\}$. Since we are assuming $a_n \leq b_n$ we know that $a_n \leq \sup\{b_k : k \geq n\}$ for a fixed n. Also, for $j \geq n$ we have $a_j \leq b_j \leq \sup\{b_k : k \geq n\}$. Hence $A_n = \sup\{a_k : $

$k \geq n\} \leq \sup\{b_k : k \geq j\} = B_n$. Since $\{A_k\}$ is decreasing, we know $\lim_{k\to\infty} A_k \leq B_n$ for any n. Thus $\lim_{k\to\infty} A_k \leq \lim_{k\to\infty} B_k$, and so $\limsup_{n\to\infty} a_n \leq \limsup_{n\to\infty} b_n$ is established. Since $a_n \leq b_n$ we know that $\inf\{a_k : k \geq n\} \leq a_n \leq b_n$ for a fixed n. Also, for $j \geq n$ we have $\inf\{a_k : k \geq n\} \leq a_j \leq b_j$. Hence $\alpha_n = \inf\{a_k : k \geq n\} \leq \inf\{b_k : k \geq j\} = B_n$. Since $\{\beta_k\}$ is increasing, we know $\alpha_n \leq \lim_{k\to\infty} \beta_k$ for any n. Thus $\lim_{k\to\infty} \alpha_k \leq \lim_{k\to\infty} \beta_k$, and so $\liminf_{n\to\infty} a_n \leq \liminf_{n\to\infty} b_n$ is established.

8. Let $a_n = \sqrt[n]{n}$. Then $a_n^n = n > 1$ and n^{th} roots of numbers greater than one are greater than one. We claim that $\{a_n\}_{n=3}^\infty$ is decreasing, and we will now show this. Observe that $a_{n+1} \leq a_n \iff a_{n+1}^{n(n+1)} \leq a_n^{n(n+1)}$, i.e $(n+1)^n \leq n^{n+1} = n \cdot n^n$. However, $(n+1)^n = n^n + n^{n-1} + \binom{n}{2}n^{n-2} + \binom{n}{3}n^{n-3} + \cdots + \binom{n}{k}n^{n-k} + \cdots + n^2 + 1$. Observe that $\binom{n}{k} = \frac{n\cdot(n-1)\cdots(n-k+1)}{k!} < n^k$, and so $\binom{n}{k}n^{n-k} < n^n$. Also, $n^2 + 1 < n^n$ for $n \geq 3$. Hence $(n+1)^n \leq n^{n+1}$, and so $a_{n+1} \leq a_n$. Since $a_n \geq 1$, we know the limit $\lim_{n\to\infty} a_n$ exists, and satisfies $\lim_{n\to\infty} a_n \geq 1$. Let $x = \lim_{n\to\infty} a_n \geq 1$. Then for all n we have $\sqrt[n]{n} > x \geq 1$, and so $n > x^n \geq 1$ for all n. This will require that $\frac{x^n}{n} < 1$ for all n. Also, $x \leq a_4 = \sqrt{2}$, so if $x = 1+\varepsilon$, we see $\varepsilon < \frac{1}{2}$. However, $\frac{(1+\varepsilon)^k}{k} < 1$ for any k also tells us $\frac{1+k\varepsilon + \frac{k(k-1)}{2}\varepsilon^2 + \cdots + \varepsilon^k}{k} < 1$ and so $\frac{1+k\varepsilon + \frac{k(k-1)}{2}\varepsilon^2}{k} < 1$. Thus $\frac{1}{k} + \varepsilon + \frac{k-1}{2}\varepsilon^2 < 1$, and so $\frac{1}{k} + \varepsilon^2 + \frac{k-1}{2}\varepsilon^2 < 1$. Hence $\varepsilon^2 < \frac{1-\frac{1}{k}}{1+\frac{k-1}{2}}$. Letting $k \to \infty$ will force $\varepsilon \to 0$. Hence we see that $\lim_{n\to\infty} a_n = 1$.

9. We proved this in the first part of our proof of the above exercise one.

10. Let $r \in \mathbb{R}$, and consider $\frac{r}{\sqrt{2}}$. Let $\{r_k\}$ be any sequence of rational numbers (whose existence is guaranteed by the above exercise) that converge to $\frac{r}{\sqrt{2}}$. Then $\{\sqrt{2}r_k\}$ is a sequence of irrational numbers converging to r.

11. Here we will make use of $\liminf_{k\to\infty} a_k = \lim_{j\to\infty}(\inf\{a_i : i \geq j\})$ and $\limsup_{k\to\infty} a_k = \lim_{j\to\infty}(\sup\{a_i : i \geq j\})$. Given a specific sequence $\{c_k\}$, since the set of subsequential limit values is a non-empty subset of $\mathbb{R} \cup \{\pm\infty\}$, we have $\liminf_{k\to\infty} c_k \leq \limsup_{k\to\infty} c_k$. Let $\{a_k\}$ and $\{b_k\}$ be any two sequences of real numbers, then $\liminf_{n\to\infty}(a_n + b_n) \leq \limsup_{n\to\infty}(a_n + b_n)$. Observe that the sequences $\alpha_j = \inf\{a_i : i \geq j\}$,

$\beta_j = \inf\{b_i : i \geq j\}$ and $\inf\{a_i + b_i : i \geq j\}$ are increasing and the sequences $A_j = \sup\{a_i : i \geq j\}$, $B_j = \sup\{b_i : i \geq j\}$ and $\sup\{a_i + b_j : i \geq j\}$ are decreasing. Let $n \in \mathbb{N}$. Then $\alpha_n + \beta_n \leq a_n + b_n$, and in fact, for any $k \geq n$, $\alpha_n + \beta_n \leq a_k + b_k$. So $\alpha_n + \beta_n \leq \inf\{a_k + b_k : k \geq n\}$. Thus $\alpha_n + \beta_n \leq \lim_{j \to \infty} \inf\{a_k + b_k : k \geq j\}$ for any n. Taking limits of the left hand side as $n \to \infty$, we get $\liminf_{n \to \infty} a_n + \liminf_{n \to \infty} b_n \leq \liminf_{n \to \infty}(a_n + b_n)$. Let $n \in \mathbb{N}$. Then $a_n + b_n \leq A_n + B_n$, and in fact, for any $k \geq n$, $a_k + b_k \leq A_n + B_n$. So $\sup\{a_k + b_k : k \geq n\} \leq A_n + B_n$. Thus $\lim_{j \to \infty} \sup\{a_k + b_k : k \geq j\} \leq A_n + B_n$ for any n. Taking limits of the right hand side as $n \to \infty$, we get $\limsup_{n \to \infty}(a_n + b_n) \leq \limsup_{n \to \infty} a_n + \limsup_{n \to \infty} b_n$.

12. If $\lim_{n \to \infty} a_n$ and $\lim_{n \to \infty} b_n$ exist. Then every subsequence of $\{a_n\}$ and every subsequence of $\{b_n\}$ converges to the respective limit of the sequence. Using this and linearity properties of limits we get $\liminf_{n \to \infty} a_n + \liminf_{n \to \infty} b_n = \lim_{n \to \infty} a_n + \lim_{n \to \infty} b_n = \lim_{n \to \infty}(a_n + b_n) = \limsup_{n \to \infty}(a_n + b_n) = \limsup_{n \to \infty} a_n + \limsup_{n \to \infty} b_n$.

B.11 Homework Solutions 3.4.1

1. $1, -1, 1, -1, \cdots, (-1)^{n+1}, \cdots$

2. $1, 2, 3, 4, \cdots, n, \cdots$

3. Let $\{a_n\}$ be a Cauchy sequence of real numbers. There is an $N \in \mathbb{N}$ so that $n, m \geq N$ implies $|a_n - a_m| < 1$. So for such an n, $a_N - 1 < a_n < a_N + 1$. Thus, for any $k \in \mathbb{N}$ we have $a_k \geq \min\{a_1, a_2, \cdots, a_{N-1}, a_N - 1\}$ and $a_k \leq \max\{a_1, a_2, \cdots, a_{N-1}, a_N + 1\}$.

4. Let $\{a_n\}$ and $\{b_n\}$ be Cauchy sequences. Then given $\varepsilon > 0$, there are $N_a, N_b \in \mathbb{N}$ so that $n, m \geq N_a$ implies $|a_n - a_m| < \frac{\varepsilon}{2}$ and $n, m \geq N_b$ implies $|b_n - b_m| < \frac{\varepsilon}{2}$. Let $n, m \geq \max\{N_a, N_b\}$. Then $|(a_n + b_n) - (a_m + b_m)| = |(a_n - a_m) + (b_n - b_m)| \leq |a_n - a_m| + |b_n - b_m| < \frac{\varepsilon}{2} + \frac{\varepsilon}{2} = \varepsilon$. The above exercise showed that there are real positive numbers A, B so that $|a_n| \leq A$ and $|b_n| \leq B$ for all n. Let M_a be so that $|a_n - a_m| < \frac{\varepsilon}{2B}$ for $n, m \geq M_a$, and let M_b be so that $|b_n - b_m| < \frac{\varepsilon}{2A}$ for any $n, m \geq M_b$. Then for $n, m \geq \max\{M_a, M_b\}$ we have $|a_n b_n - a_m b_m| = |a_n b_n - a_n b_m + a_n b_m - a_m b_m| \leq |a_n b_n - a_n b_m| + |a_n b_m - a_m b_m| = |a_n| \cdot |b_n - b_m| + |b_m| \cdot |a_n - a_m| \leq A|b_n - b_m| + B|a_n - a_m| < A \cdot \frac{\varepsilon}{2A} + B \cdot \frac{\varepsilon}{2B} = \varepsilon$.

5. Suppose that $\{a_n\}$ is a Cauchy sequence of natural numbers. Then there is an $N \in \mathbb{N}$ so that for $n, m \geq N$ we have $|a_n - a_m| < \frac{1}{3}$. Thus, for $n > N$ we have $a_N - \frac{1}{3} < a_n < a_N + \frac{1}{3}$. The interval $(a_N - \frac{1}{3}, a_N + \frac{1}{3})$ only contains one natural number, namely a_N. Thus for $n \geq N$ we have $a_n = a_N$.

6. Assume $\{a_n\}$ is a sequence which satisfies $d(x_n, x_{n+1}) < r^n$, and $r \in (0, 1)$. Observe that $d(x_n, x_{n+k})$ satisfies $d(x_n, x_{n+k}) \leq d(x_n, x_{n+1}) + d(x_{n+1}, x_{n+2}) + \cdots + d(x_{n+k-1}, x_{n+k}) \leq r^n + r^{n+1} + \cdots + r^{n+k-1} = r^n \cdot \frac{1-r^{k-1}}{1-r} < r^n \cdot \frac{1}{1-r}$. Since $b_n = r^n \to 0$ as $n \to \infty$, given $\varepsilon > 0$ we can find an N so that $r^N \cdot \frac{1}{1-r} < \varepsilon$.

7. For $a_n = \sqrt{n}$ we have $|a_{n+1} - a_n| = \frac{1}{\sqrt{n+1}+\sqrt{n}} \to 0$ as $n \to \infty$. However, $\sqrt{n+k} - \sqrt{n} = \frac{k}{\sqrt{n+k}+\sqrt{n}} > \frac{k}{2\sqrt{n+k}}$. If we pick $n = k$ we see that $|a_{2n} - a_n| > \frac{\sqrt{n}}{2\sqrt{2}} \to \infty$ as $n \to \infty$.

B.12 Homework Solutions 3.5.4

1. Here we have $\sqrt[k]{a_k} = \sqrt[k]{\frac{1}{k^k}} = \frac{1}{k} \to 0$ as $k \to \infty$, and so by the ratio test the series $\sum_{k=1}^{\infty} \frac{1}{k^k}$ converges.

2. Here we have $a_{2k+1} = \frac{1}{2^{2k+1}}$ and $a_{2k} = \frac{1}{2^{2(k-1)}}$. So $\frac{a_{2k+1}}{a_{2k}} = \frac{1}{2^3}$ and $\frac{a_{2(k+1)}}{a_{2k+1}} = 2$. So by the ratio test, no information can be drawn.

3. For the same series as in the above exercise, we have $\sqrt[2k+1]{a_{2k+1}} = \frac{1}{2}$ and $\sqrt[2k]{a_{2k}} = \frac{\sqrt[k]{2}}{2} \to \frac{1}{2}$. So $\limsup_{n \to \infty} \sqrt[n]{a_n} = \frac{1}{2}$, and so the series converges by the root test.

4. For the series $\sum \frac{2^k}{k!}$ we have $\frac{a_{k+1}}{a_k} = \frac{2}{k+1} \to 0$ as $k \to \infty$, and so the series converges by the ratio test.

5. One can show that the Fibonacci sequence, which is defined by $f_1 = 1, f_2 = 1$ and $f_{n+1} = f_n + f_{n-1}$, satisfies $f_n = \frac{1}{\sqrt{5}}(x^n - \frac{(-1)^n}{x^n})$ where $x = \frac{1+\sqrt{5}}{2}$. For the series $\sum \frac{1}{f_n}$, we have $\frac{a_{n+1}}{a_n} = \frac{f_n}{f_{n+1}} = \frac{x^n - \frac{(-1)^n}{x^n}}{x^{n+1} - \frac{(-1)^{n+1}}{x^{n+1}}}$
$= \frac{1 - \frac{(-1)^n}{x^{2n}}}{x - \frac{(-1)^{n+1}}{x^{2n+1}}} \to \frac{1}{x} < 1$. So the series converges by the ratio test.

6. For $x \in (0,1)$ we consider $\sum k x^k$. We have $\sqrt[k]{a_k} = \sqrt[k]{k} \cdot x \to x < 1$ as $k \to \infty$. So this series converges by the root test.

7. Suppose $\sum a_k$ converges. Let $b_n = \frac{a_1 + a_2 + \cdots + a_n}{n}$, we will show $\sum b_k$ diverges. Here we have $\sum_{k=1}^{n} b_k = a_1 \cdot (1 + \frac{1}{2} + \cdots + \frac{1}{n}) + a_2 \cdot (\frac{1}{2} + \frac{1}{3} + \cdots + \frac{1}{n}) + \cdots + a_{n-1} \cdot (\frac{1}{n-1} + \frac{1}{n}) + a_n \cdot \frac{1}{n}$. For j so that $a_j \neq 0$ we have $a_j \cdot (\frac{1}{j} + \frac{1}{j+1} + \cdots + \frac{1}{n}) \to \infty$ as $n \to \infty$, and so $\sum_{k=1}^{n} b_k \to \infty$ as $n \to \infty$.

8. Consider $\sum_{n=1}^{\infty} \frac{1}{n^{1+\frac{1}{n}}}$. Here we have $\frac{1}{n^{1+\frac{1}{n}}} = \frac{1}{n} \cdot \frac{1}{\sqrt[n]{n}}$. An earlier exercise's solution told us $\sqrt[n]{n} \to 1$ as $n \to \infty$, and in fact $\{\sqrt[n]{n}\}$ is strictly decreasing for $n \geq 3$. So, for $n \geq 3$ $\sqrt[n]{n} < 2$ and so $\frac{1}{n} \cdot \frac{1}{\sqrt[n]{n}} > \frac{1}{2} \cdot \frac{1}{n}$. Since $\sum \frac{1}{2} \cdot \frac{1}{n}$ diverges, we must also have $\sum \frac{1}{n} \cdot \frac{1}{\sqrt[n]{n}}$ diverges.

9. Consider $\sum \frac{n!}{n^n}$. Here, $\frac{a_{n+1}}{a_n} = (\frac{n}{n+1})^n = \frac{1}{(1+\frac{1}{n})^n} \to \frac{1}{e}$ as $n \to \infty$. So this series converges by the ratio test.

10. Consider $\sum \frac{1^2 + 2^2 + \cdots + n^2}{n!}$. Then $\frac{a_{n+1}}{a_n} = \frac{1}{n+1}[1 + \frac{(n+1)^2}{1 + \cdots + n^2}] = \frac{1}{n+1} + \frac{1}{\frac{1}{n+1} + \cdots + \frac{n^2}{n+1}} \to 0$. Thus the series converges by the ratio test.

11. Consider $\sum \frac{\sqrt{k+1}}{\sqrt[3]{k^5 + k^3 + 1}}$. $\frac{\sqrt{k+1}}{\sqrt[3]{k^5 + k^3 + 1}} \leq \frac{\sqrt{2k}}{\sqrt[3]{k^5}} = \sqrt{2} \frac{1}{k^{\frac{7}{6}}}$. Now $\sum \frac{1}{k^{\frac{7}{6}}}$ is a convergent p series, and so our series converges by the comparison test.

12. Consider $\sum (\frac{n}{n^2+1})^n$. $\sqrt[n]{a_n} = \frac{n}{n^2+1} = \frac{1}{n+\frac{1}{n}} \to 0$, and so the series converges by the root test.

13. Consider $\sum (\frac{n}{n^2+1})^{n^{1.0001}}$. Observe that for $n > 2^{10000}$ we have $n^{0.0001} - 1 > 1$, and so for such n we have $(n+\frac{1}{n})^{n[n^{0.0001} - 1]} > (n+\frac{1}{n})^n > 1$. Hence, for such an n we have $\frac{(n+\frac{1}{n})^{n^{1.0001}}}{(n+\frac{1}{n})^n} > 1$. Hence $(\frac{1}{n+\frac{1}{n}})^n > (\frac{1}{n+\frac{1}{n}})^{n^{1.0001}}$, and using the above exercise and the comparison test, we have that our given series converges.

14. Consider $\sum 2^{\frac{(-1)^n}{n}}$. Here $a_n = \frac{1}{\sqrt[n]{2}}$ if n is odd and $a_n = \sqrt[n]{2}$ if n is even. Either way, $a_n \to 1$ and so by the divergence test, the series diverges.

15. Consider $\sum (\frac{5+(-1)^n}{2})^{-n}$. So $0 < a_n \leq (\frac{1}{2})^n$. Using that $\sum \frac{1}{2^n}$ converges, we get that the given series converges by the comparison test.

16. Consider $\sum(\frac{5+(-1)^n}{2})^n$. Here we have $a^n > 2^n$, and so the series diverges by the divergence test.

B.13 Homework Solutions 3.5.11

1. Consider $\sum \frac{n!}{n^n} x^n$. Then $\frac{|a_{n+1}|}{|a_n|} = \frac{|x|}{(1+\frac{1}{n})^n} \to \frac{|x|}{e}$, and so the series converges absolutely if $|x| < e$. Note in fact that if we let $x = \pm e$, by Stirling's formula we have $n!(\frac{e}{n})^n \geq \sqrt{2\pi n}$, and so the series diverges at $x = \pm e$.

2. For $p \in \mathbb{N}$ and $q > 0$, we will determine the radius of convergence of $\sum \frac{(pn)!}{(n!)^q} x^n$. Here $\frac{|a_{n+1}|}{|a_n|} = \frac{(pn+p)(pn+p-1)\cdots(pn+1)}{(n+1)^q} |x| = p^p \frac{(n+1)}{(n+1)^{\frac{q}{p}}} \cdot \frac{(n+\frac{p-1}{p})}{(n+1)^{\frac{q}{p}}} \cdots \frac{(n+\frac{1}{p})}{(n+1)^{\frac{q}{p}}} |x|$

$= p^p [\prod_{i=1}^p \frac{n^{1-\frac{q}{p}+\frac{i}{q}}\cdot p \cdot n^{\frac{i}{p}}}{(1+\frac{1}{n})^{\frac{q}{p}}}] \cdot |x|$. If $p > q$ then the quantity $n^{1-\frac{q}{p}} \to \infty$ as $n \to \infty$, and this series does not converge for any non-zero x values, and hence the radius of convergence is 0. If $q = p$ then

$p^p [\prod_{i=1}^p \frac{n^{1-\frac{q}{p}+\frac{i}{q}} \cdot p \cdot n^{\frac{i}{p}}}{(1+\frac{1}{n})^{\frac{q}{p}}}] \cdot |x| \to p^p |x|$ and so the series converges absolutely if $|x| < p^{-p}$. If $q > p$, then $p^p [\prod_{i=1}^p \frac{n^{1-\frac{q}{p}+\frac{i}{q}} \cdot p \cdot n^{\frac{i}{p}}}{(1+\frac{1}{n})^{\frac{q}{p}}}] \cdot |x| \to 0$, and so the radius of convergence is infinite, and so the series converges absolutely for all values of x.

3. For a given power series $\sum_{n=0}^\infty a_n x^n$ we consider the series $\sum_{n=0}^\infty n a_n x^{n-1}$ and the series $\sum_{n=0}^\infty \frac{a_n}{n+1} x^{n+1}$. For absolute convergence we use the ratio test. The first series $\sum_{n=0}^\infty a_n x^n$ converges absolutely if $|x| < \frac{1}{\limsup_{n\to\infty} \frac{|a_{n+1}|}{|a_n|}} := r_1$. For the second series $\sum_{n=0}^\infty n a_n x^{n-1}$, it converges absolutely if $|x| < \frac{1}{\limsup_{n\to\infty} \frac{(n+1)|a_{n+1}|}{n|a_n|}} := r_2$. The third series $\sum_{n=0}^\infty \frac{a_n}{n+1} x^{n+1}$ converges absolutely if $|x| < \frac{1}{\limsup_{n\to\infty} \frac{n|a_{n+1}|}{(n+1)|a_n|}} := r_3$. We will show that the radius of converge of the three series is the same if we can show $\limsup_{n\to\infty} \frac{|a_{n+1}|}{|a_n|}$, $\limsup_{n\to\infty} \frac{(n+1)|a_{n+1}|}{n|a_n|}$ and $\limsup_{n\to\infty} \frac{n|a_{n+1}|}{(n+1)|a_n|}$ are all the same. Clearly, for a given n, $\frac{n}{n+1} \cdot \frac{|a_{n+1}|}{|a_n|} \leq \frac{|a_{n+1}|}{|a_n|} \leq \frac{n+1}{n} \frac{|a_{n+1}|}{|a_n|}$, and so by exercise 7 from 3.3.1 we have

$\limsup_{n\to\infty} \frac{n}{n+1} \cdot \frac{|a_{n+1}|}{|a_n|} \leq \limsup_{n\to\infty} \frac{|a_{n+1}|}{|a_n|} \leq \limsup_{n\to\infty} \frac{n+1}{n} \frac{|a_{n+1}|}{|a_n|}$. Let $\varepsilon > 0$ be given, then for $n > \frac{1}{\varepsilon} - 1$ we have $\frac{n}{n+1} = 1 - \frac{1}{n+1} > 1 - \varepsilon$. Hence, for such an n we have $\frac{n}{n+1} \cdot \frac{|a_{n+1}|}{|a_n|} > (1 - \varepsilon)\frac{|a_{n+1}|}{|a_n|}$. So $\limsup_{n\to\infty} \frac{n}{n+1} \cdot \frac{|a_{n+1}|}{|a_n|} \geq (1 - \varepsilon) \limsup_{n\to\infty} \frac{|a_{n+1}|}{|a_n|}$. For $n > \frac{1}{\varepsilon}$ we have $\frac{n+1}{n} < 1 + \varepsilon$. Hence $\frac{n+1}{n} \frac{|a_{n+1}|}{|a_n|} \leq (1+\varepsilon) \frac{|a_{n+1}|}{|a_n|}$ for such an n, and so $\limsup_{n\to\infty} \frac{n+1}{n} \frac{|a_{n+1}|}{|a_n|} \leq (1+\varepsilon) \limsup_{n\to\infty} \frac{|a_{n+1}|}{|a_n|}$. So, $(1-\varepsilon) \limsup_{n\to\infty} \frac{|a_{n+1}|}{|a_n|} \leq \limsup_{n\to\infty} \frac{n}{n+1} \cdot \frac{|a_{n+1}|}{|a_n|} \leq \limsup_{n\to\infty} \frac{|a_{n+1}|}{|a_n|} \leq \limsup_{n\to\infty} \frac{n+1}{n} \frac{|a_{n+1}|}{|a_n|} \leq (1+\varepsilon) \limsup_{n\to\infty} \frac{|a_{n+1}|}{|a_n|}$. Since $\varepsilon > 0$ was arbitrary, it must be that $\limsup_{n\to\infty} \frac{n}{n+1} \cdot \frac{|a_{n+1}|}{|a_n|} = \limsup_{n\to\infty} \frac{|a_{n+1}|}{|a_n|} = \limsup_{n\to\infty} \frac{n+1}{n} \frac{|a_{n+1}|}{|a_n|}$, and so $r_1 = r_2 = r_3$ is established.

4. Let $f(x) = \sum_{n=1}^{\infty} \frac{1}{n} x^n$. Then $\frac{|a_{n+1}|}{|a_n|} = \frac{n}{n+1}|x| \to |x|$ as $n \to \infty$. Thus the series $f(x)$ converges absolutely for $x \in (-1, 1)$. Now $f(-1) = \sum (-1)^n \frac{1}{n}$ is the alternating harmonic series, which converges conditionally, and $f(1) = \sum \frac{1}{n}$ is the harmonic series, which diverges. Hence the domain of $f(x)$ is $[-1, 1)$. Consider $\mathrm{Log}(x) = -f(1-x)$, then $\mathrm{Log}(x)$ has domain $(0, 2]$.

5. Consider $J_0(x) = \sum_{n=0}^{\infty} \frac{(-1)^n}{(n!)^2 2^{2n}} x^{2n}$. Here $\frac{|a_{n+1}|}{|a_n|} = \frac{x^2}{(4(n+1)^2)} \to 0$. Here the radius of convergence is ∞, and the series converges for any x.

6. Consider $J_1(x) = \sum_{n=0}^{\infty} \frac{(-1)^n}{2n!(n+1)! 2^{2n}} x^{2n+1}$. Here $\frac{|a_{n+1}|}{|a_n|} = \frac{x^2}{4(n+1)(n+2)} \to 0$, and so the radius of convergence is ∞ and the series converges for all values of x.

7. $\mathrm{Sin}(x)$ and $\mathrm{Cos}(x)$ converge absolutely for all values of x. So $\mathrm{Sin}(x) \cdot \mathrm{Cos}(x) = (\sum_{k=0}^{\infty} \frac{(-1)^k}{(2k+1)!} x^{2k+1}) \cdot (\sum_{m=0}^{\infty} \frac{(-1)^m}{(2m)!} x^{2m}) = \sum_{j=0}^{\infty} (\sum_{i=0}^{j} \frac{(-1)^i}{(2i+1)!} x^{2i+1} \cdot \frac{(-1)^{j-i}}{(2(j-i))!} x^{2(j-i)}) = \sum_{j=0}^{\infty} (\sum_{i=0}^{j} \frac{1}{(2i+1)!(2(j-i))!})(-1)^j x^{2j+1} = \sum_{j=0}^{\infty} (\sum_{i=0}^{j} \binom{2j+1}{2i+1}) \frac{(-1)^j}{(2j+1)!} x^{2j+1} = \sum_{j=0}^{\infty} (\sum_{i=0}^{j} (\binom{2j}{2i} + \binom{2j}{2i+1})) \frac{(-1)^j}{(2j+1)!} x^{2j+1} = \sum_{j=0}^{\infty} (\sum_{s=0}^{2j} \binom{2j}{s}) \frac{(-1)^j}{(2j+1)!} x^{2j+1} = \sum_{j=0}^{\infty} (1+1)^{2j} \cdot \frac{(-1)^j}{(2j+1)!} x^{2j+1} = \frac{1}{2} \sum_{j=0}^{\infty} \frac{(-1)^j}{(2j+1)!} (2x)^{2j+1} = \frac{1}{2} \mathrm{Sin}(2x)$.

B.14 Homework Solutions 4.1.2

1. Assume $\lim_{x \to a} f(x) = L$. Then given $\varepsilon > 0$, $\exists \delta > 0$ so that $0 < |x - a| < \delta \Rightarrow |f(x) - L| < \varepsilon$. So, for $a < x < a + \delta$ we have $|f(x) - L| < \varepsilon$. Hence $\lim_{x \to a^+} f(x) = L$. Also, $a - \delta < x < a$ implies $|f(x) - L| < \varepsilon$, and so $\lim_{x \to a^-} f(x) = L$. Next assume $\lim_{x \to a^-} f(x) = L = \lim_{x \to a^+} f(x)$. Then, given $\varepsilon > 0$, $\exists \delta_- > 0$ so that $a - \delta_- < x < a$ implies $|f(x) - L| < \varepsilon$, and $\exists \delta_+ > 0$ so that $a < x < a + \delta_+$ implies $|f(x) - L| < \varepsilon$. Thus for $0 < |x - a| < \min\{\delta_-, \delta_+\}$ we have $|f(x) - L| < \varepsilon$. Hence $\lim_{x \to a} f(x) = L$.

2. First we will establish that for $0 < a < b \iff 0 < a^3 \leq b^3$. If $0 < a < b$, then multiplying by a we get $a^2 < ab$ and multiplying by b we get $ab < b^2$, and so $a^2 < b^2$. Then multiplying by a, we get $a^3 < b^2 a$. Multiplying the original inequality by b^2, we get $ab^2 < b^3$, and so $a^3 < b^3$. To get the converse, we proceed by a proof by contraposition. Suppose now that $a, b > 0$ and $a \geq b$. Then performing the same argument as above (just switch all $<$'s to $\geq$'s), we get the desired result. Note that this also says that $0 < a < b \iff 0 < \sqrt[3]{a} < \sqrt[3]{b}$. Let $\varepsilon > 0$ be given. For $a = 0$ we use that if $|x| < \varepsilon^3$, then $|\sqrt[3]{x}| = \sqrt[3]{|x|} < \varepsilon$. If $a < 0$ and we assume $x < 0$ as well, we observe that $-\sqrt[3]{x} = \sqrt[3]{-x} = |\sqrt[3]{x}|$. So if we establish the result of positive a, if $a < 0$ then $\lim_{x \to -a} \sqrt[3]{x} = \sqrt[3]{-a}$. So $\exists \delta > 0$ so that $|x - (-a)| = |x - (-a)| < \delta$ implies $|\sqrt[3]{x} - \sqrt[3]{-a}| < \varepsilon$. If $y = -x$, then for $|y - a| = |-x - a| = |x - (-a)| < \delta$, we have $|\sqrt[3]{y} - \sqrt[3]{a}| = |-\sqrt[3]{-x} - (-\sqrt[3]{a})| = |\sqrt[3]{x} - \sqrt[3]{-a}| < \varepsilon$. Now, assume $a > 0$, and let $\varepsilon > 0$ be given. Without loss of generality, we shall also assume $x > 0$. Consider $|\sqrt[3]{x} - \sqrt[3]{a}| = \frac{|x-a|}{(\sqrt[3]{x})^2 + \sqrt[3]{x} \cdot \sqrt[3]{a} + (\sqrt[3]{a})^2} < \frac{|x-a|}{(\sqrt[3]{a})^2}$. If we assume $|x - a| < \min\{a, (\sqrt[3]{a})^2 \cdot \varepsilon\}$, then $|\sqrt[3]{x} - \sqrt[3]{a}| < \varepsilon$.

3. First we will establish that for $0 < a < b \iff 0 < a^n \leq b^n$. We shall proceed by induction. For $n = 1$ this is obvious, so we assume it is true for all $1 \leq k < n$. Taking $0 < a^{n-1} < b^{n-1}$, multiplying by a we get $0 < a^n < ab^{n-1}$. Taking $0 < a < b$ and multiplying by b^{n-1} we get $0 < ab^{n-1} < b^n$ and so we see $0 < a^n < b^n$. To get the converse, we proceed by a proof by contraposition. Suppose now that $a, b > 0$ and $a \geq b$. Then performing the same argument as above (just switch all $<$'s to $\geq$'s), we get the desired result. Note that this also says that $0 < a < b \iff 0 < \sqrt[n]{a} < \sqrt[n]{b}$. Let $\varepsilon > 0$ be given. For $a = 0$ we use that if

$|x| < \varepsilon^n$, then $|\sqrt[n]{x}| = \sqrt[n]{|x|} < \varepsilon$. In the case where n is odd and $a < 0$, we assume $x < 0$ as well, we observe that $-\sqrt[n]{x} = \sqrt[n]{-x} = |\sqrt[n]{x}|$. So if we establish the result of positive a, if $a < 0$ then $\lim_{x \to -a} \sqrt[n]{x} = \sqrt[n]{-a}$. So $\exists \delta > 0$ so that $|x - (-a)| = |x - (-a)| < \delta$ implies $|\sqrt[n]{x} - \sqrt[n]{-a}| < \varepsilon$. If $y = -x$, then for $|y - a| = |-x - a| = |x - (-a)| < \delta$, we have $|\sqrt[n]{y} - \sqrt[n]{a}| = |-\sqrt[n]{-x} - (-\sqrt[n]{a})| = |\sqrt[n]{x} - \sqrt[n]{-a}| < \varepsilon$. Now, assume $a > 0$, and let $\varepsilon > 0$ be given. Without loss of generality, we shall also assume $x > 0$. Consider $|\sqrt[n]{x} - \sqrt[n]{a}| = \frac{|x-a|}{\sum_{k=0}^{n-1}(\sqrt[n]{x})^k \cdot (\sqrt[n]{a})^{n-1-k}} < \frac{|x-a|}{(\sqrt[n]{a})^{n-1}}$. If we assume $|x - a| < \min\{a, (\sqrt[n]{a})^{n-1} \cdot \varepsilon\}$, then $|\sqrt[n]{x} - \sqrt[n]{a}| < \varepsilon$.

B.15 Homework Solutions 4.5

1. Let $\varepsilon = \frac{1}{2}$ and fix any $\delta > 0$. If $a \in \mathbb{Q}$, then $\exists x \in (a - \delta, a + \delta) \cap (\mathbb{R} \setminus \mathbb{Q})$, and so $|f(x) - f(a)| = 2 \geq \frac{1}{2} = \varepsilon$. If $a \in (\mathbb{R} \setminus \mathbb{Q})$, then $\exists x \in (a - \delta, a + \delta) \cap \mathbb{Q}$, and so $|f(x) - f(a)| = 2 \geq \frac{1}{2} = \varepsilon$.

2. Let $a \neq 0$ be fixed, let $\varepsilon = \frac{|a|}{2}$, and fix any $\delta > 0$. If $a \in \mathbb{Q}$, then $\exists x \in (a - \delta, a + \delta) \cap (\mathbb{R} \setminus \mathbb{Q})$, and so $|f(x) - f(a)| = 2|a| \geq \frac{|a|}{2} = \varepsilon$. If $a \in (\mathbb{R} \setminus \mathbb{Q})$, then $\exists x \in (a - \delta, a + \delta) \cap \mathbb{Q}$, and so $|f(x) - f(a)| = 2|a| \geq \frac{|a|}{2} = \varepsilon$. For 0, let $\varepsilon > 0$ be given and let $\delta = \varepsilon$. Then for $x \in (-\varepsilon, \varepsilon)$ we have $|f(x) - f(0)| = |f(x)| = |x| < \varepsilon$.

3. Let $f(x) = 0$ if $x \in \mathbb{R} \setminus \mathbb{Q}$, and let $f(x) = n$ if $x = \frac{m}{n} \in \mathbb{Q}$, $n > 0$ and the gcd of m, n being one. Let $\varepsilon = \frac{1}{2}$, and consider any $\delta > 0$. If $x = \frac{m}{n} \in \mathbb{Q}$ we take $y \in (x - \delta, x + \delta) \cap (\mathbb{R} \setminus \mathbb{Q})$, and in such a case $|f(x) - f(y)| = n > \frac{1}{2} = \varepsilon$. So f is not continuous at any point in $\mathbb{Q}$. Let $x \in \mathbb{R} \setminus \mathbb{Q}$, and let $y \in (x - \delta, x + \delta) \cap \mathbb{Q}$, with $y = \frac{m}{n}$. Then $|f(x) - f(y)| = n > \frac{1}{2} = \varepsilon$. Thus f is not continuous at any point in $\mathbb{R} \setminus \mathbb{Q}$.

4. Suppose $f : D \to \mathbb{R}$ and $(-r, r) \subseteq D$ with $r > 0$. Suppose $f(0) > 0$ and f is continuous at 0. Let $\varepsilon = \frac{f(0)}{2}$, and let $\delta > 0$, $\delta \leq r$ be so that $|x - 0| < \delta$ implies $|f(x) - f(0)| < \varepsilon = \frac{f(0)}{2}$. So, for $x \in (-\delta, \delta)$ we have $0 < \frac{f(0)}{2} < f(x) < \frac{3f(0)}{2}$.

5. Suppose $f(x) \geq 0$ on $\mathbb{Q}$ and f is continuous everywhere. Consider $y \in \mathbb{R} \setminus \mathbb{Q}$, with $f(y) \neq 0$. Suppose $f(y) < 0$. Let $\varepsilon = \frac{|f(y)|}{2} = \frac{-f(y)}{2}$.

Then $\exists \delta > 0$ so that $|x - y| < \delta$ implies $|f(x) - f(y)| < \varepsilon$. Let $x \in (y - \delta, y + \delta) \cap \mathbb{Q}$. Then $|f(x) - f(y)| < \frac{-f(y)}{2}$, and so $\frac{3f(y)}{2} < f(x) < \frac{f(y)}{2} < 0$, a contradiction. So it must be $f \geq 0$ on all of $\mathbb{R}$. If we assume $f > 0$ on $\mathbb{Q}$, we cannot conclude more than $f \geq 0$ on $\mathbb{R}$. To see this, consider $f(x) = (x - \sqrt{2})^2$.

6. Let $\{r_n\}$ be an enumeration of $\mathbb{Q} \cap [0,1]$, and define $f(x) = \sum_{r_n \leq x} 2^{-n}$. Then clearly $0 \leq f \leq 1$ and f is increasing. Moreover, for $x = r_k$ we have $\lim_{x \to r_k+} f(x) - \lim_{x \to r_k-} = r_k$, and so f cannot be continuous at r_k.

7. Suppose $f: \mathbb{R} \to \mathbb{R}$ satisfies $f(x+y) = f(x) + f(y)$. Suppose f is continuous at x_0. Observe that $f(0+0) = f(0)$ and so $2f(0) = f(0)$ which implies $f(0) = 0$. Thus $0 = f(0) = f(x+(-x)) = f(x) + f(-x)$ and so $f(-x) = -f(x)$. Let $n \in \mathbb{N}$, then $f(n) = f(\underbrace{1+1+\cdots+1}_{n \text{ times}}) = \underbrace{f(1) + f(1) + \cdots + f(1)}_{n \text{ times}} = nf(1)$. $f(1) = f(\underbrace{\frac{1}{n} + \cdots + \frac{1}{n}}_{n \text{ times}}) = nf(\frac{1}{n})$, and so $f(\frac{1}{n}) = \frac{f(1)}{n}$. Likewise for $m \in \mathbb{N}$ we have $f(\frac{m}{n}) = f(\underbrace{\frac{1}{n} + \cdots + \frac{1}{n}}_{m \text{ times}}) = \underbrace{f(\frac{1}{n}) + \cdots + f(\frac{1}{n})}_{m \text{ times}} = mf(\frac{1}{n}) = \frac{m}{n}f(1)$. So we see that on $\mathbb{Q}$, $f(x) = xf(1)$. Note that f continuous at x_0 implies that for a given $\varepsilon > 0$, $\exists \delta > 0$ so that $|x - x_0| < \delta$ implies $|f(x) - f(x_0)| < \varepsilon$. However $f(x - x_0) = f(x) - f(x_0)$, and so for $y \in (-\delta, \delta)$ we write $x = x_0 + y$. So $y = x - x_0 \in (-\delta, \delta)$ and thus $|f(y)| = |f(y) - f(0)| < \varepsilon$. Hence f is continuous at 0. Consider $z \in \mathbb{R}$ and suppose $|z - y| < \delta$. The $|f(y) - f(z)| = |f(y - z)| < \varepsilon$, by the continuity at 0. Assume $y > 0$ and $y \notin \mathbb{Q}$. Then for a given $n \in \mathbb{N}$, with $\frac{1}{n} < \delta$, there is an $x \in (y - \frac{1}{n}, y + \frac{1}{n}) \cap \mathbb{Q}$. Moreover, $|f(x) - f(y)| < \varepsilon$. Thus $f(x) - \varepsilon < f(y) < f(x) + \varepsilon$. Hence $xf(1) - \varepsilon < f(y) < xf(1) + \varepsilon$. If $f(1) \geq 0$ we use $(y - \frac{1}{n})f(1) - \varepsilon \leq xf(1) - \varepsilon < f(y) < xf(1) + \varepsilon \leq (y + \frac{1}{n}) + \varepsilon$, and if $f(1) < 0$ we have $(y + \frac{1}{n})f(1) - \varepsilon \leq xf(1) - \varepsilon < f(y) < xf(1) + \varepsilon \leq (y - \frac{1}{n}) + \varepsilon$. Letting $n \to \infty$ we get $yf(1) - \varepsilon < f(y) < yf(1) + \varepsilon$. Since $\varepsilon > 0$ is arbitrary, we get $f(y) = yf(1)$.

8. Suppose that $|f(x) - f(y)| < K|x - y|$ on some set D, then given $\varepsilon > 0$, by picking $\delta = \frac{\varepsilon}{K}$ we see $|x - y| < \delta \Rightarrow |f(x) - f(y)| < \varepsilon$. Hence f is uniformly continuous on D.

9. Suppose that $|f(x) - f(y)| < K|x-y|^\alpha$ on some set D, where $0 < \alpha < 1$, then given $\varepsilon > 0$, by picking $\delta = (\frac{\varepsilon}{K})^{\frac{1}{\alpha}}$ we see $|x-y| < \delta \Rightarrow |f(x) - f(y)| < \varepsilon$. Hence f is uniformly continuous on D.

10. Define $T_n(x) = (1+\frac{x}{n})^n$. First, we shall assume $x > 0$. By the binomial theorem, we have $T_n(x) = \sum_{k=0}^n \binom{n}{k}(\frac{x}{n})^k = 1 + x + \frac{1}{2!}(1-\frac{1}{n})x^2 + \frac{1}{3!}(1-\frac{1}{n})(1-\frac{2}{n})x^3 + \cdots + \frac{1}{n!}(\prod_{j=1}^{n-1}(1-\frac{j}{n}))x^n < S_n(x) := 1 + x + \frac{x^2}{2!} + \cdots + \frac{x^n}{n!}$. So $T_n(x) < S_n(x) \to e^x$ as $n \to \infty$. Hence $\limsup_{n \to \infty} T_n(x) \leq e^x$. For $n > m$ we have $T_n(x) > 1 + x + \frac{1}{2!}(1-\frac{1}{n})x^2 + \cdots + \frac{1}{m!}(\prod_{j=1}^{m-1}(1-\frac{j}{n}))x^m$. Keeping m fixed, we let $n \to \infty$ along a subsequence which attains $\liminf_{n\to\infty} T_n(x)$ and so $\liminf_{n\to\infty} T_n(x) \geq 1 + x + \frac{1}{2!}x^2 + \cdots + \frac{1}{m!}x^m = S_m(x)$. We then let $m \to \infty$, and so we get $e^x \leq \liminf_{n\to\infty} T_n(x) \leq \limsup_{n\to\infty} T_n(x) \leq e^x$. So $\lim_{n\to\infty} T_n(x) = e^x$. If $x = 0$, we observe $T_n(0) = 1 \to e^0 = 1$ as $n \to \infty$. Now, let $x > 0$ and consider $T_n(-x)$. Observe that $T_n(-x) = (1-\frac{x}{n})^n = \frac{1}{(1+\frac{x}{n-x})^n}$. By the Archimedean property, there exists a whole number k so that $k \leq x < k_1$, and so $n - (k+1) \leq n - x < n - k$. Assume that $n > k+1$ for the remainder of the argument. Thus $\frac{1}{(1+\frac{x}{n-(k+1)})^n} \leq \frac{1}{(1+\frac{x}{n-x})^n} \leq \frac{1}{(1+\frac{x}{n-k})^n}$. We may rewrite this as $\frac{(1+\frac{x}{n-(k+1)})^{-(k+1)}}{(1+\frac{x}{n-(k+1)})^n} \leq T_n(-x) \leq \frac{(1+\frac{x}{n-k})^{-k}}{(1+\frac{x}{n-k})^n}$. Letting $n \to \infty$, we get $\frac{1}{e^x} \leq \lim_{n\to\infty} T_n(-x) \leq \frac{1}{e^x} = e^{-x}$.

11. If $Z = \{x : f(x) = 0\}$ is finite, then it is clearly closed. Suppose Z has infinitely many elements, and suppose that x is a limit point for Z. Since the domain of f is closed, x is in this domain. Since f is continuous at x, given $\varepsilon > 0$, $\exists \delta > 0$ so that $|x - y| < \delta \Rightarrow |f(x) - f(y)| < \varepsilon$. However, $Z \cap [(x - \delta, x) \cup (x, x + \delta)] \neq \emptyset$. so we choose a y in this set. Hence $f(y) = 0$, and so $-\varepsilon < f(x) < \varepsilon$. Since $\varepsilon > 0$ is arbitrary, we get $f(x) = 0$.

12. Suppose $f : (a, b) \to \mathbb{R}$ satisfies for $x < y$ and $0 < \lambda < 1$ we have $f(\lambda x + (1-\lambda)y) < \lambda f(x) + (1-\lambda)f(y)$. Let $a < s < t < u < b$ be $t = \lambda s + (1-\lambda)u$. Then $t - s = (1-\lambda)(u-s)$ and $u - t = \lambda(u-s)$. So $\frac{f(u)-f(t)}{u-t} > \frac{f(u)-[\lambda f(s) + (1-\lambda)f(u)]}{u-t} = \frac{\lambda[f(u)-f(s)]}{\lambda(u-s)} = \frac{f(u)-f(s)}{u-s}$.

B.15. HOMEWORK SOLUTIONS 4.5   443

Also, $\frac{f(t)-f(s)}{t-s} < \frac{\lambda f(s)+(1-\lambda)f(u)-f(s)}{(1-\lambda)(u-s)} = \frac{(1-\lambda)[f(u)-f(s)]}{(1-\lambda)(u-s)} = \frac{f(u)-f(s)}{u-s}$. So $\frac{f(t)-f(s)}{t-s} < \frac{f(u)-f(s)}{u-s} < \frac{f(u)-f(t)}{u-t}$. If we let $m(x,y) = \frac{f(y)-f(x)}{y-x}$, we have established for $a < s < t < u < b$ that $m(s,t) < m(s,u) < m(t,u)$. Fix $[s,u] \subseteq (a,b)$. We will show that f is Lipschitz continuous on this set (s,u). Let $t < \tau$ be two points in (s,u). Consider any $r \in (a,s)$ and any $v \in (u,b)$. Then $a < r < s < t < \tau < u < v < b$. So we have $m(r,s) < m(r,t)$, $m(s,t) < m(s,\tau) < m(t,\tau)$, $m(t,\tau) < m(t,u) < m(\tau,u)$ and $m(\tau,u) < m(\tau,v) < m(u,v)$. Putting this together, we get $m(r,s) < m(t,\tau) < m(u,v)$, and so $(\tau-t)m(r,s) < f(\tau)-f(t) < (\tau-t)m(u,v)$. Hence $|f(\tau)-f(t)| < (|m(r,s)|+|m(u,v)|)|\tau-t|$. Since $(s,u) \subseteq (a,b)$ was arbitrary, we know that f is continuous on all of (a,b).

13. Suppose f is uniformly continuous on D, a bounded set in $\mathbb{R}$. Then for $\varepsilon = 1$, there is a $\delta > 1$ so that $|x-y| < \delta$, $x,y \in D$, implies $|f(x)-f(y)| < 1$. For such a δ, the collection $\{B_{\frac{\delta}{4}}(x) : x \in \bar{D}\}$ is an open cover of $\bar{D}$. $\bar{D}$ is compact, so there is a finite subcover $\{B_{\frac{\delta}{4}}(x_k) : k = 1, 2, \cdots, n\}$. Next we create a list $y_1, y_2, \cdots, y_n$, where $y_k = x_k$ if $x_k \in D$. If $x_k \notin D$, then it must be x_k is a limit point of D. So, there is a $y_k \in B_{\frac{\delta}{4}}(x_k) \cap D$. Observe as well that $\{B_{\frac{\delta}{2}}(y_k) : k = 1, 2, \cdots\}$ is an open cover of D. Now let $x \in D$, then $x \in B_{\frac{\delta}{2}}(y_k)$ for some k. Then $|f(x)| = |f(x)-f(y_k)+f(y_k)| \leq |f(x)-f(y_k)|+|f(y_k)| < 1 + \max_{1 \leq i \leq n}|f(y_i)|$.

14. Let $f : [0,1] \to [0,1]$ be continuous. Suppose $g(x) = f(x)-x$, and so g is continuous on $[0,1]$ as well. If $f(x) = x$ doesn't have a solution, then $g(x) \neq 0$ for every $x \in [0,1]$. Since $g(0) = f(0)-0 = f(0) \in [0,1]$, it must be that $g(x) > 0$ on $[0,1]$. However, $g(1) = f(1)-1 > 0$ implies $f(1) > 1$, a contradiction.

15. Let t be such that $\lim_{x \to t^+} f(x)$ and $\lim_{x \to t^-} f(x)$ both exist, however, $\lim_{x \to t^-} f(x) \neq \lim_{x \to t^+} f(x)$. Consider $S_f :=\{t : \lim_{x \to t^+} f(x) > \lim_{x \to t^-} f(x)\}$ and $T_f := \{t : \lim_{x \to t^+} f(x) < \lim_{t \to t^-} f(x)\}$. Let $t \in S_f$. Let $p \in \mathbb{Q}$ be such that $\lim_{x \to t^-} f(x) < p < \lim_{x \to t^+} f(x)$. Let $\varepsilon = \min\{p-\lim_{x \to t^-} f(x), \lim_{x \to t^+} f(x)-p\}$. Then $\exists \delta_- > 0$ so that $t-\delta_- < x < t$ implies $f(x) < \lim_{z \to t^-} f(z)+\varepsilon \leq p$. $\exists \delta_+$ such that $t < x < t+\delta_+$ implies $f(x) > \lim_{z \to t^+} f(z)-\varepsilon \geq p$. Fix $q, r \in \mathbb{R}$ so that $t-\delta_- < q < t$ and $t < r < t+\delta_+$. Then $q < x < t$ implies

$f(x) < p$, and $t < z < r$ implies $f(z) > p$. Clearly $\mathbb{Q}^3$ is countable, and so the collection of all such (p, q, r) is at most countably infinite. Suppose (p, q, r) is such a triple associated with $t, \tau \in S_f$. If $t \neq \tau$, we may assume $t < \tau$ by otherwise relabeling. Take σ so that $q < t < \sigma < \tau < r$. $q < \sigma < \tau$ implies $f(\sigma) < p$, while $t < \sigma < r$ implies $f(\sigma) > p$. This gives us a contradiction if $t \neq \tau$, and thus we see that $t = \tau$. Hence each $p, q, r \in \mathbb{Q}^3$ can be associated with at most one point in S_f. Therefore S_f is at most countably infinite. Observing that $T_f = S_{-f}$ allows us to see T_f is at most countably infinite as well.

16. Suppose $f : (a, b) \to \mathbb{R}$ is continuous and $f(\frac{x+y}{2}) < \frac{f(x)+f(y)}{2}$. Claim that for $\lambda = \frac{j}{2^k}$, where $k \in \mathbb{N}$ and $j = 1, \cdots, 2^k - 1$ we have $f(\lambda x + (1-\lambda)y) < \lambda f(x) + (1-\lambda)f(y)$. We will proceed by induction on k. For $k = 1$, we only need to consider $j = 1$, and this is the assumption about f, namely $f(\frac{x+y}{2}) < \frac{f(x)+f(y)}{2}$. Suppose for $k \in \mathbb{N}$, and any $j \in \{1, 2, \cdots 2^k - 1\}$ $f(\frac{j}{2^k}x + (1-\frac{j}{2^k})y) < \frac{j}{2^k}f(x) + (1-\frac{j}{2^k})f(y)$ is true. We need to show that for $k+1$ and any $j \in \{1, \cdots, 2^{k+1} - 1\}$, for $\lambda = \frac{j}{2^{k+1}}$ we have $f(\lambda x + (1-\lambda)y) < \lambda f(x) + (1-\lambda)f(y)$. Suppose we consider $\lambda = \frac{j}{2^{k+1}}$. We may assume j is odd, otherwise this will be of the form $\lambda = \frac{i}{2^k}$, and fall under the case of our induction hypothesis. So we have $\lambda = \frac{2m+1}{2^{k+1}}$. However, $\lambda x + 1(-\lambda)y = \frac{2m+1}{2^{k+1}}x + (1 - \frac{2m+1}{2^{k+1}})y = \frac{\tilde{x}+\tilde{y}}{2}$, where $\tilde{x} = \frac{m}{2^k}x + (1 - \frac{m}{2^k})y$, and $\tilde{y} = \frac{m+1}{2^k}x + (1 - \frac{m+1}{2^k})y$. So by our assumption about f and the induction hypothesis, we observe $f(\lambda x + 1(-\lambda)y) = f(\frac{2m+1}{2^{k+1}}x + (1 - \frac{2m+1}{2^{k+1}})y) = f(\frac{\tilde{x}+\tilde{y}}{2}) < \frac{1}{2}f(\tilde{x}) + \frac{1}{2}f(\tilde{y}) = \frac{1}{2}f(\frac{m}{2^k}x + (1 - \frac{m}{2^k})y) + \frac{1}{2}f(\frac{m+1}{2^k}x + (1-\frac{m+1}{2^k})y) < \frac{1}{2}(\frac{m}{2^k}f(x) + (1-\frac{m}{2^k})f(y)) + \frac{1}{2}(\frac{m+1}{2^k}f(x) + (1 - \frac{m+1}{2^k})f(y)) = (\frac{2m+1}{2^{k+1}})f(x) + (1 - \frac{2m+1}{2^{k+1}})f(y)) = \lambda f(x) + (1-\lambda)f(y)$. So we see that $f(\lambda x + (1-\lambda)y) < \lambda f(x) + (1-\lambda)f(y)$ for all dyadic rational numbers λ. For a general real number $\lambda \in (0, 1)$ we use that the set of dyadic rational numbers is dense in $\mathbb{R}$, so there is a sequence of dyadic rational numbers $\lambda_n \to \lambda$. Hence $f(\lambda_n x + (1-\lambda_n)y) < \lambda_n f(x) + (1-\lambda_n)f(y)$. Using continuity of f and letting n approach infinity, we see $f(\lambda x + (1-\lambda)y) < \lambda f(x) + (1-\lambda)f(y)$. Hence f is convex.

17. Suppose first that $x > 0$. For $x \neq \frac{1}{n}$ for all $n \in \mathbb{N}$, We have that $x \in (\frac{1}{m+1}, \frac{1}{m})$ for some m. Consider $y \in (\frac{1}{m+1}, \frac{1}{m})$ as well. Then given $\varepsilon > 0$ we observe that for $\delta = \min\{\frac{1}{m} - x, x - \frac{1}{m+1}, \frac{\varepsilon}{m(m+1)}\}$ we have $|x - y| < \delta$ implies $|f(x) - f(y)| = m(m+1)|x - y| < \varepsilon$. For $x < 0$ with

$|x| \neq \frac{1}{n}$ for any n, we also have for $\delta = \min\{\frac{1}{m} - |x|, |x| - \frac{1}{m+1}, \frac{\varepsilon}{m(m+1)}\}$ we have $|x - y| < \delta$ implies $|f(x) - f(y)| = |f(|x|) - f(|y|)| = m(m+1)||x| - |y|| = m(m+1)|x - y| < \varepsilon$. In fact if f is continuous at $x > 0$, we claim that f is continuous at $-x$. To see this suppose $\delta > 0$ is so that $|x - y| < \delta$ implies $|f(x) - f(y)| < \varepsilon$. Let $-y < 0$ be so that $|(-y) - (-x)| < \min\{\delta, \frac{x}{2}\}$. Then $|(-y) - (-x)| = |y - x| < \delta$ and so $|f(-y) - f(-x)| = |f(y) - f(x)| < \varepsilon$. Consider next $x = \frac{1}{2k}$. First, we consider $y < x$ with $y \in (\frac{1}{2k+1}, \frac{1}{2k})$ and $|x - y| < \frac{\varepsilon}{2k(2k+1)}$. In such a case $|f(y) - f(x)| = 2k(2k+1)|x - y| < \varepsilon$. If $y > x$ with $y \in (\frac{1}{2k}, \frac{1}{2k-1})$ and $|x - y| < \frac{\varepsilon}{2k(2k-1)}$ we have $|f(y) - f(x)| = 2k(2k-1)|x - y| < \varepsilon$. Now we consider $x = \frac{1}{2k+1}$. First, we consider $y < x$ with $y \in (\frac{1}{2k+2}, \frac{1}{2k+1})$ and $|x - y| < \frac{\varepsilon}{(2k+2)(2k+1)}$. In such a case $|f(y) - f(x)| = (2k+2)(2k+1)|x - y| < \varepsilon$. If $y > x$ with $y \in (\frac{1}{2k+1}, \frac{1}{2k})$ and $|x - y| < \frac{\varepsilon}{2k(2k+1)}$ we have $|f(y) - f(x)| = 2k(2k+1)|x - y| < \varepsilon$. Hence we see that f is continuous at any point in $(-\frac{1}{2}, 0) \cup (0, \frac{1}{2})$. For $x = \frac{1}{2}$ we assume $y \in (\frac{1}{3}, \frac{1}{2})$ and $|x - y| < \frac{\varepsilon}{6}$. Here we have $|f(x) - f(y)| = 6|x - y| < \varepsilon$. For $x = \frac{-1}{2}$ we assume $y \in (-\frac{1}{2}, -\frac{1}{3})$ and $|x - y| < \frac{\varepsilon}{6}$. Here we have $|f(x) - f(y)| = 6|x - y| < \varepsilon$. Hence f is continuous at $x = \pm\frac{1}{2}$. Consider now $x = 0$. Let $\varepsilon = \frac{1}{3}$. Let $\delta > 0$. Consider $|y| \leq \delta$ with $|y| = \frac{1}{2n+1}$, then $|f(y) - f(x)| = f(|y|) = f(\frac{1}{2n+1}) = 1 > \frac{1}{3} = \varepsilon$. So we see that f is not continuous at $x = 0$.

18. Suppose f is continuous on an interval I and $f(I) = \{r_1, r_2, \cdots, r_n\}$, and $r_1 < r_2 < \cdots < r_n$. If $n > 1$, we pick an interval $[a, b] \subseteq I$ so that $\{f(a), f(b)\} = \{r_1, r_2\}$. Then by the intermediate value theorem, there is a $c \in (a, b)$ so that $f(c) = \frac{f(a)+f(b)}{2} = \frac{r_1+r_2}{2}$. However, $r_1 < \frac{r_1+r_2}{2} < r_2$, and this is a contradiction, and so $f(I) = \{r_1\}$.

19. Take $f(x) = 1$ if $x \in \mathbb{Q}$ and $f(x) = -1$ if $x \in \mathbb{R} \setminus \mathbb{Q}$. Then $|f|$ is continuous everywhere, but f is continuous nowhere.

20. Let $g(x) := f(x + \frac{1}{2}) - f(x)$, where $g : [0, \frac{1}{2}] \to \mathbb{R}$. g is continuous on $[0, \frac{1}{2}]$, since f is continuous on $[0, 1]$. If $g(0) = 0$, then we have $f(c + \frac{1}{2}) = f(c)$ has a solution for $c = 0$. If $g(0) < 0$, then $f(\frac{1}{2}) < f(0)$ and so $g(\frac{1}{2}) = f(1) - f(\frac{1}{2}) = f(0) - f(\frac{1}{2}) > 0$. If $g(0) > 0$, then $f(\frac{1}{2}) > f(0)$ and so $g(\frac{1}{2}) = f(1) - f(\frac{1}{2}) = f(0) - f(\frac{1}{2}) < 0$. So in either of these cases the intermediate value theorem gives us the existence of a $c \in (0, \frac{1}{2})$ so that $g(c) = 0$, i.e. $f(c + \frac{1}{2}) = f(c)$.

21. Here we need to show $\{x \in D : f(x) - g(x) > 0\}$ is open in D. That is, it is an open set intersected with D. Since $f - g$ is continuous, it suffices to prove that for a continuous $h : D \to \mathbb{R}$, the set $\{x \in D : h(x) > 0\}$ is open in D. Let $x \in D$ be so that $h(x) > 0$, and let $\varepsilon = \frac{h(x)}{2}$. There exists a $\delta > 0$ so that $|x - y| < \delta$, $y \in D$, implies $|h(x) - h(y)| < \varepsilon = \frac{h(x)}{2}$. Thus, for $y \in B_\delta(x) \cap D$, we have $0 < \frac{h(x)}{2} < h(y) < \frac{3h(x)}{2}$. Hence $B_\delta(x) \cap D \subseteq \{z \in D : h(z) > 0\}$.

22. Let $\varepsilon > 0$ be given. Then we know that $f(x) = \sqrt{x}$ is uniformly continuous on $[0, 2]$. Let $\delta' > 0$ be so that $|x - y| < \delta'$, $x, y \in [0, 2]$, implies $|\sqrt{x} - \sqrt{y}| < \varepsilon$. For $x, y \in [1, \infty)$ we observe that $|\sqrt{x} - \sqrt{y}| = \frac{|x-y|}{\sqrt{x}+\sqrt{y}} < |x - y|$ and so on this interval we see that $\delta = \varepsilon$, we have $|x - y| < \varepsilon$ implies $|\sqrt{x} - \sqrt{y}| < \varepsilon$. Take $\delta = \min\{1, \delta', \varepsilon\}$, and for any $x, y \geq 0$, we have $|x - y| < \delta$ implies $|\sqrt{x} - \sqrt{y}| < \varepsilon$.

23. For $f(x) = xe^x - 1$ we have $f(x)$ is continuous everywhere. Moreover $f(0.5625) = -0.012781755 < 0$ and $f(0.5703125) = 0.008779997 > 0$, and so by the intermediate value theorem, $f(x) = 0$ has a solution on $(0.5625, 0.5703125)$.

24. For $f : D \to \mathbb{R}$ define for $\delta > 0$ the quantity $\omega(\delta) = \sup\{|f(x) - f(y)| : x, y \in D, |x - y| < \delta\}$, which is clearly non-negative if it is real. Let $\delta_1 < \delta_2$, then $|x - y| < \delta_1 \Rightarrow |x - y| < \delta_2$, and so $\{|f(x) - f(y)| : x, y \in D, |x - y| < \delta_1\} \subseteq \{|f(x) - f(y)| : x, y \in D, |x - y| < \delta_2\}$. Hence $\omega(\delta_1) \leq \omega(\delta_2)$, if $\omega(\delta_1)$ is real. Suppose f is uniformly continuous on D. Then given $\varepsilon > 0$, $\exists \Delta$ so that for $\delta < \Delta$ we have $|x - y| < \delta \Rightarrow |f(x) - f(y)| < \varepsilon$. So for $\delta < \Delta$ we have $\omega(\delta) \leq \varepsilon$. Hence $\lim_{\delta \to 0^+} \omega(\delta) \leq \varepsilon$. Since this is true for any $\varepsilon > 0$, it must be $\lim_{\delta \to 0^+} \omega(\delta) = 0$. Now suppose $\omega(\delta) \to 0$ as $\delta \to 0^+$. Given $\varepsilon > 0$ $\exists \Delta$ so that $0 < \delta < \Delta$ implies $0 \leq \omega(\delta) < \frac{\varepsilon}{2}$. Hence for $|x - y| < \delta$, $x, y \in D$, we have $|f(x) - f(y)| \leq \frac{\varepsilon}{2} < \varepsilon$. So f is uniformly continuous on D.

25. Suppose f is upper semicontinuous on a compact set E. Then for all $z \in E$ we have $\limsup_{x \to z, x \in E} f(x) \leq f(z)$. Then $L(z) := \lim_{\delta \to 0^+} \sup\{f(x) : x \in B'_\delta(z) \cap E\} \leq f(z)$ for all $z \in E$. Given $\varepsilon > 0$ there is a positive $\Delta = \Delta(\varepsilon, z)$ so that $0 < \delta < \Delta$ implies $|\sup\{f(x) : x \in B'_\delta(z) \cap E\} - L(z)| < \varepsilon$. This implies $\sup\{f(x) : x \in B'_\delta(z) \cap E\} < L(z) + \varepsilon \leq f(z) + \varepsilon$. Hence for $x \in B_\delta(z)$ we have $f(x) < f(z) + \varepsilon$. For $\varepsilon = 1$ we have that

for $\Delta(z,1)$ as above that $\{B_{\frac{\Delta(z,1)}{2}}(z) : z \in E\}$ is an open cover of E. So there is a finite subcover $\{B_{\frac{\Delta(z_1,1)}{2}}(z_1), B_{\frac{\Delta(z_2,1)}{2}}(z_2), \cdots, B_{\frac{\Delta(z_m,1)}{2}}(z_m)\}$ of E. Hence for a given $x \in E$, $x \in B_{\frac{\Delta(z_j,1)}{2}}(z_j)$ for some j and so $f(x) \leq f(z_j) + 1 \leq \max_{1 \leq i \leq m} f(z_i) + 1$.

26. Suppose f is usc at z. Then $\limsup_{x \to z, x \in E} f(x) \leq f(z)$, that is, $\lim_{\delta \to 0^+} \sup\{f(x) : x \in B'_\delta(z) \cap E\} \leq f(z)$. Hence $-\lim_{\delta \to 0^+} \sup\{f(x) : x \in B'_\delta(z) \cap E\} \geq -f(z)$, and so $\lim_{\delta \to 0^+} -\sup\{f(x) : x \in B'_\delta(z) \cap E\} \geq -f(z)$. From an earlier exercise, we know $\lim_{\delta \to 0^+} \inf\{-f(x) : x \in B'_\delta(z) \cap E\} \geq -f(z)$, and this is $\liminf_{x \to z, x \in E}[-f(x)] \geq -f(z)$. Thus $-f$ is lsc at z.

B.16 Homework Solutions 5.1.6

1. Suppose $f(x) = \sin \frac{1}{x}$ for $x \neq 0$, and $f(0) = 0$. For $x_0 \neq 0$, $\lim_{x \to x_0} \sin \frac{1}{x} = \sin(\lim_{x \to x_0} \frac{1}{x}) = \sin \frac{1}{x_0} = f(x_0)$. However f is not continuous at 0. To see this, let $0 < \varepsilon < 1$ and $\delta > 0$, the $x = \frac{1}{\frac{\pi}{2} + 2\pi n} \in (-\delta, \delta)$ if $n \in \mathbb{N}$ is sufficiently large. Moreover $|\sin \frac{1}{\frac{\pi}{2} + 2\pi n} - 0| = 1 > \varepsilon$. For $x \neq 0$, we may use the chain rule to get $f'(x) = \frac{-\cos \frac{1}{x}}{x^2}$.

2. Let $f(x) = x^2 \sin \frac{1}{x}$ for $x \neq 0$, and $f(0) = 0$. For $x \neq 0$, we have $f'(x) = 2x \sin \frac{1}{x} - \cos \frac{1}{x}$. $f'(0) = \lim_{x \to 0} x \sin \frac{1}{x} = 0$ since $0 \leq |x \sin \frac{1}{x}| \leq |x| \to 0$ as $x \to 0$. However $\lim_{x \to 0} f'(x)$ does not exist, since $\lim_{x \to 0} \cos \frac{1}{x}$ does not exist. To see this, let $0 < \varepsilon < 1$ and $\delta > 0$, and observe that $x = \frac{1}{2\pi n} \in (-\delta, \delta)$ if n is large. Moreover, for such an x, $\cos \frac{1}{x} = 1$. Thus $|\cos \frac{1}{x} - 0| = 1 > \varepsilon$.

3. Suppose $f(x) = x \sin \frac{1}{x}$ for $x \neq 0$, and $f(0) = 0$. For $x_0 \neq 0$, we have $\lim_{x \to x_0} x \sin \frac{1}{x} = [\lim_{x \to x_0} x] \cdot [\lim_{x \to x_0} \sin \frac{1}{x}] = x_0 \cdot \sin(\lim_{x \to x_0} \frac{1}{x}) = x_0 \sin \frac{1}{x_0} = f(x_0)$. Also, $\lim_{x \to 0} x \sin \frac{1}{x} = 0$ since $0 \leq |x \sin \frac{1}{x}| \leq |x| \to 0$ as $x \to 0$, so f is continuous at 0. For $x \neq 0$, we have $f'(x) = \sin \frac{1}{x} - \frac{\cos \frac{1}{x}}{x}$. Also, $\frac{f(x) - f(0)}{x} = \sin \frac{1}{x}$ does not have a limit as $x \to 0$, as we established above. Hence $f'(0)$ does not exist.

4. Observe that for $x \neq y$ we have $\frac{|f(x) - f(y)|}{|x - y|} \leq |x - y|^{\alpha - 1}$. Letting $x \to y$ gives us $f'(y) = 0$. Since this is true for any y, we know that $f' \equiv 0$.

5. $f(0) = f(0+0) = f(0)^2$, and so $f(0)^2 - f(0) = 0$. Hence $f(0) = 0$ or $f(0) = 1$. If $f(0) = 0$, the $f(x) = f(x+0) = f(x)f(0) = 0$ and so $f \equiv 0$. Thus $f'(x) = 0$ exists for all x, and satisfies $f'(x) = f(x)f'(0)$ as well. Otherwise, we have $f(0) = 1$. If $f(0) = 1$, $\lim_{h \to 0} \frac{f(x+h)-f(x)}{h} = \lim_{h \to 0} \frac{f(x)f(h)-f(x)}{h} = f(x) \lim_{h \to 0} \frac{f(h)-1}{h} = f(x)f'(0)$.

6. Recall the sum of angles formula $\cos(x+y) = \cos x \cos y - \sin x \sin y$. Let x and y both be z, and so $\cos(2z) = \cos^2 z - \sin^2 z$. Using $\sin^2 z = 1 - \cos^2 z$, we have $\cos^2(2z) = 2\cos^2 z - 1$, and so $\cos^2 z = \frac{\cos^2(2z)+1}{2}$. Letting $z = \frac{x}{2}$, and then taking square roots, and the appropriate sign gives us the desired result for $\cos \frac{x}{2}$. Then use that $\sin^2 z = 1 - \cos^2 z = 1 - \frac{\cos^2(2z)+1}{2} = \frac{1-\cos^2(2z)}{2}$. Using $z = \frac{x}{2}$, then taking square roots, and the appropriate sign establishes the other result.

B.17 Homework Solutions 5.3.4

1. Since $\lim_{x \to 1^-} g = 1 \neq 0 = \lim_{x \to 1^+} g$ and $g(1) = 1$ exists, we have that g has a simple discontinuity at a point where it is defined. We also know that if $y = f$ has $f' = g$, then g cannot have any simple discontinuities at points where g is defined.

2. Suppose $f'(x) = m$, a constant function. Then consider $g(x) = f(x) - mx$. Then $g'(x) \equiv 0$, so $g(x) \equiv b$. Hence $f(x) = g(x) + mx = mx + b$.

3. Consider $0 \leq \frac{|f(x)-f(y)|}{|x-y|} \leq |x-y|^{\alpha-1} \to 0$ as $x \to y$. Hence $f'(y) \equiv 0$ on $\mathbb{R}$. Thus $f(x)$ is a constant function.

4. Suppose f is differentiable on $[a,b]$ and $f(a) = f(b)$. By the mean value theorem, we know that there exists $c \in (a,b)$ so that $f'(c) = \frac{f(b)-f(a)}{b-a} = 0$.

5. Suppose $f(a) = \max_{[a,b]} f$ and $f(b) = \min_{[a,b]} f$. For $h > 0$ we have $\frac{f(a+h)-f(a)}{h} \leq 0$ and so $\lim_{h \to 0^+} \frac{f(a+h)-f(a)}{h} \leq 0$. For $h < 0$ we have $\frac{f(b+h)-f(b)}{h} \leq 0$ and so $\lim_{h \to 0^-} \frac{f(b+h)-f(b)}{h} \leq 0$.

6. Here we have for $f(x) = \sqrt[k]{x}$, $g(x) = f^{-1}(x) = x^k$. So for $x_0 = g(y_0) = y_0^k$ we have $f'(x_0) = \frac{1}{g'(y_0)} = \frac{1}{ky_0^{k-1}} = \frac{1}{kx_0^{\frac{k-1}{k}}} = \frac{1}{k}x_0^{\frac{1}{k}-1}$.

7. Let $x < y$ be two points in $[a, b]$. Then, by the mean value theorem, we have $f(y) - f(x) = f'(x) \cdot (y - x)$, for some $z \in (x, y)$, and thus $z \in (a, b)$. Hence $|f(y) - f(x)| = |f'(z)| \cdot |y - x| \leq \max_{(a,b)} |f'| \cdot |y - x|$. So f is Lipschitz continuous on $[a, b]$.

8. If $f'(x) = g'(x)$ for all x in $\mathbb{R}$, we have $h(x) = f(x) - g(x)$ has $h'(x) = f'(x) - g'(x) = 0$ for every $x \in \mathbb{R}$, and so $h(x) = C$ for some C. Thus $f(x) = g(x) + C$.

9. $f(x) = |x|$ does not violate the theorem since at 0 the derivative does not exist. The theorem states that if $f'(x)$ exists on $[a, b]$, then it cannot have any simple discontinuities on $[a, b]$. So for $f(x) = |x|$, such an interval $[a, b]$ where f' exists cannot contain zero.

10. By L'Hôpital's rule we have the limit $\lim_{n \to \infty} \log(1+\frac{x}{n})^n = \lim_{n \to \infty} \frac{\log(1+\frac{x}{n})}{\frac{1}{n}} = \lim_{y \to 0^+} \frac{\log(1+xy)}{y} = \lim_{y \to 0^+} \frac{x}{1+xy} = x$. So, by continuity of $y = e^x$ we have $e^x = e^{\lim_{n \to \infty} \log(1+\frac{x}{n})^n} = \lim_{n \to \infty} e^{\log(1+\frac{x}{n})^n} = \lim_{n \to \infty} (1+\frac{x}{n})^n$.

11. By L'Hôpital's rule $\lim_{x \to 0^+} \log x^x = \lim_{x \to 0^+} \frac{\log x}{\frac{1}{x}} = \lim_{x \to 0^+} (-x) = 0$. So, by continuity, we have $1 = e^0 = e^{\lim_{x \to 0^+} \log x^x} = \lim_{x \to 0^+} e^{\log x^x} = \lim_{x \to 0^+} x^x$.

12. By repeated use of L'Hôptial's rule, we have the limit $\lim_{x \to \infty} \frac{x^n}{e^x} = \lim_{x \to \infty} \frac{nx^{n-1}}{e^x} = \cdots = \lim_{x \to \infty} \frac{n!x}{e^x} = \lim_{x \to \infty} \frac{n!}{e^x} = \lim_{n \to \infty} n! e^{-x} = 0$.

13. For $0 < x < 1$ we have $|\frac{x^\alpha}{\log x}| \leq \frac{1}{|\log x|} \to 0$ as $x \to 0^+$, since $\log x \to -\infty$ as $x \to 0^+$.

B.18 Homework Solutions 5.4.1

1. Suppose $f'(x) = c_0 + c_1 x + \cdots + c_n x^n$, $c_n \neq 0$. Consider $g(x) = c_0 x + \frac{c_1}{2} x^2 + \cdots + \frac{c_n}{n+1} x^{n+1}$. Then $h(x) = f(x) - g(x)$ satisfies $h'(x) = f'(x) - g'(x) = 0$ everywhere, and so $h(x) = C$ on $\mathbb{R}$. Hence $f(x) = h(x) + g(x) = C + c_0 x + \frac{c_1}{2} x^2 + \cdots + \frac{c_n}{n+1} x^{n+1}$.

2. Here we proceed by induction. If $f'(x) = 0$ as a function, then $f(x) = C$. Suppose $f^{(n)}(x) = 0$ identically implies f is a polynomial of degree $n - 1$ or less. Let g be a function so that $g^{(n+1)} = 0$ everywhere. Then

$f = g'$ satisfies $f^{(n)} = 0$ everywhere. So $f'(x)$ is a polynomial of degree $j \le n-1$. Thus, by the above exercise, f is a polynomial of degree $j+1 \le n$.

B.19 Homework Solutions 5.5.4

1. We proceed by induction on n, for $n > 1$. For $n = 2$ we have for $p_1, p_2 >$ with $p_1 + p_2 = 1$, $f(p_1 x + p_2 y) = f(p_1 x + (1-p_1)y) < p_1 f(x) + (1-p_1)f(y) = p_1 f(x) + p_2 f(y)$. Now assume that for $p_1, p_2, \cdots, p_n > 0$ so that $p_1 + p_2 + \cdots + p_n = 1$ and any $x_1, x_2, \cdots, x_n$ that $f(p_1 x_1 + p_2 x_2 + \cdots + p_{n-1} x_{n-1} + p_n x_n) < p_1 f(x_1) + p_2 f(x_2) + \cdots + p_{n-1} f(x_{n-1}) + p_n f(x_n)$. Consider $p_1, p_2, \cdots, p_n, P_{n+1} > 0$ so that $p_1 + p_2 + \cdots + p_n + p_{n+1} = 1$ and any $x_1, x_2, \cdots, x_n, x_{n+1}$. Then $f(p_1 x_1 + p_2 x_2 + \cdots + p_n x_n + p_{n+1} x_{n+1}) = f(p_1 x_1 + (1-p_1) \sum_{k=2}^{n+1} \frac{p_k}{1-p_1} x_k) < p_1 f(x_1) + (1-p_1) f(\sum_{k=2}^{n+1} \frac{p_k}{1-p_1} x_k) < p_1 f(x_1) + (1-p_1)(\sum_{k=2}^{n+1} \frac{p_k}{1-p_1} f(x_k)) = \sum_{k=1}^{n+1} p_k f(x_k)$.

2. Suppose $f : (a,b) \to \mathbb{R}$ satisfies for $x < y$ and $0 < \lambda < 1$ we have $f(\lambda x + (1-\lambda)y) < \lambda f(x) + (1-\lambda)f(y)$. Let $a < s < t < u < b$ be $t = \lambda s + (1-\lambda)u$. Then $t - s = (1-\lambda)(u-s)$ and $u - t = \lambda(u-s)$. So $\frac{f(u)-f(t)}{u-t} > \frac{f(u) - [\lambda f(s) + (1-\lambda)f(u)]}{u-t} = \frac{\lambda[f(u)-f(s)]}{\lambda(u-s)} = \frac{f(u)-f(s)}{u-s}$. Also, $\frac{f(t)-f(s)}{t-s} < \frac{\lambda f(s) + (1-\lambda)f(u) - f(s)}{(1-\lambda)(u-s)} = \frac{(1-\lambda)[f(u)-f(s)]}{(1-\lambda)(u-s)} = \frac{f(u)-f(s)}{u-s}$. So $\frac{f(t)-f(s)}{t-s} < \frac{f(u)-f(s)}{u-s} < \frac{f(u)-f(t)}{u-t}$. If we let $m(x,y) = \frac{f(y)-f(x)}{y-x}$, we have established for $a < s < t < u < b$ that $m(s,t) < m(s,u) < m(t,u)$. Fix $[s,u] \subseteq (a,b)$. We will show that f is Lipschitz continuous on this set (s,u). Let $t < \tau$ be two points in (s,u). Consider any $r \in (a,s)$ and any $v \in (u,b)$. Then $a < r < s < t < \tau < u < v < b$. So we have $m(r,s) < m(r,t) < m(s,t)$, $m(s,t) < m(s,\tau) < m(t,\tau)$, $m(t,\tau) < m(t,u) < m(\tau,u)$ and $m(\tau,u) < m(\tau,v) < m(u,v)$. Putting this together, we get $m(r,s) < m(t,\tau) < m(u,v)$, and so $(\tau-t)m(r,s) < f(\tau) - f(t) < (\tau-t)m(u,v)$. Hence $|f(\tau) - f(t)| < (|m(r,s)| + |m(u,v)|)|\tau - t|$. Since $(s,u) \subseteq (a,b)$ was arbitrary, we know that f is continuous on all of (a,b).

3. For $g(x) = |x|$ we have $g(\lambda x + (1-\lambda)y) = |\lambda x + (1-\lambda)y| \le |\lambda x| + |(1-\lambda)y| = \lambda|x| + (1-\lambda)|y|$. This is convex provided we extend the definition to allow $\le$ instead of $<$. For $f(x) = x^2$ we recall the result $2xy \le x^2 + y^2$, which holds since $(x-y)^2 \ge 0$. Thus $f(\frac{x+y}{2}) = (\frac{x+y}{2})^2 \frac{x^2 + 2xy + y^2}{4} \le$

$\frac{x^2+(x^2+y^2)+y^2}{4} = \frac{x^2}{2} + \frac{y^2}{2} = \frac{f(x)+f(y)}{2}$. Note that there is a strict inequality unless $x = y$, which we are not assuming is the case. Thus $f(\frac{x+y}{2}) < \frac{f(x)+f(y)}{2}$. Since f is continuous, we know by an earlier exercise that f is convex.

4. Suppose f and g are convex. Then $(f+g)(\lambda x+(1-\lambda)y) = f(\lambda x+(1-\lambda)y)+g(\lambda x+(1-\lambda)y) < [\lambda f(x)+(1-\lambda)f(y)]+[\lambda g(x)+(1-\lambda)g(y)] = \lambda(f(x)+g(x))+(1-\lambda)(f(y)+g(y))$.

5. Suppose f is convex and $c > 0$. Then $f(\lambda x + (1-\lambda)y) < \lambda f(x) + (1-\lambda)f(y)$, and upon multiplication by c we get $cf(\lambda x+(1-\lambda)y) < \lambda cf(x)+(1-\lambda)cf(y)$, and so cf is convex. If $c < 0$, then multiplying $f(\lambda x+(1-\lambda)y) < \lambda f(x)+(1-\lambda)f(y)$ by c yields $cf(\lambda x+(1-\lambda)y) > \lambda cf(x)+(1-\lambda)cf(y)$, and so cf is concave.

6. Let $a < s < t < u < b$, and f is convex on (a,b). Then from an earlier exercise we know $\frac{f(t)-f(s)}{t-s} < \frac{f(u)-f(s)}{u-s} < \frac{f(u)-f(t)}{u-t}$. The second inequality tells us $\frac{f(u+h)-f(u)}{h}$ is increasing as $h < 0$ increases to 0^-. The first inequality tells us $\frac{f(u+h)-f(u)}{h}$ decreases as $h > 0$ decreases to 0^+. Moreover $\frac{f(u-h)-f(u)}{-h} \le \frac{f(u+h)-f(u)}{h}$ follows as well from these two inequalities. Thus $\lim_{h \to 0^-} \frac{f(u+h)-f(u)}{h}$ exists, $\lim_{h \to 0^+} \frac{f(u+h)-f(u)}{h}$ exists, and $\lim_{h \to 0^-} \frac{f(u+h)-f(u)}{h} \le \lim_{h \to 0^+} \frac{f(u+h)-f(u)}{h}$. If they are the same, then f is differentiable at u. Let $D^+ f(x) = \lim_{h \to 0^+} \frac{f(x+h)-f(x)}{h}$ and $D^- f(x) = \lim_{h \to 0^-} \frac{f(x+h)-f(x)}{h}$. Then $D^- f(x) \le D^+ f(x)$. For $y < x$ we have $D^- f(x) \le D^+ f(x) \le \frac{f(x)-f(y)}{x-y} \le D^- f(x) \le D^+ f(x)$, so $D^- f$ and $D^+ f$ are monotone increasing. Since they both exist everywhere on (a,b), the set of discontinuities of both are at most countably infinite. Moreover, if x is a point of continuity of $D^+ f$, letting $y \to x^-$ we get $D^+ f(x) = D^- f(x)$, so at such a point f' exists. Hence the set of discontinuities of f' are at most countably infinite.

B.20 Homework Solutions 5.6.2

1. The function $f(x) = e^x$ has $f^{(n)}(x) = e^x$, and so $f^{(n)}(0) = 1$ for each n. Thus the Taylor series for $y = e^x$ about 0 is $\sum_{n=0}^{\infty} \frac{1}{n!} x^n =$

$1 + x + \frac{x^2}{2!} + \cdots + \frac{x^n}{n!} + \cdots$. This is exactly how $y = e^x$ was defined, so by definition $y = e^x$ is analytic about zero everywhere.

2. For $f(x) = \sin x$ we have $f^{(4k)}(x) = \sin x$, $k = 1, 2, \cdots$ and $f^{(4k+1)}(x) = \cos x$, $f^{(4k+2)}(x) = -\sin x$, and $f^{(4k+3)}(x) = -\cos x$ for $k = 0, 1, 2, \cdots$. Thus, the Taylor series for $\sin x$ about 0 is $x - \frac{x^3}{3!} + \frac{x^5}{5!} - \frac{x^7}{7!} + \cdots$. This is exactly how $y = \sin x$ was defined and so by definition $y = \sin x$ is analytic about $c = 0$ everywhere.

3. For $f(x) = \cos x$ we have $f^{(4k)}(x) = \cos x$, $k = 1, 2, \cdots$ and $f^{(4k+1)}(x) = -\sin x$, $f^{(4k+2)}(x) = -\cos x$, and $f^{(4k+3)}(x) = \sin x$ for $k = 0, 1, 2, \cdots$. Thus, the Taylor series for $\cos x$ about 0 is $1 - \frac{x^2}{2!} + \frac{x^4}{4!} - \frac{x^6}{6!} + \cdots$. This is exactly how $y = \cos x$ was defined and so by definition $y = \sin x$ is analytic about $c = 0$ everywhere.

4. For $f(x) = \sin x$ we have $f^{(4k)} = \sin x$, $k = 1, 2, \cdots$ and $f^{(4k+1)}(x) = \cos x$, $f^{(4k+2)}(x) = -\sin x$, and $f^{(4k+3)}(x) = -\cos x$ for $k = 0, 1, 2, \cdots$. So we have $f^{(4k)}(\frac{\pi}{6}) = \frac{1}{2}$, $k = 1, 2, \cdots$ and $f^{(4k+1)}(\frac{\pi}{6}) = \frac{\sqrt{3}}{2}$, $f^{(4k+2)}(\frac{\pi}{6}) = -\frac{1}{2}$, and $f^{(4k+3)}(\frac{\pi}{6}) = -\frac{\sqrt{3}}{2}$ for $k = 0, 1, 2, \cdots$. Thus, the Taylor series for $\sin x$ about $\frac{\pi}{6}$ is $\frac{1}{2} + \frac{\sqrt{3}}{2}(x - \frac{\pi}{6}) - \frac{1}{2} \cdot \frac{(x-\frac{\pi}{6})^2}{2!} - \frac{\sqrt{3}}{2} \cdot \frac{(x-\frac{\pi}{6})^3}{3!} + \frac{1}{2} \cdot \frac{(x-\frac{\pi}{6})^4}{4!} + \frac{\sqrt{3}}{2} \cdot \frac{(x-\frac{\pi}{6})^5}{5!} = \frac{1}{2} \cdot \frac{(x-\frac{\pi}{6})^6}{6!} - \frac{\sqrt{3}}{2} \cdot \frac{(x-\frac{\pi}{6})^7}{7!} + \cdots$. Note that the Taylor remainder formula is $R_{n-1}(x) = \frac{f^{(n)}(\xi)}{n!}(x - \frac{\pi}{6})^n$ So $|R_{n-1}(x)| \leq \frac{|x-\frac{\pi}{6}|^n}{n!}$. Observe that $e^{|x-\frac{\pi}{6}|} = \sum_{k=0}^{\infty} \frac{|x-\frac{\pi}{6}|^k}{k!}$ converges, so we know that by the divergence theorem $\frac{|x-\frac{\pi}{6}|^n}{n!} \to 0$ as $n \to \infty$ for any fixed $x \in \mathbb{R}$. Thus $|R_{n-1}(x)| \to 0$ as $n \to \infty$. Hence $\sin x$ is analytic about $c = \frac{Pi}{6}$ for every x.

5. The Taylor series for $\cos x$ about $\frac{\pi}{4}$ is $\sum_{k=0}^{\infty} \frac{1}{\sqrt{2}}(-1)^k \frac{(x-\frac{\pi}{4})^k}{k!}$. The remainder term is $R_{n-1}(x) = f^{(n)}(\xi)\frac{(x-\frac{\pi}{4})^n}{n!}$. $|R_{n-1}(x)| \leq \frac{|x-\frac{\pi}{4}|^n}{n!}$. Using that $e^{|x-\frac{\pi}{4}|} = \sum_{k=0}^{\infty} \frac{|x-\frac{\pi}{4}|^k}{k!}$ converges, we know $\frac{|x-\frac{\pi}{4}|^n}{n!} \to 0$ as $n \to \infty$ for each x. Thus $y = \cos x$ is analytic about $c = \frac{\pi}{4}$ for every $x \in \mathbb{R}$.

6. For $y = \sqrt{x}$ and $c = 100$, $T_2(x) = 10 + \frac{x-100}{20} - \frac{(x-100)^2}{8000}$, $T_2(x) = 10 + \frac{x-100}{20} - \frac{(x-100)^2}{8000} + \frac{(x-100)^3}{1,600,000}$, $R_2(x) = \frac{1}{16}\xi^{-\frac{5}{2}}(x-100)^3$, and $R_3(x) = -\frac{5}{128}\xi^{-\frac{7}{2}}(x-100)^4$. $T_2(99) = 10 - \frac{1}{20} - \frac{1}{8000}$ and $|R_2(99)| < \frac{1}{16 \cdot 9^5}$. $R_3(99) = 10 - \frac{1}{20} - \frac{1}{8000} - \frac{1}{1,600,000}$ and $|R_3(99)| < \frac{5}{128 \cdot 9^7}$. $T_2(101) =$

$10 + \frac{1}{20} - \frac{1}{8000}$, $|R_2(101)| < \frac{1}{16 \cdot 10^5}$, $T_3(101) = 10 + \frac{1}{20} - \frac{1}{8000} + \frac{1}{1,600,000}$
and $|R_3(101)| < \frac{5}{128 \cdot 10^7}$.

B.21 Homework Solutions 5.7.1

1. Consider $f(x) = x^2 - a$. Then $x_{n+1} = x_n - \frac{f(x_n)}{f'(x_n)} = x_n - \frac{x_n^2 - a}{2x_n} = \frac{1}{2}(x_n + \frac{a}{x_n})$. Taking this condition, we get $\sqrt{a} - x_{n+1} = \sqrt{a} - \frac{x}{2} - \frac{a}{2x_n} = -\frac{1}{2x_n}(a - 2\sqrt{a}x_n + x_n^2) = -\frac{1}{2x_n}(\sqrt{a} - x_n)^2$.

2. Consider $f(x) = x^3 - 2x^2 - 5 = 0$. Then $x_{n+1} = x_n - \frac{x_n^3 - 2x_n^2 - 5}{3x_n^2 - 4x_n}$. Using $x_0 = 3$, we get $x_1 = 2.733333333$, $x_2 = 2.691624726$, $x_3 = 2.690647977$, $x_4 = 2.690647448$, and $x_5 = 2.690647448$.

3. For $f(x) = x^3 - 3x + 6$, we have $x_{n+1} = x_n - \frac{x_n^3 - 3x_n + 6}{3x_n^2 - 3}$. For $x_0 = 1$ we have $f'(x_0) = 0$, and so x_1 is not defined.

4. Consider $f(x) = -2x^4 + 3x^2 + \frac{11}{8}$. We have f is continuous everywhere. $f(-2) < 0$, $f(2) < 0$ and $f(0) > 0$. By the intermediate value theorem, we have that $f(x) = 0$ has a root on $(-2, 0)$ and $f(x) = 0$ has a root on $(0, 2)$. Using Newton's method, $x_{n+1} = x_n - \frac{-2x_n^4 + 3x_n^2 + \frac{11}{8}}{-8x_n^3 + 6x_n}$. Using $x_0 = 2$ we have $x_1 = 1.641826923$, $x_2 = 1.443401794$, $x_3 = 1.374820447$, $x_4 = 1.366861904$, $x_5 = 1.366760416$, $x_6 = 1.366760399$, and $x_7 = 1.366760399$. If $x_0 = 0.5$, then $x_1 = -0.5$, $x_3 = 0.5$, and so $x_{2n} = 0.5$ and $x_{2n+1} = -0.5$. Observe that $x_0 = 0.5$ satisfies $f''(x_0) = 0$.

B.22 Homework Solutions 6.1.9

1. Since f is continuous on $[a, b]$ it is Riemann integrable on $[a, b]$. Likewise, since f is strictly increasing, and bounded on $[a, b]$, it is Riemann integrable on $[a, b]$. A third way to see that $f(x) = x$ is Riemann integrable on $[a, b]$ is for any partition $\mathcal{P}$ we will show $U(f, [a, b], \mathcal{P}) - L(f, [a, b], \mathcal{P}) \leq 2\delta(b - a)$, where the mesh or norm of the partition $\mathcal{P}$ of $[a, b]$ is δ. Note that this can be made less than any $\varepsilon > 0$ by making δ small enough. Let $\mathcal{P}^*$ be any marked partition of $[a, b]$, and $S = S(f, [a, b], \mathcal{P}^*) = \sum_{i=1}^{n} x_i^*(x_i - x_{i-1})$, then $L = \sum_{i=1}^{n} x_{i-1}(x_i - x_{i-1}) \leq S \leq \sum_{i=1}^{n} x_i(x_i - x_{i-1}) = U$. Observe that $U - L = \sum_{i=1}^{n}(x_i - $

$x_{i-1})^2$. Let $\delta := \max_{1 \leq i \leq n}(x_i - x_{i-1})$. Now $S = \sum_{i=1}^n x_i^*(x_i - x_{i-1}) \leq \sum_{i=1}^n (x_{i-1} + \delta)(x_i - x_{i-1}) = L + \delta(b-a)$, and $S \geq \sum_{i=1}^n (x_i - \delta)(x_i - x_{i-1}) = U - \delta(b-a)$ and so $L + \delta(b-a) - (U - \delta(b-a)) \geq 0$. Thus $U - L \leq 2\delta(b-a)$.

2. For $f(x) = 0$ for $x \in \mathbb{Q}$ and $f(x) = \frac{1}{x}$ for $x \notin \mathbb{Q}$, we see that $\frac{1}{x} \to \infty$ as $x \in \mathbb{R} \setminus \mathbb{Q}$ approaches 0 from the right. Thus f is unbounded on $[0,1]$, and so f is not Riemann integrable on $[0,1]$.

3. Suppose $f = 0$ except at $\xi_1 < \xi_2 < \cdots < \xi_m$. Let $M = \max\{|f(\xi_k)| : k = 1, 2, \cdots, m\}$. Let $\mathcal{P}^*$ be any marked partition of $[a,b]$, and let $\delta = \max\{(x_i - x_{i-1}) : i = 1, 2 \cdots, n\}$, where our partition is $a = x_0 < x_1 < \cdots < x_{n-1} < x_n = b$. Then $|S(f, [a,b], \mathcal{P}^*)| = |\sum_{k=1}^n f(x_k^*)(x_k - x_{k-1})| \leq \sum_{k=1}^n M(x_k - x_{k-1}) \leq 2mM\delta$. This can be made less than ε if δ is small enough, and so we see $\int_a^b f dx = 0$.

4. Suppose $f(x) = g(x)$ except at possibly finitely many points, and suppose g is Riemann integrable on $[a,b]$. Then by the above exercise, $h(x) = f(x) - g(x)$ is Riemann integrable with $\int_a^b h dx = 0$. Then f is Riemann integrable, since $f = h + g$, and so $\int_a^b f dx = \int_a^b h dx + \int_a^b g dx = \int_a^b g dx$.

5. Let $f(x) = 0$ if x is zero or is an irrational number, and $f(x) = n$ where $x = \frac{m}{n} \in \mathbb{Q}$, in reduced form, with $n > 0$. In the discussion on the Thomae function, we showed that for any sequence of rational numbers $\frac{m_k}{n_k}$ converging to an irrational x, we must have $n_k \to \infty$. Thus, on any $[a,b]$, $b \neq a$, there is an irrational number $c \in [a,b]$. Let $\frac{m_k}{n_k} \to c$ be any sequence of rational numbers in reduced form with $\frac{m_k}{n_k} \in [a,b]$, then $f(\frac{m_k}{n_k}) = n_k \to \infty$ as $k \to \infty$. Thus $f(x)$ is not bounded on $[a,b]$, and so $f(x)$ is not Riemann integrable on $[a,b]$.

6. If $f([a,b])$ is a finite set, it doesn't have to be a step function. Consider $f(x) = \chi_\mathbb{Q}$ which has $f(x) = 1$ if $x \in \mathbb{Q}$, and is otherwise equal to zero in value.

7. Consider $f(x) = \chi_{\mathbb{Q} \cap [a,b]} - \chi_{(\mathbb{R} \setminus \mathbb{Q}) \cap [a,b]}$, which is a simple function that is not Riemann integrable on $[a,b]$.

B.22. HOMEWORK SOLUTIONS 6.1.9

8. Suppose $f(x)$ is continuous on $[a,b]$, with $f \geq 0$, and $\int_a^b f dx = 0$. Suppose there is a $c \in [a,b]$ so that $f(c) > 0$. Let $\varepsilon = \frac{f(c)}{2}$. By continuity of f at c we have that $\exists \delta > 0$ so that $x \in (c-\delta, c+\delta) \cap [a,b]$ implies $|f(x) - f(c)| < \frac{f(c)}{2}$. Hence $x \in (c-\delta, c+\delta) \cap [a,b]$ implies $\frac{f(c)}{2} < f(x) < \frac{3f(c)}{2}$. If $c \neq a, b$ let $\delta_1 \leq \delta$ also satisfies $\delta_1 < \min\{c-a, b-c\}$. Then $0 = \int_a^b f(x) dx = \int_a^{c-\delta_1} f dx + \int_{c-\delta_1}^{c+\delta_1} f dx + \int_{c+\delta_1}^b f dx \geq \int_{c-\delta_1}^{c+\delta_1} f dx \geq \frac{f(c)}{2} \cdot 2\delta > 0$, which is a contradiction. If $c = a$, then $0 = \int_a^{a+\delta} f dx + \int_{a+\delta}^b f dx \geq \int_a^{a+\delta} f dx \geq \frac{f(c)}{2}\delta > 0$, also a contradiction. If $c = b$, then observe that $0 = \int_a^{b-\delta} f dx + \int_{b-\delta}^b f dx \geq \int_{b-\delta}^b f dx \geq \frac{f(c)}{2}\delta > 0$, a contradiction. Hence $f(x)$ must be zero everywhere on $[a,b]$.

9. We know $\min_{a \leq x \leq b} f(x) \leq \frac{1}{b-a}\int_a^b f dx \leq \max_{a \leq x \leq b} f(x)$. Since f is continuous on $[a,b]$, there is $\alpha, \beta \in [a,b]$ so that $f(\alpha) = \min_{a \leq x \leq b} f(x)$ and $f(\beta) = \max_{a \leq x \leq b} f(x)$. Consider the closed interval with endpoints α and β. If $f(\alpha) = f(\beta)$, then f is constant on $[a,b]$ and so we use $c = a$. If $f(\alpha) \neq f(\beta)$, and $\frac{1}{b-a}\int_a^b f dx \neq f(\alpha)$ or $f(\beta)$. Then we apply the intermediate value theorem to the interval with endpoints α and β and $L = \frac{1}{b-a}\int_a^b f dx$, to get the existence of $c \in (a,b)$ so that $f(c) = L$.

10. Suppose there is a $c \in [a,b]$ so that $f(c) > 0$, and $f \geq 0$ on $[a,b]$. Let $\varepsilon = \frac{f(c)}{2}$. By continuity of f at c we have that $\exists \delta > 0$ so that $x \in (c-\delta, c+\delta) \cap [a,b]$ implies $|f(x) - f(c)| < \frac{f(c)}{2}$. Hence $x \in (c-\delta, c+\delta) \cap [a,b]$ implies $\frac{f(c)}{2} < f(x) < \frac{3f(c)}{2}$. If $c \neq a, b$ let $\delta_1 \leq \delta$ also satisfies $\delta_1 < \min\{c-a, b-c\}$. Then $\int_a^b f(x)dx = \int_a^{c-\delta_1} f dx + \int_{c-\delta_1}^{c+\delta_1} f dx + \int_{c+\delta_1}^b f dx \geq \int_{c-\delta_1}^{c+\delta_1} f dx \geq \frac{f(c)}{2} \cdot 2\delta > 0$. If $c = a$, then $0 = \int_a^{a+\delta} f dx + \int_{a+\delta}^b f dx \geq \int_a^{a+\delta} f dx \geq \frac{f(c)}{2}\delta > 0$. If $c = b$, then observe that $0 = \int_a^{b-\delta} f dx + \int_{b-\delta}^b f dx \geq \int_{b-\delta}^b f dx \geq \frac{f(c)}{2}\delta > 0$.

11. No, f^2 integrable does not imply f is integrable. To see this, take $f(x) = \chi_{\mathbb{Q} \cap [a,b]} - \chi_{(\mathbb{R} \setminus \mathbb{Q}) \cap [a,b]}$, which is a simple function that is not Riemann integrable on $[a,b]$.

12. Since f^3 integrable on $[a,b]$ we know f^3 is bounded on $[a,b]$. Likewise, $y = \sqrt[3]{x}$ is continuous everywhere, and we have $f = \sqrt[3]{f^3}$. So f is a

continuous function composed with a bounded integrable function, and so f is integrable on $[a,b]$.

13. Since f integrable on $[a,b]$ implies f is bounded on $[a,b]$. Likewise, $y = x^2$ is continuous everywhere, and we have $f^2 = (f)^2$. So f^2 is a continuous function composed with a bounded integrable function, and so f^2 is integrable on $[a,b]$. Likewise, $|f| = \sqrt{f^2}$ is then a continuous function composed with a bounded integrable function, and so $|f|$ is integrable on $[a,b]$ whenever f is integrable on $[a,b]$.

14. Suppose $\lim_{c \to a^+} \int_c^b f\,dx = L$. Then given $\varepsilon > 0$, there is a $\delta > 0$ so that $a < c < a + \delta$ implies $|\int_c^b f\,dx - L| < \frac{\varepsilon}{2}$. Let δ satisfy this limit condition, and also satisfy $\delta < \frac{\varepsilon}{2\max_{a \leq x \leq b}|f(x)|}$, in the case where f is not identically equal to zero on $[a,b]$. Suppose $\int_a^b f\,dx$ exists as well. Then $\int_a^c f\,dx + \int_c^b f\,dx = \int_a^b f\,dx$ for any $a < c < b$. Moreover, for such a δ as above, when $a < c < a + \delta$ we have $|\int_a^b f\,dz - L| = |\int_a^c f\,dx + \int_c^b f\,dx - L| \leq |\int_a^c f\,dx| + |\int_c^b f\,dx - L| < |\int_a^c f\,dx| + \frac{\varepsilon}{2} \leq \max_{a \leq x \leq b}|f(x)| \cdot \delta + \frac{\varepsilon}{2} < \frac{\varepsilon}{2} + \frac{\varepsilon}{2} = \varepsilon$. Since $\varepsilon > 0$ is arbitrary, it must be $\int_a^b f\,dx = L$.

15. Suppose $\int_a^b f\,dx$ exists for all $b > a$, $f \geq 0$, f is decreasing and $\lim_{x \to \infty} f(x) = 0$. So for any $c, d > a$ we have $\int_c^d f\,dx$ exists. For $k \in \mathbb{N}$ so that $k \geq a$, we have $f(k) \leq \int_k^{k+1} f\,dx \leq f(k+1)$. Thus we have $f(k) + f(k+1) + \cdots + f(k+j) \leq \int_k^{k+j+1} f\,dx \leq f(k+1) + f(k+2) + \cdots + f(k+j+1)$. If we assume $\lim_{b \to \infty} \int_a^b f\,dx$ exists, we see that $f(k) + f(k+1) + \cdots + f(k+j)$ is bounded for all j, and since $S_j := f(k) + f(k+1) + \cdots + f(k+j)$ is monotone increasing, it must be that $\lim_{j \to \infty} S_j$ exists and satisfies $\lim_{j \to \infty} S_j \leq \lim_{b \to \infty} \int_a^b f\,dx$. If $\sum_{j=1}^{\infty} f(k+j)$ exists, we know $\int_k^{k+j+1} f\,dx \leq \sum_{i=1}^{\infty} f(k+i)$. Moreover, given $\varepsilon > 0$, there is an N so that for $n, m > N$ we have $|\sum_{j=m}^n f(k+j)| < \varepsilon$. Hence $0 \leq \int_{k+m-1}^{k+n} f\,dx < \varepsilon$ for all $m, n > N$, and so $\lim_{b \to \infty} \int_a^b f\,dx$ exists.

16. Yes, since $f(x) = \sin \frac{1}{x}$ for $x \neq 0$, $f(0) = 0$, f is bounded, and f is continuous at all points except $x = 0$, hence $\int_{-1}^1 f(x)\,dx$ exists.

17. Yes, since $f(x) = x\sin\frac{1}{x}$ for $x \neq 0$, $f(0) = 0$, f is continuous on $[-1.1]$ and so also bounded on $[-1,1]$, and thus $\int_{-1}^{1} f(x)dx$ exists.

B.23 Homework Solutions 6.2.2

1. Suppose $g : [a, b] \to \mathbb{R}$ is strictly increasing, $g' \in C^0([a, b])$ and f is continuous on $[g(a), g(b)]$. We will now show $\int_{g(a)}^{g(b)} f(x)dx = \int_{a}^{b} f(g(x))g'(x)dx$. By assumptions, since f is continuous and $(f \circ g) \cdot g'$ is as well, both $\int_{g(a)}^{g(b)} f(x)dx$ and $\int_{a}^{b} f(g(x))g'(x)dx$ exist. Let $F(t) = \int_{g(a)}^{t} f(x)dx$, for $g(a) < t < g(b)$, and $G(z) = \int_{a}^{z} f(g(x))g'(x)dx$, for $a < z < b$. Then $F'(t) = f(t)$ and $G'(z) = f(g(z))g'(z)$. Also, $G'(z) = F'(g(z))g'(z) = \frac{d}{dz}F(g(z))$, and so for $a < z < b$ we have $G(z) = F(g(z)) + C$. Letting $z \to a^+$ gives us $C = 0$, and letting $z \to b^-$, and (essentially) using exercise 14 from the previous section, we obtain $G(b) = \lim_{z \to b^-} G(z) = \lim_{z \to b^-} F(g(z)) = F(g(b))$, i.e. $\int_{g(a)}^{g(b)} f(x)dx = \int_{a}^{b} f(g(x))g'(x)dx$.

2. Suppose $g : [a, b] \to \mathbb{R}$ is strictly decreasing, $g' \in C^0([a, b])$ and f is continuous on $[g(b), g(a)]$. We will now show $\int_{g(a)}^{g(b)} f(x)dx = -\int_{g(b)}^{g(a)} f(x)dx = \int_{a}^{b} f(g(x))g'(x)dx$. By assumptions, since f is continuous and $(f \circ g) \cdot g'$ is as well, both $\int_{g(b)}^{g(a)} f(x)dx$ and $\int_{a}^{b} f(g(x))g'(x)dx$ exist. Let $F(t) = \int_{g(b)}^{t} f(x)dx$, for $g(b) < t < g(a)$, and $G(z) = \int_{a}^{z} f(g(x))g'(x)dx$, for $a < z < b$. Then $F'(t) = f(t)$ and $G'(z) = f(g(z))g'(z)$. Also, $G'(z) = F'(g(z))g'(z) = \frac{d}{dz}F(g(z))$, and so for $a < z < b$ we have $G(z) = F(g(z)) + C$. Letting $z \to a^+$ and (essentially) using exercise 14 from the previous section, we obtain $0 = \int_{g(b)}^{g(a)} f(x)dx + C$. So for $a < z < b$ we have $\int_{a}^{z} f(g(x))g'(x)dx = \int_{g(b)}^{g(z)} f(x)dx - \int_{g(b)}^{g(a)} f(x)dx$. Letting $z \to b^-$, and (essentially) using exercise 14 from the previous section, we obtain $G(b) = \lim_{z \to b^-} G(z) = \lim_{z \to b^-} F(g(z)) - \int_{g(b)}^{g(a)} f(x)dx = F(g(b)) - \int_{g(b)}^{g(a)} f(x)dx = -\int_{g(b)}^{g(a)} f(x)dx$, i.e. $\int_{g(a)}^{g(b)} f(x)dx = \int_{a}^{b} f(g(x))g'(x)dx$. Hence we see that from the argument just given, and the previous exercise, we have the integral substitution result when g is strictly monotone.

3. Next we shall assume $g : [a, b] \to \mathbb{R}$, $g' \in C^0([a, b])$, and $g'(x) = 0$ only at finitely many points in $[a, b]$. Moreover, suppose f is continuous

on $[\min_{a \leq x \leq b} g(x), \max_{a \leq x \leq b} g(x)]$. We will now show that the integral substitution formula holds: $\int_{g(a)}^{g(b)} f(x)dx = \int_a^b f(g(x))g'(x)dx$. Suppose $a \leq \alpha < \beta < \gamma \leq b$. Consider first where g is strictly increasing on (α, β) and g is strictly decreasing on (β, γ). From the above exercises, we have $\int_{g(\alpha)}^{g(\beta)} f(x)dx = \int_\alpha^\beta f(g(x))g'(x)dx$, and $-\int_{g(\gamma)}^{g(\beta)} f(x)dx = \int_\beta^\gamma f(g(x))g'(x)dx$. Consider first the subcase $g(\alpha) \leq g(\gamma) < g(\beta)$. Then, $\int_{g(\alpha)}^{g(\gamma)} f(x)dx = \int_{g(\alpha)}^{g(\beta)} f(x)dx - \int_{g(\gamma)}^{g(\beta)} f(x)dx = \int_\alpha^\beta f(g(x))g'(x)dx + \int_\beta^\gamma f(g(x))g'(x)dx = \int_\alpha^\gamma f(g(x))g'(x)dx$. Consider second the subcase $g(\gamma) < g(\alpha) < g(\beta)$. Then, $\int_{g(\alpha)}^{g(\gamma)} f(x)dx = -\int_{g(\gamma)}^{g(\alpha)} f(x)dx = \int_{g(\alpha)}^{g(\beta)} f(x)dx - \int_{g(\gamma)}^{g(\beta)} f(x)dx = \int_\alpha^\beta f(g(x))g'(x)dx + \int_\beta^\gamma f(g(x))g'(x)dx = \int_\alpha^\gamma f(g(x))g'(x)dx$. Consider second where g is strictly decreasing on (α, β) and g is strictly increasing on (β, γ). From the above exercises, we have $\int_{g(\alpha)}^{g(\beta)} f(x)dx = -\int_{g(\beta)}^{g(\alpha)} f(x)dx = \int_\alpha^\beta f(g(x))g'(x)dx$, and $\int_{g(\beta)}^{g(\gamma)} f(x)dx = \int_\beta^\gamma f(g(x))g'(x)dx$. Consider first the subcase $g(\beta) < g(\gamma) \leq g(\alpha)$. Then, $\int_{g(\alpha)}^{g(\gamma)} f(x)dx = -\int_{g(\gamma)}^{g(\alpha)} f(x)dx = -\int_{g(\beta)}^{g(\alpha)} f(x)dx + \int_{g(\beta)}^{g(\gamma)} f(x)dx = \int_\alpha^\beta f(g(x))g'(x)dx + \int_\beta^\gamma f(g(x))g'(x)dx = \int_\alpha^\gamma f(g(x))g'(x)dx$. Consider second the subcase $g(\beta) < g(\alpha) < g(\gamma)$. Then, $\int_{g(\alpha)}^{g(\gamma)} f(x)dx = -\int_{g(\beta)}^{g(\alpha)} f(x)dx + \int_{g(\beta)}^{g(\gamma)} f(x)dx = \int_\alpha^\beta f(g(x))g'(x)dx + \int_\beta^\gamma f(g(x))g'(x)dx = \int_\alpha^\gamma f(g(x))g'(x)dx$. Now we suppose that for $a < c < b$ we know that $\int_{g(a)}^{g(c)} f(x)dx = \int_a^c f(g(x))g'(x)dx$, where here of course if $g(a) > g(c)$, then $\int_{g(a)}^{g(c)} f(x)dx = -\int_{g(c)}^{g(a)} f(x)dx$. Suppose $c < d \leq b$ is so that g is strictly monotone on (c, d). Suppose first that g is increasing on (c, d), and so $\int_{g(c)}^{g(d)} f(x)dx = \int_c^d f(g(x))g'(x)dx$ by the first exercise. If $g(a) < g(c)$, then $\int_{g(a)}^{g(d)} f(x)dx = \int_{g(a)}^{g(c)} f(x)dx + \int_{g(c)}^{g(d)} f(x)dx = \int_a^c f(g(x))g'(x)dx + \int_c^d f(g(x))g'(x)dx = \int_a^d f(g(x))g'(x)dx$. Next suppose $g(c) \leq g(a)$. First consider the subcase $g(c) \leq g(d) \leq g(a)$. Then we have $-\int_{g(d)}^{g(a)} f(x)dx = -\int_{g(c)}^{g(a)} f(x)dx + \int_{g(c)}^{g(d)} f(x)dx = \int_a^c f(g(x))g'(x)dx + \int_c^d f(g(x))g'(x)dx = \int_a^d f(g(x))g'(x)dx$. Second, consider the subcase $g(c) \leq g(a) < g(d)$. Then we have that $\int_{g(a)}^{g(d)} f(x)dx = -\int_{g(c)}^{g(a)} f(x)dx + \int_{g(c)}^{g(d)} f(x)dx = \int_a^c f(g(x))g'(x)dx + \int_c^d f(g(x))g'(x)dx = \int_a^d f(g(x))g'(x)dx$. Suppose now that g is decreasing on (c, d), and so

$\int_{g(c)}^{g(d)} f(x)dx = -\int_{g(d)}^{g(c)} f(x)dx = \int_c^d f(g(x))g'(x)dx$ by the second exercise. If $g(a) < g(c)$, we consider first the subcase $g(a) < g(d) < g(c)$. Then $\int_{g(a)}^{g(d)} f(x)dx = \int_{g(a)}^{g(c)} f(x)dx - \int_{g(d)}^{g(c)} f(x)dx = \int_a^c f(g(x))g'(x)dx + \int_c^d f(g(x))g'(x)dx = \int_a^d f(g(x))g'(x)dx$. We consider next the subcase $g(d) \leq g(a) < g(c)$. Then $-\int_{g(d)}^{g(a)} f(x)dx = \int_{g(a)}^{g(c)} f(x)dx - \int_{g(d)}^{g(c)} f(x)dx = \int_a^c f(g(x))g'(x)dx + \int_c^d f(g(x))g'(x)dx = \int_a^d f(g(x))g'(x)dx$. Lastly, if $g(c) \leq g(a)$, then $g(d) < g(a)$, and so $-\int_{g(d)}^{g(a)} f(x)dx = -\int_{g(c)}^{g(a)} f(x)dx - \int_{g(c)}^{g(d)} f(x)dx = \int_a^c f(g(x))g'(x)dx + \int_c^d f(g(x))g'(x)dx = \int_a^d f(g(x))g'(x)dx$.

B.24 Homework Solutions 6.3.1

1. Let $f(x) = x^2 \sin\frac{1}{x}$ for $x \neq 0$, and $f(0) = 0$. For $x \neq 0$ we have $f'(x) = 2x\sin\frac{1}{x} - \cos\frac{1}{x}$, and f is continuous at zero, since for $x \neq 0$ we have $0 \leq f(x) \leq x^2$ and $x^2 \to 0$ as $x \to 0$. Let $\mathcal{P}$ be any partition of a given interval $[a, b]$. In particular, suppose $\mathcal{P}$ is given by $a = x_0 < x_1 < \cdots < x_n = b$. If $0 \in [a, b]$, we have that $0 \in [x_{i-1}, x_i]$ for some i. By otherwise considering $\mathcal{Q} = \mathcal{P} \cup \{0\}$, we may assume that 0 is a partition point of $\mathcal{P}$. (In particular, $V(f, [a, b], \mathcal{Q}) \leq V(f, [a, b], \mathcal{P})$, holds for any refinement $\mathcal{Q}$ of $\mathcal{P}$, so this assumption is justified.) By the mean value theorem we get the bound $|f(x_j) - f(x_{j-1})| = |f'(x_j^*)|(x_j - x_{j-1})$, where $x_{j-1} < x_j^* < x_j$. Moreover, $|f'(x_j^*)| = |2x_j^* \sin\frac{1}{x_j^*} - \cos\frac{1}{x_j^*}| \leq 2\max\{|a|, |b|\} + 1$. So we have $V(f, [a, b], \mathcal{P}) \leq \sum_{k=1}^n (2\max\{|a|, |b|\} + 1)(x_k - x_{k-1}) = (2\max\{|a|, |b|\} + 1) \cdot (b - a)$. Hence $V(f, [a, b]) \leq (2\max\{|a|, |b|\} + 1)(b - a)$.

2. Suppose f is Lipschitz continuous on $[a, b]$. Then $\exists K > 0$ so that $|f(x) - f(y)| \leq K|x - y|$, for any $x, y \in [a, b]$. Let $\mathcal{P}$ be a partition given by $a = x_0 < x_1 < \cdots < x_n = b$. Then $V(f, [a, b], \mathcal{P}) = \sum_{j=1}^n |f(x_j) - f(x_{j-1})| \leq \sum_{j=1}^n K|x_j - x_{j-1}| = K\sum_{j=1}^n (x_j - x_{j-1}) = K(b-a)$. Hence $V(f, [a, b]) \leq K(b-a)$.

3. Suppose $a = y_0 \leq y_1 \leq \cdots \leq y_m = b$, and $f(x) = \sum_{k=1}^m c_k \chi_{[y_{k-1}, y_k]}(x)$. Let $\mathcal{P}$ be a partition $a = x_0 < x_1 < \cdots < x_n = b$ of $[a, b]$. Since a refinement $\mathcal{Q}$ of $\mathcal{P}$ has $V(f, [a, b], \mathcal{Q}) \geq V(f, [a, b], \mathcal{P})$, we may assume each y_k, $k = 0, 1, \cdots, m$, is a partition point of $\mathcal{P}$. Then $V(f, [a, b], \mathcal{P}) = \sum_{k=1}^n |f(x_k) - f(x_{k-1})| = \sum_{k=1}^m |c_k - c_{k-1}|$. Taking the supremum over

all such partitions does not increase this quantity, since any partition $\mathcal{P}$ will result in a $V(f,[a,b],\mathcal{P})$ which is a sum of terms of the form $|c_j - c_i|$ where $j > i$. Also, if $j \neq i+1$, then we also observe that $|c_j - c_i| = |c_j - c_{i+1} + c_{i+1} - c_i| \leq |c_j - c_{i+1}| + |c_{i+1} - c_i|$, and so we see $V(f,[a,b]) = \sum_{k=1}^{m} |c_k - c_{k-1}|$.

4. Suppose $f \in C^1([a,b])$. Let ε be given. Since $\int_a^b |f'|dx$ exists, there is a $\delta > 0$ so that if $\mathcal{P}^\dagger$ is any marked partition of $[a,b]$ whose norm is less than δ, then $|S(|f'|,[a,b],\mathcal{P}^\dagger) - \int_a^b |f'|dx| < \varepsilon$. Let $\mathcal{Q}$ be any partition of $[a,b]$ so that $V(f,[a,b],\mathcal{Q}) > V(f,[a,b]) - \varepsilon$. Let $\mathcal{P}$ be a refinement of $\mathcal{Q}$ which satisfies $\|\mathcal{P}\| < \delta$. Then $V(f,[a,b],\mathcal{P}) > V(f,[a,b]) - \varepsilon$. However, by the mean value theorem $V(f,[a,b],\mathcal{P}) = \sum_{k=1}^{n} |f(x_k) - f(x_{k-1})| = \sum_{k=1}^{n} |f'(x_k^*)|(x_k - x_{k-1})$, for $x_{k-1} < x_k^* < x_k$, for each k. Moreover, let $\mathcal{P}^\dagger$ be $\mathcal{P}$ with the marking points $\{x_k^* : k = 1, 2, \cdots, n\}$. Then $V(f,[a,b],\mathcal{P}) = S(|f'|,[a,b],\mathcal{P}^\dagger) \leq \int_a^b |f'|dx + \varepsilon$. Thus $\int_a^b |f'|dx + \varepsilon \geq S(|f'|,[a,b],\mathcal{P}^\dagger) = V(f,[a,b],\mathcal{P}) \geq V(f,[a,b]) - \varepsilon$. This implies that $V(f,[a,b]) \leq \int_a^b |f'|dx + 2\varepsilon$. Since $\varepsilon > 0$ is arbitrary, we get $V(f,[a,b]) \leq \int_a^b |f'(x)|dx$.

5. Suppose that f, g are of bounded variation on $[a,b]$, and $|g| \geq \varepsilon > 0$ on $[a,b]$. Let $\mathcal{P}$ be any partition of $[a,b]$. Then $V(\frac{f}{g},[a,b],\mathcal{P}) = \sum_{k=1}^{n} \left|\frac{f(x_k)}{g(x_k)} - \frac{f(x_{k-1})}{g(x_{k-1})}\right| = \sum_{k=1}^{n} \frac{|f(x_k)g(x_{k-1}) - g(x_k)f(x_{k-1})|}{|g(x_k)g(x_{k-1})|} \leq \frac{1}{\varepsilon^2} \sum_{k=1}^{n} |f(x_k)g(x_{k-1}) - g(x_k)f(x_{k-1})| = \frac{1}{\varepsilon^2} \sum_{k=1}^{n} |f(x_k)g(x_{k-1}) - f(x_{k-1})g(x_{k-1}) + f(x_{k-1})g(x_{k-1}) - g(x_k)f(x_{k-1})| \leq \frac{1}{\varepsilon^2} \sum_{k=1}^{n} |g(x_{k-1})||f(x_k) - f(x_{k-1})| + \frac{1}{\varepsilon^2} \sum_{k=1}^{n} |f(x_{k-1})||g(x_k) - g(x_k)| \leq \frac{\sup_{x \in [a,b]} |f(x)|}{\varepsilon^2} V(f,[a,b]) + \frac{\sup_{x \in [a,b]} |g(x)|}{\varepsilon^2} V(g,[a,b])$.

6. Since a function of bounded variation f can be written as the difference between two increasing functions $f(x) = g(x) - h(x)$, and each increasing function can have at most a countable number of discontinuities, each necessarily of the first kind, the same is true for f. Since at any point $c \in [a,b]$ we know $\lim_{x \to c^-} f(x) = \lim_{x \to c^-} g(x) - \lim_{x \to c^-} h(x) = \sup\{g(x) : x < c\} - \sup\{h(x) : x < c\}$, and $\lim_{x \to c^+} f(x) = \lim_{x \to c^+} g(x) - \lim_{x \to c^+} h(x) = \inf\{g(x) : x > c\} - \inf\{h(x) : x > c\}$. If both $\lim_{x \to c^-} g(x) = \lim_{x \to c^+} g(x) = g(c)$ and $\lim_{x \to c^-} h(x) = \lim_{x \to c^+} h(x) = h(c)$, then clearly f will be continuous at $x = c$. However, if either $\lim_{x \to c^-} g(x) = \lim_{x \to c^+} g(x) = g(c)$ or $\lim_{x \to c^-} h(x) = \lim_{x \to c^+} h(x) = h(c)$ fails to hold, then f is not continuous at c. In such a case f has a

B.25 Homework Solutions 6.4.1

1. Let $f(x) = \sqrt{1-x^2}$, for $x \in [0, r]$. Here we have that the arclength is $\int_0^r \sqrt{1 + (\frac{-x}{\sqrt{1-x^2}})^2} dx = \int_0^r \frac{1}{\sqrt{1-x^2}} dx = \sin^{-1} r$. This exists for any $0 < r < 1$, and $\sin^{-1} r \to \frac{\pi}{2}$ as $r \to 1^-$, and so an exercise from a previous section (essentially 6.1.9 number 15), we may also define the arclength of $f(x)$ on $[0, 1]$ to be this limiting value.

2. Let $\vec{\gamma}(t) = (\gamma_1(t), \gamma_2(t), \cdots, \gamma_n(t))$ be a parametrized curve, with $a \leq t \leq b$. Let $\mathcal{P}$ be a partition of $[a, b]$ given by $a = t_0 < t_1 < \cdots < t_{m-1} < t_m = b$. Then for any $1 \leq j \leq n$, $V(\gamma_j, [a, b], \mathcal{P}) = \sum_{k=1}^m |\gamma_j(t_k) - \gamma_j(t_{k-1})| \leq \lambda(\vec{\gamma}, [a, b], \mathcal{P}) = \sum_{k=1}^m \|\vec{\gamma}(t_k) - \vec{\gamma}(t_{k-1})\|$
$= \sum_{k=1}^m \sqrt{\sum_{j=1}^n (\gamma_j(t_k) - \gamma_j(t_{k-1}))^2} \leq \sum_{k=1}^m \sum_{j=1}^n |\gamma_j(t_k) - \gamma_j(t_{k-1})|$
$= \sum_{j=1}^n V(\gamma_j, [a, b], \mathcal{P})$. Certainly if each γ_j is of bounded variation on $[a, b]$, then $\lambda(\vec{\gamma}, [a, b], \mathcal{P})$ is bounded from above, and hence $\Lambda(\vec{\gamma}, [a, b])$ exists, and is bounded from above by $\sum_{j=1}^n V(\gamma_j, [a, b])$. Conversely, if $\Lambda(\vec{\gamma}, [a, b])$ exists, then for each $1 \leq j \leq n$ we have that $V(\gamma_j, [a, b], \mathcal{P}) \leq \Lambda(\vec{\gamma}, [a, b])$. So in this case, $V(\gamma_j, [a, b])$ will also exist, and satisfy this same bound, for $1 \leq j \leq n$. For the second part, suppose that $\vec{\gamma}$ is $C^1([a, b])$. Let $\varepsilon > 0$ be given. Let $\mathcal{P}$ be a partition given by $a = t_0 < t_1 < \cdots < t_{m-1} < t_m = b$. Then $|\lambda(\vec{\gamma}, [a, b], \mathcal{P}) - \int_a^b \|\frac{d}{dt} \vec{\gamma}(t)\| dt|$
$= |\sum_{k=1}^m \|\vec{\gamma}(t_k) - \vec{\gamma}(t_{k-1})\| - \int_a^b \|\frac{d}{dt}\vec{\gamma}(t)\| dt| = |\sum_{k=1}^m [\|\vec{\gamma}(t_k) - \vec{\gamma}(t_{k-1})\| - \int_{t_{k-1}}^{t_k} \|\frac{d}{dt}\vec{\gamma}(t)\| dt]| \leq \sum_{k=1}^m |\|\vec{\gamma}(t_k) - \vec{\gamma}(t_{k-1})\| - \int_{t_{k-1}}^{t_k} \|\frac{d}{dt}\vec{\gamma}(t)\| dt|$
$= \sum_{k=1}^m |\int_{t_{k-1}}^{t_k} \|\frac{\vec{\gamma}(t_k) - \vec{\gamma}(t_{k-1})}{t_k - t_{k-1}}\| - \|\frac{d}{dt}\vec{\gamma}(t)\| dt| \leq \sum_{k=1}^m \int_{t_{k-1}}^{t_k} \|\frac{\vec{\gamma}(t_k) - \vec{\gamma}(t_{k-1})}{t_k - t_{k-1}}\| - \|\frac{d}{dt}\vec{\gamma}(t)\|| dt \leq \sum_{k=1}^m \int_{t_{k-1}}^{t_k} \|\frac{\vec{\gamma}(t_k) - \vec{\gamma}(t_{k-1})}{t_k - t_{k-1}} - \frac{d}{dt}\vec{\gamma}(t)\| dt$
$= \sum_{k=1}^m \int_{t_{k-1}}^{t_k} \frac{\|\vec{\gamma}(t_k) - \vec{\gamma}(t_{k-1}) - \frac{d}{dt}\vec{\gamma}(t) \cdot (t_k - t_{k-1})\|}{t_k - t_{k-1}} dt$. However, the vector $\vec{\gamma}(t_k) - \vec{\gamma}(t_{k-1}) - \frac{d}{dt}\vec{\gamma}(t) \cdot (t_k - t_{k-1})$ has n components $\gamma_j(t_k) - \gamma_j(t_{k-1}) - \frac{d}{dt}\gamma_j(t) \cdot (t_k - t_{k-1})$. By the fundamental theorem of calculus, each

one can be written as $\int_{t_{k-1}}^{t_k} \frac{d}{dt}\vec{\gamma}_j(\tau) - \frac{d}{dt}\vec{\gamma}_j(t)dt$. Moreover, if we assume that $\|\mathcal{P}\| < \delta$ is so that for each j (using the uniform continuity of $\frac{d}{dt}\vec{\gamma}_j$) on $[t_{k-1}, t_k]$ the function $\frac{d}{dt}\vec{\gamma}_j(\tau) - \frac{d}{dt}\vec{\gamma}_j(t)$ is in absolute value less than $\frac{\varepsilon}{\sqrt{n}(b-a)}$. So $\int_{t_{k-1}}^{t_k} \frac{d}{dt}\vec{\gamma}_j(\tau) - \frac{d}{dt}\vec{\gamma}_j(t)dt$ is in absolute value less that $\frac{\varepsilon}{\sqrt{n}(b-a)}(t_k - t_{k-1})$. Then $\frac{\|\vec{\gamma}(t_k) - \vec{\gamma}(t_{k-1}) - \frac{d}{dt}\vec{\gamma}(t)\cdot(t_k - t_{k-1})\|}{t_k - t_{k-1}} < \frac{\varepsilon}{b-a}$, and so $\sum_{k=1}^m \int_{t_{k-1}}^{t_k} \frac{\|\vec{\gamma}(t_k) - \vec{\gamma}(t_{k-1}) - \frac{d}{dt}\vec{\gamma}(t)\cdot(t_k - t_{k-1})\|}{t_k - t_{k-1}} dt < \varepsilon$. So it must be that $\lambda(\vec{\gamma}, [a, b], \mathcal{P})$ is bounded, in particular, is bounded above by $\int_a^b \|\frac{d}{dt}\vec{\gamma}(t)\|dt + \varepsilon$. So Λ exists as well, and by otherwise taking a refinement of the existing $\mathcal{P}$, which only increases the value of λ, we may assume that $\lambda(\vec{\gamma}, [a, b], \mathcal{P})$ is greater than $\Lambda(\vec{\gamma}, [a, b]) - \varepsilon$. Thus we see that since $\lambda(\vec{\gamma}, [a, b], \mathcal{P})$ is within ε of $\Lambda(\vec{\gamma}, [a, b])$, it must be that $\Lambda(\vec{\gamma}, [a, b])$ is within 2ε of $\int_a^b \|\frac{d}{dt}\vec{\gamma}(t)\|dt$. Since ε is arbitrary, we get the result.

3. Here $f(x) = \frac{1}{3}(x^2 + 2)^{\frac{3}{2}}$, for $x \in [0, 3]$. The arclength is
$\int_0^2 \sqrt{1 + (x\sqrt{x^2 + 2})^2} dx = \int_0^2 \sqrt{1 + 2x^2 + x^4} dx = \int_0^3 1 + x^2 dx = 12$.

4. Here $f(x) = \frac{1}{4}x^4 + \frac{1}{8x^2}$, for $x \in [1, 2]$. The arclength is
$\int_1^2 \sqrt{1 + (x^3 - \frac{1}{4x^3})^2} dx = \int_1^2 \sqrt{x^6 + \frac{1}{16x^6} + \frac{1}{2}} dx = \int_1^2 x^3 + \frac{1}{4x^3} dx = \frac{123}{32}$.

B.26 Homework Solutions 6.5.4

1. We may view the Cantor set C as $\bigcap_{k=0}^\infty C_k$, where $C_0 = [0, 1]$, $C_1 = [0, \frac{1}{3}] \cup [\frac{2}{3}, 1]$, $C_2 = [0, \frac{1}{9}] \cup [\frac{2}{9}, \frac{1}{3}] \cup [\frac{2}{3}, \frac{7}{9}] \cup [\frac{8}{9}, 1]$, and in general, C_k is the disjoint union of 2^k intervals of the type $[\frac{j}{3^k}, \frac{j+1}{3^k}]$. Let $\varepsilon > 0$ be given, and take $k > \frac{\log \frac{\varepsilon}{2}}{\log \frac{2}{3}}$. Cover each interval $[\frac{j}{3^k}, \frac{j+1}{3^k}]$ by the open interval $(\frac{j}{3^k} - \frac{\varepsilon}{2^{k+2}}, \frac{j+1}{3^k} + \frac{\varepsilon}{2^{k+2}})$. The length of this open interval is $\frac{1}{3^k} + \frac{\varepsilon}{2^{k+1}}$, and so the sum of all the lengths of the open intervals that cover C_k is $\frac{2^k}{3^k} + \frac{\varepsilon}{2} < \varepsilon$, by our choice of k. Since $C \subseteq C_k$, we have that C is also covered by this collection of open sets, whose sum of their lengths is less than ε. Since this can be done for any $\varepsilon > 0$, we see that C has measure zero.

2. Thomae's function is clearly bounded and it is continuous at all points in $(\mathbb{R} \setminus \mathbb{Q}) \cap [a, b]$, and so its set of discontinuities is $\mathbb{Q} \cap [a, b]$, which is a

countable set, and thus it has measure zero. Hence Thomae's function is integrable on $[a,b]$.

3. The set of discontinuities of $f = \chi_\mathbb{Q}$ on $[a,b]$ is the entire interval, which is not a set of measure zero. Hence Lebesgue's criterion is not satisfied, and so $\chi_\mathbb{Q}$ is not Riemann integrable on any $[a,b]$, with $a < b$.

4. Suppose f,g are Riemann integrable on $[a,b]$, and $\int_a^b |f(x) - g(x)| dx = 0$. Let $h(x) = |f(x) - g(x)|$. Then by Lebesgue's criterion, we know that the set of discontinuities of $h(x)$ on $[a,b]$ has measure zero. Let $x \in (a,b)$ be a point where $f(x)$ is continuous and suppose $h(x) > 0$. In such a case, we know that there is a $\delta > 0$ so that $|x - y| < \delta$ implies $|h(y) - h(x)| < \frac{h(x)}{2}$, i.e. $0 < \frac{h(x)}{2} < h(y) < \frac{3h(x)}{2}$. Suppose $\delta \leq \min\{x-a, b-x\}$ as well. Then, $0 = \int_a^b h dx = \int_a^{x-\delta} h dx + \int_{x-\delta}^{x+\delta} h dx + \int_{x+\delta}^b h dx \geq \int_{x-\delta}^{x+\delta} h dx > 2\delta \cdot \frac{h(x)}{2} = \delta h(x) > 0$, which is a contradiction. Thus it must be that h continuous at $x \in (a,b)$ implies $h(x) = 0$, i.e $f(x) = g(x)$. Since the set of x's for which this doesn't hold has measure zero, we have the desired result.

5. If f is integrable on $[a,b]$ and $F(x) = \int_a^x f(t) dt$. By Lebesgue's criterion we know f is continuous on $[a,b] \setminus Z$, where Z has measure zero. At the points where f is continuous, we know $F'(x) = f(x)$ holds. Thus the values of x for which this condition doesn't hold lie in a set of measure zero.

B.27 Homework Solutions 7.1.1

1. $f(x) = 0$ if $0 \leq x < 1$ and $f(1) = 1$. f is continuous and differentiable on $[0,1)$ provided we use right-hand limits at zero. $\int_0^1 f dx = 0$, and $\int_0^1 x^n dx = \frac{1}{n+1} \to 0$ as $n \to \infty$. $\frac{d}{dx} f_n = nx^{n-1} \to 0$ on $[0,1)$ (since $\sum nx^{n-1}$ converges for $|x| < 1$). So $\frac{d}{dx} f_n \to \frac{d}{dx} f$ on $[0,1)$. For $x_0 \in [0,1)$ we have $\lim_{x \to x_0} \lim_{n \to \infty} f_n(x) = \lim_{x \to x_0} 0 = 0$. Also, $\lim_{n \to \infty} \lim_{x \to x_0} f_n(x) = \lim_{n \to \infty} x_0^n = 0$. $\lim_{x \to 1^-} \lim_{n \to \infty} f_n(x) = \lim_{x \to 1^-} 0 = 0$ and $\lim_{n \to \infty} \lim_{x \to 1^-} f_n(x) = \lim_{n \to \infty} 1 = 1$.

2. Let $f_n(x) = \frac{\sin nx}{n}$. For any fixed x $\lim_{n \to \infty} f_n(x) = 0$, and so $f(x) = 0$ for all values of x. Hence f is continuous and differentiable (with $f' = 0$)

everywhere, and the integral of f exists on every interval $[a, b]$, and has a value of zero. $\frac{d}{dx}f_n(x) = \cos nx$, and except for $x = 2\pi k$, $k \in \mathbb{Z}$, we have $\lim_{n\to\infty} \frac{d}{dx}f_n(x)$ does not exist. However, for $x = 2\pi k$ we do have $\lim_{n\to\infty} \frac{d}{dx}f_n(x) = 0 = \frac{d}{dx}f(x)$. $\int_a^b f_n(x)dx = \frac{\cos na - \cos nb}{n^2} \to 0 = \int_a^b f(x)dx$. $\lim_{n\to\infty} \lim_{x\to x_0} f_n(x) = \lim_{n\to\infty} \frac{\sin nx_0}{n} = 0 = \lim_{x\to x_0} 0 = \lim_{x\to x_0} \lim_{n\to\infty} f_n(x)$.

3. Consider $f_n : [0, 1] \to \mathbb{R}$, with $f_n(x) = 2n - 2n^2 x$ on $(0, \frac{1}{n}]$, and zero elsewhere on $[0, 1]$. Then $f_n(x) \to 0 \equiv f(x)$ point-wise on $[0, 1]$. $\int_0^1 f_n(x)dx = \int_0^{\frac{1}{n}} 2n - 2n^2 x\, dx = 1$. However, $\int_0^1 f(x)dx = 0$.

4. Consider $f_n : [0, 1] \to \mathbb{R}$, with $f_n(x) = 3n + 1$ for $x \in (0, \frac{1}{n}]$, and zero elsewhere on $[0, 1]$. Ten $f_n(x) \to 0 \equiv f(x)$ point-wise on $[0, 1]$. $\int_0^1 f_n(x)dx = 3 + \frac{1}{n} \to 3$ as $n \to \infty$. However, $\int_0^1 f(x)dx = 0$.

5. Observe that $x^{\frac{1}{2n-1}} = -|x|^{\frac{1}{2n-1}}$ for $x < 0$. So, it suffices to consider $x > 0$. Observe that $g_n(x) = \sqrt[2n-1]{x}$ is continuous on $[-1, 1]$, for each n. Clearly $g_n(0) = 0$ regardless of the value of n. $\ln g_n(x) = \frac{\ln x}{2n-1} \to 0$ as $n \to \infty$ for any fixed positive x. Since $\ln x$ and e^x are continuous on their domains, we have $\ln g_n(x) \to 0$ implies $g_n(x) \to 1$ as $n \to \infty$, for any fixed positive x. It follows that $\sqrt[2n-1]{x} \to -1$ for $x < 0$. Also, $x^{\frac{2n}{2n-1}} = x \cdot \sqrt[2n-1]{x}$ converges to x as $n \to \infty$ if $x > 0$ is fixed, and converges to $-x$ as $n \to \infty$ if $x < 0$ is fixed. That is $f_n(x) = x^{\frac{2n}{2n-1}} \to |x| =: f(x)$ for x fixed, as $n \to \infty$. Clearly $\frac{d}{dx}f_n(x) = \frac{2n}{2n-1}\sqrt[2n-1]{x}$ for all $x \in [-1, 1]$, and this converges to 1 for $x > 0$, as $n \to \infty$, and it converges to -1 for $x < 0$, as $n \to \infty$, and it converges to 0 for $x = 0$, as $n \to \infty$. However, $f(x) = |x|$ is not differentiable at $x = 0$. Thus $\frac{d}{dx}f_n(0) \to 0$ as $n \to \infty$, but $\frac{d}{dx}f(0)$ does not exist. Hence $\frac{d}{dx}f_n \not\to \frac{d}{dx}f$ on $[-1, 1]$.

B.28 Homework Solutions 7.2.4

1. Consider $f_n(x) = x + \frac{1}{n}$. Then for $f(x) = x$ we have $|f_n(x) - f(x)| = \frac{1}{n} \to 0$ as $n \to \infty$, regardless of the value of x.

2. For f_n in the above exercise, we now consider $f_n^2 = x^2 + \frac{2x}{n} + \frac{1}{n^2}$. Then $|f_n^2(x) - f^2(x)| = \frac{|2x + \frac{1}{n}|}{n}$. For each fixed x this goes to zero as n ap-

proaches infinity, and so $f_n^2(x)$ converges to $f^2(x)$ point-wise. However, suppose that given $\varepsilon > 0$ we require $\frac{|2x+\frac{1}{n}|}{n} < \varepsilon$, then $-\frac{\varepsilon n^2+1}{2n} < x < \frac{\varepsilon n^2-1}{2n}$. This shows that $f_n^2(x)$ cannot converge to $f^2(x)$ uniformly on $\mathbb{R}$.

3. Consider $f_n(x) = x^2 e^{-nx}$ on $[0, \infty)$. We will now show that $f_n(x)$ converges uniformly to $f(x) \equiv 0$ on this interval. Clearly $f_n(x) \geq 0$. Observe that $e^{nx} \geq 1 + nx + \frac{n^2 x^2}{2}$ for $x \geq 0$. Hence $f_n(x) \leq \frac{x^2}{\frac{n^2 x^2}{2}+nx+1}$. For $x > 0$ we actually have $f_n(x) < \frac{x^2}{\frac{n^2 x^2}{2}+nx} = \frac{1}{n(\frac{n}{2}+\frac{1}{x})} < \frac{2}{n^2}$. This shows that $f_n(x) \to 0$ uniformly on $(0, \infty)$. Since $f_n(0) = 0$ for all n, we have $f_n(x) \to 0$ uniformly on $[0, \infty)$.

4. Suppose that $f_n \to f$ and $g_n \to g$ uniformly on A. Consider $h_n = f_n + g_n$ and $k_n = c \cdot f_n$ for some constant c. Let $\varepsilon > 0$ be given. Since $f_n \to f$ uniformly on A and $g_n \to g$ uniformly on A, there is an N so that $n > N$ implies $|f_n(x) - f(x)| < \frac{\varepsilon}{2}$ and $|g_n(x) - g(x)| < \frac{\varepsilon}{2}$ for all $x \in A$. Thus, for such $n > N$ we have $|h_n(x) - (f(x) + g(x))| = |(f_n(x) - f(x)) + (g_n(x) - g(x))| \leq |f_n(x) - f(x)| + |g_n(x) - g(x)| < \varepsilon$. Thus $h_n \to f + g$ uniformly on A. If $c = 0$, then $cf, cf_n \equiv 0$ and so $k_n \to cf$ uniformly on A. Otherwise, there is an $M > 0$ so that $n > M$ implies $|f_n(x) - f(x)| < \frac{\varepsilon}{|c|} \ \forall x \in A$. Hence for such $n > M$ we have $|k_n(x) - cf(x)| = |c| \cdot |f_n(x) - f(x)| < |c| \cdot \frac{\varepsilon}{|c|} = \varepsilon$. Thus $k_n \to cf$ uniformly on A.

5. Suppose that $f_n \to f$ and $g_n \to g$ uniformly on A, f_n and g_n are bounded on A for each n, we will now show that $f_n \cdot g_n \to f \cdot g$ uniformly on A. By the assumption of uniform convergence on A, we have that there is an N so that for $n > N$ and $x \in A$, we have both $|f(x) - f_n(x)| < 1$ and $|g(x) - g_n(x)| < 1$. Let's call this result (*). Hence, by (*), we have $f_{N+1}(x) - 1 < f(x) < f_{N+1}(x) + 1$ and $g_{N+1}(x) - 1 < g(x) < g_{N+1}(x) + 1$ for every $x \in A$. From this we obtain both $-\max_{y \in A} |f_{N+1}(y)| - 1 < f(x) < \max_{y \in A} |f_{N+1}(y)| + 1$ and $-\max_{y \in A} |g_{N+1}(y)| - 1 < g(x) < \max_{y \in A} |g_{N+1}(y)| + 1$ for every $x \in A$. Thus we have $\max_{y \in A} |f(y)| \leq \max_{x \in A} |f_{N+1}(x)| + 1$ and $\max_{y \in A} |g(y)| \leq \max_{x \in A} |f_{N+1}(x)| + 1$. Also, from (*), we have for any $n > N$ and any $x \in A$, $-1 + g(x) < g_n(x) < g(x) + 1$. So we obtain the inequality $-1 - \max_{y \in A} |g(y)| < g_n(x) < 1 + \max_{y \in A} |g(y)|$ for any $n > N$ and any $x \in A$. From this observation, it follows

that $\max_{x \in A} |g_n(x)| \leq \max_{y \in A} |g(y)| + 1$ for $n > N$. Now we are ready to establish the uniform convergence of $f_n \cdot g_n$ to $f \cdot g$ on A. Since $f_n \to f$ and $g_n \to g$ uniformly on A, given $\varepsilon > 0$, there is an L so that $n > L$ implies both $|f_n(x) - f(x)| < \frac{\varepsilon}{2[\max_{y \in A}|g(y)|+1]}$ and $|g_n(x) - g(x)| < \frac{\varepsilon}{2[\max_{y \in A}|f(y)|+1]}$ for every $x \in A$. Suppose now $n > \max\{L, N\}$ where N is as given above. Then for any $x \in A$, we have $|f_n(x)g_n(x) - f(x)g(x)| = |f_n(x)g_n(x) - f(x)g_n(x) + f(x)g_n(x) - f(x)g(x)| \leq |g_n(x)| \cdot |f_n(x) - f(x)| + |f(x)| \cdot |g_n(x) - g(x)| \leq |g_n(x)| \cdot |f_n(x) - f(x)| + [|f(x)| + 1] \cdot |g_n(x) - g(x)| \leq [\max_{y \in A} |g(y)| + 1] \cdot |f_n(x) - f(x)| + [\max_{y \in A} |f(y)| + 1] \cdot |g_n(x) - g(x)| < [\max_{y \in A} |g(y)| + 1] \cdot \frac{\varepsilon}{2[\max_{y \in A}|g(y)|+1]} + [\max_{y \in A} |f(y)| + 1] \cdot \frac{\varepsilon}{2[\max_{y \in A}|f(y)|+1]} = \varepsilon$. Hence we get the desired result.

6. Suppose $f_n \to f$ uniformly on A, and $\|f_n\|_{\sup,A} \leq M$ for all n. Suppose g is continuous on $[-M, M]$. We will now show that $g \circ f_n \to g \circ f$ uniformly on A. Since g is continuous on $[-M, M]$, we know that g is uniformly continuous on $[-M, M]$. So, given $\varepsilon > 0$, there exists $\delta > 0$ so that for any $x, y \in [-M, M]$ satisfying $|x - y| < \delta$ we have $|g(x) - g(y)| < \varepsilon$. Call this condition (*). Since $f_n \to f$ uniformly on A we have that given $\eta > 0$, there is an N so that $n > N$ implies $|f_n(x) - f(x)| < \eta$ for any $x \in A$. Call this condition (**). Using (**) we see that for every $x \in A$ and any $n > N$ we have $f_n(x) - \eta < f(x) < f_n(x) + \eta$. This implies $-M - \eta < f(x) < M + \eta$ for any $x \in A$. Here we used $-M \leq f_n(x) < M$. Thus we obtain $\max_{y \in A} |f(y)| \leq M + \eta$. Since this is true for all $\eta > 0$, we get $\|f\|_{\sup,A} \leq M$. Now in (**) we use $\eta = \delta$, where the δ comes from (*). Thus obtaining for $n > N$ and any $x \in A$ that $|f_n(x) - f(x)| < \delta$, and so by (**) we have that $|g(f_n(x)) - g(f(x))| < \varepsilon$.

7. Suppose that $f_n : [0, \infty) \to \mathbb{R}$, with $f_1(x) = \sqrt{x}$ and $f_{n+1}(x) = \sqrt{x + f_n(x)}$. We will now show that $f_n(x) \to f(x)$ point-wise, where $f(0) = 0$ and $f(x) = \frac{1+\sqrt{1+4x}}{2}$. Since $f_1(0) = 0$, we get $f_2(0) = \sqrt{0+0} = 0$. Likewise, if $f_n(0) = 0$, we get $f_{n+1}(0) = \sqrt{0+0} = 0$. Hence $f_n(0) \to 0 =: f(0)$. For $x > 0$ we will now show $f_n(x) \to \frac{1+\sqrt{1+4x}}{2} =: f(x)$ as $n \to \infty$. By definition, we have $f_n(x) \geq 0$ for any positive x and any natural number n. We will now show that $f_n(x) \leq \frac{1+\sqrt{1+4x}}{2}$ for all positive numbers x and all natural numbers n. Since $0 \leq 2 + 2\sqrt{1+4x}$ we have $4x \leq 2 + 4x + 2\sqrt{1+4x} =$

B.28. HOMEWORK SOLUTIONS 7.2.4

$(1 + \sqrt{1+4x})^2$, and since $y = \sqrt{x}$ is an increasing function, we obtain $2\sqrt{x} \leq 1 + \sqrt{1+4x}$. Thus $f_1(x) \leq f(x)$. Now suppose that $f_n(x) \leq \frac{1+\sqrt{1+4x}}{2}$, then $x + f_n(x) \leq \frac{2x+1+\sqrt{1+4x}}{2}$. Taking square roots, we obtain $f_{n+1}(x) = \sqrt{x + f_n(x)} \leq \sqrt{\frac{2x+1+\sqrt{1+4x}}{2}} = \frac{\sqrt{4x+2+2\sqrt{4x+1}}}{2} = \frac{1+\sqrt{1+4x}}{2}$. Now we will show that for $x > 0$ we have that $\{f_n(x)\}_{n=1}^\infty$ is monotone increasing. $f_2(x) = \sqrt{x + \sqrt{x}} > \sqrt{x}$. Now suppose $f_{n-1}(x) < f_n(x)$. From this we obtain $x + f_{n-1}(x) < x + f_n(x)$, and so $f_n(x) = \sqrt{x + f_{n-1}(x)} < \sqrt{x + f_n(x)} = f_{n+1}(x)$. Thus, for each positive x, the sequence $\{f_n(x)\}_{n=1}^\infty$ converges. Let $f(x)$ denote the limit of the sequence. Taking the condition $f_{n+1}(x) = \sqrt{x + f_n(x)}$, we get $f(x) = \lim_{n\to\infty} f_{n+1}(x) = \lim_{n\to\infty} \sqrt{x + f_n(x)} = \sqrt{x + \lim_{n\to\infty} f_n(x)} = \sqrt{x + f(x)}$. Hence $f(x) = \sqrt{x + f(x)}$. From this we obtain $[f(x)]^2 - f(x) - x = 0$. Since $f(x) > 0$, it follows that $f(x) = \frac{1+\sqrt{1+4x}}{2}$. To show that $f_n(x)$ does not converge uniformly to $f(x)$ on $[0, \infty)$ we assume it does. Observing that each $f_n(x)$ is continuous on $[0, \infty)$ we see that $0 = \lim_{n\to\infty} 0 = \lim_{n\to\infty} f_n(0) = \lim_{n\to\infty} \lim_{x\to 0} f_n(x) = \lim_{x\to 0} \lim_{n\to\infty} f_n(x) = \lim_{x\to 0} \frac{1+\sqrt{1+4x}}{2} = 1$, a contradiction.

8. Suppose that $f_n : [a, b] \to \mathbb{R}$, f_n is increasing, and $f_n \to f$ point-wise. We have that f must be increasing as well, since for any $x < y$ in $[a, b]$ we have $f_n(x) \leq f_n(y)$ for every n. Thus $a_n := f_n(y) - f_n(x) \geq 0$ and $a_n \to f(y) - f(x)$. Hence $f(y) - f(x) \geq 0$. Suppose we assume that each f_n is strictly increasing, all that we can conclude is that f is increasing. To see this, take $[a, b] = [0, 1]$, and $f_n(x) = x^n$. Then $f(x)$ is identically zero on $[0, 1)$, but $f(1) = 1$.

9. We will now produce a sequence of functions $f_n : [a, b] \to \mathbb{R}$ that are Riemann integrable on $[a, b]$, and converge point-wise to f which is not Riemann integrable. To do this we first define $g_n : [0, 1] \to \mathbb{R}$, given by $g_n(x) = 1$ if $x = \frac{m}{k}$, $m, k \in \mathbb{N}$, with m, k relatively prime, $m \leq k$, and $k \leq n$. Otherwise, we let $g_n(x) = 0$. Then $g_n(x)$ is zero on all but finitely many points in $[0, 1]$. Hence $g_n(x)$ is Riemann integrable for each n. In particular $\int_0^1 g_n dx = 0$. Moreover, $g_n \to g$, where $g(x) = 1$ if $x \in \mathbb{Q} \cap [0, 1]$, and otherwise $g(x) = 0$. Clearly $g(x)$ is not Riemann integrable. For the general interval $[a, b]$ we just compose $g_n(x)$ with a linear mapping that takes a to zero and b to one. That is, define $f_n(x) = g_n(\frac{x-a}{b-a})$. By properties of u-substitution established in an

earlier homework exercise, we get the desired result.

10. We will provide an example of a uniformly convergent sequence of functions on $(0,1)$ whose sequence of derivative functions do not converge. For this we take $f_n(x) = \frac{\sin nx}{\sqrt{n}}$. Here we have $f_n \to f$, where $f \equiv 0$. This follows since $|f_n(x) - f(x)| = \frac{|\sin nx|}{\sqrt{n}} = \frac{1}{\sqrt{n}} \to 0$ independent of the value of x. However, $\frac{df_n}{dx}(x) = \sqrt{n}\cos nx$ does not as converge as $n \to \infty$ for any x in $(0,1)$.

11. Suppose that $\{f_n\}_{n=1}^\infty$ is bounded in A, with $f_n \to f$ uniformly on A. We will now show $\lim_{n\to\infty} \|f_n\|_{\sup,A} = \|f\|_{\sup,A}$. Given $\varepsilon > 0$, $\exists N$ such that $\forall n > N$ and $\forall x \in A$ we have $|f_n(x) - f(x)| < \varepsilon$. Let us call this condition (*). By (*), for any $n > N$ and $x \in A$ we know $f_n(x) - \varepsilon < f(x) < f_n(x) + \varepsilon$ and so $-\|f_n\|_{\sup,A} - \varepsilon \leq f_n(x) - \varepsilon < f(x) < f_n(x) + \varepsilon \leq \|f_n\|_{\sup,A} + \varepsilon$. This gives us $\|f\|_{\sup,A} \leq \|f_n\|_{\sup,A} + \varepsilon$. Also, by (*), for any $n > N$ and $x \in A$ we know $f(x) - \varepsilon < f_n(x) < f(x) + \varepsilon$ and so $-\|f\|_{\sup,A} - \varepsilon \leq f(x) - \varepsilon < f_n(x) < f(x) + \varepsilon \leq \|f\|_{\sup,A} + \varepsilon$. This gives us $\|f_n\|_{\sup,A} \leq \|f\|_{\sup,A} + \varepsilon$. Hence we have $\|f_n\|_{\sup,A} - \varepsilon \leq \|f\|_{\sup,A} \leq \|f_n\|_{\sup,A} + \varepsilon$. For all $n > N$. This gives us the desired result.

12. Take $f_n : [0,1] \to \mathbb{R}$ with $f_n(x) = \frac{\chi_{\mathbb{Q} \cap [0,1]}(x)}{n}$. Then $f_n(x) = 0$ except for $f_n(x) = \frac{1}{n}$ when $x \in \mathbb{Q} \cap [0,1]$. Moreover, f_n is not continuous anywhere on $[0,1]$ for each n. Clearly $f_n(x) \to 0$ uniformly, and the zero function is differentiable.

B.29 Homework Solutions 7.3.2

1. From theorem 7.3.4 and lemmas 7.3.2 and 3 we know that if $f(x) = \sum_{n=0}^\infty a_n(x-c)^n$, then $f^{(k)}(x) = \sum_{n=k}^\infty n(n-1)\cdots(n-k+1)a_n(x-c)^{n-k}$.

2. Using the above exercise and evaluating at $x = c$ we have $f(c) = a_0$ and $f^{(k)}(c) = k(k-1)\cdots 1 a_k$.

3. Suppose that $f(x) = \int_0^x e^{-t^2} dt$. Then by the fundamental theorem of calculus, $f'(x) = e^{-x^2}$. Using the definition of e^x as a power series we have $f'(x) = e^{-x^2} = \sum_{k=0}^\infty \frac{(-x^2)^k}{k!}$. From this we obtain $f(x) = C + \sum_{k=0}^\infty \frac{(-1)^k}{k!} \cdot \frac{x^{2k+1}}{2k+1}$. Using that $f(0) = 0$, we see that $C = 0$.

4. Let $\alpha \in \mathbb{R}$. Define $f(x) = (1+x)^\alpha$ for $x > -1$. We may use the fact that $(1+x)^\alpha = e^{\alpha \log(1+x)}$, and so $f'(x) = \frac{d}{dx}(1+x)^\alpha = \alpha \cdot (1+x)^{\alpha-1}$. Continuing inductively, we see that $f^{(k)}(x) = \alpha \cdot (\alpha-1) \cdots (\alpha-k+1) \cdot (1+x)^{\alpha-k}$. From this we see that $f(0) = 1$, $f'(0) = \alpha$ and in general $f^{(k)}(0) = \alpha \cdot (\alpha-1) \cdots (\alpha-k+1)$. For $k \in \mathbb{N}$ we define $\binom{\alpha}{k} := \frac{\alpha \cdot (\alpha-1) \cdots (\alpha-k+1)}{k!}$ and $\binom{\alpha}{0} := 1$. Thus the power series (Taylor series) expansion for $f(x)$ about 0 is given by $g(x) := \sum_{k=0}^{\infty} \binom{\alpha}{k} x^k$. Using the ratio test, we see that the series converges absolutely if $1 > \lim_{k \to \infty} |\frac{\binom{\alpha}{k+1} x^{k+1}}{\binom{\alpha}{k} x^k}| = \lim_{k \to \infty} |\frac{\alpha+1}{k+1} - 1| \cdot |x| = |x|$. Thus the series converges absolutely for $|x| < 1$. By term by term differentiation of the power series, for $-1 < x < 1$, we see that g satisfies $(1+x) \cdot g'(x) = \alpha g(x)$. Also, we see that f also satisfies $(1+x) \cdot f'(x) = \alpha f(x)$. For $-1 < x < 1$ we consider the function $\frac{g(x)}{f(x)} = \frac{g(x)}{(1+x)^\alpha}$. $\frac{d}{dx} \frac{g(x)}{f(x)} = \frac{g'(x) \cdot f(x) - g(x) \cdot f'(x)}{(f(x))^2} = \frac{g'(x) \cdot (1+x)^\alpha - g(x) \cdot \alpha (1+x)^{\alpha-1}}{(1+x)^{2\alpha}} = \frac{\alpha g(x) \cdot (1+x)^{\alpha-1} - g(x) \cdot \alpha (1+x)^{\alpha-1}}{(1+x)^{2\alpha}} \equiv 0$. Thus $\frac{g(x)}{f(x)} \equiv Constant$ on $-1 < x < 1$. Using $x = 0$ we see that $Constant = 1$. Hence on $-1 < x < 1$ we have $g(x) = f(x)$.

B.30 Homework Solutions 7.4.3

1. Let $f : \mathbb{R} \to \mathbb{R}$ is continuous and define $f_n(x) = f(nx)$. Suppose that $\{f_n\}_{n=1}^{\infty}$ is uniformly equicontinuous on $[0, 1]$. We claim that f is a constant function. By uniform equicontinuity, given $\varepsilon > 0$ there is a $\delta > 0$ so that for any $x, y \in [0, 1]$ satisfying $|x - y| < \delta$ and any $n \in \mathbb{N}$ we have $|f_n(x) - f_n(y)| < \varepsilon$. That is, $|f(nx) - f(ny)| < \varepsilon$. Consider $i \in \mathbb{N}$, then for $k \geq i$ we have $x = \frac{i}{k} \in [0, 1]$. Let $v \in \mathbb{R}$. Consider $y = x + z$ where $|z| < \delta$ and $y \in [0, 1]$. Let $w \in \mathbb{R}$ be such that $v = i + w$. We fix $k \in \mathbb{N}$, and we require that $kz = w$. Recall that $|z| < \delta$ is needed and $|w| = k|z| < k\delta$. This is possible since we may choose $k > i$ sufficiently large so that $|z| = \frac{|w|}{k} < \delta$. Having made such a choice of k, we have $x = \frac{i}{k}$, and $y = x + z$, and if necessary make k larger so that $y \in [0, 1]$ is satisfied. Then $|x - y| = |z| < \delta$, and so $|f_k(x) - f_k(y)| < \varepsilon$. That is, $|f(kx) - f(ky)| < \varepsilon$. However, $f(kx) = f(i)$ and $f(ky) = f(i + w) = f(v)$. Hence $|f(i) - f(v)| < \varepsilon$. Since $\varepsilon > 0$ was arbitrary, and our choice of $i \in \mathbb{N}$ and $v \in \mathbb{R}$ were independent of ε, we have $f(i) = f(v)$. Thus f is a constant function.

2. Let $\{f_n\}_{n=1}^\infty$ be a uniformly equicontinuous sequence of functions on a compact set K, and suppose $f_n \to f$ point-wise on K. Then for each $x \in K$, $\{f_n(x)\}_{n=1}^\infty$ is bounded since it is a convergent sequence. Thus, by the Ascoli-Arzelà theorem we have that $\{f_n\}_{n=1}^\infty$ is uniformly bounded, and contains a uniformly convergent subsequence $\{f_{n_k}\}_{k=1}^\infty$. For a fixed $x \in K$, since $f_n(x) \to f(x)$ as $n \to \infty$, and $\{f_{n_k}(x)\}_{k=1}^\infty$ converges uniformly, it must be $f_{n_k}(x) \to f(x)$. Moreover, since $\{f_{n_k}\}_{k=1}^\infty$ converges uniformly to f, we have that $f(x)$ is continuous on K. We will now show that $\{f_n\}_{n=1}^\infty$ converges uniformly to f on K. Since $\{f_n\}_{n=1}^\infty$ is uniformly equicontinuous, given $\varepsilon > 0$ there is a $\delta_1 > 0$ so that for any $x, y \in K$ with $|x - y| < \delta_1$ we have that $|f_n(x) - f_n(y)| < \frac{\varepsilon}{3}$. Since f is continuous on K, it is uniformly continuous. Thus there is a $\delta_2 > 0$ so that for $x, y \in K$ satisfying $|x - y| < \delta_2$ we have $|f(x) - f(y)| < \frac{\varepsilon}{3}$. Let $\delta = \min\{\delta_1, \delta_2\}$. The collection $\{B_\delta(x) : x \in K\}$ is an open cover for K, and by compactness, there is a finite subcover $\{B_\delta(x_1), B_\delta(x_2), \cdots, B_\delta(x_J)\}$. For $i = 1, 2, \cdots, J$ we have that there is an N_i so that for $n > N_i$ we have $|f_n(x_i) - f(x_i)| < \frac{\varepsilon}{3}$. Let $N = \max\{N_1, N_2, \cdots, N_J\}$. Let $x \in K$, then there is a $k \in \{1, 2, \cdots, J\}$ so that $x \in B_\delta(x_k)$. Thus, for $n > N$ we have $|f_n(x) - f(x)| = |f_n(x) - f_n(x_k) + f_n(x_k) - f(x_k) + f(x_k) - f(x)| \leq |f_n(x) - f_n(x_k)| + |f_n(x_k) - f(x_k)| + |f(x_k) - f(x)| < \frac{\varepsilon}{3} + \frac{\varepsilon}{3} + \frac{\varepsilon}{3} = \varepsilon$. Thus $\{f_n\}_{n=1}^\infty$ converges uniformly to f on K.

3. Suppose $\|f_n\|_{C^{0,\alpha}([a,b])} \leq M$, $M > 0$, for any $n \in \mathbb{N}$. Then by definition of $\|\cdot\|_{C^{0,\alpha}([a,b])}$ we have $\|f_n\|_{\sup,[a,b]} \leq M$ for and $\sup_{x \neq y; x, y \in [a,b]} \frac{|f(x) - f(y)|}{|x-y|^\alpha} < M$ for all n. The first observation gives us that $\{f_n\}_{n=1}^\infty$ is uniformly bounded. The second condition tells us that for any n we have $|f_n(x) - f_n(y)| \leq M \cdot |x-y|^\alpha$. If we require that $M \cdot |x-y|^\alpha < \varepsilon$ we can attain this if $|x-y| < \delta = (\frac{\varepsilon}{M})^{\frac{1}{\alpha}}$. So for $|x-y| < \delta$ we obtain $|f_n(x) - f_n(y)| < \varepsilon$ for all n. This gives us that the family of functions is uniformly equicontinuous on $[a, b]$. So by Ascoli-Arzelà, there is a uniformly convergent subsequence $\{f_{n_k}\}_{k=1}^\infty$.

4. Let $f_n(x) = \sin nx$. We will show that this family of functions is not uniformly equicontinuous on any $[a, b] \subset \mathbb{R}$, with $b > a$. Suppose the contrary. By Ascoli-Arzelà, there is a subsequence $f_{n_k}(x) = \sin n_k x$, so that f_{n_k} converges uniformly to some function f on $[a, b]$. Since each f_{n_k} is continuous on $[a, b]$, we have f is continuous on $[a, b]$. More-

over, $\lim_{k\to\infty} \int_\alpha^\beta f_{n_k}(x)dx = \int_\alpha^\beta f(x)dx$ for any $[a,b] \subseteq [a,b]$. However, $|\int_\alpha^\beta \sin n_k x \, dx| = \frac{|\cos n_k \alpha - \cos n_k \beta|}{n_k} \leq \frac{2}{n_k} \to 0$ as $k \to \infty$. Thus, $\int_\alpha^\beta f(x)dx = 0$. Since f is continuous, and this is true for every $[\alpha,\beta] \subseteq [a,b]$, we have that $f \equiv 0$ on $[a,b]$. Since $\{f_{n_k}\}$ is uniformly bounded, we also know that $f_{n_k}^2 \to f^2$ uniformly on $[a,b]$. Moreover, $\lim_{k\to\infty} \int_a^b (f_{n_k}(x))^2 dx = \int_a^b (f(x))^2 dx = \int_a^b 0 \, dx = 0$. However, $\lim_{k\to\infty} \int_a^b \sin^2 n_k x \, dx = \lim_{k\to\infty} \int_a^b (\frac{1}{2} - \frac{1}{2}\cos 2n_k x) dx = \lim_{k\to\infty}[\frac{b-a}{2} - \frac{\sin 2n_k \beta - \sin 2n_k \alpha}{4n_k}] = \frac{b-a}{2} \neq 0$, a contradiction. Thus $\{\sin(nx)\}_{n=1}^\infty$ cannot contain a uniformly convergent subsequence on any non-trivial interval $[a,b]$.

5. Let $m \neq 0$ be any real number. Then $f_n : \mathbb{R} \to \mathbb{R}$, $f_n(x) := mx + n$ is uniformly equicontinuous, since given $\varepsilon > 0$, we have $|x - y| < \frac{\varepsilon}{|m|}$ implies $|f_n(x) - f_n(y)| = |m| \cdot |x - y| < \varepsilon$. However, $\{f_n\}_{n=1}^\infty$ is not uniformly bounded on any set in $\mathbb{R}$.

6. Consider $f_n : \mathbb{R} \to [-1,1]$ given by $f_n(x) = \sin x$ if $2\pi n \leq x \leq 2\pi(n+1)$ and otherwise $f_n(x) = 0$. Now $f_n(x) \to f(x)$ point-wise, where $f(x) = 0$ for all x. Moreover, $\|f_n\|_{\sup,\mathbb{R}} \leq 1 \, \forall n \in \mathbb{N}$, and $|f_n(x) - f_n(y)| \leq |x - y|$ for all n, hence $\{f_n\}_{n=1}^\infty$ is uniformly bounded and uniformly equicontinuous. However, there is no uniformly convergent subsequence, since where $f_n(x)$ is not equal to zero depends on n explicitly.

B.31 Homework Solutions 8.1.2

1. Let $\|\vec{x}\|_1 = \|(x_1, x_2, \cdots, x_n)\|_1 = \sum_{k=1}^n |x_k|$. Since $|x_k| \geq 0$ for all k, we have $\|\vec{x}\|_1 \geq 0$. Moreover, $\|\vec{x}\|_1 = 0$ if and only if $|x_k| = 0$ for all k, and this is equivalent to $\vec{x} = \vec{0}$. Let $c \in \mathbb{R}$ and $\vec{x} \in \mathbb{R}^n$. Then $\|c \cdot \vec{x}\|_1 = \|(cx_1, cx_2, \cdots, cx_n)\|_1 = \sum_{k=1}^n |cx_k| = \sum_{k=1}^n |c| \cdot |x_k| = |c| \sum_{k=1}^n |x_k| = |c| \cdot \|\vec{x}\|_1$. Let $\vec{x} = (x_1, x_2, \cdots, x_n)$ and $\vec{y} = (y_1, y_2, \cdots, y_n)$. Then $\|\vec{x} + \vec{y}\|_1 = \|(x_1 + y_1, x_2 + y_2, \cdots, x_n + y_n)\|_1 = \sum_{k=1}^n |x_k + y_k| \leq \sum_{k=1}^n (|x_k| + |y_k|) = \sum_{k=1}^n |x_k| + \sum_{k=1}^n |y_k| = \|\vec{x}\|_1 + \|\vec{y}\|_1$. Thus we see that $\|\cdot\|_1$ is a norm on $\mathbb{R}^n$.

2. For $\vec{x} = (x_1, x_2, \cdots, x_n)$, we define $\|\vec{x}\|_p = (\sum_{k=1}^n |x_k|^p)^{\frac{1}{p}}$. Since $|a|^p \geq 0$ for any real number a and equal to zero if and only if $a = 0$, we have $\|\vec{x}\|_p^p \geq 0$, and so $\|\vec{x}\|_p \geq 0$ for every $\vec{x}$. $f(x) = x^{\frac{1}{p}} \geq 0$ on $[0, \infty)$ and

equals zero if and only if $x = 0$. Thus $\|\vec{x}\|_p = 0$ implies $\sum_{k=1}^n |x_k|^p = 0$ and this holds if and only if each $|x_k|^p = 0$, which in turn holds if and only if each $|x_k| = 0$. Finally, this holds if and only if $\vec{x} = \vec{0}$. Now, for $c = \mathbb{R}$, $\|c \cdot \vec{x}\|_p = \|(cx_1, cx_2, \cdots, cx_n)\|_p = (\sum_{k=1}^n |cx_k|^p)^{\frac{1}{p}} = (|c|^p \cdot \sum_{k=1}^n |x_k|^p)^{\frac{1}{p}} = |c| \cdot (\sum_{k=1}^n |x_k|^p)^{\frac{1}{p}} = |c| \cdot \|\vec{x}\|_p$. Finally, we must establish the triangle inequality for the norm $\|\cdot\|_p$. This result is often called Minkowski's inequality. Here we restrict ourselves to $p > 1$, since the $p = 1$ case is established in the above exercise. First we observe that since $f(x) = -\log x$ is convex, we have $\frac{1}{p}\log a + \frac{1}{q}\log b \leq \log(\frac{a}{p} + \frac{b}{q})$ where $a, b > 0$ and $\frac{1}{p} + \frac{1}{q} = 1$. Taking exponentials of both sides of this inequality, we obtain $a^{\frac{1}{p}} \cdot b^{\frac{1}{q}} \leq \frac{a}{p} + \frac{b}{q}$. Letting $\alpha = a^{\frac{1}{p}}$ and $\beta = b^{\frac{1}{q}}$, we obtain $\alpha \cdot \beta \leq \frac{\alpha^p}{p} + \frac{\beta^q}{q}$. Note that this holds for any $p > 1$ and any pair $\alpha, \beta > 1$, and this result is often called Young's inequality. Next we will establish what is often called Hölder's inequality. Hölder's inequality says that $\sum |x_i y_i| \leq (\sum |x_i|^p)^{\frac{1}{p}} \cdot (\sum |y_i|^q)^{\frac{1}{q}}$, where as above $\frac{1}{p} + \frac{1}{q} = 1$, $p > 1$, and in this case either $\vec{x} = (x_1, x_2, \cdots, x_n)$, $\vec{y} = (y_1, y_2, \cdots, y_n)$, or $\{x_i\}$ and $\{y_i\}$ are sequences. To prove Hölder's inequality, we will first assume $(\sum |X_i|^p)^{\frac{1}{p}} = 1 = (\sum |Y_j|^q)^{\frac{1}{q}}$. Then by Young's inequality, $\sum |X_i Y_i| \leq \sum (\frac{|X_i|^p}{p} + \frac{|Y_i|^q}{q}) = \frac{\sum |X_i|^p}{p} + \frac{\sum |Y_i|^q}{q} = \frac{1}{p} + \frac{1}{q} = 1$. The general case follows by taking $X_i = \frac{x_i}{(\sum |x_j|^p)^{\frac{1}{p}}}$ and $Y_i = \frac{y_i}{(\sum |Y_j|^q)^{\frac{1}{q}}}$. We will now establish the triangle inequality. Consider $\sum |x_i + y_i|^p = \sum |x_i + y_i|^{p-1} \cdot |x_i + y_i| \leq \sum |x_i + y_i|^{p-1} \cdot (|x_i| + |y_i|) = \sum |x_i + y_i|^{p-1} \cdot |x_i| + \sum |x_i + y_i|^{p-1} \cdot |y_i|$. We then apply Hölder's inequality, to obtain $\sum |x_i + y_i|^{p-1} \cdot |x_i| + \sum |x_i + y_i|^{p-1} \cdot |y_i| \leq (\sum (|x_i + y_i|^{p-1})^{\frac{p}{p-1}})^{\frac{p-1}{p}} \cdot (\sum |x_i|^p)^{\frac{1}{p}} + (\sum (|x_i + y_i|^{p-1})^{\frac{p}{p-1}})^{\frac{p-1}{p}} \cdot (\sum |y_i|^p)^{\frac{1}{p}} = (\sum |x_i + y_i|^p)^{\frac{p-1}{p}} \cdot ((\sum |x_i|^p)^{\frac{1}{p}} + (\sum |y_i|^p)^{\frac{1}{p}})$. Thus we get $(\sum |x_i + y_i|^p)^{\frac{1}{p}} \leq (\sum |x_i|^p)^{\frac{1}{p}} + (\sum |y_i|^p)^{\frac{1}{p}}$.

3. Let $\|\vec{x}\|_\infty = \|(x_1, x_2, \cdots, x_n)\|_\infty = \max\{|x_1|, |x_2|, \cdots, |x_n|\}$. Clearly $|x_k| \geq 0$ for every k, and so $\|\vec{x}\|_\infty \geq 0$ for any $\vec{x}$. Moreover, $\|\vec{x}\|_\infty = 0$ if and only if $\max\{|x_1|, |x_2|, \cdots, |x_n|\} = 0$. Now, for each $k \in \{1, 2, \cdots, n\}$, we have that $0 \leq \|x_k\| \leq \max\{|x_1|, |x_2|, \cdots, |x_n|\} = 0$. Thus, $\|\vec{x}\|_\infty = 0$ if and only if each $|x_k| = 0$, which holds if and only if $\vec{x} = \vec{0}$. Let $c \in \mathbb{R}$ and $\vec{x} \in \mathbb{R}^n$. Then $\|c \cdot \vec{x}\|_\infty = \|(cx_1, cx_2, \cdots, cx_n)\|_\infty = \max\{|cx_1|, |cx_2|, \cdots, |cx_n|\} = |c| \max\{|x_1|, |x_2|, \cdots, |x_n|\} = |c| \cdot \|\vec{x}\|_\infty$. Let $\vec{x} = (x_1, x_2, \cdots, x_n)$ and $\vec{y} = (y_1, y_2, \cdots, y_n)$. Then $\|\vec{x} + \vec{y}\|_\infty =$

$\max\{|x_1+y_1|, |x_2+y_2|, \cdots, |x_n+y_n|\} \leq \max\{|x_1|+|y_1|, |x_2|+|y_2|, \cdots, |x_n|+|y_n|\} \leq \max\{|x_1|, |x_2|, \cdots, |x_n|\} + \max\{|y_1|, |y_2|, \cdots, |y_n|\} = \|\vec{x}\|_\infty + \|\vec{y}\|_\infty$. Note that the last inequality follows since the maximum of the $|x_k|$'s and the maximum of the $|y_j|$'s do not have to occur on the same subscript value.

4. Given $\vec{x} = (x_1, x_2, \cdots, x_n)$, we have $\|\vec{x}\|_1^2 = \sum_{k=1}^n |x_k|^2 + 2\sum_{1 \leq i < j \leq n} |x_i| \cdot |x_j| \geq \sum_{k=1}^n |x_k|^2 = \|\vec{x}\|^2$. Thus, $\|\vec{x}\|_1 \geq \|\vec{x}\|$. Moreover, $\|\vec{x}\|_1^2 = \|\vec{x}\|^2 + 2\sum_{1 \leq i < j \leq n} |x_i| \cdot |x_j| \leq \|\vec{x}\|^2 + \sum_{1 \leq i < j \leq n}(|x_i|^2 + |x_j|^2) = \|\vec{x}\|^2 + \sum_{j=1}^n \sum_{i=1}^{j-1}(|x_i|^2 + |x_j|^2) \leq \|\vec{x}\|^2 + \sum_{j=1}^n \sum_{i=1}^n(|x_i|^2 + |x_j|^2) = \|\vec{x}\|^2 + \sum_{j=1}^n(\|\vec{x}\|^2 + n|x_j|^2) = (2n+1)\|\vec{x}\|^2$. So we may take $K = \sqrt{2n+1}$, and we have $\frac{1}{K}\|\vec{x}\| \leq \|\vec{x}\|_1 \leq K\|\vec{x}\|$, and from this we also get $\frac{1}{K}\|\vec{x}\|_1 \leq \|\vec{x}\| \leq K\|\vec{x}\|_1$.

5. Given $\vec{x} = (x_1, x_2, \cdots, x_n)$, we have that $\|\vec{x}\|_p = (\sum_{i=1}^n |x_i|^p)^{\frac{1}{p}}$. Now Using exercises 6 and 9 in this problem set, $\|\vec{x}\|_p \leq (n\|\vec{x}\|_\infty^p)^{\frac{1}{p}} = n^{\frac{1}{p}} \cdot \|\vec{x}\|_\infty \leq n^{\frac{1}{p}}\sqrt{n}\|\vec{x}\|$. Moreover, $\|\vec{x}\|_p \geq (\|\vec{x}\|_\infty^p)^{\frac{1}{p}} = \|\vec{x}\|_\infty \geq \frac{1}{n^{\frac{1}{p}}}\|\vec{x}\|_\infty \geq \frac{1}{n^{\frac{1}{p}}\sqrt{n}}\|\vec{x}\|$. This gives the desired result with $K = n^{\frac{1}{p}}\sqrt{n}$.

6. Given $\vec{x} = (x_1, x_2, \cdots, x_n)$, we have that $\|\vec{x}\|_\infty = \max_{1 \leq i \leq n} |x_i| = |x_k|$ for some k. Thus, $\|\vec{x}\|_\infty^2 = |x_k|^2 \leq \sum_{i=1}^n |x_i|^2 = \|\vec{x}\|^2$. So we see $\|\vec{x}\|_\infty \leq \|\vec{x}\|$. Observe as well that $\|\vec{x}\| = \sqrt{|x_1|^2 + |x_2|^2 + \cdots + |x_n|^2} \leq \sqrt{n \cdot \max\{|x_1|^2, |x_2|^2, \cdots, |x_n|^2\}} = \sqrt{n \cdot (\max\{|x_1|, |x_2|, \cdots, |x_n|\})^2} = \sqrt{n} \cdot \|\vec{x}\|_\infty$. Thus $\frac{1}{\sqrt{n}}\|\vec{x}\| \leq \|\vec{x}\|_\infty \leq \sqrt{n}\|\vec{x}\|$, and so it follows that $\frac{1}{\sqrt{n}}\|\vec{x}\|_\infty \leq \|\vec{x}\| \leq \sqrt{n}\|\vec{x}\|_\infty$.

7. This follows directly from exercise 4 with exactly the same $K = \sqrt{2n+1}$.

8. The result follows from what was established in exercise 6, using the same K.

9. This follows directly from exercise 6 with exactly the same $K = \sqrt{n}$.

10. Consider the collection $M_{m,n}(\mathbb{F})$ of $m \times n$ matrices with coefficients in a field $\mathbb{F}$. We let $[a_{ij}]$ represent a matrix whose coefficient in the ith row and j^{th} column is $a_{ij} \in \mathbb{F}$. Consider the matrix Δ_{ij} whose entries are all $0_\mathbb{F}$ except in the ith row and j^{th} column the entry is $1_\mathbb{F}$. Suppose

that $\sum_{1\leq i\leq n, 1\leq j\leq m} c_{ij}\Delta_{ij} = 0_{m\times n}$, the $m\times n$ matrix whose coefficients are all $0_\mathbb{F}$. Then we know that $c_{ij} = 0_\mathbb{F}$ for all i,j. Any matrix $[a_{ij}]$ can be represented by $\sum_{i,j} a_{ij}\Delta_{ij}$, and so $\{\Delta_{ij}\}$ spans $M_{m,n}(\mathbb{F})$. Thus the dimension of $M_{m,n}(\mathbb{F})$ is at most $m\cdot n$. However, we showed that that $\{\Delta_{ij}\}$ is a linearly independent collection, and so the dimension must be at least $m\cdot n$.

11. Consider $V = \{a_0 + a_1 x + a_2 x^2 : a_0, a_1, a_2 \in \mathbb{R}\}$. Let $c \in \mathbb{R}$. We will now show that $\{1, x+c, (x+c)^2\}$ is a basis for V. $c_0\cdot 1 + c_1\cdot (x+c) + c_2\cdot(x+c)^2 = 0$ if and only if $c_0 + c_1 c + c_2 c + (c_1 + 2cc_2)x + c_2 x^2 = 0$ if and only if $c_0 + c_1 c + c_2 c = 0, c_1 + 2cc_2 = 0, c_2 = 0$. Since $c_2 = 0$, we get $c_1 = c_1 + 2cc_2 = 0$, and thus $c_0 = c_0 + c_1 c + c_2 c = 0$. Thus $\{1, x+c, (x+c)^2\}$ is linearly independent. Now, given $a_0 + a_1 x + a_2 x^2$, we wish to find b_0, b_1, b_2 so that $b_0 + b_1(x+c) + b_2(x+c)^2 = a_0 + a_1 x + a_2 x^2$. This holds if and only if $b_0 + b_1 c + b_2 c = a_0, b_1 + 2cb_2 = a_1, b_2 = a_2$. This holds if and only if $b_2 = a_2, b_1 = a_1 - 2ca_2$ and $b_0 = a_0 - (a_1 - 2ca_2)c - a_2 c$. Thus $\{1, x+c, (x+c)^2\}$ spans V. Hence it is a basis for V.

B.32 Homework Solutions 8.2.2

1. Let V be the space of $n \times 1$ matrices over a field $\mathbb{F}$, and W be the space of $m \times 1$ matrices over $\mathbb{F}$. Let $A = [a_{ij}] \in M_{m,n}(\mathbb{F})$ be an $m \times n$ matrix with coefficients in $\mathbb{F}$. We define $T: V \to W$, by $T(\vec{x}) = A\vec{x}$. We will now prove that $T\vec{x} = \vec{0}$ for all $\vec{x}$ if and only if $a_{ij} = 0_\mathbb{F}$ for all i,j. Then if $a_{ij} = 0_\mathbb{F}\ \forall i,j$ then $A\vec{x} = \vec{0}$ is clear, since the jth in $A\vec{x}$ is $\sum_{i=1}^n a_{ij}x_i$. Now suppose $A\vec{x} = \vec{0}$ for all $\vec{x}$. Take $\vec{x}$ with $x_1, x_2, \cdots, x_{j-1}, x_{j+1}, \cdots, x_n = 0_\mathbb{F}$ and $x_j = 1_\mathbb{F}$. Then $A\vec{x} = \vec{y}$, where $y_i = a_{ij} = 0_\mathbb{F}$ for $i = 1, 2, \cdots m$ by assumption. Since $1 \leq j \leq n$ is arbitrary, we get $a_{ij} = 0_\mathbb{F}$ for all i, j.

2. Let T, V, W be as in the above exercise. Let $m < n$, we will now show that the kernel of T, that is the subset of the domain that maps to zero, is non-trivial. Clearly $T(\vec{0}) = \vec{0}$. this follws by linearity, since $T(\vec{0}) + T(\vec{0}) = T(\vec{0} + \vec{0}) = T(\vec{0})$. If $T(\vec{x}) = \vec{0}$, then $T(c\vec{x}) = cT(\vec{x}) = c\vec{0} = \vec{0}$. Moreover, if $T(\vec{x})$ and $T(\vec{y}) = \vec{0}$, then $\vec{0} = T(\vec{x}) + T(\vec{y}) = T(\vec{x} + \vec{y})$. Thus we see that the kernel of T is non-empty, and is a linear subspace of V. Let $\vec{n_1}, \cdots, \vec{n_k}$ be a basis for the kernel of T (not that we are not assuming that the kernel is non-trivial, if it is

indeed trivial this collection can be empty). Then there are vectors $\{\vec{x}_{k+1}, \cdots, \vec{x}_n\}$ so that $\{\vec{n}_1, \cdots, \vec{n}_k, \vec{x}_{k+1}, \cdots, \vec{x}_n\}$ is a basis for V. Then the vectors $\{T(\vec{n}_1), \cdots, T(\vec{n}_k), T(\vec{x}_{k+1}), \cdots, T(\vec{x}_n)\}$ span the range of T. This follows since to be in the range of T the vector must be of the form $T(\vec{y})$. However, $\vec{y} = c_1\vec{n}_1 + \cdots + c_k\vec{n}_k + c_{k+1}\vec{x}_{k+1} + \cdots + c_n\vec{x}_n$, and so $T(\vec{y}) = T(c_1\vec{n}_1 + \cdots + c_k\vec{n}_k + c_{k+1}\vec{x}_{k+1} + \cdots + c_n\vec{x}_n) = c_1 T(\vec{n}_1) + \cdots + c_k T(\vec{n}_k) + c_{k+1}T(\vec{x}_{k+1}) + \cdots + c_n T(\vec{x}_n)$. Since $T(\vec{n}_i) = \vec{0}$, we actually have $T(\vec{y}) = c_{k+1}T(\vec{x}_{k+1}) + \cdots + c_n T(\vec{x}_n)$. Observe that the range is a linear subspace of W, since for $T(\vec{y})$ in the range, $cT(\vec{y}) = T(c\vec{y})$ and $T(c\vec{y})$ is also in the range. For $T(\vec{y}), T(\vec{z})$ in the range, we have $T(\vec{y}) + T(\vec{z}) = T(\vec{y} + \vec{z})$, and $T(\vec{y}+\vec{z})$ is in the range. Now, the dimension of the range is at most m, and $m < n$. Moreover $\{T(\vec{n}_1), \cdots, T(\vec{n}_k), T(\vec{x}_{k+1}), \cdots, T(\vec{x}_n)\}$ span the range of T. In fact, $\{T(\vec{x}_{k+1}), \cdots, T(\vec{x}_n)\}$ span the range of T, since $T(\vec{n}_i) = \vec{0}$. We will now show that $\{T(\vec{x}_{k+1}), \cdots, T(\vec{x}_n)\}$ are linearly independent. Suppose $\sum_{i=k+1}^{n} c_i T(\vec{x}_i) = \vec{0}$. Then $T(\sum_{i=k+1}^{n} c_i\vec{x}_i) = \vec{0}$. Thus $\sum_{i=k+1}^{n} c_i\vec{x}_i = \vec{\alpha}$ is in the kernel of T, and in this case, it must be that there are constants $b_1, \cdots b_k$ so that $\alpha = \sum_{i=1}^{k} b_i\vec{n}_i$. Thus, $\sum_{i=1}^{k} b_i\vec{n}_i - \sum_{i=k+1}^{n} c_i\vec{x}_i = \vec{0}$. By the linear independence of the collection $\{\vec{n}_1, \cdots, \vec{n}_k, \vec{x}_{k+1}, \cdots, \vec{x}_n\}$, we get that $b_1, \cdots, b_k, c_{k+1}, \cdots, c_n = 0_\mathbb{F}$. Thus we have that $n - k \leq m < n$, and so $k \geq 0$. That is, the kernel cannot be trivial, it must be a linear subspace of dimension at least one.

3. Let $V = \{a_0 + a_1 x + a_2 x^2 : a_0, a_1, a_2 \in \mathbb{R}\}$. In an earlier exercise, it was shown that $B = \{1, x + c, (x+c)^2\}$ is a basis for V. Let $f(x) = a_0 + a_1 x + a_2 x^2$. Then $[f(x)]_B = (a_0 - c(a_1 - 2ca_2) - ca_2, a_1 - 2ca_2, a_2)$.

4. Suppose that $L : V \to W$ be a linear transformation between vector spaces V and W that is a bijection. Consider $L^{-1}(\vec{y} + \vec{w}) = \vec{z}$. Then $\vec{y} + \vec{w} = L(\vec{z})$. Likewise, $L^{-1}(\vec{y}) + L^{-1}(\vec{w}) = \vec{u} + \vec{v}$ implies $\vec{y} + \vec{w} = L(\vec{u}) + L(\vec{v}) = L(\vec{u} + \vec{v})$. Thus $L(\vec{z}) = L(\vec{u}+\vec{v})$ and so $\vec{z} = \vec{u}+\vec{v}$, since L is injective. That is, $L^{-1}(\vec{y}+\vec{w}) = L^{-1}(\vec{u}) + L^{-1}(\vec{w})$. Now, consider $L^{-1}(c\vec{y}) = \vec{x}$, for c a non-zero scalar. Then $c\vec{y} = L(\vec{x})$. Moreover, consider $cL^{-1}(\vec{y}) = \vec{z}$. Then $L^{-1}(\vec{y}) = c^{-1} \cdot \vec{z}$, and so $\vec{y} = L(c^{-1}\vec{z}) = c^{-1}L(\vec{z})$ which implies $c\vec{y} = L(\vec{z})$. Hence, $\vec{x} = \vec{z}$. That is $L^{-1}(c\vec{y}) = cL^{-1}(\vec{y})$. The case where $c = 0$ is obvious, since $c\vec{x} = \vec{0}$.

B.33 Homework Solutions 8.3.4

1. Let $f : \mathbb{R}^2 \to \mathbb{R}$ be given by $f(x,y) = |xy|$. Since $f(x,0) = 0$, we have $f_x(0,0) = \lim_{x \to 0} \frac{f(x,0)-f(0,0)}{x} = \lim_{x \to 0} \frac{0}{x} = 0$. Likewise, since $f(0,y) = 0$, $f_y(0,0) = \lim_{y \to 0} \frac{f(0,y)-f(0,0)}{y} = 0$. In this case, f differentiable at $(0,0)$ is equivalent to $0 = \lim_{(x,y) \to (0,0)} \frac{|xy|}{\sqrt{x^2+y^2}}$. However, $|xy| \leq \frac{x^2+y^2}{2}$ and so $\frac{|xy|}{\sqrt{x^2+y^2}} \leq \frac{1}{2}\sqrt{x^2+y^2} \to 0$ as $(x,y) \to (0,0)$. Thus we see that f is differentiable at $(0,0)$.

2. Let $f(x,y) = x^2 \arctan \frac{y}{x} - y^2 \arctan \frac{x}{y}$ if $x,y \neq 0$ and otherwise equal to 0. We will now show $f_{xy}(0,0) = 1$, but $f_{yx}(0,0) = -1$. $f(x,0) = 0$ and $f(0,y) = 0$, and so as in the above exercise we have $f_x(0,0)$ and $f_y(0,0)$ are both 0. Moreover, for $x \neq 0$, $f_y(x,0) = \lim_{y \to 0} \frac{f(x,y)-f(x,0)}{y} = \lim_{y \to 0} \frac{f(x,y)}{y} = \lim_{y \to 0} \frac{x^2 \arctan \frac{y}{x} - y^2 \arctan \frac{x}{y}}{y} = \lim_{y \to 0} x \frac{\arctan \frac{y}{x}}{\frac{y}{x}} = x \lim_{z \to 0} \frac{\arctan z}{z} = x \lim_{z \to 0} \frac{1}{1+z^2} = x$. For $y \neq 0$, $f_x(0,y) = \lim_{x \to 0} \frac{f(x,y)-f(0,y)}{x} = \lim_{x \to 0} \frac{f(x,y)}{x} = \lim_{x \to 0} (x \arctan \frac{y}{x} - \frac{y^2}{x} \arctan \frac{x}{y}) = \lim_{x \to 0} -y \frac{\arctan \frac{x}{y}}{\frac{x}{y}} = -y \lim_{z \to 0} \frac{\arctan z}{z} = -y$. Now $f_{xy}(0,0) = \lim_{y \to 0} \frac{f_x(0,y)-f_x(0,0)}{y} = \lim_{y \to 0} \frac{f_x(0,y)}{y} = \lim_{y \to 0} \frac{-y}{y} = -1$. $f_{yx}(0,0) = \lim_{x \to 0} \frac{f_y(x,0)-f_y(0,0)}{x} = \lim_{x \to 0} \frac{f_y(x,0)}{x} = \lim_{x \to 0} \frac{x}{x} = 1$.

3. Consider $f(x,y) = \frac{xy^2}{x^2+y^2}$ for $(x,y) \neq (0,0)$ and $f(0,0) = 0$. We will now compute $\nabla_{\vec{u}} f(0,0)$ for any given $\vec{u}$. Let $\vec{u} = (\cos\theta, \sin\theta)$. Then $\nabla_{\vec{u}} f(0,0) = \lim_{t \to 0} \frac{f(t\cos\theta, t\sin\theta)-f(0,0)}{t} = \lim_{t \to 0} \frac{\frac{t^3 \cos\theta \sin\theta}{t^2(\cos^2\theta+\sin^2\theta)}}{t} = \lim_{t \to 0} \cos\theta \sin\theta = \cos\theta \sin\theta$. If f is differentiable at $(0,0)$, then $\nabla_{\vec{u}} f(0,0) = f_x(0,0) \cos\theta + f_y(0,0) \sin\theta$. The mapping $\vec{u} \mapsto \nabla_{\vec{u}} f(0,0)$ is linear in $\vec{u}$ whenever f is differentiable at $(0,0)$. In our case, $\nabla_{\vec{u}} f(0,0) = \nabla_{(\cos\theta, \sin\theta)} f(0,0) = \cos\theta \cdot \sin\theta$ is non-linear. Hence we see that f is not differentiable at $(0,0)$.

4. Consider the function $f(x,y) = \frac{xy}{x^4+y^4}$ if $(x,y) \neq (0,0)$ and $f(0,0) = 0$. If $(x,y) \neq (0,0)$, then $f_x = \frac{y(x^4+y^4)-xy(4x^3)}{(x^4+y^4)^2}$ and $f_y = \frac{x(x^4+y^4)-xy(4y^3)}{(x^4+y^4)^2}$. Since $f|_{x=0}, f|_{y=0} \equiv 0$ we have $f_x(0,0), f_y(0,0) = 0$. Hence f_x and f_y exist at all points in $\mathbb{R}^2$. However, $f(x,x) = \frac{1}{2x^2} \to \infty$ as $x \to 0$. Thus f is not continuous at $(0,0)$, and so f cannot be differentiable at $(0,0)$.

5. Let $f(x,y) = \frac{xy}{\sqrt{x^2+y^2}} \sin \frac{1}{\sqrt{x^2+y^2}}$ if $(x,y) \neq (0,0)$ and $f(0,0) = 0$. By the chain rule for functions of one variable, for $(x,y) \neq (0,0)$ we have $f_x(x,y) = \frac{y\sqrt{x^2+y^2} - xy \cdot \frac{x}{\sqrt{x^2+y^2}}}{x^2+y^2} \cdot \sin \frac{1}{\sqrt{x^2+y^2}} - \frac{x^2 y}{(x^2+y^2)^2} \cos \frac{1}{\sqrt{x^2+y^2}}$ and

$f_y(x,y) = \frac{x\sqrt{x^2+y^2} - xy \cdot \frac{y}{\sqrt{x^2+y^2}}}{x^2+y^2} \cdot \sin \frac{1}{\sqrt{x^2+y^2}} - \frac{xy^2}{(x^2+y^2)^2} \cos \frac{1}{\sqrt{x^2+y^2}}$. Since $f|_{y=0}, f|_{x=0} \equiv 0$, we have $f_x(0,0), f_y(0,0) = 0$. However, $\nabla_{(\cos\theta,\sin\theta)} f(0,0) = \lim_{t\to 0} \frac{f(t\cos\theta, t\sin\theta) - f(0,0)}{t} = \lim_{t\to 0} \cos\theta \sin\theta \sin\frac{1}{t}$, which does not exist unless $\cos\theta$ or $\sin\theta = 0$. If f is differentiable at $(0,0)$, then by the chain rule we would have $\nabla_{\vec{u}} f(0,0) = (f_x(0,0), f_y(0,0)) \cdot \vec{u}$ for every $\vec{u}$.

6. Consider $f(x,y) = a_{0,0} + a_{1,0}x + a_{0,1}y + a_{2,0}x^2 + a_{1,1}xy + a_{0,2}y^2 + a_{3,0}x^3 + a_{2,1}x^2 y + a_{1,2}xy^2 + a_{0,3}y^3$. We will require f to satisfy $f_{xx} + f_{yy} = 0$. A direct computation shows $f_{xx} = 2a_{2,0} + 6a_{3,0}x + 2a_{2,1}y$ and $f_{yy} = 2a_{0,2} + 2a_{1,2}x + 6a_{0,3}y$. Thus $f_{xx} + f_{yy} = 0$ is equivalent to $2a_{2,0} + 2a_{0,2} + (6a_{3,0} + 2a_{1,2})x + (2a_{2,1} + 6a_{0,3})y$. This is equivalent to $a_{2,0} = -a_{0,2}$, $a_{1,2} = -3a_{3,0}$ and $a_{2,1} = -3a_{0,3}$.

7. Suppose that $f : D \to \mathbb{R}^m$, where $D \subseteq \mathbb{R}^n$ is convex. Suppose f is differentiable in D and that there is an M so that $\|Df\|_{\mathcal{L}(\mathbb{R}^n, \mathbb{R}^m)} \leq M$ for f restricted to D of course. We will show $\|f(\vec{x}) - f(\vec{y})\| \leq M\|\vec{x} - \vec{y}\|$ holds on D. First we shall prove the following: $g : [a,b] \to \mathbb{R}^k$, and g differentiable on (a,b) and continuous on $[a,b]$, then there is an x in (a,b) so that $\|g(b) - g(a)\| \leq (b-a)\|Dg(x)\|$. Let $\vec{z} = g(b) - g(a)$, and consider the function $h(t) = \vec{z} \cdot g(t)$, for $a \leq t \leq b$. The mean value theorem guarantees the existence of $a < x < b$ so that $h(b) - h(a) = (b-a)h'(x) = (b-a)\vec{z} \cdot Dg(x)$. However, $h(b) - h(a) = \vec{z} \cdot g(b) - \vec{z} \cdot g(a) = \vec{z} \cdot (g(b) - g(a)) = \vec{z} \cdot \vec{z} = \|\vec{z}\|^2$. Thus, $\|\vec{z}\|^2 = (b-a)\vec{z} \cdot \|Dg(x)\| \leq (b-a)\|\vec{z}\| \cdot \|Dg(x)\|$. Hence $\|\vec{z}\| \leq (b-a)\|Dg(x)\|$. Now, let f be as above. Then for any $\vec{x}, \vec{y} \in D$, by convexity $\{(1-t)\vec{x} + t\vec{y} : 0 \leq t \leq 1\} \subseteq D$. We define $g(t) = f((1-t)\vec{x} + t\vec{y})$ for $t \in [0,1]$. Then by the chain rule, $Dg(t) = Df((1-t)\vec{x} + t\vec{y})(\vec{y} - \vec{x})$. Moreover, $\|Dg(t)\| \leq \|Df((1-t)\vec{x} + t\vec{y})\| \cdot \|\vec{y} - \vec{x}\| \leq M\|\vec{y} - \vec{x}\|$. Hence $\|f(\vec{y}) - f(\vec{x})\| = \|g(1) - g(0)\| \leq M\|\vec{y} - \vec{x}\|$.

8. Suppose f is differentiable on an open convex set $D \subseteq \mathbb{R}^n$, and here $Df \equiv 0$. Given any points $\vec{x}, \vec{y} \in D$ for $M = 0$ and the above exercise we know that $\|f(\vec{x}) - f(\vec{y})\| \leq M \cdot \|\vec{x} - \vec{y}\| = 0$. Thus f is a constant

function on D.

9. Let f and g be real-valued differentiable functions sharing a common domain in $\mathbb{R}^n$. We know that $D(f \cdot g)\vec{u}$ may be expressed as $\sum_{k=1}^n \frac{\partial (f\cdot g)}{\partial x_k} u_k$, where $\vec{u} = \sum_{k=1}^n u_k \vec{e}_k$, we will get the result (product rule) if we establish $\frac{\partial (f\cdot g)}{\partial x_k} = \frac{\partial f}{\partial x_k} \cdot g + f \cdot \frac{\partial g}{\partial x_k}$ for any fixed k. Since $\frac{\partial (f\cdot g)(\vec{x}_0)}{\partial x_k} = \frac{d}{dt}|_{t=0}(f\cdot g)(\vec{x}_0 + t\vec{e}_k)$ we get $\frac{\partial}{\partial x_k}(f\cdot g) = \frac{\partial f}{\partial x_k} \cdot g + f \cdot \frac{\partial g}{\partial x_k}$. Likewise, we get the corresponding quotient rule since it is true for partial derivatives. That is, $\frac{\partial (\frac{f}{g})(\vec{x}_0)}{\partial x_k} = \frac{d}{dt}|_{t=0}(\frac{f}{g})(\vec{x}_0 + t\vec{e}_k)$ we get $\frac{\partial}{\partial x_k}(\frac{f}{g}) = \frac{\frac{\partial f}{\partial x_k}\cdot g - f\cdot \frac{\partial g}{\partial x_k}}{g^2}$.

10. Find the local extrema of the following functions, and classify them as relative maximums, minimums, or neither. Since each f is $C^3(\mathbb{R}^2)$, the results of section 8.3.3 apply for each.

    (a) $f(x,y) = (4x^2 + y^2)e^{-x^2-4y^2}$. $f_x = (8x - 8x^3 - 2xy^2)e^{-x^2-4y^2}$, and $f_y = (2y - 8y^3 - 32x^2 y)e^{-x^2-4y^2}$. $f_x, f_y = 0 \iff x(4-4x^2-y^2) = 0$ and $y(1 - 4y^2 - 16x^2) = 0$. The first condition says $x = 0$ or $x^2 + (\frac{y}{2})^2 = 1$. The second condition says $y = 0$ or $x^2 + (\frac{y}{2})^2 = \frac{1}{16}$. This yields $(0,0), (1,0), (-1,0), (0, \frac{1}{2}) (0, -\frac{1}{2})$ are the critical points for f. Now, $f_{xx} = [8 - 24x^2 - 2y^2 + (8 - 8x^3 - 2xy^2)(-2x)]e^{-x^2-4y^2}$, $f_{yy} = [2 - 24y^2 - 32x^2 + (2y - 8y^3 - 32x^2 y)(-8y)]e^{-x^2-4y^2}$ and $f_{xy} = f_{yx} = [-4xy + (8x - 8x^3 - 2xy^2)(-8y)]e^{-x^2-4y^2}$. At $(0,0)$ the Hessian matrix of f has entries $f_{xx} = 8$, $f_{yy} = 2$ on the diagonal, and $f_{xy} = f_{xy} = 0$ on the off-diagonal entries. Thus the Hessian matrix is positive definite at $(0,0)$, telling us that f has a relative minimum at $(0,0)$. In actuality, $f_{xy} = 0$ at all the critical points. At $(0, \pm\frac{1}{2})$ we have that $f_{xx} \approx 2.759095809$ and $f_{yy} \approx -1.471517765$, and so f has neither a relative minimum nor a relative maximum at $(0, \pm\frac{1}{2})$. At $(\pm 1, 0)$, $f_{xx} \approx -43.49250926$ and $f_{yy} \approx -81.54845485$, and so at $(\pm 1, 0)$, the Hessian matrix is negative definite, and so f has a relative maximum at $(\pm 1, 0)$.

    (b) $f(x,y) = x^2 - y^2 + 10$. $f_x = 2x$, $f_y = -2y$. $f_x, f_y = 0$ if and only if $(x,y) = (0,0)$. $f_{xx} = 2$, $f_{yy} = -2$, and $f_{xy} = 0$. The Hessian matrix is neither positive definite or negative definite at $(0,0)$. Hence at $(0,0)$ f has neither a relative minimum nor a relative maximum.

(c) $f(x,y) = x^3 + y^3 - x - y$. $f_x = 3x^2 - 1$, $f_y = 3y^2 - 1$, and so the critical points of f are $(\frac{1}{\sqrt{3}}, \frac{1}{\sqrt{3}})$, $(\frac{-1}{\sqrt{3}}, \frac{1}{\sqrt{3}})$, $(\frac{1}{\sqrt{3}}, \frac{-1}{\sqrt{3}})$ and $(\frac{-1}{\sqrt{3}}, \frac{-1}{\sqrt{3}})$. Moreover, $f_{xx} = 6x$, $f_{yy} = 6y$ and $f_{xy} = 0$. Hence the Hessian matrix is positive definite at $(\frac{1}{\sqrt{3}}, \frac{1}{\sqrt{3}})$, and so here f has a relative minimum. The Hessian matrix is negative definite at $(\frac{-1}{\sqrt{3}}, \frac{-1}{\sqrt{3}})$, and so here f has a relative maximum. At the other two critical values, it has neither a relative maximum nor a relative minimum.

(d) $f(x,y) = \frac{x^4}{32} + x^2y^2 - x - y^2$. $f_x = \frac{x^3}{8} + 2xy^2 - 1$, $f_y = 2y(x^2-1)$. The critical points are $(2,0)$, $(1, \frac{\sqrt{7}}{4})$, $(-1, \frac{\sqrt{7}}{4})$, $(1, \frac{-\sqrt{7}}{4})$ and $(-1, \frac{-\sqrt{7}}{4})$. $f_{xx} = \frac{3}{8}x^2 + 2y^2 \geq 0$, $f_{yy} = 2(x^2-1)$ and $f_{xy} = 4xy$. At $(2,0)$ we have $f_{xx} = \frac{3}{2}$, $f_{yy} = 6$ and $f_{xy} = 0$. So the Hessian matrix is positive definite, and so f has a relative minimum at this critical value. At $\pm(1, \frac{\sqrt{7}}{4})$ we have $f_{xx} = \frac{5}{4}$, $f_{yy} = 0$ and $f_{xy} = \sqrt{7}$. Thus we see that f will have one positive eigenvalue, and one negative eigenvalue, and so the Hessian matrix is neither positive definite nor negative definite. Thus f has neither a relative maximum nor a relative minimum at either of these critical values. At $\pm(-1, \frac{\sqrt{7}}{4})$ we have $f_{xx} = \frac{5}{4}$, $f_{yy} = 0$ and $f_{xy} = -\sqrt{7}$. Thus we see that f will have one positive eigenvalue, and one negative eigenvalue, and so the Hessian matrix is neither positive definite nor negative definite. Thus f has neither a relative maximum nor a relative minimum at either of these critical values.

(e) $f(x,y) = x^4 + x^2y^2 - y$. $f_x = 4x^3 + 2xy^2$, $f_y = 2x^2y - 1$. $f_x = 0$ implies $x = 0$. Using this in f_y, we get $f_y = -1$. Thus there are no critical points for f. Since f is C^1, we know that this implies that f cannot have any relative extrema.

(f) $f(x,y) = \frac{x}{1+x^2+y^2}$. $f_x = \frac{1+y^2-x^2}{(1+x^2+y^2)^2}$, $f_y = \frac{-2xy}{(1+x^2+y^2)^2}$, and so the critical points are $(1,0)$ and $(-1,0)$. $f_{xx} = \frac{-2x(1+x^2+y^2)^2 - 4x(1+y^2-x^2)(1+x^2+y^2)}{(1+x^2+y^2)^4}$, $f_{yy} = \frac{-2x(1+x^2+y^2)^2 + 8xy^2(1+x^2+y^2)}{(1+x^2+y^2)^4}$, and $f_{xy} = \frac{-2y(1+x^2+y^2)^2 + 8x^2y(1+x^2+y^2)}{(1+x^2+y^2)^4}$. $f_{xy}(\pm 1, 0) = 0$, $f_{xx}(1,0) = -\frac{1}{2}$, $f_{xx}(-1,0) = \frac{1}{2}$, $f_{yy}(1,0) = -\frac{1}{2}$ and $f_{yy}(-1,0) = \frac{1}{2}$. Hence the Hessian matrix for f is negative definite at $(1,0)$, and positive definite at $(-1,0)$. Hence f has a relative maximum at $(1,0)$ and a relative minimum at $(-1,0)$.

(g) $f(x,y) = x^4 + y^4 - x^3 = x^3(x-1) + y^4$. $f_x = 4x^3 - 3x^2 = x^2(4x-3)$, $f_y = 4y^3$. So the critical points for f are $(0,0)$ and $(\frac{3}{4}, 0)$. We observe that the function $f|_{y=0}$ is decreasing in a neighbourhood of $x = 0$, and so f has neither a relative maximum nor a relative minimum at $(0,0)$. $f_{xx} = 12x^2 - 6x$, $f_{yy} = 12y^2$ and $f_{xy} = 0$. At $(\frac{3}{4}, 0)$ we have $f_{xx} > 0$, $f_{yy} = 0$ and $f_{xy} = 0$. Consider the function $g(t) = f(\frac{3}{4} + t\cos\theta, t\sin\theta) = (\frac{3}{4} + t\cos\theta)^3(t\cos\theta - \frac{1}{4}) + (t\sin\theta)^4$. We observe that if $\theta = \pm\frac{\pi}{2}$, $g(t) = \frac{-3^3}{4^4} + t^4$, which has a minimum at $t = 0$. Moreover, in general, $g'(t) = 3\cos\theta(\frac{3}{4} + t\cos\theta)^2(t\cos\theta - \frac{1}{4}) + \cos\theta(\frac{3}{4} + t\cos\theta)^3 + 4\sin\theta(t\sin\theta)^3$. $g'(0) = 0$. $g''(t) = 6\cos^2\theta(\frac{3}{4} + t\cos\theta)(t\cos\theta - \frac{1}{4}) + 6(\frac{3}{4} + t\cos\theta)^2\cos^2\theta + 12(t\sin\theta)^2\sin^2\theta$. Thus we see $g''(0) > 0$ for any $\theta \neq \pm\frac{\pi}{2}$. Thus we see that f has a relative minimum at $(\frac{3}{4}, 0)$.

11. Consider $f(x, y) = xy$ subject to $x + y = 20$. Let $f(x, y) = xy$ and $g(x, y) = x + y$. Then $\nabla f = (y, x)$ and $\nabla g = (1, 1)$, and so by the method of Lagrange multipliers, we require $(y, x) = \lambda(1, 1)$. This condition yields $x = y$. Using $g = 20$ yields $x = 10$. Thus $f = 10 \cdot 10 = 100$ is an extremum of f subject to $g = 20$. Since $f|_{g=20}$ can be expressed by $f(x, 20 - x) = x(20 - x)$, we see that this extremum is a maximum.

12. Consider $f(x, y) = x^2 + y^2$ subject to $g(x, y) = xy = 1$. Then, by the method of Lagrange multipliers we consider $\nabla f = (2x, 2y) = \lambda(y, x) = \lambda(y, x)$. From this, we get $y^2 = x^2$, and thus $x = \pm y$. Using $g = xy = 1$, we see that $x = y = 1$ or $x = y = -1$. Both of these yield $f = 2$. Moreover, we see that this is a minimum, since our condition $f|_{g=1}$ can be expressed as $f(x, \frac{1}{x}) = x^2 + \frac{1}{x^2}$, which clearly has no maximum on $\{x : x \neq 0\}$.

B.34 Homework Solutions 8.4.4

1. Let $f(x) = x + 2x^2 \sin\frac{1}{x}$ for $x \neq 0$ and $f(0) = 0$. In the example after the Inverse Function Theorem, we showed that the function $g(x) = \alpha x + x^2 \sin\frac{1}{x}$ for $x \neq 0$, $g(0) = 0$ and $\alpha \in (0, 1)$ cannot be invertible in an neighbourhood of 0. Hence, $g(x)$ is not injective in any neighbourhood of zero. For $\alpha = \frac{1}{2}$, we have $f(x) = 2g(x)$, and so the same is true for $f(x)$. Note that $g'(0) = \alpha$ was also established, and this shows that $f'(0) = 1$. Moreover, for $x \neq 0$ we have $f'(x) = 1 + 4x \sin\frac{1}{x} - 2\cos\frac{1}{x}$ and so $|f'| \leq 7$ on $(-1, 1)$.

2. Consider the system $3x+y-z+u^2 = 0$, $x-y+2z+u = 0$ and $2x+2y-3z+2u = 0$. We may think of this symbolically as $f_1(x,y,z,u) = 0$, $f_2(x,y,z,u) = 0$ and $f_3(x,y,z,u) = 0$. f_1, f_2, f_3 are all C^1 functions since they are polynomials. Moreover, for $(x_0, y_0, z_0, u_0) = (0,0,0,0)$, we have $f_1(x_0, y_0, z_0, u_0) = 0$, $f_2(x_0, y_0, z_0, u_0) = 0$ and $f_3(x_0, y_0, z_0, u_0) = 0$. However, let (x_0, y_0, z_0, u_0) be any vector so that $f_1(x_0, y_0, z_0, u_0) = 0$, $f_2(x_0, y_0, z_0, u_0) = 0$ and $f_3(x_0, y_0, z_0, u_0) = 0$. Writing the system $f_1 = 0$, $f_2 = 0$ and $f_3 = 0$ symbolically as $\vec{f}(x,y,z,u) = \vec{0}$. We see that
$$D_{(x,y,u)} \vec{f}(x_0, y_0, u_0) = \begin{bmatrix} 3 & 1 & 2u_0 \\ 1 & -1 & 1 \\ 2 & 2 & 2 \end{bmatrix}.$$ This matrix's determinant has value $-12 + 8u_0 \neq 0$ unless $u_0 = \frac{3}{2}$, and so this matrix is invertible. So, by the Implicit Function Theorem, there is an open neighbourhood of (x_0, y_0, u_0) and a function $g(x,y,u)$ so that $f_1(x, y, g(x,y,u), u) = 0$, $f_2(x, y, g(x,y,u), u) = 0$ and $f_3(x, y, g(x,y,u), u) = 0$ in this neighbourhood. We see that $D_{(x,z,u)} \vec{f}(x_0, z_0, u_0) = \begin{bmatrix} 3 & -1 & 2u_0 \\ 1 & 2 & 1 \\ 2 & -3 & 2 \end{bmatrix}$. This matrix's determinant has value $21 - 14u_0 \neq 0$ unless $u_0 = \frac{3}{2}$, and so this matrix is invertible. So, by the Implicit Function Theorem, there is an open neighbourhood of (x_0, z_0, u_0) and a function $g(x,z,u)$ so that $f_1(x, g(x,z,u), z, u) = 0$, $f_2(x, g(x,z,u), z, u) = 0$ and $f_3(x, g(x,z,u), z, u) = 0$ in this neighbourhood. We see that $D_{(y,z,u)} \vec{f}(y_0, z_0, u_0) = \begin{bmatrix} 1 & -1 & 2u_0 \\ -1 & 2 & 1 \\ 2 & -3 & 2 \end{bmatrix}$. This matrix's determinant has value $3 - 2u_0 \neq 0$ unless $u_0 = \frac{3}{2}$, and so this matrix is invertible. So, by the Implicit Function Theorem, there is an open neighbourhood of (y_0, z_0, u_0) and a function $g(y, z, u)$ so that $f_1(g(y,z,u), y, z, u) = 0$, $f_2(g(y,z,u), y, z, u) = 0$ and $f_3(g(y,z,u), y, z, u) = 0$ in this neighbourhood. Now, fix any $u \in \mathbb{R}$ and consider system $3x + y - z = -u^2$, $x - y + 2z = -u$ and $2x + 2y - 3z = -2u$. We see that this may be written as $\begin{bmatrix} 3 & 1 & -1 \\ 1 & -1 & 2 \\ 2 & 2 & -3 \end{bmatrix} \cdot \begin{bmatrix} x \\ y \\ z \end{bmatrix} = \begin{bmatrix} -u^2 \\ -u \\ -2u \end{bmatrix}$. This 3 by 3 matrix on the left-hand-side of the equation has

determinant value 0, and so this matrix is not invertible. So the system cannot be solved for a unique x, y, z in terms of any u.

3. Consider $f(x, y, z) = x^2 y + e^x + z$. Then f is C^1 since it is the sum of a polynomial plus the exponential function composed with a polynomial. $f(0, 1, -1) = 0$, $D_x f = 2xy + e^x$, and so $D_x f(0, 1, -1) = 1$. So, by the Implicit Function Theorem, there is a C^1 function $g(y, z)$ defined on an open neighbourhood of $(1, -1)$ so that $f(g(y, z), y, z) = 0$ on this open set. That is $[g(y, z)]^2 \cdot y + e^{g(x,y)} + z = 0$ on this open set. Differentiating with respect to y we obtain $2g \cdot y \cdot D_y g + g^2 + e^g \cdot D_y g = 0$, and so $D_y g = \frac{-g^2}{2gy + e^g}$. Hence $D_y g(1, -1) = 0$. Likewise, $2g \cdot y D_z g + e^g \cdot D_z g + 1 = 0$, and so $D_z g = \frac{-1}{2gy + e^g}$. Hence $D_z g(1, -1) = -1$.

B.35 Homework Solutions 9.1.3

1. Let C denote the Cantor set as discussed earlier in the second chapter. Let $C_0, C_1, \cdots$ be the collection of compact sets which have: $\cdots \subseteq C_k \subseteq \cdots \subseteq C_3 \subseteq C_2 \subseteq C_1 \subseteq C_0$; $C_0 := [0, 1] \subseteq \mathbb{R}$, and define $C_1 = C_0 \setminus (\frac{1}{3}, \frac{2}{3}) = [0, \frac{1}{3}] \cup [\frac{2}{3}, 1]$, $C_2 := [0, \frac{1}{9}] \cup [\frac{2}{9}, \frac{3}{9}] \cup [\frac{6}{9}, \frac{7}{9}] \cup [\frac{8}{9}, 1]$, etc.; C_k is a union of 2^k disjoint closed intervals, each of length $\frac{1}{3^k}$; $C = \bigcap_{k=0}^{\infty} C_k$. Then since C and all of the C_k's are measurable (since they are all compact sets), and $C_k \searrow C$, $\mu(C_k) = (\frac{2}{3})^k$, and we have $\mu(C) = \lim_{k \to \infty} \mu(C_k) = \lim_{k \to \infty} (\frac{2}{3})^k = 0$.

2. Let E be a set with a non-empty interior E°. Then $\exists \vec{x}_0 = (x_1^0, x_2^0, \cdots, x_n^0) \in E$ and $r > 0$ so that $B_r(\vec{x}) = \{\vec{y} : \|\vec{x} - \vec{y}\| < r\} \subseteq E$. Thus $S = [x_1^0 - \frac{r}{\sqrt{n}}, x_1^0 + \frac{r}{\sqrt{n}}] \times [x_2^0 - \frac{r}{\sqrt{n}}, x_2^0 + \frac{r}{\sqrt{n}}] \times \cdots \times [x_n^0 - \frac{r}{\sqrt{n}}, x_n^0 + \frac{r}{\sqrt{n}}] \subseteq E$. Moreover, $0 < (\frac{2r}{\sqrt{n}})^n = \mu(S) = \mu_e(S) < \mu_e(E)$.

3. Suppose $E \subseteq [a, b]$ is so that $\mu(E) = 0$. Then given $x \in [a, b]$, either $x \in E$ or $x \in E^c$. If $x \in E$, then from the above exercise we know that $x \in \partial E$, since E° is empty. Thus, for any $r > 0$ there is an $y \in B_r(x) \cap E^c$. Thus we see that E^c is dense in $[a, b]$.

4. The following is a construction of a two dimensional Cantor set: Consider the unit square $\{(x, y) : 0 \leq x, y \leq 1\}$ and subdivide this set into 9 equal parts and keep only the 4 closed corner squares, removing the remaining regions (in the shape of a cross). Repeat this process with

the remaining 4 closed squares, resulting in a collection of 16 closed squares. Continue along these lines, ad infinitum. We will show that the remaining region is a perfect set, has $\mathbb{R}^2$ Lebesgue measure zero, and is the cross product of the $\mathbb{R}^1$ Cantor set with itself. In the construction of the one-dimensional Cantor set, we have the collection of compact sets $\cdots \subset C_k \subset \cdots \subset C_2 \subset C_1 \subset C_0 = [0,1]$, where C_k is the disjoint union of 2^k closed intervals, each of length $\frac{1}{3^k}$. Then the Cantor set is $C = \cap_{k=0}^\infty C_k$. In the two-dimensional case, we have the sequence $\cdots \subset C_k^2 \subset \cdots \subset C_2^2 \subset C_1^2 \subset C_0^2 = [0,1]^2$. Each C_k^2 is the disjoint union of 4^k closed intervals, each of volume (measure) $\frac{1}{9^k}$. Thus each C_k^2 is measurable, and compact with $\mu(C_k^2) = (\frac{4}{9})^k$. $C_k^2 \searrow \cap_{k=0}^\infty C_k^2$, which is non-empty, compact and measurable. Hence $\mu(\cap_k C_k^2) = \lim_{k \to \infty} \mu(C_k^2) = \lim_{k \to \infty} (\frac{4}{9})^k = 0$. In fact, $\cap_k C_k^2 = C^2$. So every point looks like (x,y), where $x,y \in C$. Moreover, there are $x_n \to x$ and $y_n \to y$, each x_n, y_n in C, and each (x_n, y_n) distinct and never equal to (x,y). Thus we see C^2 is a perfect set.

5. If $\{E_k\}_{k=1}^\infty$ is a collection of measurable sets so that $\sum_{k=1}^\infty \mu(E_k) < \infty$, then we define $\limsup_{k \to \infty} E_k := \cap_{j=1}^\infty (\cup_{k=j}^\infty E_k)$, and $\liminf_{k \to \infty} E_k := \cup_{j=1}^\infty (\cap_{k=j}^\infty E_k)$. We will now show that both are measurable, and have measure zero. Since $\{E_k\}_{k=1}^\infty$ is a countable collection of measurable sets, we know that $I_k = \cap_{j=k}^\infty E_j$ and $U_k = \cup_{j=k}^\infty E_j$ are measurable for any k. Thus the sets $\limsup_{k \to \infty} E_k = \cap_{j=1}^\infty U_j$ and $\liminf_{k \to \infty} E_k = \cup_{j=1}^\infty I_j$ are measurable. Observe that $U_j \searrow \limsup_{k \to \infty} E_k$ and $I_j \nearrow \liminf_{k \to \infty} I_k$. Let $\varepsilon > 0$ be given. Since $\sum \mu(E_k) < \infty$, there exists J so that $j \geq J$ implies $\sum_{k=j}^\infty \mu(E_k) < \varepsilon$. So, for such j we have $\mu(\limsup_{k \to \infty} E_k) \leq \mu(U_j) = \mu(\cup_{k=j}^\infty E_j) \leq \sum_{k=j}^\infty \mu(E_k) < \varepsilon$. Since $\varepsilon > 0$ is arbitrary, we see that $\mu(\limsup_{k \to \infty} E_k) = 0$. Moreover, for any $j \geq J$ we have $\mu(I_j) = \mu(\cap_{k=j}^\infty E_k) \leq \mu(\cup_{k=j}^\infty E_k) < \varepsilon$. Thus $\mu(I_j) < \varepsilon \, \forall j \geq J$, and so we have $\liminf_{k \to \infty} E_k \leq \varepsilon$. Since $\varepsilon > 0$ is arbitrary, we have $\mu(\liminf_{k \to \infty} E_k) = 0$.

6. We will now show that the outer measure of a set is unchanged if in the definition of outer measure we use coverings of the set by parallelepipeds with a fixed orientation (with edges parallel to a fixed set of n linearly independent, pairwise orthogonal vectors), rather than by intervals. Note that the parallelepipeds mentioned here are the images of intervals under an orthogonal transformation of $\mathbb{R}^n$. In this case,

we will refer to such parallelepipeds as rectangles, and we will use the notation I^* for such a rectangle. That is, the edges of I^* are parallel to the fixed collection of linearly independent, pairwise orthogonal vectors mentioned above. For such a rectangle I^*, we define $v^*(I^*)$ to be the products of the lengths of the edges which all share a single vertex on this rectangle. Suppose we create an outer measure μ_e^* for $\mathbb{R}^n$ by $\mu_e^*(E) = \inf\{\sum_k v^*(I_k^*) : E \subseteq \cup_k I_k^*\}$, where the summation and union is over countable collections. We will now show that $\mu_e(E) = \mu_e^*(E)$ for any $E \subseteq \mathbb{R}^n$. Let $\varepsilon > 0$ be given. Given any I^*, we may find another rectangle I_1^* so that $I^* \subseteq (I_1^*)^o$ and $v^*((I_1^*)^o) \leq v^*(I^*) + \varepsilon$. It was shown that any open set can be written as a countable collection of non-overlapping closed cubes. Let $\{I_k\}$ be such a collection of cubes so that $I^* \subseteq (I_1^*)^o = \cup_k I_k$. Since $\{I_k\}$ are non-overlapping, and for any m, $\sum_{k=1}^m v(I_k) \leq v^*(I_1^*)$. Therefore, $\sum_{k=1}^\infty v(I_k) \leq v^*(I_1^*) \leq v^*(I^*) + \varepsilon$. Now, let $E \subseteq \mathbb{R}^n$, and $\varepsilon > 0$. Choose $\{I_k\}_{k=1}^\infty$ so that $E \subseteq \cup_k I_k$ and $\sum_k v(I_k) \leq \mu_e(E) + \frac{\varepsilon}{2}$. For each k, find a collection of rectangles $\{I_{k,j}^*\}_{j=1}^\infty$ so that $I_k \subseteq \cup_j I_{k,j}^*$ and $\sum_j v^*(I_{k,j}^*) \leq v(I_k) + \frac{\varepsilon}{2^{k+1}}$. Thus, $\sum_{k,j} v^*(I_{k,j}^*) \leq \sum_k v(I_k) + \frac{\varepsilon}{2} \leq \mu_e(E) + \varepsilon$. Since $E \subseteq \cup_{k,j} I_{k,j}^*$, we obtain $\mu_e^*(E) \leq \mu_e(E) + \varepsilon$. Since $\varepsilon > 0$ is arbitrary, $\mu_e^*(E) \leq \mu_e(E)$. By symmetry of this argument (or reversing the rolls of the coordinate systems in $\mathbb{R}^n$) we get the reverse inequality. Thus $\mu_e(E) = \mu_e^*(E)$.

7. Let E be a set and define $E + \vec{x} := \{\vec{e} + \vec{x} : \vec{e} \in E\}$. Claim that $E + \vec{x}$ satisfies: $\mu_e(E + \vec{x}) = \mu_e(E)$, and E measurable implies $E + \vec{x}$ is measurable. Let $\varepsilon > 0$ be given. First we observe that if $I = [a_1, b_1] \times [a_2, b_2] \times \cdots \times [a_n, b_n]$, then $I + \vec{x} = [a_1 + x_1, b_1 + x_1] \times [a_2 + x_2, b_2 + x_2] \times \cdots \times [a_n + x_n, b_n + x_n]$, where $\vec{x} = (x_1, x_2, \cdots, x_n)$. $V(I) = \prod_{k=1}^n (b_k - a_k) = \prod_{k=1}^n [(b_k + x_k) - (a_k + x_k)] = V(I + \vec{x})$. Suppose that $\{I_k\}$ is a collection of intervals so that $E \subseteq \cup_k I_k$ and $\sum V(I_k) < \mu_e(E) + \varepsilon$. Since $\{I_k + \vec{x}\}$ satisfies $E + \vec{x} \subseteq \cup_k (I_k + \vec{x})$ we have $\mu_e(E + \vec{x}) \leq \sum_k V(I_k + \vec{x}) = \sum_k V(I_k) < \mu_e(E) + \varepsilon$. Thus $\mu_e(E + \vec{x}) \leq \mu_e(E) + \varepsilon$ for any ε. Hence $\mu_e(E + \vec{x}) \leq \mu_e(E)$. Since this is true for any E and any $\vec{x}$ and $E = (E + \vec{x}) + (-\vec{x})$ we get $\mu_e(E) = \mu_e((E + \vec{x}) + (-\vec{x})) \leq \mu_e(E + \vec{x})$. So we obtain $\mu_e(E + \vec{x}) = \mu_e(E)$. Let $\varepsilon > 0$ be arbitrary. If E is measurable, then $\exists$ open O_ε so that $E \subseteq O_\varepsilon$ and $\mu_e(O_\varepsilon \setminus E) < \varepsilon$. Then $O_\varepsilon + \vec{x}$ is open, and $E + \vec{x} \subseteq O_\varepsilon + \vec{x}$. Also, $\mu_e((O_\varepsilon + \vec{x}) \setminus (E + \varepsilon)) < \varepsilon$. Hence $E + \vec{x}$ is measurable.

8. Let E be a non-measurable subset of $[0,1)$ as developed in the section the Existence of Non-measurable Sets. Let $\{r_k\}_{k=1}^{\infty}$ be an enumeration of the rational numbers in $(0,1)$ and define $E_k = E + r_k$. $\{E_k\}$ are pairwise disjoint. To see this note that if $x \in E$, then $x + q \notin E$ $\forall q \in \mathbb{Q} \setminus \{0\}$. So $(E + r_k) \cap E$ is empty unless $r_k = 0$, and in such a case it is E. By the above exercise have $\mu_e(E_k) = \mu_e(E)$. Clearly $\cup_k E_k \subseteq [0, 2)$. $\mu_e(\cup_k E_k) \leq 2 < \sum_k \mu_e(E_k) = \sum_k \mu_e(E) = \infty$.

B.36 Homework Solutions 9.2.2

1. Suppose f is a real-valued measurable function. Let $c \in \mathbb{R}$ and $\alpha \in \mathbb{R}$. We will now show cf and $f + c$ are measurable functions. If $c = 0$ then $f + c = f$, and so it is measurable. Likewise, cf is identically zero in this case, and so $\{\vec{x} : cf(\vec{x}) > \alpha\}$ is either empty, or the entire domain, and hence is a measurable set. Thus cf is measurable. In $c \neq 0$, then since f is measurable we know $\{\vec{x} : f(\vec{x}) > \alpha - c\}$ and $\{\vec{x} : f(\vec{x}) > \frac{\alpha}{c}\}$ are both measurable sets. Hence $\{\vec{x} : f(\vec{x}) + c > \alpha\}$ and $\{\vec{x} : cf(\vec{x}) > \alpha\}$ are measurable sets. Thus we see that $f + c$ and cf are measurable functions.

2. Let $f = \sum_{k=1}^{n} c_k \chi_{E_k}$ be a simple function. Suppose that each E_k is a measurable set. Let $\alpha \in \mathbb{R}$. Then $\{\vec{x} : f(\vec{x}) > \alpha\} = \cup_{k:c_k > \alpha} E_k$ is either empty or a union of some of the E_j's in $\{E_1, E_2, \cdots, E_n\}$. Either way, this set is measurable. Hence f is a measurable function. Now suppose that f is a measurable function. Then $\{\vec{x} : f(\vec{x}) = c_k\} = E_k$ is measurable for any k. Hence each E_k is a measurable set.

3. Suppose that f and g are measurable functions defined on a measurable set D. Let $M(\vec{x}) = \max\{f(\vec{x}), g(\vec{x})\}$ and $m(\vec{x}) = \min\{f(\vec{x}), g(\vec{x})\}$. Let $\alpha \in \mathbb{R}$. Then $\{\vec{x} : f(\vec{x}) > \alpha\}$ and $\{\vec{x} : g(\vec{x}) > \alpha\}$ are measurable sets. So $\{\vec{x} : M(\vec{x}) > \alpha\} = \{\vec{x} : f(\vec{x}) > \alpha\} \cup \{\vec{x} : g(\vec{x}) > \alpha\}$ is measurable and $\{\vec{x} : m(\vec{x}) > \alpha\} = \{\vec{x} : f(\vec{x}) > \alpha\} \cap \{\vec{x} : g(\vec{x}) > \alpha\}$. Hence we see that M and m are measurable functions.

4. Suppose that f is upper or lower semicontinuous on a measurable set D, we will now show f is measurable. Since f is upper semicontinuous if and only if $-f$ is lower semicontinuous, we only need to consider the case where f is upper semicontinuous. Let $\alpha \in \mathbb{R}$. First, suppose that

f is usc on D. Let $\vec{x}_0$ be a limit point for $\{\vec{x} \in D : f(\vec{x}) \geq \alpha\}$. Then there is a sequence $\{\vec{x}_k\}_{k=1}^{\infty}$, $\vec{x}_k \in D$ for each k, so that $f(\vec{x}_k) \geq \alpha$ for each k. Since f is usc at $\vec{x}_0$ we have $f(\vec{x}_0) \geq \limsup_{k \to \infty} f(\vec{x}_k) \geq \alpha$. Therefore, $\vec{x}_0 \in \{\vec{x} \in D : f(\vec{x}) \geq \alpha\}$. So $\{\vec{x} \in D : f(\vec{x}) \geq \alpha\}$ is relatively closed in D. Conversely, let $\vec{x}_0$ be a limit point of D which is in D. If f is not usc at $\vec{x}_0$, then $f(\vec{x}_0) < \infty$, and $\exists M$ and a sequence $\{\vec{x}_k\}$ so that $f(\vec{x}_0) < M$, $\vec{x}_k \in D$, $\vec{x}_k \to \vec{x}_0$ and $f(\vec{x}_k) \geq M$. Then $\{\vec{x} \in D : f(\vec{x}) \geq M\}$ is not relatively closed, since it does not contain all its limit points. Finally, we will show that f is measurable if and only if $\{\vec{x} \in D : f(\vec{x}) \geq \alpha\}$ is measurable for all α. The one direction of this biconditional was already established in the text. For the other direction, we simply use that $\{\vec{x} : f(\vec{x}) \geq \alpha + \frac{1}{n}\}$ is measurable implies $\cup_{n \in \mathbb{N}} \{\vec{x} : f(\vec{x}) \geq \alpha + \frac{1}{n}\}$ is measurable, but this set is $\{\vec{x} : f(\vec{x}) > \alpha\}$, and it is measurable.

5. Suppose $\{f_k\}$ is a sequence of measurable functions on a set D. We will now show $f_k \to_m f$ on $D \iff \forall \varepsilon > 0 \; \exists N \in \mathbb{N}$ so that $n > N$ implies $\mu(\{\vec{x} \in D : |f_n(\vec{x}) - f(\vec{x})| > \varepsilon\}) < \varepsilon$. Suppose $f_k \to_m f$ on D. Then for all $\varepsilon > 0$ we have $\lim_{n \to \infty} \mu(\{\vec{x} \in D : |f_n(\vec{x}) - f(\vec{x})| > \varepsilon\}) = 0$. Thus $\forall \varepsilon > 0$ and $\forall \eta > 0$, there is an $N \in \mathbb{N}$ so that $n > N$ implies $\mu(\{\vec{x} \in D : |f_n(\vec{x}) - f(\vec{x})| > \varepsilon\}) < \eta$. Taking $\eta = \varepsilon$ we get the one direction of the biconditional. Conversely, suppose $\forall \varepsilon > 0 \; \exists N \in \mathbb{N}$ so that $n > N$ implies $\mu(\{\vec{x} \in D : |f_n(\vec{x}) - f(\vec{x})| > \varepsilon\}) < \varepsilon$. Let $\{\varepsilon_k\}_{k=1}^{\infty}$ be a sequence so that $\varepsilon_k \searrow 0$. Without loss of generality, suppose $\varepsilon_k < \varepsilon$ as well. Then $\forall k$ we have $\exists N_k$ so that $n > N_k$ implies $\mu(\{\vec{x} \in D : |f_n(\vec{x}) - f(\vec{x})| > \varepsilon_k\}) < \varepsilon_k$. Observe that $\{\vec{x} \in D : |f_n(\vec{x}) - f(\vec{x})| > \varepsilon\} \subseteq \{\vec{x} \in D : |f_n(\vec{x}) - f(\vec{x})| > \varepsilon_k\}$, and so for $n > N_k$ we have $\mu(\{\vec{x} \in D : |f_n(\vec{x}) - f(\vec{x})| > \varepsilon\}) < \varepsilon_k$. Hence $\lim_{n \to \infty} \mu(\{\vec{x} \in D : |f_n(\vec{x}) - f(\vec{x})| > \varepsilon\}) = 0$.

6. A necessary and sufficient condition for $\{f_k\}$ to converge in measure on D is: $\forall \varepsilon > 0 \; \exists N$ so that for $n, m > N$ we have $\mu(\{\vec{x} \in D : |f_n(\vec{x}) - f_m(\vec{x})| > \varepsilon\}) < \varepsilon$. Suppose first that $f_k \to_m f$. Then from the above exercise we know that $\exists N$ so that $n > N$ implies $\mu(\{\vec{x} \in D : |f_n(\vec{x}) - f(\vec{x})| > \frac{\varepsilon}{2}\}) < \frac{\varepsilon}{2}$. Also, $\{\vec{x} : |f_k(\vec{x}) - f_j(\vec{x})| > \varepsilon\} \subseteq \{\vec{x} : |f_k(\vec{x}) - f(\vec{x})| > \frac{\varepsilon}{2}\} \cup \{\vec{x} : |f_j(\vec{x}) - f(\vec{x})| > \frac{\varepsilon}{2}\}$. The result follows from these facts. Conversely, suppose $\forall \varepsilon > 0 \; \exists N$ so that for $n, m > N$ we have $\mu(\{\vec{x} \in D : |f_n(\vec{x}) - f_m(\vec{x})| > \varepsilon\}) < \varepsilon$. We will now show $\{f_k\}$

converges in measure. For each $j \in \mathbb{N}$, choose N_j so that $n, m \geq N_j$ implies $\mu(\{\vec{x} : |f_n(\vec{x}) - f_m(\vec{x})| > \frac{1}{2^j}\}) < \frac{1}{2^j}$. Moreover, we may assume that $N_1 < N_2 < N_3 < \cdots$. Let $D_j = \{\vec{x} : |f_n(\vec{x}) - f_m(\vec{x})| > \frac{1}{2^j}\}$. Then for all $\vec{x} \in D \setminus D_j$ we have $|f_{N_{j+1}}(\vec{x}) - f_{N_j}(\vec{x})| \leq \frac{1}{2^j}$, where $\mu(D_j) < \frac{1}{2^j}$. Since $\sum \mu(D_j) < \infty$, from an earlier exercise we have $Z = \limsup_{k \to \infty} D_k = \bigcap_{k=1}^{\infty} \bigcup_{j=k}^{\infty} D_k$ has measure zero. Let $U_k = \bigcup_{j=k}^{\infty} D_j$. Note that for $\vec{x} \notin Z$ we have $\vec{x} \notin U_k$ for k sufficiently large. For such a k and for $j \geq k$ and $\vec{x} \notin U_k$ we have $|f_{N_{j+1}}(\vec{x}) - f_{N_j}(\vec{x})| \leq \frac{1}{2^j}$. Thus $\sum_{j=k}^{\infty}(f_{N_{j+1}}(\vec{x}) - f_{N_j}(\vec{x}))$ converges uniformly outside of U_k. Thus for $\vec{x} \notin Z$ we have $f(\vec{x}) = \lim_{j \to \infty} f_{N_j}(\vec{x}) = f_{N_1}(\vec{x}) + \sum_{j=1}^{\infty}(f_{N_{j+1}}(\vec{x}) - f_{N_j}(\vec{x}))$. We define f to be zero on Z. Hence $f_{N_j} \to f$ pointwise a.e. as $j \to \infty$. It remains to show that $f_j \to_m f$ on D. Observe that for $x \notin Z$ we have $f(\vec{x}) - f_{N_j}(\vec{x}) = \sum_{i=j}^{\infty}(f_{N_{j+1}}(\vec{x}) - f_{N_j}(\vec{x}))$, and so for $x \notin U_j$ we have $|f(\vec{x}) - f_{N_j}(\vec{x})| \leq \sum_{i=j}^{\infty} |f_{N_{j+1}}(\vec{x}) - f_{N_j}(\vec{x})| < \frac{1}{2^{j-1}}$. In particular, $\mu(\{\vec{x} : |f(\vec{x}) - f_{N_j}(\vec{x})| \geq \frac{1}{2^{j-1}}\}) \leq \mu(U_j) \leq \frac{1}{2^{j-1}}$. Hence $\{f_{N_j}\}$ converges in measure to f on D. To see that $f_k \to_m f$ on D, we observe that for a given $\varepsilon > 0$, $\{\vec{x} : |f_n(\vec{x}) - f(\vec{x})| > \varepsilon\} \subseteq \{\vec{x} : |f_k(\vec{x}) - f_{N_j}(\vec{x})| > \frac{\varepsilon}{2}\} \cup \{\vec{x} : |f_{N_j}(\vec{x}) - f(\vec{x})| > \frac{\varepsilon}{2}\}$ for any N_j. We may pick N_j large enough to make the $\mu(\{\vec{x} : |f_{N_j}(\vec{x}) - f(\vec{x})| > \frac{\varepsilon}{2}\}) < \frac{\varepsilon}{2}$. Because $\{f_j\}$ is Cauchy in measure, by picking k sufficiently large (and if necessary N_j still larger) we obtain $\mu(\{\vec{x} : |f_k(\vec{x}) - f_{N_j}(\vec{x})| > \frac{\varepsilon}{2}\}) < \frac{\varepsilon}{2}$.

7. Suppose $f_k \to_m f$ and $g_k \to_m g$ on D. Then given $\varepsilon > 0$ there is an N so that $n \geq N$ implies $\mu(\{\vec{x} : |f_n(\vec{x}) - f(\vec{x})| > \frac{\varepsilon}{2}\}) < \frac{\varepsilon}{2}$ and $\mu(\{\vec{x} : |g_n(\vec{x}) - g(\vec{x})| > \frac{\varepsilon}{2}\}) < \frac{\varepsilon}{2}$. Observe that $\{\vec{x} : |f_k(\vec{x}) + g_k(\vec{x}) - f(\vec{x}) - g(\vec{x})| > \varepsilon\} \subseteq \{\vec{x} : |f_n(\vec{x}) - f(\vec{x})| > \frac{\varepsilon}{2}\} \cup \{\vec{x} : |g_n(\vec{x}) - g(\vec{x})| > \frac{\varepsilon}{2}\}$, and so for $n \geq N$ we have $\mu(\{\vec{x} : |f_k(\vec{x}) + g_k(\vec{x}) - f(\vec{x}) - g(\vec{x})| > \varepsilon\}) \leq \mu(\{\vec{x} : |f_n(\vec{x}) - f(\vec{x})| > \frac{\varepsilon}{2}\}) + \mu(\{\vec{x} : |g_n(\vec{x}) - g(\vec{x})| > \frac{\varepsilon}{2}\}) < \varepsilon$.

8. Suppose f is measurable on a measurable set D with $\mu(D) < \infty$. Suppose also that f is finite a.e. We know that $|f|$ is also measurable. The set $\bigcup_{n=1}^{\infty}\{\vec{x} : n-1 < |f(\vec{x})| \leq n\} \subset D$, and so $\sum_{n=1}^{\infty} \mu(\{\vec{x} : n-1 < |f(\vec{x})| \leq n\}) = \mu(\bigcup_{n=1}^{\infty}\{\vec{x} : n-1 < |f(\vec{x})| \leq n\}) \leq \mu(D) < \infty$. Let $\varepsilon > 0$. Since $\sum_{n=1}^{\infty} \mu(\{\vec{x} : n-1 < |f(\vec{x})| \leq n\})$ converges, we know that given $\varepsilon > 0$, there is an N so that $\sum_{n=N}^{\infty} \mu(\{\vec{x} : n-1 < |f(\vec{x})| \leq n\}) < \varepsilon$. Hence $\mu(\bigcup_{n=N}^{\infty}\{\vec{x} : n-1 < |f(\vec{x})| \leq n\}) < \varepsilon$. That is, $\mu(\{\vec{x} : |f(\vec{x})| > N\}) < \varepsilon$. Let a be any number $\in [N, \infty)$.

9. Suppose that $f_n \to_m f$ and $g_n \to_m g$ on D, with $\mu(D) < \infty$ and f and g

finite a.e. Let $\varepsilon > 0$ be given. From the above exercise, there is an N so that $\mu(\{\vec{x} : |f(\vec{x})| > N\}) + \mu(\{\vec{x} : |g(\vec{x})| > N\}) < \frac{\varepsilon}{2}$. Observe that $\{\vec{x} : |f_n(\vec{x})g_n(\vec{x}) - f(\vec{x})g(\vec{x})| > \varepsilon\} \subseteq \{\vec{x} : |f_n(\vec{x}) - f(\vec{x})| \cdot |g_n(\vec{x})| > \frac{\varepsilon}{2}\} \cup \{\vec{x} : |g_n(\vec{x}) - g(\vec{x})| \cdot |f(\vec{x})| > \frac{\varepsilon}{2}\}$. Let $F = \{\vec{x} : |f(\vec{x})| \leq N\} \cap \{\vec{x} : |g(\vec{x})| \leq N\}$. Then we know $\mu(F^c) < \frac{\varepsilon}{2}$. Moreover, $\mu(\{\vec{x} : |f_n(\vec{x})g_n(\vec{x}) - f(\vec{x})g(\vec{x})| > \varepsilon\}) \leq \mu(F^c) + \mu(\{\vec{x} : |f_n(\vec{x}) - f(\vec{x})| \cdot |g_n(\vec{x})| > \frac{\varepsilon}{2}\} \cap F) + \mu(\{\vec{x} : |g_n(\vec{x}) - g(\vec{x})| \cdot |f(\vec{x})| > \frac{\varepsilon}{2}\} \cap F)$. Since $\{\vec{x} : |g_n(\vec{x}) - g(\vec{x})| \cdot |f(\vec{x})| > \frac{\varepsilon}{2}\} \cap F \subseteq \{\vec{x} : |g_n(\vec{x}) - g(\vec{x})| \cdot N > \frac{\varepsilon}{2}\} \cap F \subseteq \{\vec{x} : |g_n(\vec{x}) - g(\vec{x})| \cdot N > \frac{\varepsilon}{2}\}$, we get $\mu(\{\vec{x} : |f_n(\vec{x})g_n(\vec{x}) - f(\vec{x})g(\vec{x})| > \varepsilon\}) \leq \frac{\varepsilon}{2} + \mu(\{\vec{x} : |f_n(\vec{x}) - f(\vec{x})| \cdot |g_n(\vec{x})| > \frac{\varepsilon}{2}\} \cap F) + \mu(\{\vec{x} : |g_n(\vec{x}) - g(\vec{x})| \cdot N > \frac{\varepsilon}{2}\})$. Also, observe that $\mu(\{\vec{x} : |f_n(\vec{x}) - f(\vec{x})| \cdot |g_n(\vec{x})| > \frac{\varepsilon}{2}\} \cap F) \leq \mu(\{\vec{x} : |f_n(\vec{x}) - f(\vec{x})| \cdot |g_n(\vec{x}) - g(\vec{x})| > \frac{\varepsilon}{2}\} \cap F) + \mu(\{\vec{x} : |f_n(\vec{x}) - f(\vec{x})| \cdot |f(\vec{x})| > \frac{\varepsilon}{2}\} \cap F) \leq \mu(\{\vec{x} : |f_n(\vec{x}) - f(\vec{x})| \geq \sqrt{\frac{\varepsilon}{2}}\}) + \mu(\{\vec{x} : |g_n(\vec{x}) - g(\vec{x})| \geq \sqrt{\frac{\varepsilon}{2}}\}) + \mu(\{\vec{x} : |f_n(\vec{x}) - f(\vec{x})| \geq \frac{1}{N} \cdot \sqrt{\frac{\varepsilon}{2}}\})$. Since $g_n \to_m g$ we have $\mu(\{\vec{x} : |g_n(\vec{x}) - g(\vec{x})| \geq \sqrt{\frac{\varepsilon}{2}}\}) + \mu(\{\vec{x} : |g_n(\vec{x}) - g(\vec{x})| \cdot N > \frac{\varepsilon}{2}\}) = \mu(\{\vec{x} : |g_n(\vec{x}) - g(\vec{x})| \geq \sqrt{\frac{\varepsilon}{2}}\}) + \mu(\{\vec{x} : |g_n(\vec{x}) - g(\vec{x})| > \frac{\varepsilon}{2N}\}) < \frac{\varepsilon}{4}$ if $n > N_1$. Also, since $f_n \to_m f$ we have $\mu(\{\vec{x} : |f_n(\vec{x}) - f(\vec{x})| \geq \sqrt{\frac{\varepsilon}{2}}\}) + \mu(\{\vec{x} : |f_n(\vec{x}) - f(\vec{x})| \geq \frac{1}{N} \cdot \sqrt{\frac{\varepsilon}{2}}\}) < \frac{\varepsilon}{4}$ if $n > N_2$. So, if $n > \max\{N_1, N_2\}$ we have $\mu(\{\vec{x} : |f_n(\vec{x})g_n(\vec{x}) - f(\vec{x})g(\vec{x})| > \varepsilon\}) < \varepsilon$. Hence $f_n \cdot g_n \to_m f \cdot g$ on D.

10. Suppose $\{f_k\}$ is a sequence of measurable functions defined on a measurable set D. For a fixed $\varepsilon > 0$ and any $m, n \in \mathbb{N}$ we have $\{\vec{x} \in D : -\varepsilon < f_m(\vec{x}) - f_n(\vec{x}) < \varepsilon\}$ is measurable. For a fixed $\varepsilon > 0$ and any $n \in \mathbb{N}$ we have $I(\varepsilon, n) := \cap_{m=n+1}^{\infty} \{\vec{x} \in D : -\varepsilon < f_m(\vec{x}) - f_n(\vec{x}) < \varepsilon\}$ is measurable. Hence $U(\varepsilon) := \cup_{n \in \mathbb{N}} I(\varepsilon, n)$ is measurable for any $\varepsilon > 0$. Consider $\cap_{j=1}^{\infty} U(\frac{1}{j})$. This set is measurable. Moreover, this is the set of points $\vec{x}$ for which $\{f_k(\vec{x})\}_{k=1}^{\infty}$ converges.

11. Suppose $f_k \to_m f$. Then $\forall \varepsilon > 0$ we have $\lim_{k \to \infty} \mu(\{\vec{x} : |f_k(\vec{x}) - f(\vec{x})| \geq \varepsilon\}) = 0$. This implies $(f_k - f) \to_m 0$ since the condition is exactly the same. Likewise, if $(f_k - f) \to_m 0$ we see that $\lim_{k \to \infty} \mu(\{\vec{x} : |f_k(\vec{x}) - f(\vec{x})| \geq \varepsilon\}) = 0$, and so $f_k \to_m f$.

12. Let $D = \mathbb{R}$, $f(x) \equiv 0$, $f_n(x) = \frac{\chi_{(0,n)}}{n}$, $g(x) = g_n(x) = x$. Clearly $g_n \to_m g$ on $\mathbb{R}$. Moreover, given $\varepsilon > 0$, $\{x : |f_n(x) - f(x)| > \varepsilon\} = \{x : \chi_{(0,n)}(x) > n\varepsilon\}$. This set is a subset of $(0, n)$ for which $\frac{1}{n} > \varepsilon$. That is $n < \frac{1}{\varepsilon}$. Clearly as $n \to \infty$ this condition cannot be satisfied. That

is $\{x : |f_n(x) - f(x)| > \varepsilon\} = \emptyset$ if $n > \frac{1}{\varepsilon}$. Hence $f_n \to_m f$. Whereas $\{x : |f_n(x)g_n(x) - f(x)g(x)| > \varepsilon\} = \{x : \frac{x \cdot \chi_{(0,n)}(x)}{n} > \varepsilon\} = (0, n) \cap (n\varepsilon, \infty) = (n\varepsilon, n)$. Hence $\mu(\{x : |f_n(x)g_n(x) - f(x)g(x)| > \varepsilon\}) = n - n\varepsilon = n(1 - \varepsilon) \to \infty$ as $n \to \infty$ if $0 < \varepsilon < 1$. So $f_n g_n \not\to_m fg$ on $\mathbb{R}$.

13. Suppose $f_n \to_m f$. Observe that $\{\vec{x} \in D : ||f_n(\vec{x})| - |f(\vec{x})|| \geq \varepsilon\} \subseteq \{\vec{x} \in D : |f_n(\vec{x}) - f(\vec{x})| \geq \varepsilon\}$. Thus $\mu(\{\vec{x} \in D : ||f_n(\vec{x})| - |f(\vec{x})|| \geq \varepsilon\}) \leq \mu(\{\vec{x} \in D : |f_n(\vec{x}) - f(\vec{x})| \geq \varepsilon\})$. Hence $\lim_{n \to \infty} \mu(\{\vec{x} \in D : |f_n(\vec{x}) - f(\vec{x})| \geq \varepsilon\}) = 0$ implies $\lim_{n \to \infty} \mu(\{\vec{x} \in D : ||f_n(\vec{x})| - |f(\vec{x})|| > \varepsilon\}) = 0$. Thus $|f_n| \to_m |f|$.

14. Suppose $f_n \to_m f$ then $\forall \varepsilon > 0$ we have $\lim_{n \to \infty} \mu(\{\vec{x} : |f_n(\vec{x}) - f(\vec{x})| \geq \varepsilon\}) = 0$. Thus for any $n_1 < n_2 < \cdots < n_k < \cdots$, we have $\lim_{k \to \infty} \mu(\{\vec{x} : |f_{n_k}(\vec{x}) - f(\vec{x})| \geq \varepsilon\}) = 0$. Hence $f_{n_k} \to_m f$.

B.37 Homework Solutions 9.3.4

1. Suppose $g(x) = 0$ if $0 \leq x \leq \frac{1}{2}$, and $g(x) = 1$ if $\frac{1}{2} < x \leq 1$. Define $f_{2k}(x) = g(x)$ and $f_{2k+1} = g(1-x)$. $\int_0^1 g(x)dx = \int_{\frac{1}{2}}^1 1 dx = \frac{1}{2}$. $\int_0^1 g(1-x)dx = \int_0^{\frac{1}{2}} 1 dx = \frac{1}{2}$. So $\int_0^1 f_k dx = \frac{1}{2}$, which implies $\lim_{k \to \infty} \int_0^1 f_k dx = \frac{1}{2}$. For any $x \in [0, 1]$ we have $\{f_k(x)\}$ is an alternating sequence of 0's and 1's. Thus $\liminf_{k \to \infty} f_k(x) = 0$ for every x.

2. Suppose that f is Riemann integrable on $[a, b]$, and define $F(x) = \int_{[a,x]} f d\mu$. Since f is also Riemann integrable on $[a, x]$ for every $a \leq x \leq b$, we have $\int_{[a,x]} f d\mu = \int_a^x f(t) dt$. Since f is Riemann integrable on $[a, b]$ we know that f is continuous a.e. on $[a, b]$. Let $x \in [a, b]$ be a point where f is continuous. Then by the fundamental theorem of calculus we know that $F(x) = \int_a^x f(t) dt$ is differentiable at x, and $F'(x) = f(x)$.

3. Let $f_n(x) = \frac{1}{2n}$ for $-n \leq x \leq n$, and let f equal zero otherwise. Then $f_n \to f$, $f \equiv 0$ pointwise, and in fact uniformly pointwise, since $|f_n(x)| \leq \frac{1}{2n} \to 0$ as $n \to \infty$. However, $\int_{\mathbb{R}} f_n dx = \int_{-n}^n \frac{1}{2n} dx = 1 \to 1$ as $n \to \infty$

4. Suppose that $f(x) = \frac{1}{1+|x|}$. $f \geq 0$, and f is even and continuous on $\mathbb{R}$. Let $g_n(x) = f(x) \cdot \chi_{[-n,n]}(x)$. Then $g_n \nearrow f$ as $n \to \infty$, and $g_n(x) \leq f(x)$

everywhere. By the monotone convergence theorem, $\int_{\mathbb{R}} g_n d\mu \to \int_{\mathbb{R}} f d\mu$. $\int_{\mathbb{R}} g_n d\mu = \int_{-n}^{n} f dx = 2\int_{0}^{n} f dx = 2\int_{0}^{n} \frac{1}{1+x} dx \geq 2\int_{1}^{n} \frac{1}{1+x} dx \geq \int_{1}^{n} \frac{1}{x} dx = \log(n) \to \infty$ as $n \to \infty$. Thus $\infty = \lim_{n \to \infty} \int_{\mathbb{R}} g_n dx = \int_{\mathbb{R}} f dx$. Thus we see that $f \notin L^1(\mathbb{R})$. Next we consider $h = f^2$. h is even, positive and continuous on $\mathbb{R}$. Define $h_n = h \cdot \chi_{[-n,n]}$. Then $h_n \leq h$ and $h_n \nearrow h$ as $n \to \infty$. So by the monotone convergence theorem $\int_{\mathbb{R}} h_n d\mu \to \int_{\mathbb{R}} h d\mu$. $\int_{\mathbb{R}} h_n d\mu = \int_{-n}^{n} \frac{1}{1+2|x|+x^2} dx = 2\int_{0}^{n} \frac{1}{1+2x+x^2} dx \leq 2\int_{0}^{1} 1 dx + 2\int_{1}^{n} \frac{1}{x^2} dx = 2 + 2(1 - \frac{1}{n}) \leq 4$. Thus $\{\int_{\mathbb{R}} h_n d\mu\}$ is a bounded monotone sequence, and so it has a finite limit. Thus we see $f \in L^2(\mathbb{R})$.

5. Let D be a measurable set of positive measure. Then for $f, g \in L^1(D)$, and $\alpha \in \mathbb{R}$ we have $\|f\|_{L^1(D)} = 0$ if and only if $|f| = 0$ a.e. on D. That is, $f = 0$ a.e. on D. $\|\alpha f\|_{L^1(D)} = \int_{D} |\alpha f| d\mu = |\alpha| \int_{D} |f| d\mu = |\alpha| \cdot \|f\|_{L^1(D)}$. $\|f + g\|_{L^1(D)} = \int_{D} |f + g| d\mu \leq \int_{D} |f| + |g| d\mu = \int_{D} |f| d\mu + \int_{D} |g| d\mu = \|f\|_{L^1(D)} + \|g\|_{L^1(D)}$. Suppose $\{f_k\}$ is a Cauchy sequence in $L^1(D)$. Then $\mu(\{\vec{x} \in D : |f_k(\vec{x}) - f_j(\vec{x})| > \varepsilon\}) \leq \frac{1}{\varepsilon} \int_{D} |f_k - f_j| d\mu$. Hence $\{f_k\}$ is a Cauchy sequence in measure. Thus, there is a subsequence $\{f_{k_n}\}$ and a function f so that $f_{k_n} \to f$ a.e. in D. Given $\varepsilon > 0$, there is an N so that $k, k_n > N$ implies $\int_{D} |f_{k_n} - f_k| d\mu < \varepsilon$. Fixing k, and letting $n \to \infty$, by Fatou's lemma we get $\int_{D} |f - f_k| d\mu = \int_{d} \liminf_{n \to \infty} |f_{k_n} - f_k| d\mu \leq \liminf_{n \to \infty} \int_{D} |f_{n_k} - f_k| d\mu < \varepsilon$. Thus $\|f - f_k\|_{L^1(D)} \to 0$ as $k \to \infty$. Moreover, $\|f\|_{L^1(D)} \leq \|f - f_k\|_{L^1(D)} + \|f_k\|_{L^1(D)}$, and so $f \in L^1(D)$.

6. Consider $C := \{\sin nx : n \in \mathbb{Z}\}$ as a set in $L^2([-\pi, \pi])$. $\|s \in nx\|^2_{L^2([-\pi,\pi])} = \sqrt{\pi}$. $\|\sin nx - \sin mx\|_{L^2([-\pi,\pi])} = \sqrt{2\pi}$ if $n \neq m$. Clearly C is bounded. Moreover, the distance between two distinct elements in C is $\sqrt{2\pi}$. Hence C has no limit points. Thus C is closed and bounded. However, $\{B_{\frac{1}{3}\sqrt{\pi}}(\sin nx) : n \in \mathbb{Z}\}$ is an open cover of C. This cannot have a finite subcover, since $B_{\frac{1}{3}\sqrt{\pi}}(\sin nx) \cap C = \{\sin nx\}$.

7. Suppose that $\int_{D} |f_k - f|^p d\mu \to 0$ as $k \to \infty$. We will now show $f_k \to_m f$, and so that there is a subsequence $\{f_{n_k}\}$ so that $f_{n_k} \to f$ a.e. Since we are assuming $\int_{D} |f_k|^p d\mu < \infty$, it must be that f_k is finite a.e. Suppose $f_k \not\to_m f$. Thus, there exists $\varepsilon > 0$ so that $\lim_{k \to \infty} \mu(\{\vec{x} \in D : |f_k(\vec{x}) - f(\vec{x})| \geq \varepsilon\}) \neq 0$. So, $\limsup_{k \to \infty} \mu(\{\vec{x} \in D : |f_k(\vec{x}) - f(\vec{x})| \geq \varepsilon\}) = \rho > 0$. Therefore, for any $k \in \mathbb{N}$ $\sup_{j \geq k} \mu(\{\vec{x} \in D : |f_j(\vec{x}) - f(\vec{x})| \geq \varepsilon\}) \geq \rho$. Let $0 < \eta < \rho$. Since $\int_{D} |f_k - f|^p d\mu \to 0$,

$\exists N$ so that $\forall n \geq N$ we have $\int_D |f_n - f|^p d\mu < \varepsilon^p(\rho - \eta)$. Moreover, $\sup_{j \geq N} \mu(\{\vec{x} \in D : |f_j(\vec{x}) - f(\vec{x})| \geq \varepsilon\}) \geq \rho$, so there is $n \geq N$ so that $\mu(\{\vec{x} \in D : |f_n(\vec{x}) - f(\vec{x})| \geq \varepsilon\}) \geq \rho - \eta$. Let $F_n = \{\vec{x} \in D : |f_n(\vec{x}) - f(\vec{x})| \geq \varepsilon\}$. $\forall \vec{x} \in F_n$ we know $|f_n(\vec{x}) - f(\vec{x})|^p \geq \varepsilon^p$, and so $\varepsilon^p(\rho - \eta) > \int_D |f_n - f|^p d\mu \geq \int_{F_n} |f_n - f|^p d\mu \geq \varepsilon^p \mu(F_n) \geq \varepsilon^p(\rho - \eta)$. This yields a contradiction. Hence $f_n \to_m f$. Hence, there is a subsequence which converges to f a.e.

8. Suppose $f_k \in L^p(D)$, $\int_D |f_k - f|^p d\mu \to 0$ and $\int_D |f_k|^p d\mu \leq M$ for every k. From the above exercise, we know that there is a subsequence $\{f_{n_k}\}$ so that $f_{n_k} \to f$ a.e. Thus $|f_{n_k}|^p \to |f|^p$ pointwise a.e. By Fatou's lemma, $\int_D \liminf_{k \to \infty} |f_{n_k}|^p d\mu \leq \liminf_{k \to \infty} \int_D |f_{n_k}|^p d\mu \leq M$. Since $\liminf |f_{n_k}|^p = f$ a.e. we have $\int_D |f|^p d\mu \leq M$.

9. Suppose f is measurable on D and for every measurable $A \subset D$ we have $\int_A f d\mu = 0$. Let $\varepsilon > 0$. Let $A_\varepsilon = \{\vec{x} \in D : f(\vec{x}) \geq \varepsilon\}$ and $B_\varepsilon = \{\vec{x} \in D : f(\vec{x}) \leq -\varepsilon\}$. Then $0 = \int_{A_\varepsilon} d\mu \geq \varepsilon \mu(A_\varepsilon)$ and so $\mu(A_\varepsilon) = 0$. Since this is true for any ε, we have $\mu(\{\vec{x} \in D : f(\vec{x}) > 0\}) = 0$. (To see this use $\{f > 0\} = \cup_{n \in \mathbb{N}} \{f \geq \frac{1}{n}\}$). Likewise, $0 = \int_{B_\varepsilon} d\mu \leq -\varepsilon \mu(A_\varepsilon)$ and so $\mu(B_\varepsilon) = 0$. Since this is true for any ε, we have $\mu(\{\vec{x} \in D : f(\vec{x}) < 0\}) = 0$. Hence $f = 0$ a.e. on D.

10. Let $a, b \in \mathbb{R}$. $(|a| - |b|)^2 \geq 0$, and so $a^2 - 2|a| \cdot |b| + b^2 \geq 0$. Hence $2|a| \cdot |b| \leq a^2 + b^2$. Thus $a \cdot b \leq |a| \cdot |b| \leq \frac{a^2}{2} + \frac{b^2}{2}$. Likewise, given any $\varepsilon > 0$, and A, B any real numbers, letting $a = A \cdot \sqrt{2\varepsilon}$ and $b = \frac{B}{\sqrt{2\varepsilon}}$, we get $AB = ab \leq \frac{a^2}{2} + \frac{b^2}{2} = \varepsilon A^2 + \frac{B^2}{4\varepsilon}$.

11. Let $f(x) = e^x$. Since f is convex, we know $e^{\frac{x}{p} + \frac{y}{q}} \leq \frac{e^x}{p} + \frac{e^y}{q}$, where $\frac{1}{p} + \frac{1}{q} = 1$. Let $a, b > 0$. Then $ab = e^{\log(ab)} = e^{\log(a) + \log(b)} = e^{\frac{\log(a^p)}{p} + \frac{\log(b^q)}{q}} \leq \frac{e^{\log(a^p)}}{p} + \frac{e^{\log(b^q)}}{q} = \frac{a^p}{p} + \frac{b^q}{q}$.

12. Let $a, b > 0$ and $\varepsilon > 0$. Let $A = (p\varepsilon)^{\frac{1}{p}} a$ and $B = \frac{b}{(p\varepsilon)^{\frac{1}{p}}}$. Then $ab = AB \leq \frac{A^p}{p} + \frac{B^q}{q} = \varepsilon \cdot a^p + \frac{b^q}{(p\varepsilon)^{\frac{q}{p}} q}$.

13. Let $1 < p, q < \infty$ be so that $\frac{1}{p} + \frac{1}{q} = 1$. Suppose f, g are measurable functions so that $\|f\|_{L^p(D)} = \|g\|_{L^q(D)} = 1$. Then by Young's inequality we have $\int_D |fg| d\mu \leq \int_D (\frac{|f|^p}{p} + \frac{|f|^q}{q}) d\mu = \frac{\|f\|_{L^p(D)}^p}{p} + \frac{\|g\|_{L^q(D)}^q}{q} = \frac{1}{p} + \frac{1}{q} = 1 =$

$\|f\|_{L^p(D)} \|g\|_{L^q(D)}$. In the general case, we may assume neither $\|f\|_{L^p(D)}$ nor $\|g\|_{L^q(D)}$ is zero, since otherwise either f or g is zero a.e., and so fg is zero a.e., and the result follows. We may also assume that neither $\|f\|_{L^p(D)}$ nor $\|g\|_{L^q(D)}$ is infinite. Because if so, we have $\int_D |fg| d\mu \leq \infty$ already holds, and hence we are done. Hence we may assume $\|f\|_{L^p(D)}$ and $\|g\|_{L^q(D)}$ are positive and finite. For $\alpha \geq 0$ we have $\|\alpha f\|_{L^p(D)} = \alpha \|f\|_{L^p(D)}$. We consider $F = \frac{1}{\|f\|_{L^p(D)}} f$ and $G = \frac{1}{\|g\|_{L^q(D)}} g$. Then $\|F\|_{L^p(D)} = 1 = \|G\|_{L^q(D)}$, and so $\int_D |FG| d\mu \leq \|F\|_{L^p(D)} \|G\|_{L^q(D)} = 1$. Thus $\frac{1}{\|f\|_{L^p(D)} \|g\|_{L^q(D)}} \int_D |fg| d\mu = \int_D |FG| d\mu \leq 1$, and so the result follows.

14. Let $1 < p < \infty$. Then by Hölder's inequality, $\|f+g\|_{L^p(D)}^p = \int_D |f+g|^p d\mu = \int_D |f+g|^{p-1} |f+g| d\mu \leq \int_D |f+g|^{p-1} (|f|+|g|) d\mu = \int_D |f+g|^{p-1} |f| d\mu + \int_D |f+g|^{p-1} |g| d\mu \leq (\int_D (|f+g|^{p-1})^{\frac{p}{p-1}} d\mu)^{\frac{p-1}{p}} \cdot (\int_D |f|^p d\mu)^{\frac{1}{p}} + (\int_D (|f+g|^{p-1})^{\frac{p}{p-1}} d\mu)^{\frac{p-1}{p}} \cdot (\int_D |g|^p d\mu)^{\frac{1}{p}} = (\int_D |f+g|^p d\mu)^{\frac{p-1}{p}} \cdot (\|f\|_{L^p(D)} + \|g\|_{L^p(D)}) = \|f+g\|_{L^p(D)}^{p-1} \cdot (\|f\|_{L^p(D)} + \|g\|_{L^p(D)})$.

15. Let $G(f; p) = (\frac{1}{\mu(D)} \int_D |f|^p d\mu)^{\frac{1}{p}}$. Let $1 \leq p < q < 1$. By Minkowski's inequality, $G(f+g; p) = (\frac{1}{\mu(D)})^{\frac{1}{p}} \|f+g\|_{L^p(D)} \leq (\frac{1}{\mu(D)})^{\frac{1}{p}} \|f\|_{L^p(D)} + (\frac{1}{\mu(D)})^{\frac{1}{p}} \|g\|_{L^p(D)} = G(f; p) + G(g; p)$ ($p = 1$ case we simply use $|f+g| \leq |pf| + |g|$ and linearity properties of the integral). Let $r = \frac{q}{p} > 1$ and $s = \frac{q}{q-p}$. By Hölder's inequality, we have $G(f; p)^p = \int_D |f|^p d\mu \leq \frac{1}{\mu(D)} (\int_D |f|^p \cdot r d\mu)^{\frac{1}{r}} \cdot (\int_D 1 d\mu)^{\frac{1}{s}} = \frac{1}{\mu(D)} (\int_D |f|^q d\mu)^{\frac{p}{q}} \mu(D)^{\frac{q-p}{q}} = (\int_D |f|^q d\mu)^{\frac{p}{q}} \mu(D)^{-\frac{p}{q}}$. Hence $G(f; p) \leq (\frac{1}{\mu(D)} \int_D |f|^q d\mu)^{\frac{1}{q}} = G(f; q)$.

16. Suppose $f, h \in L^1(D)$. Suppose $f \leq g \leq h$ holds as well, and g is measurable. Then $f = f^+ - f^-$, $g = g^+ - g^-$, $h = h^+ - h^-$, $|f| = f^+ + f^-$, $|g| = g^+ + g^-$ and $|h| = h^+ + h^-$. Observe that $g^+ \leq h^+$ and so $\int_D g^+ d\mu \leq \int_D h^+ d\mu \leq \|h\|_{L^1(D)}$. Also, $-f^- \leq -g^-$ and so $g^- \leq f^-$. Hence $\int_D g^- d\mu \leq \int_D f^- d\mu \leq \|f\|_{L^1(D)}$. Thus $\|g\|_{L^1(D)} = \int_D g^+ d\mu + \int_D g^- d\mu \leq \|f\|_{L^1(D)} + \|h\|_{L^1(D)}$.

17. Suppose $f \in L^1(D)$, and $\{f_n\}$ is a sequence such that $f_n \in L^1(D)$ for each n. Suppose for each $\vec{x} \in D$ $f_n(\vec{x}) \nearrow f(\vec{x})$. Then $f_1 \leq f_n \leq f$. Hence $|f_n| \leq |f| + |f_1|$, and so, by the dominated convergence theorem, $\int_D |f_n| d\mu \to \int_D |f| d\mu$.

18. Let $D = [0,1]$ and $f_n = n \cdot \chi_{[0,\frac{1}{n}]}$. Then $\int_D f_n d\mu = 1$ $\forall n$. Moreover $f_n(x) \to 0$ for all $x \in (0,1]$. Thus $f_n \to 0$ a.e. on $[0,1]$.

19. Let $f_{k+2^n} = \chi_{[\frac{k}{2^n}, \frac{k+1}{2^n}]}$ for $n \in \{0\} \cup \mathbb{N}$ and $k \in \{0, 1, 2, \cdots, 2^n - 1\}$. Then $\int_D f_{k+2^n} d\mu = \frac{1}{2^n} \to 0$. However, $f_j \not\to 0$ a.e.

20. Suppose $f, f_1, f_2, \cdots$ are Lebesgue integrable, and suppose $\int_D |f_n - f| d\mu \to 0$. Then $|\int_D f_n d\mu - \int_D f d\mu| = |\int_D (f_n - f) d\mu| \leq \int_D |f_n - f| d\mu \to 0$. Also, $|\int_D |f_n| d\mu - \int_D |f| d\mu| = |\int_D |f_n| - |f| d\mu| \leq \int_D ||f_n| - |f|| d\mu \leq \int_D |f_n - f| d\mu \to 0$.

21. Suppose f, and $\{f_n\}$ are bounded by M on $[a,b]$. Suppose $f_n \to f$ pointwise on $[a,b]$. Let $\varepsilon > 0$. By Egorov's theorem there is $F \subseteq [a,b]$ such that $\mu(F) < \frac{\varepsilon}{4M}$ and $f_n \to f$ uniformly on $[a,b] \setminus F$. Then. $\exists N$ so that $n \geq N$ and $\vec{x} \in [a,b] \setminus F$ we have $|f_n(\vec{x}) - f(\vec{x})| < \frac{\varepsilon}{2(b-a)}$. Then for $n \geq N$ we have $\int_{[a,b]} |f_n - f| d\mu = \int_{[a,b]\setminus F} |f_n - f| d\mu + \int_F |f_n - f| d\mu$
$\leq \mu([a,b] \setminus F) \cdot \frac{\varepsilon}{2(b-a)} + 2M\mu(F) < \varepsilon$.

22. Suppose $\int_D |f| d\mu < \infty$. Then we know f is finite a.e. In particular, $D = (\cup_{n=0}^\infty \{\vec{x} : n \leq |f(\vec{x})| < n+1\}) \cup Z$, where $Z = \{\vec{x} : |f(\vec{x})| = \infty\}$, and $\mu(Z) = 0$. So $\sum_{n=0}^\infty \int_{\{\vec{x}: n \leq |f(\vec{x})| < n+1\}} |f| d\mu = \int_D |f| d\mu < \infty$. Hence, given $\varepsilon > 0$ there is an N so that $k \geq N$ has $\sum_{n=k}^\infty \int_{\{\vec{x}: n \leq |f(\vec{x})| < n+1\}} |f| d\mu < \frac{\varepsilon}{2}$. Now, $\int_{\{\vec{x}: |f(\vec{x})| \geq k\}} |f| d\mu = \sum_{n=k}^\infty \int_{\{\vec{x}: n \leq |f(\vec{x})| < n+1\}} |f| d\mu \leq \frac{\varepsilon}{2}$. Let $h = f\chi_{\{\vec{x}: |f(\vec{x})| < k\}}$. Then $\int_D |f - h| d\mu = \int_{\{\vec{x}: |f(\vec{x})| \geq k\}} |f| d\mu < \frac{\varepsilon}{2}$. Moreover, $\int_A |h| d\mu \leq k\mu(A)$. Suppose $A \subseteq D$ is so that $\mu(A) < \frac{\varepsilon}{2k}$. Then $\int_A |f| d\mu \leq \int_A |f - h| d\mu + \int_A |h| d\mu \leq \int_D |f - h| d\mu + k\mu(A) < \frac{\varepsilon}{2} + k\frac{\varepsilon}{2k} = \varepsilon$. So $\delta = \frac{\varepsilon}{2k}$ will do.

23. Suppose $f \in L^1(\mathbb{R})$. Let $F(x) = \int_{(-\infty, x]} f\mu$. Then, by the above exercise, given $\varepsilon > 0$ there is a $\delta > 0$ so that for measurable $D \subseteq \mathbb{R}$ with $\mu(D) < \delta$ we have $\int_D f d\mu < \varepsilon$. Then for $y \in [x - \delta, x]$ we have $0 \geq F(x) - F(y) = \int_{(y, x]} f\mu < \varepsilon$. Likewise, for $y \in [x, x + \delta]$ we have $0 \leq F(y) - F(x) = \int_{(x, y]} f d\mu < \varepsilon$.

24. Consider $f_n = \frac{\chi_{[0,n]}}{n}$ on $\mathbb{R}$. Then $f_n \to 0$ uniformly on $\mathbb{R}$, and $\int_\mathbb{R} f_n = 1$ for every n.

25. Let $1 \leq p < q < \infty$, and consider $\alpha \in (0,1)$. Let $r = \alpha p + (1-\alpha)q$. Let $u = \frac{1}{\alpha} > 1$ and $v = \frac{1}{1-\alpha} > 1$. Then by Hölder's inequality,

$\|f\|_{L^r(D)} = \int_D |f|^{\alpha p}|f|^{(1-\alpha)q}d\mu \leq (\int_D (|f|^{\alpha p})^u d\mu)^{\frac{1}{u}}(\int_D (|f|^{(1-\alpha)q})^v d\mu)^{\frac{1}{v}}$
$= (\int_D |f|^p d\mu)^{\alpha}(\int_D |f|^q d\mu)^{1-\alpha} = \|f\|_{L^p(D)}^{\alpha p}\|f\|_{L^q(D)}^{(1-\alpha)q}$. The case where $p = q$ or $\alpha = 0, 1$ is trivial, since if $p = q$ then $r = p = q$ as well. If $\alpha = 0$, then $r = q$, and if $\alpha = 1$, then $r = p$. In all these special cases, $\|f\|_{L^r(D)}^r = \|f\|_{L^p(D)}^{\alpha p}\|f\|_{L^q(D)}^{(1-\alpha)q}$.

26. Consider first $p = 1$. Take $D = [0, 1]$ and $f = g = \frac{1}{\sqrt{x}}$. Here we have $\int_0^1 f dx = 2$, and so f, g are in $L^1(D)$. However, $\int_0^1 fg dx = \int_0^1 \frac{1}{x}dx = \infty$. Thus $fg \notin L^1(D)$. In the general case, for $p > 1$, consider $f = g = \frac{1}{x^{\frac{p}{2}}}$.

27. Consider $f_n = \frac{\sin\frac{x}{n}}{(1+\frac{x}{n})^n}$ on $[0, \infty)$. For a fixed $x \geq 0$, consider $y(n) = (1 + \frac{x}{n})^n$, for $n \in (1, \infty)$. Then $\log y(n) = n\log(1 + \frac{x}{n})$. $\frac{d}{dn}\log y(n) = \log(1 + \frac{x}{n}) + \frac{1}{1+\frac{x}{n}} > 0$. Hence $y(n)$ is an increasing function of n for each x fixed. Thus we see that for $n \geq 2$ we have $|f_n| \leq \frac{1}{(1+\frac{x}{2})^2}$. Moreover, $\int_{[0,\infty)} \frac{1}{(1+\frac{x}{2})^2} d\mu \leq \int_{[0,1]} 1 d\mu + \int_{[1,\infty)} \frac{4}{x^2} d\mu < \infty$. Hence, by the dominated convergence theorem, we have $\lim_{n\to\infty} \int_{[0,\infty)} f_n d\mu = \int_{[0,\infty)} \lim_{n\to\infty} f_n d\mu$. However, $\lim_{n\to\infty} f_n = \frac{\sin 0}{e^x} = 0$. Thus $\lim_{n\to\infty} \int_{[0,\infty)} f_n d\mu = 0$.

B.38 Homework Solutions 9.4.2

1. Consider $I = \int_{-\infty}^{\infty} e^{-x^2} dx$. Since $e^{-x^2-y^2}$ is non-negative, we know that $I^2 = \int_{-\infty}^{\infty} e^{-x^2} dx \cdot \int_{-\infty}^{\infty} e^{-y^2} dy = \int_{-\infty}^{\infty}[\int_{-\infty}^{\infty} e^{-x^2-y^2} dy]dx = \int_{\mathbb{R}^2} e^{-x^2-y^2} d\mu$. However, $\int_{-\infty}^{\infty}[\int_{-\infty}^{\infty} e^{-x^2-y^2} dy]dx = \int_0^{\infty}(\int_0^{2\pi} e^{-r^2} r dr)d\theta = \int_0^{\infty} 2\pi e^{-r^2} r dr$. By a substitution $u = -r^2$, $du = -2r dr$, we have $\int_0^{\infty} 2\pi e^{-r^2} r dr = \int_{-\infty}^{0} \pi e^u du = \pi$. Hence $I = \sqrt{\pi}$.

2. Let $V_n = \mu(\{\vec{x} \in \mathbb{R}^n : \|\vec{x}\| \leq 1\})$. We will now prove the following recursion formula $V_n = 2V_{n-1}\int_0^1 (1-t^2)^{\frac{n-1}{2}} dt$ by induction on n. $V_1 = \int_{[-1,1]} d\mu = \int_0^1 (1-t^2)^0 dt = 2$. $V_2 = 4\int_{[0,1]} \sqrt{1-x^2} dx = 2V_1 \int_{[0,1]} \sqrt{1-t^2} dt = \pi$. Suppose that for some $n \in \mathbb{N} \setminus \{1, 2\}$ we have $V_{n-1} = 2V_{n-2}\int_0^1 (1-t^2)^{\frac{n-1}{2}} dt$. However, by Fubini's theorem, $V_n = \int_{B_1(\vec{0})} d\mu =$

$\int_{-1}^{1}[\int_{-\sqrt{1-x_1^2}}^{\sqrt{1-x_1^2}}[\int_{-\sqrt{1-x_1^2-x_2^2}}^{\sqrt{1-x_1^2-x_2^2}}[\cdots[\int_{-\sqrt{1-x_1^2-x_2^2-\cdots-x_{n-2}^2-x_{n-1}^2}}^{\sqrt{1-x_1^2-x_2^2-\cdots-x_{n-2}^2-x_{n-1}^2}} 1 dx_n]\cdots]dx_3]dx_2]dx_1$.

Let $y_j = \frac{x_j}{\sqrt{1-x_1^2}}$, for $j = 2, 3, \cdots, n$, and $y_1 = x_1$. Then $\sqrt{1-x_1^2}dx_j = dy_j$ for $j = 2, 3, \cdots, n$. Observe that $x_1^2 + x_2^2 + \cdots + x_k^2 \leq 1 \iff x_2^2 + \cdots + x_k^2 \leq 1 - x_1^2 \iff y_2^2 + \cdots + y_k^2 \leq 1$. So, $V_n = \int_{-1}^{1} [\int_{\{\vec{y} \in \mathbb{R}^{n-1} : \|\vec{y}\| \leq 1\}} (\sqrt{1-y_1^2})^{n-1} dy_n \cdots dy_2] dy_1 = V_{n-1} \int_{-1}^{1} (\sqrt{1-y_1^2})^{n-1} dy_1$
$= 2V_{n-1} \int_0^1 (1-t^2)^{\frac{n-1}{2}} dt$.

3. Let $a > 0$ and consider $\int_{\mathbb{R}^n} e^{-a\|\vec{x}\|^2} dV$. Since $e^{-a\|\vec{x}\|^2} > 0$, we know $\int_{\mathbb{R}^n} e^{-a\|\vec{x}\|^2} dV = \int_{-\infty}^{\infty} (\int_{-\infty}^{\infty} \cdots (\int_{-\infty}^{\infty} e^{-a(x_1^2+x_2^2+\cdots+x_n^2)} dx_n) \cdots dx_2) dx_1$
$= \int_{-\infty}^{\infty} e^{-ax_1^2} dx_1 \cdots \int_{-\infty}^{\infty} e^{-ax_2^2} dx_2 \cdots \int_{-\infty}^{\infty} e^{-ax_n^2} dx_n = (\int_{-\infty}^{\infty} e^{-ax^2} dx)^n$. Now, letting $u = \sqrt{a} \cdot x$, we have $\int_{-\infty}^{\infty} e^{-ax^2} dx = \frac{1}{\sqrt{a}} \int_{infty}^{\infty} e^{-u^2} du = \sqrt{\frac{\pi}{a}}$ by the above exercise. Hence, $\int_{\mathbb{R}^n} e^{-a\|\vec{x}\|^2} dV = (\sqrt{\frac{\pi}{a}})^n = (\frac{\pi}{a})^{\frac{n}{2}}$.

4. Letting $\sigma(\mathbb{S}^{n-1})$ be the hypersurface area of $\mathbb{S}^{n-1} = \{\vec{x} : \|\vec{x}\| = 1\}$ in $\mathbb{R}^n$, we will establish that $\sigma(\mathbb{S}^{n-1}) = \frac{2\pi^{\frac{n}{2}}}{\Gamma(\frac{n}{2})}$, where $\Gamma(x) = \int_0^{\infty} t^{x-1} e^{-t} dt$ is the so-called **Gamma function** defined for $x \in \mathbb{C}$ with positive real parts. Here we will assume the so-called polar coordinates or co-area formula that allows one to compute $\int_{\mathbb{R}^n} f dV = \int_0^{\infty} [\int_{\partial B_r(\vec{x}_0)} f d\sigma] dr$. So, $\sigma(\mathbb{S}^{n-1}) = \int_{\partial B_1(\vec{0})} 1 d\sigma$ as a surface integral in the calculus sense. For simplicity of notation, we let $V_n = \mu(\{\vec{x} \in \mathbb{R}^n : \|\vec{x}\| \leq 1\}) = \mu(B_1(\vec{0}))$, $V_n(R) = \mu(\{\vec{x} \in \mathbb{R}^n : \|\vec{x}\| \leq R\}) = \mu(B_R(\vec{0}))$, S_{n-1} to be the $n-1$ dimensional hypersurface area of $\mathbb{S}^{n-1} = \{\vec{x} \in \mathbb{R}^n : \|\vec{x}\| = 1\} = \partial B_1(\vec{0})$, and $S_{n-1}(R)$ is the $n-1$ dimensional hypersurface area of $\{\vec{x} \in \mathbb{R}^n : \|\vec{x}\| = R\} = \partial B_R(\vec{0})$. Then, we have $V_n(R) = \int_{B_R(\vec{0})} 1 dx_1 \cdots dx_n$. Letting $R\vec{y} = \vec{x}$ for $\vec{y} \in B_1(\vec{0})$, we have $R^n dy_1 \cdots dy_n = dx_1 \cdots dx_n$, and so $V_n(R) = \int_{B_1(\vec{0})} R^n dy_1 \cdots dy_n = R^n V_n$. If dS_{n-1} is the differential hypersurface area of $\partial B_1(\vec{0})$ and $dS_{n-1}(R)$ is the differential hypersurface area of $\partial B_R(\vec{0})$, then $dS_{n-1}(R) = R^{n-1} dS_{n-1}$ and so $S_{n-1}(R) = R^{n-1} S_{n-1}$. By polar coordinates, we have $R^n V_n = V_n(R) = \int_0^R \int_{\partial B_r(\vec{0})} dS_{n-1}(r) dr$
$= \int_0^R r^{n-1} \int_{\partial B_1(\vec{0})} dS_{n-1} dr = \int_0^R r^{n-1} dr \cdot S_{n-1} = \frac{R^n}{n} S_{n-1}$. Hence, we see that $V_n = \frac{S_{n-1}}{n}$. Recall that $\int_{\mathbb{R}^n} e^{-\|\vec{x}\|^2} dx_1 \cdots dx_n = (\sqrt{\pi})^n = (\pi)^{\frac{n}{2}}$. Also, we may compute this integral using polar coordinates as $\int_{\mathbb{R}^n} e^{-\|\vec{x}\|^2} dx_1 \cdots dx_n = \int_0^{\infty} \int_{\partial B_r(\vec{0})} e^{-r^2} dS_{n-1}(r) dr$
$= \int_0^{\infty} \int_{\partial B_1(\vec{0})} e^{-r^2} r^{n-1} dS_{n-1} dr = S_{n-1} \int_0^{\infty} e^{-r^2} r^{n-1} dr$. Next, we make the substitution $z = r^2$, and so $\frac{1}{2\sqrt{z}} dz = dr$. Hence, $S_{n-1} \int_0^{\infty} e^{-r^2} r^{n-1} dr =$

$S_{n-1} \int_0^\infty e^{-z} z^{\frac{n}{2}-\frac{1}{2}} \frac{1}{2\sqrt{z}} dz = \frac{1}{2} S_{n-1} \int_0^\infty e^{-z} z^{\frac{n}{2}-1} dz = \frac{S_{n-1}}{2} \Gamma(\frac{n}{2})$.

5. Using that $\Gamma(\frac{1}{2}n+1) = \frac{n}{2}\Gamma(\frac{1}{2}n)$, we will show that $\mu(\{\vec{x} : \|\vec{x}\| < 1\}) = \frac{\pi^{\frac{n}{2}}}{\Gamma(\frac{1}{2}n+1)}$. Moreover, we will show this by establishing that $\mu(\{\vec{x} : \|\vec{x}\| < 1\}) = n^{-1}\sigma(\mathbb{S}^{n-1})$. From the above exercise, we have $V_n = \frac{S_{n-1}}{n} = \frac{2\pi^{\frac{n}{2}}}{n\Gamma(\frac{n}{2})}$
$= \frac{\pi^{\frac{n}{2}}}{\frac{n}{2}\Gamma(\frac{n}{2})} = \frac{\pi^{\frac{n}{2}}}{\Gamma(\frac{n}{2}+1)}$ The last identity follows since $\Gamma(x+1) = \int_0^\infty t^x e^{-t} dt$, which by integration by parts (use $u = t^x$, $dv = e^{-t}dt$) is $-t^x e^{-t}|_{t=0}^\infty + \int_0^\infty xt^{x-1}e^{-t}dt = x\int_0^\infty t^{x-1}e^{-t}dt = x\Gamma(x)$.

6. Suppose f is measurable on $\mathbb{R}^{n+m}$ and $\int_{\mathbb{R}^n}[\int_{\mathbb{R}^m}|f|d\vec{y}]d\vec{x} < \infty$. Then, by Tonelli's theorem, $\int_{\mathbb{R}^{n+m}}|f|d\mu = \int_{\mathbb{R}^n}[\int_{\mathbb{R}^m}|f|d\vec{y}]d\vec{x} < \infty$. So, $f \in L^1(\mathbb{R}^{n+m})$, and so by Fubini's theorem, $\int_{\mathbb{R}^{n+m}} d\vec{y} = \int_{\mathbb{R}^n}[\int_{\mathbb{R}^m} f d\vec{y}]d\vec{x} = \int_{\mathbb{R}^m}[\int_{\mathbb{R}^n} f d\vec{x}]d\vec{y}$.

7. Let $f(x,y) = y^{-2}$ if $0 < x < y < 1$, $f(x,y) = -x^{-2}$ if $0 < y < x < 1$, and $f = 0$ otherwise. Then $\int_0^1 \int_0^1 f dx dy = \int_0^1 [\int_0^y y^{-2} dx + \int_y^1 -x^{-2} dx] dy$
$= \int_0^1 y^{-1} + 1 - y^{-1} dy = \int_0^1 1 dy = 1$. Whereas, $\int_0^1 \int_0^1 f dy dx = \int_0^1 [\int_0^x -x^{-2} dy + \int_x^1 y^{-2} dy] dx = \int_0^1 -x^{-1} - 1 + x^{-1} dx = \int_0^1 -1 dx = -1$.

Bibliography

[1] Eves, Howard. *Foundations and Fundamental Concepts of Mathematics.* 3rd ed. Mineola, New York: Dover Publications, Inc., 1990.

[2] Gelbaum, Bernard R., and Olmsted, John M. H. *Counterexamples in Analysis.* Mineola, New York: Dover Publications, Inc., 2003.

[3] Jacobson, Nathan. *Basic Algebra I,* 2nd ed. Mineola, New York: Dover Publications, Inc., 1985.

[4] Jost, Jürgen. *Postmodern Analysis,* 2nd ed. New York, New York: Springer, 2003.

[5] Krantz, Steven G., and Parks, Harold R. *The Implict Function Theorem: History, Theory, and Applications.* Boston: Birkhäuser, 2003.

[6] Kreyszig, Erwin. *Introductory Functional Analysis with Applications.* Hoboken, New Jersey: John Wiley and Sons, Inc., 1978.

[7] Landau, E.G.H. *Foundations of Analysis: The Arithmetic of Whole, Rational, Irrational and Complex Numbers.* Trans. by F. Steinhardt. New York: Chelsea, 1951.

[8] Reed, Michael, and Simon, Barry. *Methods of Modern Mathematical Physics 1: Functional Analysis.* New York: Academic Press, 2003.

[9] Robison, Gerson B. *A New Approach to Circular Functions, II and lim (sin x)/x*, Mathematics Magazine, 41 (1968) 66-70.

[10] Rudin, Walter. *Principles of Mathematical Analysis.* 3rd ed. New York: McGraw-Hill, Inc., 1976.

[11] Shilov, Georgi E. *Elementary Real and Complex Analysis*. Mineola, New York: Dover Publications, Inc., 1996.

[12] Spivak, Michael. *Calculus*. Houston: Publish or Perish, 1994.

[13] Strichartz, Robert S. *The Way of Analysis*. Boston: Jones and Bartlett Publishers, 2000.

[14] Wheeden, Richard, and Zygmund, Antoni. *Measure and Integral: An Introduction to Real Analysis*. New York: Marcel Dekker, Inc., 1977.

Index

C^1, 328
$C^{0,\alpha}([a,b])$, 292
G_δ sets, 342
$L^1(D)$, 382
$L^p(D)$, 383
$L^\infty(D)$, 385
σ-algebra, 348

a.e., 354
absolute convergence of a series, 125
absolute convergence of a series of functions, 283
absolute extremum, 191
absolute maximum, 191
absolute minimum, 191
absolute value, 26
accumulation point, 72
additive inverse, 12
affine function, 311
algebra, σ, 348
algebraic numbers, 45
almost everywhere, 354
alternating harmonic series, 124
alternating series, 122
analytic function, 211
angle between vectors, 66
antiderivative, 244, 246
Archimedean property of $\mathbb{R}$, 34
arclength of a curve, 261
Ascoli-Arzelà theorem, 290
associativity of addition, 12

associativity of multiplication, 12
axiom of induction, 14

ball, punctured, 157
balls and disks, 68
Banach space, 312
basis of a vector space, 300
basis, canonical, 302
Bessel function, 137
binomial theorem, 98
Bolzano-Weierstrass theorem, 83
boundary of a set, 72
bounded monotone convergence theorem, 96
bounded sequence, 90
boundedness, 32
boundedness for functions, 35
boundedness, uniformly, 288

cancellation property for fields, 18
cancellation property of the integers, 15
cancellation property of the naturals, 14, 410
canonical basis, 302
Cantor Set, 86
Cantor set, 86
Cauchy condensation test, 113
Cauchy convergence criterion, 105
Cauchy convergence criterion for uniform convergence, 275

Cauchy criterion for integrability, 222, 366
Cauchy product of series, 127
Cauchy sequence, 103
Cauchy's inequality, 390
Cauchy-Schwarz inequality, 66
center of power series expansion, 133
characteristic function of $\mathbb{Q}$, 144
closed in a set D, 152
closed set, 71
closure of a set, 72
commutativity of addition, 12
commutativity of multiplication, 12
compact metric space, 76
compactness, 76
comparison test for series, 111
complement of a set, 71
complete metric space, 105
completeness axiom of the set of real numbers, 32
complex conjugate, 62
complex numbers, 60
concavity, 204
conditionally convergent series, 126
connected set, 84
continuity, 143
continuity, Hölder, 292
continuity, modulus of, 165
continuous at a point, 143
continuous function, 144
continuous on a set, 144
continuous on a set D, 154
continuous, Hölder, 163
continuous, Lipschitz, 163
continuous, uniformly, 155
contraction mapping, 327
convergence in measure, 359
convergent sequence, 89

convex function, 163
convex sets, or convexity, 69
convexity, 204
convolution of functions, 397
coordinates, 302
coordinates of a vector with respect to a basis, 309
cosecant function, 183
cosine function, 183
cotangent function, 183
countability of the algebraic numbers, 45
countability of the rational numbers, 42
countable sets, 42
critical value, 191
cusp, 171
cuts, Dedekind, 47

Darboux integral, 232, 369
decimal expansions for real numbers, 40
decreasing function, 158
decreasing sequence, 95
Dedekind cuts, 47
DeMorgan's laws, 73
dense set, 72
derivative, 167, 313
diameter of a set, 70
differentiable, 167
differentiable at a point, 167
differentiable in Banach spaces, 314
dimension of a vector space, 300
direction of steepest ascent, 324
direction of steepest descent, 324
directional derivative, 316
disconnected set, 84
discontinuity, 156

INDEX

discontinuity of the first kind, 156, 164
discontinuity of the second kind, 156
discontinuity, simple, 156, 164
discontinuous at a point, 156
distance function, 70
distance in Euclidean space, 66
distributive property, 13
divergence theorem for series, 108
divergent sequence, 89
division, 16
dot product, 65

e, 120
equicontinuous, 289
equivalent norms, 304
essential suppremum, 385
Euclidean algorithm, or division algorithm, 41
Euclidean space, 66
Euler's number, 120
even integers, 38
exponential notation for fields, 16

factorial, 118
Fibonacci sequence, 90
field, 13
finite intersection property of compact sets, 77
finite sets, 42
finite subcover, 76
first derivative test, 202
fixed point, 327
Fourier series, 283
Fubini's theorem, 392
function, convex, 163
function, decreasing, 158
function, increasing, 158

function, monotone, 158
function, strictly decreasing, 158
function, strictly increasing, 158
function, strictly monotone, 158
function,measurable, 353
functions of bounded variation, 249

Gamma function, 398
geometric series, 110
global maximum, 191
global minimum, 191
gradient, 316
greatest lower bound, 32

Hölder continuity, 163, 292
Hölder continuous, 201
Hölder's inequality, 391, 472
half-angle formulas for sine and cosine, 191
harmonic function, 326
harmonic series, 108
Heine-Borel theorem, 80
Hessian, 322
Hessian matrix, 322
Hilbert space, 316

implicit function theorem, 333
increasing function, 158
increasing sequence, 95
infimum, or inf, 32
infinite series, 106
infinity, 33
inner product, 65
inner product space, 67
integers, 15
integral, 218, 363
integral test, 244
integral, Darboux, 232, 369
interior of a set, 71

interior point, 71
intermediate value theorem, 154
intermediate value theorem for derivatives, 197
interval, 67
interval of convergence of a power series, 133
inverse function theorem, 328
inverse trigonometric functions, 190
isolated point, 72
isomorphic, 56
isomorphism of fields, 56
iterated integral, 392

Jensen's inequality, 208

kernel of a linear transformation, 306

L'Hôpial's rule, 198
L-p spaces, 383
Lagrange multipliers, method of, 325
least upper bound, 32
Lebesgue integrable function, 382
Lebesgue measurable set, 342
Lebesgue measure of a set, 342
Lebesgue outer measure, 340
Lebesgue sum, 373
left-hand limit, 143, 156
length of a curve, 261
length of a rectifiable curve in $\mathbb{R}^n$, 263
level set, 324
Liapounov's inequality, 392
liminf of a function, 157
limit comparison test for series, 112
limit inferior, 101
limit infimum of a function, 157
limit laws, 92, 140
limit of a function, 139
limit of a sequence, 89

limit point, 72
limit superior, 101
limit supremum of a function, 157
limit, left-hand limit, 143, 156
limit, right-hand limit, 143, 156
limsup, 101
limsup of a function, 157
linear combination of vectors, 300
linear mapping, 306
linear operator, 307
linear transformation, 306
linearly independent set of vectors, 299
Lipschitz continuity, 163
local maximum, 191
local minimum, 191
logarithm, 150
lower semicontinuous, 157
lower sum, 232, 368

magnitude of a vector in Euclidean space, 66
marked partition, 218, 362
mean value theorem for integrals, 243
measurable function, 353
measurable set, Lebesgue, 342
measure zero, 264
measure, Lebesgue, 342
mesh of a partition, 218, 362
metric space, 70
metric space, compact, 76
metric space, complete, 105
Minkowski's inequality, 391, 472
modulus of continuity, 165
monotone function, 158
monotone sequence, 96
multiples notation for fields, 16
multiplicative inverse, 12

n^{th} roots, 35

INDEX

natural numbers, 13
negative definite, 323
negative element of an ordered field, 31
negative integers, 25
negative part of a real number, 254
negative variation of a function on an interval, 255
neighbourhood of a point, 72
Newton's method, 211
Newton-Raphson method, 211
non-negative, 25
non-positive, 25
norm of a linear transformation, 307
norm of a partition, 218, 362
norm of a vector in Euclidean space, 66
norm or absolute value of a complex number, 62
normed linear space, 67, 303
norms, equivalent, 304

odd integers, 38
one, 12
open cover, 76
open in a set D, 152
open set, 71
order of operations, 16
order on the set of integers, 25
order on the set of rational numbers, 25
order properties of the naturals, 14, 410
ordered field, 29
oscillation of a function, 265
outer measure of a set, 340

p-series, 114

partial derivative, 316
partition of a measurable set, 373
partition of an interval, 217
partition of an interval in $\mathbb{R}^n$, 362
perfect set, 72
point-wise converge of a series of functions, 283
point-wise convergence, 271
polynomial, 97
positive definite, 323
positive element of an ordered field, 30
positive integers, 25
positive part of a real number, 254
positive variation of a function on an interval, 255
power series, 133
powers, 16
product rule, 326
properties of exponents, 27
punctured ball, 157
Pythagorean identity, 180, 183

quotient rule, 326

radius of convergence of a power series, 135
ratio test for series, 115
rational numbers, 15, 22
rearrangement of a series, 129
rectifiable curve, 261
refinement of a partition, 232, 369
relative extremum, 191
relative maximum, 191
relative minimum, 191
Riemann integrable, 218, 363
Riemann integral, 218, 363
Riemann sum, 218, 362

right-hand limit, 143, 156
Rolle's theorem, 192
root test for series, 114

scalar multiplication, 65
secant function, 183
second derivative, 201
second derivative test, 203
semicontinuous, lower, 157
semicontinuous, upper, 157
sequence of partial sums, 106
sequence, bounded, 90
sequence, Cauchy, 103
sequence, convergent, 89
sequence, decreasing, 95
sequence, divergent, 89
sequence, Fibonacci, 90
sequence, increasing, 95
sequence, monotone, 96
sequence, strictly decreasing, 95
sequence, strictly increasing, 95
sequence, subsequence, 98
sequence, subsequential limit, 98
sequences, 89
series, 106
series, absolute convergence, 125
series, alternating, 122
series, alternating harmonic, 124
series, Cauchy condensation test, 113
series, Cauchy product, 127
series, comparison test, 111
series, conditional convergent, 126
series, convergent, 106
series, divergent, 106
series, geometric, 110
series, limit comparison test, 112
series, p-series, 114
series, ratio test, 115

series, rearrangement, 129
series, root test, 114
series, telescoping, 110
set of measure zero, 264
set, of type G_δ, 342
sigma algebra, 348
simple discontinuity, 156, 164
simple function, 374
simple functions, 227
sine function, 183
span of a set of vectors, 300
square roots, 35
step function, 226
strictly decreasing function, 158
strictly decreasing sequence, 95
strictly increasing function, 158
strictly increasing sequence, 95
strictly monotone function, 158
subcover, 76
subsequence, 98
subsequential limit, 98
subtraction, 15
supremum, or sup, 32

tangent function, 183
tangent line, 194
Taylor polynomial, 208
Taylor remainder term, 209
Taylor series, 209
Taylor's theorem, 209, 321
telescoping series, 110
ternary expansion, 87
Thomae function, 145
Thomae function Riemann integrability, 225
total variation of a function, 249
total variation of a function on an interval, 249

INDEX

transcendental number, 45
triangle inequality for complex numbers, 63
triangle inequality for metric spaces, 70
triangle inequality for real numbers, 26
triangle inequality in Euclidean space, 66
trichotomy property, 29
trigonometric functions, 183

u-substitution, 248
uncountability of the real numbers, 42
uncountable sets, 42
uniform boundedness, 288
uniform convergence, 274
uniform convergence of a series of functions, 283
uniform convergence, Cauchy convergence criterion, 275
uniform equicontinuity, 289
uniformly continuous on a set D, 155
uniformly equicontinuous, 289
uniqueness of inverses in fields, 17
uniqueness of one in fields, 17
uniqueness of zero in fields, 17
unity, 12
upper semicontinuous, 157
upper sum, 232, 368

variation, total, 249
vector addition, 65
vector space, 297
volume of an interval, 339

Weierstrass approximation theorem, 293

Weierstrass M-test, 283

Young's inequality, 390, 472

zero, 12

CPSIA information can be obtained
at www.ICGtesting.com
Printed in the USA
BVOW11s0734200717
489685BV00003B/34/P